Nanomaterials for the Life Sciences
Volume 8
Nanocomposites

Edited by
Challa S. S. R. Kumar

Related Titles

Kumar, C. S. S. R. (ed.)

Nanotechnologies for the Life Sciences

10 Volume Set

ISBN: 978-3-527-31301-3

Kumar, C. S. S. R. (Ed.)

Nanomaterials for the Life Sciences (NmLS)

Book Series, 10 Volumes

Vol. 6

Semiconductor Nanomaterials

2010

ISBN: 978-3-527-32166-7

Vol. 7

Biomimetic and Bioinspired Nanomaterials

2010

ISBN: 978-3-527-32167-4

Vol. 8

Nanocomposites

2010

ISBN: 978-3-527-32168-1

Vol. 9

Carbon Nanomaterials

2011

ISBN: 978-3-527-32169-8

Vol. 10

Polymeric Nanomaterials

2011

ISBN: 978-3-527-32170-4

Nanomaterials for the Life Sciences
Volume 8

Nanocomposites

Edited by
Challa S. S. R. Kumar

WILEY-VCH Verlag GmbH & Co. KGaA

The Editor

Dr. Challa S. S. R. Kumar
CAMD
Louisiana State University
6980 Jefferson Highway
Baton Rouge, LA 70806
USA

Library of Congress Card No.: applied for

British Library Cataloguing-in-Publication Data
A catalogue record for this book is available from the British Library.

Bibliographic information published by the Deutsche Nationalbibliothek
The Deutsche Nationalbibliothek lists this publication in the Deutsche Nationalbibliografie; detailed bibliographic data are available on the Internet at <http://dnb.d-nb.de>.

Composition Toppan Best-set Premedia Ltd., Hong Kong
Printing and Binding betz-druck GmbH, Darmstadt
Cover Design Schulz Grafik-Design, Fußgönheim

Printed in the Federal Republic of Germany
Printed on acid-free paper

ISBN: 978-3-527-32168-1

Contents

Nanomaterials for the Life Sciences Vol.8: Nanocomposites. Edited by Challa S. S. R. Kumar

ISBN: 978-3-527-32168-1

Preface

In the NmLS series, we have so far covered a wide range of nanomaterials with potential applications in life sciences. With the publication of the 8th volume, we come closer to the completion of the ten volume NmLS series. In this volume, we are focusing on nanocomposites; which have been attracting enormous interest for a variety of applications ranging from packaging, automotive, electrical, biomedical and other applications due to their superior thermal, electrical conductive and other properties. Surprisingly, till recently, their utility in life sciences has been slow to be recognized and this volume captures the first ever compendium on application of nanocomposites in life sciences. The eighth volume has eleven chapters correlating the life science applications with a wide variety of nanocomposites. The range of nanocmposites presented includes those based on titania, gold, silver, chitin, quantum dots, and so on. The life science applications provided consists of biosensing, tissue engineering, medical diagnosis and therapy. The most exciting and unique aspect of the book is that it brings out the fact that conversion of nanomaterials to their corresponding nanocomposites results a dramatic change in their fundamental science as well as properties; which are significant from the point of view technological applications.

The first chapter is from the laboratories of Prof. Gayle E. Woloschak, Northwestern University, Chicago, USA. The chapter provides an overview of TiO2 nanoparticles, nanocomposites and nanoconjugates demonstrating variable degrees of photo catalytic reactivity, photo response and surface reactivity which in turn influences their interactions with biological systems; the applications ranging from diagnosis to therapy and water purification to self-sterilizing surgical devices. The chapter entitled, Titanium Dioxide Nanocomposites, is a first ever comprehensive review on titania nanocomposites for applications in life sciences. The second chapter, Chitin nanocomposites for medical applications, by Prof. Aji P Mathew demonstrates chitosan-based nanocomposites as a novel group of biomaterials with a potential to support and facilitate cell growth, for controlled drug delivery and as biosensors to detect glucose, creatine, etc. in the body. This contribution from the Luleå University of Technology, Sweden, covers chitin nanocomposites obtained from carbon nanotubes, inorganic metal nanoparticles or

Nanomaterials for the Life Sciences Vol.8: Nanocomposites. Edited by Challa S. S. R. Kumar

ISBN: 978-3-527-32168-1

montmorillonite and biobased nanowhiskers (chitin or cellulose) are used as reinforcement in chitosan matrices. The next chapter overviews the state-of-the-art of silver nanocomposites for applications related to wound healing, inflammations and biological sensing. The chapter, contributed by Prof. Andrea Travan, University of Trieste, Italy brings out the message that silver nanocomposites will play a key role in life science applications thanks to their unique properties such as scattering brightness and biocompatibility. Moving from silver nano composites to gold nanocomposites, Prof. María Isabel Pividori from Universitat Autónoma de Barcelona, Spain, reviews specific application of Gold Nanocomposite in Biosensing. In this chapter the main features of nanostructured gold composites as transducer materials for electrochemical biosensing is covered. More specifically, novel approaches based on rigid carbon–polymer composites materials modified with gold nanoparticles for the improved electrochemical biosensing of DNA is discussed.

It is now very well established that quantum dots are being applied as sensitive and reliable probes for sensing and imaging applications. This is not surprising due to their unique electronic, optical, magnetic, and mechanical properties coupled with sizes comparable with biomacromolecules in addition to their ability to resist photo bleaching, denaturation of biomolecules, and alterations in pH and temperature. The fifth chapter focusing on Quantum Dot Nanocomposites, presents recent developments on the construction, characterization, and representative applications of the quantum dot nanocomposites, with emphasis placed on the use of these nanocomposites as FRET-based biosensors for investigations of donor-acceptor distance distribution, macromolecule conformation, receptor-ligand binding, DNA hybridization, antigen/antibody recognition, and DNA/protein interaction. Prof. Feimeng Zhou from California State University, Los Angeles, USA addressed every facet of QD nano composites emphasizing their applications for life sciences. In the sixth chapter entitled, Gold-Polymer nanocomposites for Bio-imaging and Biosensing, Prof. Nobuo Uehara reiterates the fact that while excellent studies on the application of gold nanoparticles in the life sciences have recently been published, there are not many reviews on the functionality of gold nanocomposites in the life sciences despite their versatility. The authors, therefore, fill this much needed gap by reviewing in this chapter recent development in gold nanocomposites for bio-sensing and bio-imaging. While the chapter four stresses the importance of electrochemical bio sensing using gold nanocomposites, this chapter's intent is on non electrochemical based applications.

So far, we have covered specific metal, metal oxide and biopolymer-based nanocomposites. From the seventh chapter onwards, the book contains information on broad range of nanocomposites with specific applications ranging from drug delivery to tissue engineering. The seventh chapter covers an overview of the structure, biology and genetic engineering of genetically modifiable bio-macromolecules their nanocomposites, followed by examples of their applications in device fabrications, medicine and sensing. Prof. Chuanbin Mao from University of Oklahoma, Norman, USA, gave an overview of the processes used for constructing genetically engineered nanocomposites using bacteriophage and flagella as examples. This emerging field of incorporation of genetically modifiable bio-macromolecules into nanotechnology will enable production of novel nanocomposites in foreseeable future. Moving to the next chapter, the reader will be exposed to multifunctional aspects of nanocomposites. In this chapter, Prof. Seungjoo Haam of Yonsei University, Republic of Korea, describe various functional components in the bio/medicinal nanocomposites. The readers are also introduced to recent advances in the formulation of multifunctional nanocomposites which led to great technological thrusts in the field of biomedicine.

The two most important applications of nanocomposites are in drug delivery and tissue engineering. The last three chapters in this book ensure that these applications are covered completely. While the ninth chapter, Nanocomposites for drug delivery, by Prof Shou-Cang Shen, from the National University of Singapore, Singapore, emphasizes on drug delivery, the tenth chapter by Prof. Kalpana S. Katti, North Dakota State University, Fargo, USA, throws light on bone tissue engineering. The final chapter, Nanocomposites for Tissue Engineering, by Michael Gelinsky, Technische Universität Dresden, Germany, provides an exhaustive overview about all types of nanocomposites that can be manufactured into tissue engineering scaffolds and also explain natural nanocomposites like extracellular matrices and their utilisation as scaffolds. This chapter is complementary to the 10th chapter in which the emphasis is on the material components of bone tissue engineering scaffolds as well as their fabrication and characterization routes. Over all, the last two chapters cover the important components of tissue engineering-scaffolds, cells and growth factors.

Nanotechnology, more so sub disciplines such as nanocomposites, embody the spirit of interdisciplinary approaches and teams. I am, therefore, very grateful to all the authors who have shared my enthusiasm and vision by contributing high quality manuscripts keeping in tune with the theme of this volume. It is primarily due to their scholarly contributions, this book comes into existence. I am thankful to my employer, the Center for Advanced Microstructures and Devices (CAMD), for providing me with an opportunity to undertake this enormous project. No words can express the support I have been receiving from my family, friends, mentors and most importantly the readers in ensuring the quality of the editorial efforts and I am indebted to them. Finally, Wiley VCH publishers continue to do a remarkable job and I am grateful for their support.

Challa S. S. R. Kumar

List of Contributors

Gopal Abbineni
University of Oklahoma
Department of Chemistry &
Biochemistry
Norman, OK 73019
USA

Salvador Alegret
Universitat Autónoma de
Barcelona
Departament de Química
Grup de Sensors i Biosensors
08193 Barcelona
Spain

Avinash H. Ambre
North Dakota State University
Department of Civil Engineering
Fargo, ND 58105
USA

Hans Arora
Northwestern University
Feinberg School of Medicine
303 E. Chicago Ave.
Chicago, IL 60611
USA

John Boyle
Northwestern University
Feinberg School of Medicine
303 E. Chicago Ave.
Chicago, IL 60611
USA

Leonard Chia
A*STAR (Agency for Science,
Technology and Research)
Institute of Chemical and
Engineering Sciences
1 Pesek Road
Jurong Island
Singapore 627833
Singapore

Ivan Donati
University of Trieste
Department of Life Sciences
Via Giorgieri 1
34127 Trieste
Italy

Nanomaterials for the Life Sciences Vol.8: Nanocomposites. Edited by Challa S. S. R. Kumar

ISBN: 978-3-527-32168-1

Yuan-Cai Dong
A*STAR (Agency for Science, Technology and Research)
Institute of Chemical and Engineering Sciences
1 Pesek Road
Jurong Island
Singapore 627833
Singapore

Caroline Doty
Northwestern University
Feinberg School of Medicine
303 E. Chicago Ave.
Chicago, IL 60611
USA

Michael Gelinsky
Technische Universität Dresden
The Max Bergmann Center of Biomaterials and Institute for Materials Science
01069 Dresden
Germany
Technische Universität Dresden
DFG Research Center and Cluster of Excellence for Regenerative Therapies Dresden (CRTD)
01069 Dresden
Germany

Seungjoo Haam
Yonsei University
Department of Chemical and Biomolecular Engineering
Seoul 120-749
Republic of Korea

Sascha Heinemann
Technische Universität Dresden
The Max Bergmann Center of Biomaterials and Institute for Materials Science
01069 Dresden
Germany

Yong-Min Huh
Yonsei University
Department of Radiology
Seoul 120-752
Republic of Korea

Dinesh R. Katti
North Dakota State University
Department of Civil Engineering
Fargo, ND 58105
USA

Kalpana S. Katti
North Dakota State University
Department of Civil Engineering
Fargo, ND 58105
USA

Kwangyeol Lee
Korea University
Department of Chemistry
Seoul 136-701
Republic of Korea

Yunfei Long
Central South University
College of Chemistry and Chemical Engineering
Changsha
Hunan 410083
China
California State University
Department of Chemistry and Biochemistry
Los Angeles, CA 90032
USA

Chuanbin Mao
University of Oklahoma
Department of Chemistry & Biochemistry
Norman, OK 73019
USA

Eleonora Marsich
University of Trieste
Department of Life Sciences
Via Giorgieri 1
34127 Trieste
Italy

Aji P. Mathew
Luleå University of Technology
Division of Manufacturing and
Design of Wood and
Bionanocomposites
97187 Luleå
Sweden

Tsutomu Nagaoka
Osaka Prefecture University
Frontier Science Innovation
Center
1-2 Gakuen-Cho
Nakaku, Sakai
Osaka Prefecture, 599-8570
Japan

Wai Kiong Ng
A*STAR (Agency for Science
Technology and Research)
Institute of Chemical and
Engineering Sciences
1 Pesek Road
Jurong Island
Singapore 627833
Singapore

Kristiina Oksman
Luleå University of Technology
Division of Manufacturing and
Design of Wood and
Bionanocomposites
97187 Luleå
Sweden

Sergio Paoletti
University of Trieste
Department of Life Sciences
Via Giorgieri 1
34127 Trieste
Italy

Tatjana Paunesku
Northwestern University
Feinberg School of Medicine
303 E. Chicago Ave.
Chicago, IL 60611
USA

Katarina Petras
Northwestern University
Feinberg School of Medicine
303 E. Chicago Ave.
Chicago, IL 60611
USA

María Isabel Pividori
Universitat Autónoma de
Barcelona
Departament de Química
Grup de Sensors i Biosensors
08193 Barcelona
Spain

Bryan Rabatic
Rush Medical College at Rush
University
Medical Center
1653 W. Congress Parkway
Chicago
Chicago, IL 60612
USA

Shoucang Shen
A*STAR (Agency for Science Technology and Research)
Institute of Chemical and Engineering Sciences
1 Pesek Road
Jurong Island
Singapore 627833
Singapore

Reginald Beng Hee Tan
A*STAR (Agency for Science Technology and Research)
Institute of Chemical and Engineering Sciences
1 Pesek Road
Jurong Island
Singapore 627833
Singapore
The National University of Singapore
Department of Chemical and Biomolecular Engineering
4 Engineering Drive
Singapore 117576
Singapore

Andrea Travan
University of Trieste
Department of Life Sciences
Via Giorgieri 1
34127 Trieste
Italy

Nobuo Uehara
Utsunomiya University
Graduate School of Engineering
7-1-2 Yoto
Utsunomiya
Tochigi Prefecture, 321-8585
Japan

Jianxiu Wang
Central South University
College of Chemistry and Chemical Engineering
Changsha
Hunan 410083
China

Gayle Woloschak
Northwestern University
Feinberg School of Medicine
303 E. Chicago Ave, Ward 13-002
Chicago, IL 60611
USA

Jaemoon Yang
Yonsei University
Department of Radiology
Seoul 120-752
Republic of Korea

Ye Yuan
Northwestern University
Feinberg School of Medicine
303 E. Chicago Ave.
Chicago, IL 60611
USA

Feimeng Zhou
Central South University
College of Chemistry and Chemical Engineering, Changsha
Hunan 410083
China
California State University
Department of Chemistry and Biochemistry
Los Angeles, CA 90032
USA

1
Titanium Dioxide Nanocomposites

Hans Arora, Caroline Doty, Ye Yuan, John Boyle, Katarina Petras, Bryan Rabatic, Tatjana Paunesku, and Gayle Woloschak

1.1
Introduction

In Nature, titanium dioxide exists in three primary phases – anatase, rutile, and brookite – with different sizes of crystal cells in each case [1]. The popularity of titanium dioxide in materials sciences began with the first photocatalytic splitting of water in 1972 [2]. However, in recent years TiO_2 has been used widely for the preparation of different types of nanomaterials, including nanoparticles, nanorods, nanowires, nanotubes, and mesoporous and nanoporous TiO_2-containing materials [3]. Regardless of scale, TiO_2 maintains its photocatalytic abilities, and in addition, nanoscale TiO_2 has a surface reactivity that fosters its interactions with biological molecules, such as phosphorylated proteins and peptides [4], as well as some nonspecific binding with DNA [5]. Nano-anatase TiO_2, which is smaller than 20 nm, has surface corner defects that alter the size of the crystal cell [6, 7] (Table 1.1).

The surface molecules of nanoscale TiO_2 particles are "on the corner" of the particle, and are forced by confinement stress into a pentacoordinated, square-pyramidal orientation. Such molecules have a propensity for stable nanoparticle conjugation to *ortho*-substituted bidentate ligands such as 3,4-dihydroxyphenethylamine (dopamine) [7, 8]. This binding with enediol ligands "heals" the surface corner defects and returns the surface TiO_2 molecules into an octahedral geometry. As a consequence, the stability of the chemical bonds formed on the nanoparticle surface precludes further modifications of the nanoparticle surface, which may aid in reducing nanoparticle aggregation and nonspecific interactions with cellular components [9, 10].

The methods used for the synthesis of TiO_2 nanoparticles have included sol, sol–gel, solvothermal, hydrothermal and other approaches [3], although new methods and modifications of the existing methods have been attempted with great frequency. Among such efforts are included the use of different dopants in the synthesis of TiO_2 nanocomposites, such as noble metals [11] (the use of core

Nanomaterials for the Life Sciences Vol.8: Nanocomposites. Edited by Challa S. S. R. Kumar

ISBN: 978-3-527-32168-1

Table 1.1 Phases of TiO_2.

Phase (Reference)	Crystal system	a (Å)	b (Å)	c (Å)
Rutile [1]	Tetragonal	4.594	4.594	2.959
Anatase [1]	Tetragonal	3.789	3.789	9.514
Brookite [1]	Orthorhombic	9.166	5.436	5.135
Anatase with corner defects [7]	–	3.96	3.96	2.7

materials such as iron oxide–silicon dioxide–titanium dioxide core-corona-shell nanoparticles [12]) and the use of different nanoparticle surface-coating molecules and photosensitizing dyes [5, 9, 10, 13–17].

Although, at present, no systematic nomenclature is used to codify nanostructures, a proposal has been made recently to develop a "nano nomenclature" [18], and it is hoped that this may aid in making any review (including the present chapter) more systematic. For example, if an attempt were made to apply this nomenclature to the 5 nm TiO_2 nanoparticles with DNA oligonucleotide and gadolinium–DOTA conjugated to its surface [13], the formula for this nanoconjugate would be 2-5B- TiO_2-(DNA,Gd, DOTA), where the "2" indicates a metallic nanoparticle, "5B" a size of 5 nm and a spherical shape, and TiO_2 the material of the particle and the (DNA,Gd, DOTA) molecules conjugated to the surface. Obviously, this designation would require a further determination of DNA sequence, as well as information such as the anticipated strength of the chemical bonds between the nanoparticle and the conjugated materials. With TiO_2 in particular, the nomenclature would also have to include information on the crystal polymorph of the nanoparticle, the presence of "corner defects" on the nanoparticle surface, and so on.

In this chapter, the most recent applications of nanoscale TiO_2 will briefly be summarized in: (i) Photocatalysis for chemical degradation and antimicrobial activity; (ii) phosphopeptide enrichment from biological materials *in vitro*; (iii) the uptake and effects of nanoscale TiO_2 and nanocomposites in cells; (iv) the use of TiO_2 and its composites for implants and tissue engineering; and (v) toxicology studies of TiO_2 nanomaterials in animals.

A comprehensive list of the nanoparticulate TiO_2 materials reviewed in the chapter is provided in Table 1.2.

1.2
Photocatalysis by TiO_2 Nanoparticles, Nanocomposites and Nanoconjugates for Chemical Degradation and Antimicrobial Activity

The photocatalytic activity of TiO_2 molecules has been widely studied and utilized in biological, chemical, and industrial applications. The term "photocatalytic" refers to the ability of a material to form electron-hole pairs upon absorbing elec-

Table 1.2 The nanoparticulate TiO_2 materials reviewed in this chapter.

Nanoparticle composition	Size	Method(s) used for sizing	Shape	Crystal structure (%)			Manufacturer	Surface area per mass ($m^2 g^{-1}$)	Dispersity	Reference
				Anatase	Rutile	Brookite				
TiO_2	25–70 nm						Sigma-Aldrich			
TiO_2	20 nm									[19]
TiO_2	21 nm			25	75	0	DeGussa-Hüls AG (Frankfurt, Germany)			[20]
TiO_2	<100 nm						Sigma-Aldrich			[21]
TiO_2	19–21 nm			0	100	0	DeGussa (Frankfurt, Germany)	50 ± 15		[22]
TiO_2	80–110 nm	TEM, XRD	Round	100		0	Self			[23]
TiO_2	<25 nm (10 nm via MFR)	TEM	Elongated and round	100		0	Sigma-Aldrich	145		[24]
TiO_2	<75 nm (40 nm via MFR)	TEM	Round	Mix	Mix	0	Sigma-Aldrich	40		[24]
TiO_2	20–30 nm			70	30	0	CAS no. 13463-67-7 (commercial)	48.6		[25]
TiO_2	21 nm (24.1 ± 2.8)	TEM		25	75	0	DeGussa (“Aeroxide” P25)	50 ± 15		[26]

Table 1.2 *Continued*

Nanoparticle composition	Size	Method(s) used for sizing	Shape	Crystal structure (%)			Manufacturer	Surface area per mass ($m^2 g^{-1}$)	Dispersity	Reference
				Anatase	Rutile	Brookite				
TiO_2	5 nm (3.5 ± 1)	TEM		100	0	0	Nanostructured and Amorphous Materials (Los Alamos, NM)	210 ± 10 (219 ± 3)		[27]
TiO_2	20.5 ± 6.7 nm	SEM	Round				Degussa (P25)	45.41		[28]
TiO_2	25–70 nm						Sigma-Aldrich			[29]
TiO_2	15 nm			100	0	0	Sigma-Aldrich	190–290		[30]
TiO_2	7, 20 nm						Sigma-Aldrich			[31]
TiO_2	3 nm	AFM		100	0	0	Self	299.1		[32]
TiO_2	20 nm	SEM		100	0	0	Shanghai Huijing Sub-Nanoscale New Material Co., Ltd., Shanghai, China	120 (105.0)		[32]
TiO_2	5 nm	XRD		100	0	0	Self			[33]
TiO_2	30 nm	TEM					Self			[34]

TiO_2	Diameter = 4–6 nm	TEM	Rods	0	100 (XRD)	0	Self	1.646		[35]
TiO_2	21 nm						DeGussa (“Aeroxide” P25) (New Jersey, USA)			[36]
TiO_2	20 nm			100	0	0				[37]
TiO_2	21 nm (20 nm)	TEM					DeGussa Korea (P25)	50		[38]
TiO_2	120 nm (hydrodynamic diameter)	TEM		80	20	0	DeGussa AG (New Jersey, USA)	50 ± 15	0.131–0.138 (polydispersity index)	[39]
TiO_2	21 nm			80	20	0	DeGussa (“Aeroxide” P25)			[40]
TiO_2	21 nm			80	20	0	DeGussa (“Aeroxide” P25)			[41]
TiO_2	34.2 ± 26.1 nm	TEM	Spherical	Mix	Mix	0	Sigma-Aldrich	18.6 ± 1.2		[42]
TiO_2	25–70 nm	FE-SEM		100	0	0	Sigma-Aldrich	18.6 ± 1.2		[43]
TiO_2	20–30 nm	TEM		70	30	0	CAS No. 13463-67-7	48.6		[44]
TiO_2	25, 80 nm	TEM					Hangzhou Dayang Nanotechnology Co. Ltd.			[45]

Table 1.2 *Continued*

Nanoparticle composition	Size	Method(s) used for sizing	Shape	Crystal structure (%)			Manufacturer	Surface area per mass ($m^2 g^{-1}$)	Dispersity	Reference
				Anatase	Rutile	Brookite				
TiO_2	80 nm (71.43 ± 23.53 nm)	SEM		0	100	0	Hangzhou Dayang Nanotechnology Co. Ltd.			[46, 47]
TiO_2	200 × 35 nm	TEM	Rods	100	0	0	Self	26.5		[48]
TiO_2	10 nm	TEM	Dots	100	0	0	Self	169.4		[48]
98% TiO_2 (core), 2% alumina (coating)	136.0 ± 35	HR-SEM		0	100	0	DuPont	18.2		[49, 50]
88% TiO_2 (core), 7% amorphous silica, 5% alumina (coating)	149.4 ± 50	HR-SEM		0	100	0	DuPont	35.7		[49, 50]
90% TiO_2 (core), 7% alumina, 1% amorphous silica (coating)	140.0 ± 44	HR-SEM		21	79		DeGussa (P25)	38.5		[49, 50]
N–TiO_2	NA	NA	NA	NA	NA	NA	NA	NA	NA	[51]

MnO_2/TiO_2	4.53–9.17 (pore size)	XRD	NA	100	0	0	NA	98.3–142.6	NA	[52]
Rare earth oxide-TiO_2	9	XRD and TEM	Square	100	0	0	NA	NA	NA	[53]
TiO_2-carbon nanotubes	9.6–11.5	XRD	TiO_2 spheres attached to carbon tubes	84.5–99.5		0.5–15.5	NA	75–156	NA	[54]
Al/TiO_2	25–45	XRD and TEM	NA	100			NA	134–361	NA	[55]
Ag–TiO_2	12.3–32.3	XRD and TEM	NA	All three phases	All three phases	All three phases	NA	NA	NA	[56]
TiO_2–Ni/Fe_2O_4	10–15 and 25–30	XRD	Sphere	100% for 10–15 nm		100% for 25–30 nm	NA	NA	NA	[57]
TiO_2/montmorillonite	50–100	Dynamic light scattering	Sphere	100			NA	122–248	NA	[58]
Polypyrrole/TiO_2	15–20	XRD and TEM	Sphere	100			NA	NA	100% uniform dispersion	[59]
Carotenoid/TiO_2	7	NA	NA	100			Ishihara Co.	NA	NA	[60]

Table 1.2 *Continued*

Nanoparticle composition	Size	Method(s) used for sizing	Shape	Crystal structure (%)			Manufacturer	Surface area per mass ($m^2 g^{-1}$)	Dispersity	Reference
				Anatase	Rutile	Brookite				
CdS/TiO_2	8–10 × 150–300	XRD and TEM	Tube	NA	NA	NA	NA	214–245	NA	[61]
CNT/TiO_2	100–120	TEM	Tube	NA	NA	NA	NA	NA	NA	[62]
Ag/TiO_2	100	TEM	NA	100			NA	11.5	NA	[63]
$Au–TiO_2$	7–8	XRD and TEM	Sphere	100			NA	179.6–184.5	NA	[64]
$ZnO–TiO_2$	9	XRD	Hexagonal	100			NA	NA	NA	[65]
Polyaniline-AMTES-TiO_2	NA	XRD	NA	NA	NA	NA	NA	NA	NA	[66]
TiO_2/Ag	15–25	TEM	Rod	NA	NA	NA	NA	NA	NA	[67]
Au/TiO_2	10–15	XRD and TEM	Sphere	100			NA	NA	NA	[68]
Ag/TiO_2	16–20	XRD	Sphere	100	NA	NA	NA	NA	NA	[69]
Ag/TiO_2	10	XRD	NA	100	NA	NA	NA	157	NA	[70]
Ag/TiO_2 thin film	20	SEM	NA	NA	NA	NA	NA	NA	NA	[71]
$Ag–TiO_2$/Ag/a-TiO_2 thin film	35	SEM and TEM	Spheres in film	100			NA	NA	NA	[72]

Sn^{4+}/TiO_2	9	XRD and SEM	NA	100			NA	100	NA	[73]
$MWNT/TiO_2$	3–24 nm MWNT coated with 3 nm TiO_2	XRD	Needle-like structure	100			NA	172	NA	[74]
S/TiO_2	8.5–11.2	XRD and TEM	Sphere	100			NA	72.9–113.4	NA	[75]
Nd/TiO_2, W/TiO_2, Zn/TiO_2	6–8	XRD and TEM	Tetragonal	100			NA	NA	NA	[76]
EVOH-TiO_2	90	TEM	NA	100			NA	NA	NA	[77]
iPP/TiO_2	80	TEM	NA	100			NA	NA	NA	[78]
$Fe_3O_4@TiO_2$	NA	NA	NA	NA	NA	NA	NA	NA	NA	[79]
Au/TiO_2 film	5–10 nm Au particle in 250–300 nm TiO_2 film	TEM and SEM	Sphere	NA	NA	NA	NA	NA	NA	[80]
TiO_2	38 nm			100	0	0	Alfa Aesar			[81]
TiO_2	5 um						Sigma			[82]
$Fe_3O_4@TiO_2$ core-shell microspheres	Fe_3O_4 microspheres had diameter of 280 nm								10 mg/ml	[83]

Table 1.2 *Continued*

Nanoparticle composition	Size	Method(s) used for sizing	Shape	Crystal structure (%)			Manufacturer	Surface area per mass ($m^2 g^{-1}$)	Dispersity	Reference
				Anatase	Rutile	Brookite				
TiO_2	32 nm	Sintering	NA	60	40	0	Nanophase Technologies Corporations	NA	NA	[84]
TiO_2	49 mm	NA	NA	NA	NA	NA	Nanophase Technologies Corporations	NA	NA	[85]
TiO_2	32 nm	NA	NA	40	60	0	Nanophase Technologies Corporations	NA	NA	[86]
TiO_2	32 nm	Sintering	NA	NA	NA	NA	Nanophase Technologies Corporations	NA	NA	[87]
TiO_2	20, 26, 32, and 56 nm	Sintering	NA	90	10	0	Nanophase Technologies Corporations	NA	NA	[88]

tromagnetic radiation. TiO_2 is a wide gap semi band conductor with a band gap energy of 3.2 eV for the anatase crystal structure, and 3.0 eV for the rutile structure [87]. When TiO_2 absorbs photons of electromagnetic radiation with energy greater than its band gap, valence band electrons are promoted to the conduction band of the TiO_2 molecule, which leaves an electropositive hole in the valence band [2]. Thus, the absorption of electromagnetic radiation by TiO_2 produces electron-hole pairs (e^- h^+) that can be transferred through the material to the surface of the bulk TiO_2. At the surface, the charged electrons (e^-) are spatially separated from the electropositive holes (h^+), thus forming separate reducing and oxidizing centers [2]. When TiO_2 is in an oxygenated aqueous environment, the charged electrons can reduce O_2 to form superoxide ($O_2^{\cdot-}$) whereas, the electropositive holes oxidize water to form hydroxyl radicals ($OH^{\cdot}$) [88]. Thus, the reductive and oxidative abilities of the electron-hole pairs can lead to the production of strong oxidizing agents applicable for many purposes, from chemical to microbial decontamination.

In the nanoparticle regime, TiO_2 preserves its photocatalytic properties; moreover, as the reaction efficiency increases in line with the surface-to-volume ratio of the material, it has been microparticle and nanoparticle formulations TiO_2 rather than the bulk material that have been used in biological, chemical, and industrial applications. Nevertheless, whilst the TiO_2-driven photocatalytic degradation of chemicals and microorganisms has been applied to decontamination and environmental purification, many drawbacks have emerged to prevent its even wider use. The first of the two main issues is centered around the high energy requirements for triggering a photocatalytic response (photoresponse). As the band gap of TiO_2 is 3.2 eV, the anatase crystal can only absorb photons of wavelengths shorter than 388 nm, primarily in the ultraviolet (UV) light spectrum. Hence, TiO_2 nanoparticles can only absorb approximately 2–3% of solar light energies, which makes the commercial applications of TiO_2 nanoparticles ineffective with natural light sources. The second major drawback to using TiO_2 nanoparticles in industrial applications is that the photocatalytic efficiency of TiO_2 is often low, due to charge recombination. During photocatalysis, the reactive electron and the electropositive hole of the electron-hole pairs can recombine within the material before they are transferred to its surface. However, when such recombination occurs, the catalytic efficiency of the nanoparticle decreases.

In an attempt to avoid difficulties associated with TiO_2 use, investigations have been made into the applications of TiO_2 nanocomposites and nanoconjugates. Hence, the primary goals for new TiO_2-based nanoscale materials are to:

- increase the photoresponse (the energetic range at which the nanoparticles can be excited) so that the TiO_2 nanoconjugates can be excited by energies in the visible light spectrum;
- increase the photocatalytic efficiency (the ability of a photocatalytic material to overcome charge recombination and allow separated charges to interact with molecules at the surface of the material).

The major strategies to attain these goals include: conjugation to a charge transfer catalyst; noble metal deposition; doping with metal and nonmetal ions;

blending with metal oxides; coating with photosensitizing dyes; compositing with polymers; and coupling with semiconductors. Although each of the modifications has different mechanisms by which they can either increase photoresponse or photocatalytic efficiency, all were found to aid the overall photocatalytic properties of TiO_2.

1.2.1 Methods for Evaluating Photocatalysis

Before providing descriptions of the different approaches used to increase the photocatalytic ability of TiO_2 nanocomposites, there is a need to outline the various methods generally used in its evaluation. The majority of such methods involve the breakdown of various molecules, typically dyes that change color in a predictable, dose-dependent manner in response to oxidation or reduction. An exception is the use of electron paramagnetic (spin) resonance (EPR), which provides a direct measurement of the production of reactive oxygen species (ROS). The most common dyes used for quantification of photocatalysis include methylene blue and methyl orange, in addition to assays for the degradation of phenol, formaldehyde, salicylic acid and various other oxidizable molecular targets. For example, methylene blue has a blue color in an oxidizing environment, but turns clear in a reducing environment. During photocatalysis, the dyes undergo decomposition that in turn causes a change in color which can be quantified spectrophotometrically, by measuring the absorption of the dye at a specific wavelength.

In contrast, EPR provides a direct measurement of the production of radicals (typically hydroxyl radicals) created by the oxidation of water by electropositive holes.

1.2.2 Different Types of TiO_2 Nanocomposites

1.2.2.1 Modifying TiO_2 with Charge-Transfer Catalysts

Charge-transfer catalysts (CTCs) are molecules that have the ability to trap reactive electrons (e^-) and electropositive holes (h^+) [89]. In the case of TiO_2, the addition of a CTC to the nanoparticle allows for a more efficient trapping of electropositive holes on the surface hydroxyl sites of the TiO_2 molecule [89]. The improved trapping ability of the nanocomposite decreases the charge recombination in TiO_2 and leads to an overall increase in photocatalytic efficiency. The most common CTCs used with TiO_2 are Al_2O_2, Al_2O_3 and SiO_2 [89].

1.2.2.2 Coating with Photosensitizing Dyes

The process of coating creates nanoconjugates rather than nanocomposites; however, these hybrid structures also often have improved photocatalytic abilities compared to the bare nanoparticles. The primary focus of adding a photosensitizing dye to a nanoparticle is most often not to reduce the charge recombination, but rather to change the photoresponse of the TiO_2 nanoparticle. For example, alizarin can lower the photon energy required to excite the nanoparticles, thus decreasing the band gap energy of alizarin-coated TiO_2 nanoparticles to 1.4 eV, in the white light range [8]. In studies conducted by Konovalova *et al.*, the addition of carotenoids to TiO_2 led to the formation of ROS on the surface of nanoconjugates under red light irradiation [60]. One of the most effective photocatalysis reactions with TiO_2 was accomplished by the Gratzel laboratory, when a 10.6% solar light efficiency was achieved by using a dye-sensitized solar cell technology [90].

1.2.2.3 Noble Metal Deposition or Coupling

A noble metal is commonly defined as an element that can resist oxidation, even at high temperatures. Noble metals include rhenium, ruthenium, rhodium, palladium, silver, osmium, iridium, platinum, and gold; of these, the most commonly used in combination with TiO_2 nanoparticles are gold and silver. As noble metals are resistant to oxidation, they are thought to act as an electron sink, promoting the movement of reactive electrons away from the TiO_2 molecule onto the surface of the noble metal [67]. The noble metal surface then acts as a site where redox reactions occur, thus preventing charge recombination within the TiO_2 nanoparticle and increasing its photocatalytic reactivity.

In many studies, silver has been deposited onto TiO_2, primarily because it is more cost-effective than gold and platinum, but also because it has an intrinsic ability to prevent bacterial growth, as well as an effective photocatalytic ability at nanoscale [11]. Previously, silver has been added to TiO_2 nanoparticles, TiO_2 nanorods, and TiO_2 nanofilms. In fact, studies conducted by Li and colleagues have shown that silver-deposited TiO_2 anatase nanoparticles have an improved photoresponse compared to that of anatase TiO_2 nanoparticles, Degussa P25 TiO_2 nanoparticles, and mixed anatase–rutile TiO_2 nanoparticles [63]. The use of Ag–TiO_2 nanocomposite films has also been shown to have an increased photocatalytic reactivity compared to the nonmodified material. For example, UV-illuminated Ag–TiO_2 nanocomposite films are 6.3-fold more effective at photodegrading methyl orange than are UV-illuminated pure TiO_2 films [56].

The deposition of gold and platinum onto TiO_2 nanoparticles has also demonstrated an increase in the photocatalytic reactivity of TiO_2. Yu and coworkers have reported an improved photocatalytic reactivity of Au–TiO_2 nanocomposite microspheres compared to TiO_2 microspheres and Degussa P25 TiO_2 nanoparticles [64]. In addition, UV-illuminated TiO_2 nanofilms embedded with gold nanostructures have a better photonic efficiency than UV-illuminated pure TiO_2 films [80].

1.2.2.4 Doping and Grafting

The purpose of doping TiO_2 nanoparticles with metals is to create a heterojunction – a space which ranges from 10 to 100 nm in size, located between the surface of the doping metal and that of the TiO_2 nanoparticle. Within this space an interior electric field forms that aids in the separation of electron-hole pairs [91]. As a consequence, reduced electrons are driven to different surface sites away from the electropositive holes, which in turn results in a reduction of the charge recombination. A special case of heterojunctions involves the use of carbon nanotubes (CNTs) with TiO_2 nanoparticles or TiO_2 nanofilms [91]. These arrays showed up to 99.1% degradation of phenol, compared to 78.7% degradation by pure TiO_2 nanotubes [91]. Similarly, multi-walled carbon nanotubes (MWCNTs) anchored to TiO_2 nanoparticles improved the photocatalytic degradation of methylene blue [54]. Ni-deposited, CNT-grafted TiO_2 nanocomposites have demonstrated photocatalytic reactivity both with UV irradiation and with exposure to an electric field of 500 V DC, with an enhanced photocatalytic degradation of NO gas molecules in comparison to pure TiO_2 nanoparticles [62]. Other molecules which have been used successfully for grafting include nitrogen, Ag_2S, and Pd–PdO [51, 92, 93]. Nitrogen-doped TiO_2-layered/isosterate nanocomposites showed an increased photocatalytic reactivity when illuminated with visible light between 380 and 500 nm [51].

UV-irradiated TiO_2 nanocomposites doped with rare earth oxides (oxides of Eu^{3+}, Pr^{3+}, Gd^{3+}, Nd^{3+} and Y^{3+}) showed a higher extent of degradation of partially hydrolyzed polyacrylamide (HPAM) [53].

1.2.2.5 **Coupling with Semiconductors, Blending with Metal Oxides**

The coupling of TiO_2 to a narrow-gap semiconductor material can result in an increase in photocatalytic reactivity, as well as an increase in photoresponse. When a narrow-gap semiconductor coupled to a TiO_2 nanoparticle is exposed to visible light, it produces reactive electrons that can travel through the semiconductor to the nonactivated TiO_2 nanoparticle [61]. This process extends the photoresponse of the TiO_2 to visible light wavelengths. Coupling TiO_2 to a semiconductor also decreases charge recombination, because the heterojunction space between the two semiconductors allows for a more efficient separation of reactive electrons and electropositive holes [61]. Common semiconductors and metal oxides coupled to TiO_2 include: CdS, WO_3 and $Cs_xH_{3-x}PW_{12}O_{40}$ [61, 89, 94, 95]. Zhu *et al.* have reported that the bamboo-like CdS/TiO_2-nanotube nanocomposites, when activated with visible light, demonstrated a methylene blue degradation of 83.7% compared to only 9.5% and 41.1% by TiO_2-nanotubes and pure CdS, respectively [61].

The metal oxides ZnO, MnO_2 and In_2O_3, when coupled to TiO_2 nanoparticles, have also been shown as efficient in increasing the photocatalytic capability of TiO_2 nanoparticles [52, 54, 96]. The addition of MnO_2 to TiO_2 nanoparticles broadens the excitation spectrum of TIO_2 to visible light ranges, as demonstrated by methylene blue degradation [52].

A different crystal structure of the TiO_2 may also have an effect on photoresponse. For example, under UV light conditions ZnO–TiO_2 nanotubes can oxidize rhodamine B with a higher efficiency than either pure TiO_2 nanoparticles, pure TiO_2 nanotubes, pure ZnO, or ZnO–TiO_2 nanoparticles [65].

1.2.2.6 **Modifying with Polymers or Clays**

The functionalization of TiO_2 nanoparticles with polymers with good conducting properties can be used to direct the charged electrons (e^-) and electropositive holes

(h^+) away from the surface of TiO_2. Moreover, the addition of polymers that allow for a large internal interface area between the polymer and the TiO_2 particle aids in charge segregation and also prevents charge recombination [59]. Similarly, the addition of clays can aid in charge segregation by providing a large internal interface between the clay and the TiO_2 molecule. In the past, a multitude of polymer/TiO_2 nanocomposites have been used on the basis of their ability to increase photocatalytic reactivity. Most notably, polypyrrole, kaolinite, polyaniline, poly amide and poly-lactic acid (PLA) have each been added to TiO_2 nanoparticles, and all have shown enhanced photocatalytic reactivity of the nanocomposites [59, 66, 97, 98]. For example, visible polypyrrole–TiO_2 nanocomposites degraded 95.54% of methyl orange compared to 40% degradation by visible light with pure TiO_2 nanoparticles [59]. Similarly, Shi-xiong *et al.* demonstrated the ability of UV- and solar light-irradiated anilinomethytriethoxysilane–TiO_2 nanoparticles composited with polyaniline (PANI/AMTES-TiO_2) to increase the photoresponse and reduce charge recombination [66].

Novel methods for developing TiO_2–clay nanocomposites have involved the use of heterocoagulation to create TiO_2–montmorillonite (MMT) nanocomposites, in which a silicate layer is used to support the TiO_2 nanoparticle. Kun and coworkers subsequently showed that these nanocomposites had a higher photocatalytic reactivity than pure TiO_2, due to the added catalytic activity of the silicate support layer of the nanocomposites [58].

1.2.3 Antimicrobial Uses of Nanocomposites

The antibacterial and decontamination applications of TiO_2 nanoparticles have undergone extensive investigation since 1985, when Matsunaga *et al.* first demonstrated the microbicidal effect of TiO_2 [99]. Subsequently, numerous studies have documented the bactericidal activity of TiO_2 nanoparticles, founded on their photocatalytic reactivity under UV-illumination [177]. The illumination of TiO_2 leads to the generation of ROS that oxidize membrane lipids and cause disruption to the outer and cytoplasmic membranes of the bacteria by lipid peroxidation; this leads in turn to the death of the bacterial cells [100].

Since the microbicidal ability of TiO_2 nanoparticles depends on their photoresponse and photocatalytic reactivity, all of the issues pertinent to the improvement of photocatalysis will aid with the killing of bacteria, viruses, and fungi. Again, the creation of nanocomposites to circumvent the partial photocatalytic reactivity of the TiO_2 due to charge recombination, and to increase the photoresponse of TiO_2, would permit nanocomposites to be used for sterilization with sunlight, in a most energy-effective way. Multiple additions can be made to TiO_2 nanoparticles, TiO_2 nanofilms and TiO_2 nanorods so as to create TiO_2 nanocomposites. These additions include noble metal deposition, doping with metal and nonmetal ions, compositing with a polymer, and the creation of core–shell magnetic nanoparticles. Currently, TiO_2 nanocomposites are used in multiple antimicrobial, antifungal, and waste decontamination applications, and can also be used to sterilize medical devices such as catheters and dental implants [71, 101]. TiO_2 nanocomposites have also been tested for sterilization of food packaging and food preparation surfaces, to prevent the bacterial contamination of food [68, 102, 103]. However, perhaps the most often used application of TiO_2 nanocomposites in this area has been for the purification of drinking water and decontamination of waste water [88].

1.2.3.1 Antimicrobial Nanocomposites with Noble Metal Deposition

The most frequently used noble metals in antimicrobial applications are silver and gold. For example, gold-capped TiO_2 nanocomposites have a strong oxidizing ability and showed a 60–100% killing efficacy of *Escherichia coli* [68]. Likewise, silver has long been studied and recognized for its potential as an antimicrobial agent, with silver ions and nanoparticles having been shown capable of killing bacteria, viruses, and fungi [104]. Recently, Ag–TiO_2 nanocomposite powders, Ag–TiO_2 nanofilms, and Ag-deposited Ag–TiO_2 nanocomposite films were all shown to exhibit enhanced photocatalytic reactivities and bactericidal activities compared to TiO_2 nanoparticles and TiO_2 nanofilms. For example, when Zhang *et al.* used a one-pot sol–gel approach to produce 10 nm TiO_2 nanocomposites with a high Ag-loading ability, the nanocomposites showed a complete inhibition of *E. coli* growth at silver concentrations of only $2.4\,\mu g\,ml^{-1}$ [69]. The compositing of Ag into TiO_2 films has been met with similar success; as an example, Liu *et al.* used the Ag doping of a TiO_2 nanofilm to kill silver-resistant *E. coli* when the nanocomposite films were UV light-irradiated [70]. In this case, the bacterial survival rate on the nanocomposite was only 7%, compared to 53.7% on UV light-irradiated pure TiO_2 nanofilms [70]. Similarly, silicon catheters coated with Ag–TiO_2 nanofilms with embedded nanocomposites demonstrated a self-sterilizing effect, with a 99% sterilization of *E. coli*, *Pseudomonas aeruginosa* and *Staphylococcus aureus* after UV illumination [71]. A similar doping of Ag–TiO_2 nanofilms with Ag–TiO_2 nanocomposite particles led, under solar light conditions, to a photocatalytic killing of *E. coli* that was 6.9-fold more effective than with TiO_2 nanofilms, and 1.35-fold more effective than with Ag/a-TiO_2 nanofilms [72]. Finally, UV-illuminated platinum nanoparticles embedded in a TiO_2 nanofilm demonstrated an increase in the photocatalysis-driven killing of *Micrococcus lylae* cells, compared to UV-illuminated pure TiO_2 nanofilms [105].

1.2.3.2 Antimicrobial Doped and Grafted Nanocomposites

The doping of TiO_2 nanoparticles with metals and nonmetals has been shown to be an effective way of increasing the photocatalytic reactivity of TiO_2. The applications for doped TiO_2 nanocomposites range from antimicrobial coatings on textiles, the inactivation of endospores, solid-surface antimicrobial coatings, and aqueous system-based biocides [74, 102, 106]. Another practical application of TiO_2 nanocomposites has been the use of tin (Sn^{4-})-doped TiO_2 nanofilms on glass surfaces, so as to confer a self-cleaning function [73]. In line with this, Sayikan *et al.* showed that Sn^{4-}-doped TiO_2 nanofilms on UV-illuminated glass surfaces had an antibacterial effect against both Gram-negative *E. coli* and Gram-positive *Staph. aureus*, whereas the TiO_2 films alone had no antibacterial effect [73, 75].

The grafting of MWCNTs to TiO_2 was used to inactivate bacterial endospores under UV light conditions, demonstrating a biocidal efficiency (LD_{90}) of 90% for the inactivation of *Bacillus cereus* endospores. Under the same conditions, pure TiO_2 nanoparticles showed no significant biocidal capabilities [74].

In addition to the nonmetal doping of TiO_2 nanoparticles, many groups have focused on the metal and metal-ion doping of TiO_2 nanoparticles. For example, when Venkatsubramanian *et al.* compared the antibacterial and photocatalytic reactivities of W^{4+}, Nd^{3+}, and Zn^{2+}-doped TiO_2 nanocomposites [76], the antibacterial activities of the nanocomposites were rated as follows: $W^{4+}/TiO_2 > Nd^{3+}/TiO_2 > Zn^{2+}/TiO_2 >$ pure TiO_2 nanoparticles [76]. It is believed that tungsten has the greatest effect on photocatalytic reactivity and antimicrobial activity due to its ability to reduce the band gap of TiO_2 and to aid in charge separation, which makes it highly photoresponsive and photocatalytically reactive [76]. Studies conducted by Lui Li-Fen *et al.* showed iron-doped TiO_2 nanocomposites to have a higher capacity for the UV photocatalytic disinfection of *E. coli* (20% survival) than did pure TiO_2 (40% survival) [102].

As an alternative, iron oxide–silicon dioxide–titanium dioxide core-corona-shell nanoparticles showed less photocatalytic reactivity than the pure Degussa P25 TiO_2; however, the ability to "recycle" such nanoconjugate constructs on the basis of their magnetic core outweighed the relative disadvantage of their lesser photocatalytic reactivity [12].

1.2.3.3 Modifying with Polymers

Polymers are commonly used as materials for food packaging, mainly because the addition of TiO_2 nanoparticles to polymer sheets can provide an antibacterial approach to sterilization. This was demonstrated by Cerrada *et al.*, who by incorporating TiO_2 nanoparticles into an ethylene–vinyl copolymer matrix (EVOH) were able to maintain an antimicrobial capacity at the interface of the TiO_2–EVOH nanocomposite; in fact, the UV irradiated TiO_2-EVOH killed 99.9% of all Gram-positive and Gram-negative bacteria tested [77]. In a similar fashion, TiO_2-embedded polymer oxide thin films also showed an enhanced antimicrobial activity. For this, TiO_2 was incorporated into an isotactic polypropylene (iPP) polymeric matrix, so as to create an iPP–TiO_2 thin film nanocomposite which, when illuminated with UV light, showed an 8- to 9-fold log increase in bactericidal effect against *P. aeruginosa* and *Enterococcus faecalis* when compared to pure TiO_2 nanoparticles [78].

1.2.3.4 Creation of Magnetic Nanocomposites

One major problem facing those industries that use TiO_2 nanoparticles to purify water has been to separate the TiO_2 from the system after use. The removal of TiO_2 from water decontamination applications is, in fact, often very difficult, and will involve the use of a slurry system. In order to overcome this problem, a core–shell magnetic nanoparticle was created that consisted of a magnetic core encapsulated by a TiO_2 shell. Thus, in a TiO_2–$NiFe_2O_4$ core–shell magnetic nanoparticle, the core would retain the magnetic properties, and the TiO_2 shell the photocatalytic reactivity. When exposed to UV light, such nanoconjugates would cause a reduction in the growth of *E. coli* [57]. A similar Fe_3O_4@TiO_2 core–shell nanocomposite also exhibited antimicrobial activity. In practice, the nanocomposites were conjugated to an immunoglobulin G (IgG) antibody, which allowed the direct targeting of pathogenic bacteria such as *Staphylococcus saprophyticus, Streptococcus pyogenes,* methicillin-resistant *Staph. aureus,* and multi-antibiotic-resistant *S. pyogenes.* Whilst, after targeting, UV irradiation of the nanocomposites led to a reduction in bacterial survival compared to negative controls [79], the main benefit was that the Fe_3O_4 core allowed them to be separated from solution simply by applying a magnetic field [79].

With the advent of combining different materials with TiO_2, a multitude of light-activated antimicrobial applications has been developed. Typical strategies used to increase the antimicrobial activity of TiO_2 nanoparticles and nanofilms by increasing their photoresponse and photocatalytic reactivity have included noble metal deposition, doping with metal and nonmetal ions, compositing with a polymer, and the creation of core–shell magnetic nanoparticles. Clearly, these newly developed nanocomposites will continue to show promise as antimicrobial agents for both industrial and biological applications.

1.3 Use of TiO_2 Nanoparticles, Nanocomposites, and Nanoconjugates for Phosphopeptide Enrichment from Biological Materials In Vitro

In addition to its photocatalytic capabilities, TiO_2 nanoparticles also demonstrate a surface reactivity that has been harnessed for use in basic science research. As the nanoparticle surface has a high affinity for phosphate groups, this can lead to various nonspecific interactions between TiO_2 nanoparticles and biological materials, such as proteins and DNA [4, 5]. Currently phosphorylated proteins are of major interest in biomedical research, and the application of TiO_2 nanoparticles in this area will be discussed at this point.

Reversible phosphorylation is a critical cellular tool that is used to control key processes such as signal transduction, gene expression, cell cycle progression, cytoskeletal regulation, and apoptosis [81]. The most common phosphorylation targets in proteins are the amino acids serine, threonine and tyrosine, and almost 30% of all proteins in mammalian cells are phosphorylated at some point during their processing [107]. It is assumed that in diseases such as cancer, AIDS, diabetes and neuronal disorders, protein phosphorylation patterns and cell signaling networks have become disturbed, causing the negative health effects of the disease [108]. In recent years, it was shown that TiO_2 nanoparticles could be used to trap and identify phosphopeptides of interest [4], and for this purpose a range of TiO_2 columns has been created that are capable of enriching certain phosphopeptides from a complex peptide mixture. Moreover, this method is not only very effective but also introduces a greater efficiency into the subsequent steps of the total research process.

The most interesting phosphorylated proteins are usually of low abundance, such that the process of phosphorylation is typically sub-stoichiometric in nature. Yet, phosphorylation is also a highly dynamic event, and can occur at multiple sites on a protein when not all of the potential phosphorylation sites are fully occupied. Nonetheless, results with even a phosphopeptide-rich sample may be seriously affected by the presence of nonphosphorylated peptides (as demonstrated using mass spectrometry). As a result, it is very difficult to generate a sample that is suitable for mass spectrometric analysis, as the phosphorylated proteins must first be enriched and separated from their original complex sample. Although phosphorylated proteins are chemically stable, many enzymes can alter their phosphorylation status. For example, the human genome contains about 500 kinases and over 100 phosphatases; hence, when tissues or cells are lysed and samples are extracted, there is a high likelihood that further enzymatic reactions will occur and that the samples will be compromised. The benefit of TiO_2 materials in this respect is that they show a great specificity with regards to the types of oligomer that they bind, and so often are used in conjunction with the technique of immobilized metal affinity chromatography (IMAC). Notably, as TiO_2 can selectively adsorb organic phosphates, it is an ideal material for capturing phosphopeptides [4]. Moreover, TiO_2 has *amphoteric* properties, which allow it to behave as either a Lewis acid or a Lewis base, depending on the pH of the solution used to wash the TiO_2 material. Acidic conditions cause the TiO_2 to be positively charged and to exhibit anion-exchange properties [108]. Consequently, samples prepared for analysis on TiO_2 columns are often dissolved in an acidic solution so as to promote electrostatic binding between the positively and negatively charged groups [109], after which they can be desorbed under alkaline conditions [110]. An additional point is that nanosized materials have high surface area-to-volume ratios, which allows them to bind to a much greater number of targets [4]. One problem here is that TiO_2 can bind multi-phosphorylated peptides so strongly that their elution becomes difficult; consequently, it is more suited to the isolation of mono-phosphorylated peptides [109]. The overall process is very effective and about 50–75% more phosphopeptides will be detected if a TiO_2-enrichment stage precedes the mass spectrometry analysis [110].

TiO_2 nanocomposites can be synthesized via photopolymerization and adapted into stationary chromatography phases for use in either microchannels or micro tips. The composites are prepared by crosslinking TiO_2 nanoparticles with organic groups, which helps to prevent the loss of particles during washing through TiO_2 composite cartridges. In this way, a TiO_2-packed pipette tip may serve as an offline first-dimension separation step in a two-dimensional (2-D) chromatography system [107]. It has in fact been found that certain agents can improve the binding of phosphorylated peptides while blocking the attachment of nonphosphorylated peptides that are not of interest. Loading the samples in 2,5-dihydroxybenzoic acid (DHB) can reduce the binding of nonphosphorylated peptides to TiO_2, while maintaining the high binding affinity for phosphorylated peptides [110].

TiO_2 nanocomposites with a high loading capacity and high capture efficiency were formed by first silanizing nanoparticles with 3-mercaptopropyltrimethoxysilane (MPTMS), and then photopolymerizing them in the presence of a diacrylate crosslinker [4]. Scanning electron microscopy (SEM) images of the nanocomposites revealed an agglomeration of particles, which helped them to be retained within the cartridge when used as a chromatographic packing material. Further investigations revealed that the TiO_2 nanocomposites had twice as much phosphate binding capacity, and a fivefold larger capture efficiency compared to 5 μm TiO_2 particles. Overall, the results of these studies indicated the need to identify a size balance for TiO_2 nanoparticles – they must not be so small as to be lost during the enrichment process, but not too large as to have any significant phenyl phosphate adsorption [4].

Titanium dioxide can also be incorporated into microspheres which consist of an iron oxide core and a titanium dioxide shell, where the iron core imparts magnetic properties on the sphere, and allows the material–target conjugate to be isolated from the solution simply by using a magnet. However, this ultimately will result in a trade-off between process efficiency and accuracy. It has been shown that magnetic core microspheres have an ill-defined structure, and a decreased selective affinity for phosphopeptides [81]. Li *et al.* attempted to overcome these problems by first synthesizing Fe_3O_4@C microspheres, and then attaching titanium via calcination to form the Fe_3O_4@TiO_2 microspheres. Following an enrichment with iron–titanium microspheres, all three potential phosphopeptides could be identified in the tryptic β-casein samples, whereas the non-enriched sample showed, very weakly, one of the potential peptides [81].

1.4 Uptake and Effects of Nanoscale TiO_2 and Nanocomposites in Cells

The prevalence of bulk TiO_2 in common household items such as toothpaste and sunscreens, as well as its importance in industrial syntheses, has led to many studies of how bulk and nanoscale TiO_2 interact with cells. In particular, intensive studies have been undertaken to determine how particle size [111], surface area, and surface chemistry can impact the ability of TiO_2 to enter cells. Some studies have also identified the phagocytic or endocytic pathways by which TiO_2 may enter eukaryotic cells, and how the surface modification or conjugation of biomolecules to its surface can impact on the uptake pathways and nanoparticle retention dynamics [9, 15, 112, 113]. A number of toxicology studies have been conducted to determine how nanoscale/ultrafine TiO_2 can induce inflammatory or apoptotic responses from immune and epithelial cells of the lung [112–117]. More recently, interest has been expressed in using the photocatalytic properties of TiO_2, coupled with the ability to conjugate biomolecules to nanoscale TiO_2, as a novel method for delivering therapeutic or diagnostic agents to malignant cells [9, 13, 16, 118].

The following section is organized according to the sequence of events culminating with the intracellular effects of TiO_2 nanocomposites. The uptake mechanisms important for the internalization of nanocomposites will first be discussed, followed by details of the intracellular localization of internalized nanocomposites. Finally, the molecular and cellular responses of cells to internalized TiO_2 will be outlined.

1.4.1
Uptake of TiO_2 Nanomaterials

The precise mechanism by which TiO_2 crosses the selectively permeable barrier of the plasma membrane is a question that must be considered on a case-by-case basis, because different cells "favor" different uptake pathways for nanoparticles of different size, charge, and surface reactivity. On the other hand, different uptake pathways are associated with different intracellular fates of the nanomaterials internalized into cells. *Phagocytosis* is the predominant method of internalization employed by dedicated immune cells such as macrophages and neutrophils. *Endocytosis,* on the other hand, is used in almost every cell type, and can proceed through four distinct pathways: (i) clathrin-mediated endocytosis; (ii) caveolin-mediated endocytosis; (iii) macropinocytosis; and (iv) the clathrin/caveolin independent pathway. Not every cell possesses each one of these uptake mechanisms, however. In addition, *passive uptake* is a possible mechanism of cellular entry for different small molecules.

The predominant endocytic pathway in cells is that of clathrin-mediated endocytosis, which proceeds via the formation of clathrin-coated membrane invaginations that eventually pinch off to form clathrin-coated vesicles and *endosomes.* Endosomes formed from the clathrin pathway undergo acidification, and are eventually sorted for degradation in lysosomes. The pathways of caveolin-mediated endocytosis and macropinocytosis both have slower kinetics than the clathrin-mediated pathway, but the endosomes formed from these two pathways are not directed to the lysosomes. The final pathway is not well characterized, and is referred to simply as the clathrin/caveolin-independent pathway [119–121].

Studies using transmission electron microscopy (TEM) and energy dispersive X-ray spectroscopy (EDS) have shown that, while bulk TiO_2 was predominantly phagocytosed, TiO_2 nanoparticles were taken up via clathrin-coated pits [113]. Phagocytosis does not seem to be a major contributor to nanoparticle uptake, because the inhibition of phagocytosis (using cytochalasin D; cytD) in macrophages abolished the uptake of micrometer-sized particles, but not of 0.2 μm and 0.1 μm particles [112]. The study authors also noted that the intracellular nanoparticles were not membrane-enclosed, and concluded that non-phagocytic and non-endocytic mechanisms might also be responsible for their uptake. Others have proposed that nanoparticles would be sequestered in endosomes, the origin of which could be attributed to all three major pathways of endocytosis [113, 122]. Here, as in other nanoparticle uptake studies, there was a clear correlation between the nanoparticle localization in the cells, the "availability" of uptake pathways, and the cell type. As not all cell types have every endocytic mechanism [121], identical nanoparticles might be found in different cell compartments of near-isogenic cell lines [123].

Early endosomes formed by the clathrin, caveolin, and macropinocytic processes pursue a defined pathway that leads to the formation of late endosomes, followed by sorting within multivesicular bodies and, finally, fusion with degradative lysosomes [178, 179]. Notably, TiO_2 nanoparticles have been localized to endosomes as well as to multivesicular bodies [113].

Finally, nanoparticles can also penetrate epithelial cells (in particular) by the process of *transcytosis*, where particles are endocytosed from the apical surface of the cell, trafficked through the cytoplasm, and released from the basal cell surface. In an experimental set-up using Caco-2 cells that simulated intestinal epithelial cells *in vitro*, TiO_2 nanoparticles were able to pass through the cells by transcytosis, without disrupting the intercellular junctions or compromising cellular integrity [124].

1.4.2
Intracellular Localization of TiO_2 Nanomaterials

Non-functionalized TiO_2 nanoparticles are not found within the nucleus, nor in other subcellular organelles such as the mitochondria, endoplasmic reticulum, or Golgi apparatus [113]. In one study, nanoparticles were found to have aggregated in the perinuclear regions of cultured bronchial epithelial cells [116], whereas in another study they were seen to be enriched within the lysosomes of mouse fibroblast cells [125]. These differences might be attributed to the different uptake mechanisms that dominate in mouse fibroblasts and bronchial epithelial cells. In fibroblasts, where the nanoparticles were endocytosed, they ultimately appeared in the lysosomes. However, nanoparticles in the bronchial epithelial cells were not taken up by membrane-bound vesicles and could diffuse freely throughout the cytoplasm.

Recently, the surface reactivity of nanoscale TiO_2 has been used to functionalize nanoparticles with biomolecules in the form of dopamine-modified deoxyribonucleic acid (DNA) and peptide nucleic acid (PNA) oligonucleotides capable of hybridizing with cellular DNA (see also below) [5, 9, 10, 13–17]. The oligonucleotide-modified nanoconjugates which had been electroporated into the cells were shown to be retained inside the cells for up to several days post-transfection. Moreover, the bound oligonucleotides that had hybridized to nucleolar or mitochondrial sequences were seen to have aided nanoconjugate retention in the appropriate subcellular compartments such as the nucleolus and the mitochondria, respectively [9, 15]. It is important to note that electroporated nanoconjugates have free access to the cytoplasm, whereas endocytosed nanoconjugates must escape the endosome in order to reach subcellular locations.

The addition of other types of surface moieties can also be used to modulate the mechanisms of nanoparticle uptake. For example, the conjugation of specific antibodies to the nanoparticle surface fostered uptake by cells expressing antigenic cell-surface receptors, as shown in numerous other nanoparticle–cell combinations [118].

In most of these nanoparticle localization studies, TEM or X-ray fluorescence microscopy were used to interrogate the intracellular nanoparticle localization.

However, the addition of a fluorescent molecule such as Alizarin Red S to the nanoparticle as a surface-modifying moiety transformed the TiO_2 nanoconjugates into fluorescent nanoconjugates, suitable for investigations using fluorescence confocal microscopy and flow cytometry [10]. This type of surface modification may facilitate further investigations of nanoconjugate uptake, retention, and subcellular localization.

One unique application of surface-functionalized TiO_2 nanoconjugates is to surface-coat them both with a targeting moiety, such as a DNA oligonucleotide, and with dopamine-modified gadolinium (Gd)-based contrast agents. By using this approach, improved uptake and retention of Gd was obtained in targeted cell population harboring sequences that hybridized with the attached oligonucleotides [13, 16]. Such nanoconjugates would be suitable for magnetic resonance imaging (MRI) studies, and may yet find their place among biomedical diagnostics. When combined with potential therapeutic applications inherent to TiO_2 nanoparticles and nanoconjugates, this might represent a new approach towards the creation of so-called "theranostic nanomaterials," with combined applications in therapy and diagnostics.

1.4.3 Intracellular Interactions of TiO_2 Nanomaterials

Once they have been internalized, TiO_2 nanoparticles, nanocomposites and nanoconjugates can interact either passively or actively with cells. Passive interactions in this context are defined as cellular responses to nanomaterials as a foreign material. For the purpose of this section, active interactions will be defined as cellular events triggered by TiO_2-mediated photocatalysis, leading to the production of ROS and free electrons (e^-) and electropositive holes (h^+) within the intracellular milieu. Active interactions would also include such cases when the surface reactivity of the TiO_2 nanomaterial induced a cellular reaction. Whilst such interactions might have cytotoxic effects in cells, inasmuch as the activity can be targeted only to the population of cells that cause harm to the organism, such cytotoxicity might have therapeutic effects.

Several groups have shown that TiO_2 nanoparticles can induce DNA damage and apoptosis in cultured human peripheral blood lymphocytes [114, 115]. Indeed, not only has DNA damage been demonstrated (through single-cell gel electrophoresis) but also an accumulation of p53, a major regulator of DNA damage-induced cell cycle arrest. In addition, not only have increased levels of p38-MAPK and JNK (both of which are considered to be indirect activators of pro-apoptotic caspases) been demonstrated, but also an activation of caspase-8 [114, 115].

The toxicity of TiO_2 nanoparticles in an absence of photoactivation was also noted in numerous cell-based assays [126]. For this, TiO_2 nanoparticles of different sizes and crystal phases (or their mixtures) were used to treat cells *in vitro* at concentrations up to $10\,\mu g\,ml^{-1}$. In most cases, anatase TiO_2 was shown to induce some signs of cytotoxicity at concentrations above $5–10\,\mu g\,ml^{-1}$, though according to others the cytotoxic effects commenced only at concentrations above $100\,\mu g\,ml^{-1}$. In addition, the consensus was that among all particles smaller than 1 μm in size, TiO_2 was the least toxic, especially when compared to Al_2O_3 and SiO_2.

In a more recent study, NIH 3T3 cells were maintained and treated with TiO_2 particles for 11 consecutive weeks. The nanoparticle sizes ranged between 2 and 30 nm, while within the treated cells increased numbers of multinucleated cells and micronuclei were found, as well as increased levels of polyploidy. In short, these data suggested that the TiO_2 nanoparticles might lead to chromosomal instability and cellular transformation [180]. However, the authors did not elaborate on any possible contribution of ambient light to their findings.

The introduction of potential stressors such as nanoparticles can trigger the expression of cellular stress signals and inflammation by activating leukocytes. The exposure of U937 human monocytes to TiO_2 nanoparticles led to an increased expression of matrix metalloproteases (MMPs)-2 and -9, both of which are involved in tissue remodeling and typically are secreted by monocytes that have been exposed to metal oxides [117]. Results from the same study suggested that nanoscale cobalt was a better activator of MMP-2 and MMP-9 than TiO_2 nanoparticles. Macrophage inhibiting factor (MIF), a pro-inflammatory cytokine, has also been shown to be upregulated in bronchial epithelial cells exposed to bovine serum albumin (BSA)-coated TiO_2 nanoparticles. Furthermore, the oxidative stress caused by the nanoparticles led to increased levels of cytoprotective proteins such as TALDO1, an enzyme that produces reducing equivalents.

One possible way to reduce the toxicity of TiO_2 nanoparticles might be through surface modifications to decrease the reactivity of the nanoparticle surface, and/or modulate its photocatalytic reactivity and photoresponse. One such surface modifier is glycidyl isopropyl ether (GIE), which has been used to mask the reactive surface of nanoscale TiO_2, without significantly affecting either the photocatalytic properties or the internalization of nanoconjugates [10].

1.4.4 Photocatalytic Uses of TiO_2 Nanomaterials to Induce DNA Cleavage and Cytotoxicity

The photocatalytic properties of TiO_2 nanomaterials have been discussed at length in previous sections, in addition to ways in which the efficiency of the process can be improved. Just as the degradation of methylene blue can be used to gauge the photocatalytic reactivity of TiO_2, the degradation of DNA has been used for the same purpose. In experiments conducted as long as 20 years ago, plasmid DNA degradation was used as a parameter of DNA degradation by photoactivated TiO_2 [181]. In a more recent study [182], the effects of 10–20 nm and 50–60 nm anatase and rutile TiO_2 on the formation of 8-hydroxydeoxyguanosine (8-OHdG) in an *in vitro* plasmid assay were found to decrease in the order 10–20 nm anatase > 50–60 nm anatase > 50–60 nm rutile. ROS generated at the surfaces of the nanomaterials as a result of TiO_2 photoactivation were seen to play a role in DNA cleavage [127–129]. Hence, the inclusion of ROS scavengers in TiO_2–DNA nanoconjugate cleavage experiments was shown to lead to a partial, if not complete, loss of DNA cleavage.

Similar to DNA *in vitro*, cellular DNA can sustain damage in cells that contain TiO_2 nanomaterials and have been exposed to UV light. Whilst cells are able to cope with a certain amount of ROS and can protect the integrity of their nuclear and mitochondrial genomes, it is possible to overwhelm the cellular antioxidant machinery and to induce DNA cleavage *in situ*. This in turn, can lead to cell cycle arrest, senescence, or cell death. As noted above, the overpowering of cells with ROS forms the basis of many current anti-cancer treatments, and may lead (potentially) to the therapeutic use of TiO_2 nanoparticles, nanoconjugates, and nanocomposites.

The earliest example of the anticancer use of TiO_2 was the triggering of photocatalysis in media-containing cultured cells [130], while several groups have used intercellular TiO_2 nanoparticles as photosensitizers to cause oxidative damage to malignant cells [131, 132]. In one study, bladder cells treated with TiO_2 were irradiated with UV-A light, after which an increased oxidative stress (as measured by major oxidative products) and increased apoptosis were identified [131].

Currently, most examples of the deliberate use of TiO_2 nanomaterials to induce cytotoxicity are based essentially on the same principle of action as the very first experiments, although the nanoparticles of today are generally smaller than those used in the past. In the following examples, the nanoparticles or nanorods used were less than 20 nm in size, which enabled an efficient nanoparticle uptake. In studies conducted at Cheon's laboratory [183], nanorods of 3.5×10 nm were used to treat A-375 melanoma cells. Having penetrated the cells, the nanorods were then excited by UV lamp illumination, such that the products of photocatalysis caused abundant apoptotic death at UV doses that normally would be harmless to this cell type. In a more recent example, TiO_2 nanoparticles were conjugated to a monoclonal antibody against interleukin (IL)-13αR, a receptor which is overexpressed in glioblastoma multiforme. This functionalization improved nanoparticle uptake by glioma cells (GBM and A172) in an antigen presentation-dependent fashion [118]. The cytotoxicity was found to be impaired by the addition of ROS quenchers, however, and particularly by those with singlet oxygen ($^1O_2$) and hydroxyl radical ($OH^•$) traps.

While random DNA scission caused by ROS has its place in therapeutic approaches, it would be desirable to control this activity and to induce DNA degradation in specific locations in the genome; a preferred example would be at oncogene loci. In order to achieve such precision with the help of nanoparticles, several exploratory studies were conducted by Woloschak and coworkers [5]. As noted above, at the nanoscale, TiO_2 molecules on the surface of nanoparticles form "corner defects" that create a propensity for stable nanoparticle conjugation to *ortho*-substituted bidentate ligands such as 3,4-dihydroxyphenethylamine(dopamine) and 3,4-dihydroxyphenylacetic acid (dopac). These anchor molecules allow for further conjugation with new molecules via amide linkages; such covalent binding to the inorganic surface is energetically favorable because it allows the TiO_2 to regain the native octahedral geometry [7]. Dopamine is used as a linker for the attachment of oligonucleotides made from DNA and PNA, as well as peptides, antibodies and MRI contrast agents [5, 9, 10, 13–17]. When oligonucleotides are bound to the nanoparticle it is possible for h^+ to be transferred from the TiO_2 nanoparticle through the dopamine linker onto the attached biomolecule. It has been shown that TiO_2 nanoparticles can act as an electron sink, thereby allowing an accumulation of h^+ on the attached DNA molecules, leading in turn to strand cleavage at the site where the electropositive holes accumulate. Paunesku *et al.* have shown that DNA oligonucleotides bound to TiO_2 nanoparticles can participate in enzyme reactions and retain the ability to hybridize with complementary DNA sequences. In the same study, the photoactivation of TiO_2 led to site-specific cleavage of the hybridized oligonucleotides [9]. The cleavage event is thought to occur by the accumulation of multiple h^+ at guanine residues, leading to the formation of guanine cation radicals that can react with neighboring water molecules. The attachment to the nanoparticle of a DNA strand that is complementary to a sequence of cellular DNA will then allow for a targeted cleavage of the genetic material. The results of several studies have indicated that cleavage efficiency is heavily dependent on the degree of nucleic acid strand hybridization. Indeed, Rajh and colleagues have demonstrated an *in vitro* DNA sequence-specific cleavage [133]. In a study conducted by Tachikawa *et al.*, the presence of mismatches between the oligonucleotide attached to the nanoparticle and the complementary oligonucleotide target was shown to modulate the extent of DNA cleavage [134]. To explain this phenomenon, the authors hypothesized that in a perfectly hybridized DNA–TiO_2 nanoconjugate the h^+ were more efficiently trapped on the nucleic acid moiety and did not undergo charge recombination with electrons on the surface of the nanoparticle. In a more recent study, the same group hypothesized that strand cleavage would most likely require both the transfer and accumulation of h^+ across the DNA strand, as well as the presence of free photogenerated ROS, by showing that cleavage could be quenched in the presence of ROS scavengers [135].

1.5
Use of Titania Oxide and Its Composites for Implants and Tissue Engineering

For many years, titanium and Ti-based alloys have been used extensively in permanent implants for orthopedic, dental, and prosthetic applications. More recently, however, titanium has been used on the nanoscale for implant surface modifications and tissue-engineering applications. With the details of the biological response to an implant placement having been elucidated at the subcellular level, nanotechnology has been utilized for the surface modification of titanium implants to maximize the natural tissue response, to achieve implant integrity, and to prevent implant failure. For tissue-engineering applications, TiO_2 has been integrated into bioactive glass composites for use as scaffolds for bone tissue generation. Tissue engineering has already superseded autologous bone grafts in the repair of fractures, bone defects, the resolution of long-bone nonunions, total joint revision surgery, repair of tumor resection, and spine fusion. Clearly, as technology improves and the techniques are refined, tissue-engineering applications will undoubtedly become much more prevalent.

1.5.1 Osseointegration

Osseointegration, as originally defined by Brånemark, is a direct structural and functional connection between ordered living bone and the surface of a load-carrying implant. An implant is considered to be osseointegrated when there is no progressive relative movement between the implant and the bone tissue [136, 137]. Clinically, osseointegration refers to the successful implant installation and healing, with the correct stress distribution when in function [138]. In order to achieve implant osseointegration, it is first necessary to achieve primary mechanical stability of the implant, as this will allow for biological fixation by the osteoid tissue and trabecular bone at an early stage [139]. Titanium serves as a substrate for bone formation, and provides a stable interface for the formation of new bone. Notably, TiO_2 is biologically inert and allows for hydroxyapatite (HAp) to form on its surface and act as a bonding layer for further bone development [140]. The healing process begins with an inflammatory response when the implant is inserted into the bone cavity. Within the first day, mesenchymal cells, pre-osteoblasts, and osteoblasts are recruited to the implant surface, and begin to produce collagen fibrils of osteoid tissue; this is followed initially by woven bone and then mature bone formation with trabeculae rich in capillaries. Trabecular bone is a calcified tissue that forms the initial architectural network within the space between the implant and bone, providing a high resistance to implant loading. Its architecture is a complex three-dimensional (3-D) network, with arches and bridges providing a biological scaffold for cell attachment and bone deposition on the implant surface. Thus, the development of novel methods to enhance implant function will depend heavily on knowledge of these processes.

1.5.2 Implantation Methods

Many of the novelties in implantation methodology have been built upon an established understanding of conventional implant–biological interactions and the recent development of nanotechnology, with emphasis placed on the modification of implant surfaces at the nanoscale level. Typical examples include nanograined surface modification, nanophase ceramics, and nano-rough poly-lactic-*co*-glycolic acid (PLGA)-coated-nanostructured titanium. Other developments include TiO_2-doped phosphate-based glasses for applications in bone tissue engineering. The use of nanotechnology in the realm of implant engineering is based on the fact that naturally occurring proteins are of nanoscale dimensions. Bone, for example, is a nanostructured material composed of proteins such as collagen type I, which contains linear fibrils that are 300 nm long and 0.5 nm in diameter, in addition to HAp crystals that range from 2 to 5 nm in thickness and from 20 to 80 nm in length [82, 141, 142]. It has been postulated that, because cells interact with nanostructured surface substrates in a biological setting, nanophase materials are able to duplicate the natural surface behavior of cells.

1.5.3
Osteoblast Adhesion

The initial osteoblast adhesion is a rapid response to implant placement, and involves short-term chemical interactions between the cells and the implant surface that initially involve ionic and van der Waals forces. The subsequent regenerative phase is a longer process, with continued cell migration to the adhesive site and the production of extracellular matrix (ECM) proteins, cell membrane proteins, and cytoskeleton proteins. This sequence of events comprises the natural healing process, and results from a signal cascade leading to the production of transcription factors and subsequent genetic induction [139].

It has been shown that osteoblast attachment, proliferation, viability, and morphology are promoted, together with an enhanced cytocompatability, on nanograined and nanophase materials such as alumina, titania, and HAp with surface pore sizes of less than 100 nm [82, 86, 142, 143]. Ceramics such as titania and alumina have long been used in implants on the basis of their favorable biocompatibility with osteoblasts. These ceramics have been produced via the sintering of TiO_2 powder at 600 °C to generate a surface roughness of less than 20 nm. Composites prepared in this manner have demonstrated an enhanced osteoblast function when compared to conventional ceramics [82, 86, 142]. Nanophase ceramics have the added benefit of physical properties that closely mimic those of physiological bone, thus promoting enhanced calcium and phosphate precipitation.

Ceramics with nanoscale topography demonstrate an increased specificity and selectivity for the adhesion of osteoblasts, with reduced adhesion of fibroblasts and endothelial cells, most likely because nanophase ceramics adsorb higher levels of vitronectin whereas conventional ceramics are more selective for laminin. Osteoblasts demonstrate increased adhesion and proliferation on nanoceramics, while endothelial cells show increased activity on conventional ceramics. Furthermore, the stereochemistry of vitronectin may make its adsorption more selective for the smaller pores of nanoceramic surfaces [82, 142].

1.5.4
Modification of Surface Chemistry

The modification of surface chemistry has been shown to increase the biological response at implant surfaces. Both, hydroxyl and carbonyl groups on the implant material surfaces demonstrate good support for osteoblasts, thus promoting adhesion, proliferation, and differentiation [144, 145]. It has been established that hydroxylated and hydrated titanium surfaces increase the amount of free surface energy, and also induce osteoblast differentiation; they have also been shown to be associated with the generation of an osteogenic microenvironment through the production of prostaglandin (PG) E_2 and transforming growth factor-beta (TGF-β) [146]. Studies of Ti polymers with surface pore sizes of 32 nm, when treated with NaOH, have provided evidence of the ability to simulate the surface and chemical properties of bone and cartilage. The combined effects of surface nanotopography and chemical modulation serve to maximize the biological responses [83, 84].

1.5.5
Artificial Bone Substitute Materials

Various types of bioactive ceramics, glasses, and glass ceramics have been developed for use as artificial bone substitute materials. These neither damage healthy tissue, nor pose any viral or bacterial risk to patients, and can be supplied at any time, in any quantity. Examples include Bioglass® in the Na_2O–CaO–SiO_2–P_2O_5 system [147], sintered HAp; $Ca_{10}(PO_4)_6(OH)_2$ [148], sintered β-tricalcium phosphate (TCP) ($Ca_3(PO_4)_2$ [149], HAp/TCP bi-phase ceramics [150], and apatite-containing glass ceramics [151]. Phosphate glasses represent a unique class of biomaterials that are biodegradable, biocompatible, and for which the rate of degradation can be modulated from days to several months. The primary advantage of using bioactive glasses is that there is no need for surgical removal once their function has been fulfilled. Rather, they are broken down and harmlessly expelled from the body. *In vivo* studies, which included a bone-healing model of marrow ablation of the rat tibia, have shown that filling the intramedullary space with bioactive glass microspheres results in new bone formation and a high turnover of the local bone [152]. During the primary biological response to bioactive glass, relatively undifferentiated mesenchymal tissue surrounds the microspheres. Initially, the tissue differentiates into an immature woven bone structure, followed by bone remodeling and the formation of lamellar bone (though this is limited by the presence of the bioactive glass microspheres). As the microspheres begin the dissolution process, new bone is continually formed in their place. It has been proposed that there is a balance between the gradual dissolution of bioactive glass matrix, and the synthesis of new bone on its surface [153]. The rate of dissolution is controlled by increasing the covalent character of the bonds within the glass structure; this is achieved by doping the degradable composition with modified metal oxides that have a greater field strength, such as Fe_2O_3, CuO, Al_2O_3, and TiO_2 [154–158]. The ability to modulate the properties of bioactive glass allows for many potential applications as reinforcing agents for composites, and also as scaffolds for tissue engineering [159].

In addition to bone regeneration, TiO_2 nanoparticles have been used recently to create an extracorporeal culture system for hepatocytes that would mimic the *in vivo* microenvironment. For this, TiO_2 nanoparticles and nanorods of 25–75 nm in size were dispersed uniformly on the surface of the 120–300 μm pores of a highly porous chitosan structure. When the attachments of hepatic cell line HL-7702 to a chitosan scaffold and to a scaffold decorated by nanoparticles were compared, the overall difference between the two cell systems was insignificant. The most notable difference was the fact that the TiO_2/chitosan structures contained more spheroids, whereas the chitosan-only structures contained more single HL-7702 cells [160].

1.6 Toxicology Studies of TiO_2 Nanomaterials in Animals

Although no whole-animal studies appear to have been conducted with nanoconjugates and nanocomposites of TiO_2, many have described the toxicity of TiO_2 itself. Neither have there been any demonstrations of the diagnostic and/or therapeutic applications of TiO_2. In considering the potential toxicity of TiO_2 *in vivo*, it can be assumed that there will be no light exposure and activation of TiO_2 photocatalysis of cellular components; however, there is still a need to consider the surface reactivity of TiO_2 nanomaterials, their ability to bind to proteins and nucleic acids, and a general propensity of nanomaterials to interact with molecules within a similar size range [161]. In addition to the fact that biological responses caused by nanoscale materials may be very different from those caused by bulk materials [39, 162], the increased surface reactivity of TiO_2 at the nanoscale may, potentially, lead to an aggregation of nanoparticles, and to the triggering of immune responses, the clogging of vessels and ducts, and an accumulation in organs associated with the filtration of blood and lymph. At present, the conditions that either increase or decrease the rate of aggregation of TiO_2 nanomaterials have not yet been widely studied in biological systems, in part because there are so many factors to consider, such as crystal structure and size, pH, and the ionic strength of the colloid. Moreover, in a study of microscale TiO_2 particles, anatase TiO_2 was found to produce more ROS than its rutile counterpart, adding crystal polymorphology as yet another potential factor to be controlled when evaluating the toxicity of TiO_2 nanoparticles *in vivo* [163].

It has been argued that the total surface area-per-unit-mass is the best predictor of TiO_2 toxicity *in vivo*, whereas others have maintained that this characteristic is not indicative of toxicity at all [164]. Whilst photocatalytic reactivity of TiO_2 *in vivo* is unlikely, concern has been expressed that ROS may be generated at the particle surface when TiO_2 nanoparticles interact with biological processes, potentially causing oxidative damage to adjacent tissues.

The expected clearance of TiO_2 nanomaterials is via the reticuloendothelial system (RES) and, because of their small size, they would be expected to locate to the liver, spleen, lymph nodes, and kidneys. In the field of cancer diagnostics and therapeutics, nanoscale molecules would potentially accumulate in tumors because of the poorly developed and largely fenestrated blood vessels of the neovasculature. Poor lymphatic drainage would reduce the clearance of these nanoscale structures from the tumor. This phenomenon, which is referred to as the enhanced permeability and retention (EPR) effect, is a mainstay of most so-called "passive targeting" approaches to concentrate nanoparticles within tumors.

While little doubt exists regarding the clearance route of TiO_2 nanomaterials, there are many possible routes for nanoparticle entry into the organism, including inhalation, intratracheal instillation, oral administration, oral, nasal and bronchial lavage, and intraperitoneal and intravenous delivery. Not all of these approaches have been used in animals, or were used only sporadically. In the interest of focus and brevity, the following sections are organized in terms of the most common routes of entry used to study TiO_2 nanoparticle toxicity in mammalian models; moreover, only nanomaterials <100 nm in size were considered.

1.6.1 Inhalation and Intratracheal Instillation

The vast majority of studies of the effects of TiO_2 nanomaterials in mammals have focused on their interactions with the respiratory system. Such studies have been conducted in two general ways:

- *Inhalation* requires the use of nebulizers or other devices to create an aerosol of TiO_2 nanoparticles, and occasionally a closed housing unit to fit the test subjects. Knowledge of the animal's respiratory rate, lung volume, and controlling for exposure time and air flow rate allows the relatively accurate measurement of the quantity of material to which the subject is being, or has been, exposed.
- *Intratracheal instillation* not only closely simulates environmental conditions but is also relatively inexpensive, as it can be conducted with a syringe and the dosage of test material much more easily defined. Intratracheal instillation also avoids the possibility of nanoparticles being absorbed through the skin, which is a concern during inhalation [32].

1.6.1.1 Intratracheal Instillation

Results on the long-term toxicity of TiO_2 nanoparticles administered via intratracheal instillation have varied greatly. One group used mice to compare 3 nm and 20 nm nanoparticles, varying the concentrations from 0.4, 4, and 40 mg kg^{-1} body weight. After three days, the 3 nm particles showed a dose-dependent toxicity comparable to that of the 20 nm nanoparticles, which had only about one-third of the total surface area, indicating that size and surface area were not factors of toxicity with this experimental set-up and delivery route. In both cases, the nanoparticles caused slight increases in total protein, albumin, alkaline phosphatase, acid phosphatase and lactate dehydrogenase (LDH) in the bronchoalveolar lavage fluid (BALF), and were present in macrophages on histological examination [32]. However, another one-week study showed 5 nm anatase nanoparticles to cause more severe pulmonary toxicity than 21 nm and 50 nm particles at 5 and 50 mg kg^{-1}, respectively, but relatively no toxicity at 0.5 mg kg^{-1}. Inflammatory infiltrates and interstitial thickening were observed on histological examination. At the highest dose (50 mg kg^{-1}), it was suggested that the 5 nm anatase nanoparticles inhibited the phagocytic ability of alveolar macrophages, although no mechanism was suggested [165]. Several studies of this type have used 21 nm nanoparticles (80/20, anatase/rutile) that were purchased commercially. It appeared that a single dose (5, 20, 50 mg kg^{-1}) of these nanoparticles could cause a dose-dependent increase in a number of pro-inflammatory and T-cell-derived (Th1- and Th2-type) factors in the BALF at one day post-treatment, although ultimately these particles might exert chronic inflammatory damage on the lungs via a Th2 mediated pathway

[38]. Over one to two weeks, 21 nm rutile particles were found to induce pulmonary emphysema, to disrupt the septal walls of the alveoli, and to cause epithelial wall apoptosis – altogether a much more drastic response than had been observed by others with comparable doses (3–16 $mg\,kg^{-1}$ body weight) [22]. A longer-duration study in rats (1, 7, and 42 days) at lower doses showed an increased ROS production (by immunohistochemical staining), and persistence of inflammatory and cytotoxic factors, when compared to fine TiO_2 particles. In this case, it was suggested that the smaller size of the nanoparticles had allowed them to pass more easily into the interstitium, as they were also found in much greater frequency in adjacent lymph nodes. However, by controlling for equal surface area exposure, TiO_2 nanoparticles were found to be only slightly more immunogenic and cytotoxic than their fine-particle counterparts – a finding which conflicted with conclusions drawn by others [40]. A comparison between 10 nm TiO_2 anatase "dots", 200 × 35 nm anatase rods, and 300 nm rutile particles administered in a single intratracheal instillation at either 1 or 5 $mg\,kg^{-1}$ to rats, concluded that the different particle types were no more cytotoxic or immunogenic in the lung than their counterparts, and indicating that surface area was not a factor in pulmonary toxicity [48]. However, others considered the surface area and particle size to be more important determinants of toxicity than the mass of material administered [166].

1.6.1.2 **Inhalation**

Studies using inhalation as an exposure method also showed varied results. In a comparison of the inhalation of ultrafine versus fine TiO_2 nanoparticles, the ultrafine material caused a more significant microvascular dysfunction [36]. Comparisons between nanoparticles (20 nm) and pigment-grade particles (250 nm) of TiO_2 in rats during inhalation exposure over three months (6 hours per day, five days per week) showed mildly severe focal alveolitis after six months. However, at one year post-exposure the levels of alveolitis had largely returned to those of the untreated controls, although more alveolar macrophages persisted. One group concluded that, pending cessation, chronic exposure to TiO_2 nanoparticles over a limited period of time had no significantly toxic lasting effects on the lungs [19]. Rats were shown to develop a much more severe inflammatory response to 21 nm TiO_2 nanoparticles than mice, but the fibroproliferative and epithelial changes in lungs were accompanied by increased macrophage and neutrophil numbers in the BALF, particularly at an aerosol concentration of 10 $mg\,m^{-3}$. Over the course of one year post-treatment, however, there was a dose-dependent decrease in toxicity markers, even at this concentration [20]. A study of 2–5 nm particles at an aerosol concentration of 0.77 and 7.22 $mg\,m^{-3}$ showed moderate inflammatory responses in lung sections at up to two weeks after exposure, although by the third week only minimal lung toxicity or inflammation was observed, indicating a return to baseline [27]. In most cases, alveolar macrophages and neutrophils persisted in the BALF, but levels of inflammatory cytokines and markers had generally returned to normal [167].

Despite the copious data generated on the topic of TiO_2 nanoparticle toxicity, it is not entirely clear whether the nanoparticle dispersity, dose, form, or some other factor can be used as the definitive predictor for determining lung toxicity. It is possible that a combination of factors may be the true determinant to evaluate the effects on the mammalian respiratory system, although differing responses to the same treatments have also been found across different species. What does appear to be true is that, in low doses (whether via intratracheal instillation or inhalation), TiO_2 nanoparticles are relatively nontoxic and do not carry any lasting effects. Higher doses, however, tend to be retained and may increase inflammatory mediators, macrophage and neutrophil infiltration, and possibly emphysema-like reactions in the lungs. Alveolar macrophages may be impaired in their ability to phagocytose these nanoparticles if their capacity is overloaded, resulting in a decreased clearance [168]. From this, it is apparent that it is important to continue studying the effects of TiO_2 nanoparticles on the lungs, and to develop acceptable standards for inhalation exposure to TiO_2 nanoparticles, not only from a single encounter but also over extended periods of time, in order to simulate chronic low-dose exposure conditions.

1.6.2 Dermal Exposure

Recent reviews of nanoparticle dermal toxicity have included an overview of studies conducted with TiO_2 nanoparticles [169], though all of these utilized microfine TiO_2 that mostly originated from sunscreen lotions. One study that included 10/15 and 20 nm TiO_2 nanoparticles showed that, in human subjects, exposure to such materials led to an accumulation of TiO_2 no deeper than the outermost layers of the striatum corneum [170]. Other studies [171] with 10.1, 5.2 and 3.2 nm TiO_2 nanoparticles of different crystal structure demonstrated some cytotoxicity and inflammation at and above concentrations of 100 mg ml^{-1}, in conjunction with the anatase crystal phase.

1.6.3 Ingestion and Oral Lavage

The accumulation of TiO_2 nanomaterials in the environment might eventually become distributed in the food and water supply of natural wildlife and, eventually, of humans. It has been suggested that nanosized materials can cross the intestinal epithelium and exert effects on other organs in the body, notably the spleen and kidneys [172]. An interesting study using simulated intestinal epithelial cells *in vitro* (cell line Caco2) showed that TiO_2 nanoparticles at concentrations above 100 μg ml^{-1} would accumulate at the apical surface of the cells, pass through the junctions, and traverse through the cells by transcytosis [124].

In a subsequent study, male and female mice were administered a single large oral dose (5 g kg^{-1} body weight) of a TiO_2 nanoparticle suspension via a syringe [45]. The TiO_2 nanoparticles were 25 and 80 nm in size, and any effects were compared to a dose of 155 nm TiO_2 particles over the course of two weeks. The animals were sacrificed after two weeks and the internal organs analyzed for their titanium content. Any histopathological changes were also evaluated, and blood serum collected to monitor any changes in biochemical markers of organ-specific damage or inflammation. No major changes were found in the body weights of all treatment groups compared to controls, although female mice showed a significant increase in liver weight relative to total body weight. The alanine aminotransferase: aspartate aminotransferase (ALT/AST) ratio, which serves as a marker of liver damage, was significantly increased in the 25 nm nanoparticle-treated group compared to controls, but possible reasons for this gender-specific effect were not discussed. Pathological changes were identified around the central vein of the liver, and hepatocyte necrosis was evident. Serum levels of LDH and alpha-hydroxybutyrate dehydrogenase (markers of cardiovascular damage) were increased in the 25 nm- and 80 nm-nanoparticle treated mice, compared to control and 155 nm-nanoparticle treated mice. Although the 80 nm group showed the highest increases, no pathological changes were identified on histological examination of the heart. The 25 nm group also showed increased blood urea nitrogen (BUN) levels, but no significant changes in creatine kinase. Histology showed proteinaceous liquid in the renal tubules of the 80 nm mice, with significant swelling in the renal glomerulus of the 155 nm-treated group, but no obvious pathology in the 25 nm-treated group. In fact, no abnormal histological changes were found in any of the tissues of mice treated with 25 nm TiO_2 nanoparticles. An examination of the heart, lungs, testicles/ovaries, and spleen showed no overt pathology on histological examination. The titanium content analyses showed an accumulation in the liver, kidneys, spleen, and lungs (in decreasing order), which confirmed the expectation of an accumulation in the RES with eventual excretion via the kidneys [45]. Overall, the 80 nm TiO_2 nanoparticles were found to be more toxic to mice after oral ingestion, particularly to the liver, while 25 nm TiO_2 nanoparticles had no appreciable effects on the liver compared to untreated and 155 nm TiO_2 particle-treated controls. The 25 nm nanoparticles were deposited in the spleen, kidneys, and lungs of test subjects, although no overt pathology was identified. Some kidney dysfunction was noted, possibly due to the high load of TiO_2 being excreted.

1.6.4 Intravenous Injection

The majority of studies of the toxic effects of TiO_2 and TiO_2-containing nanomaterials have focused on mimicking environmental exposure (inhalation, ingestion, etc.). However, in order to study the pharmacokinetics of TiO_2 nanomaterials more directly, it is necessary to recreate a situation of 100% bioavailability, namely intravenous injection [25]. In this way, information obtained from studies in which other routes of administration have been evaluated can be placed into context, and also aid in extrapolating the data acquired from animal models to the human situation [25]. Studies have also been conducted with TiO_2 nanomaterials as a tool for cancer diagnosis and therapy [10, 13, 16], further supporting the need for a thorough evaluation of the effects of TiO_2 nanomaterials in the bloodstream. Unfortunately, very few studies have focused on the potential toxicity of TiO_2

nanomaterials in mammals in the setting of intravenous injection. It is to be expected that TiO_2 nanomaterials would be scavenged by the RES [173]. When bulk anatase TiO_2 (0.2–0.4 um) was administered intravenously (via a tail vein) to female Sprague-Dawley rats, no overt pathology was observed on histological examination. However, over a 24 h period, bulk TiO_2 was localized primarily to the celiac lymph nodes, liver and spleen, in decreasing order. After one year, accumulation in the celiac lymph nodes still far exceeded (over 18-fold) that in the mediastinal lymph nodes, where the next largest accumulation of TiO_2 was found, followed by the spleen, liver, and lungs. Accumulation in the celiac lymph nodes was deemed appropriate, as these particular lymph nodes filter lymph from the liver, indicating that the majority of the TiO_2 had traversed the liver before their deposition in the celiac lymph nodes [173]. Further investigations of the intravenous administration of TiO_2 have been conducted based on the results of this study.

A study in male Wistar rats used a single bolus injection of TiO_2 nanoparticles into the tail vein, at a concentration of 5 mg kg^{-1} body weight (0.5%). These nanoparticles were 20–30 nm in diameter, and had a surface area of 48.6 $m^2 g^{-1}$. In 200–300 g rats this translates to 1–1.5 mg TiO_2 per rat; the equivalent average dose in humans would be a single bolus of 350–450 mg TiO_2. The rats were sacrificed at 1, 14, and 28 days post-injection, and the TiO_2 content analyzed in blood cells, plasma, kidneys, spleen, brain, lymph nodes, liver, and lungs. Only the liver, spleen, lungs, and kidneys showed detectable levels of TiO_2 accumulation (in decreasing order), whilst only the liver retained TiO_2 at the final 28-day time point [25]. However, bearing in mind that the nanoparticles were smaller and did not aggregate, a much more efficient renal clearance could have been expected [174].

For nanoparticles of 20–30 nm diameter, persistent liver accumulation is to be expected when considering the principles behind the EPR effect and the profuse fenestrations of the liver. Just as nanoparticle treatments take advantage of the poorly formed and unnaturally fenestrated blood supply of tumors, those organs designed to filter the blood may also selectively acquire untargeted nanomaterials, regardless of their nature. Although a further analysis of cytokines and enzymes was conducted, there was no indication of any inflammation or cytotoxicity. Thus, in rats at 100% bioavailability, a dose level of 5 mg kg^{-1} body weight of TiO_2 nanomaterial showed neither toxic nor adverse effects [25].

One key component that was not addressed in this study was the aggregation potential of TiO_2 nanomaterials under biological conditions. A menagerie of components would be involved in mimicking an *in vitro* model of TiO_2 nanomaterial interactions with biological conditions. While pretreatment must always involve a disaggregation step (most commonly by prolonged sonication), when the nanomaterials had been administered the biological milieu (blood, peritoneum, lung) would immediately enhance the aggregation status of TiO_2 [39, 162]. By using dynamic light scattering (DLS) techniques it is possible to assess the aggregation status of TiO_2 nanomaterials under different conditions. Likewise, the use of EDS allows the presence of TiO_2 to be detected in tissue sections. DLS experiments *in vitro* showed a dramatic increase in aggregation (as characterized by hydrodynamic diameter) when the TiO_2 nanoparticles were moved from pure water to saline. In this case, three dilutions were evaluated: 100×, 1000×, and 10 000×. Whilst there was no appreciable difference between aggregation status (hydrodynamic size) in Milli-Q H_2O versus 10 mM NaCl, the *Z*-avg when nanoparticles were solvated in phosphate-buffered saline (PBS) rose dramatically, by more than 20-fold. Although the pH of Milli-Q H_2O and 10 mM NaCl were equivalent at each dilution (6.83–5.35), the pH of PBS ranged from 7.43 to 7.4. The effects of controlled pH changes at constant ionic strength were not evaluated in this study [39, 162]. This change was attributed not so much to the increase in pH as to the increase in ionic strength of the surrounding solution. It was postulated that the electric double layer surrounding the surface of the charged nanoparticles would be diminished in the presence of an increasing ionic strength of the solution, thus screening the electrostatic repulsion that would hinder aggregation in low-ionic strength solutions. For this reason, the electrostatic repulsive forces would be overcome by van der Waals attractions and the nanoparticles would aggregate [39, 162].

These same nanoparticles were injected either intravenously ($560\,mg\,kg^{-1}$) or subcutaneously ($5600\,mg\,kg^{-1}$) into female Balb/c mice, which were sacrificed at three days post-treatment and their organs examined histopathologically. By using a combination of transmission electron microscopy (TEM) and scanning electron microscopy (SEM)-EDS, TiO_2 aggregates were identified in the liver, kidneys, lungs, lymph nodes, and spleen in the setting of intravenous injection, while none were identified in sections of the heart and brain. This indicated that TiO_2 nanoparticles of this size do not accumulate appreciably in the heart, nor are they able to cross the blood–brain barrier. Liver sections showed the greatest accumulation of TiO_2 aggregate material, with the larger aggregates accumulating within the vacuoles of Kupffer cells (the monocytes of the liver), and smaller aggregates being distributed more widely between the sinusoidal spaces and the liver parenchyma. The uptake by Kupffer cells was expected, based on their role as phagocytic cells in the RES [39, 162]. While there was occasional evidence of macrophage activation and lymphocyte infiltration, there was no overt cellular degeneration of inflammation in response to infiltration and the deposition of TiO_2 aggregates. Lung sections showed a profuse distribution of large (up to 200 μm) aggregates along the alveolar walls, accompanied by an activation of intravascular macrophages, although no material was found in the alveolar lumens or the intravascular macrophages themselves. Fewer aggregates were found in the kidneys, lymph nodes, and spleen; however, beyond the colocalization of a few phagocytic cells, no obvious pathology was evident in tissue sections of these organs.

Subcutaneous injection revealed aggregates only in the liver, lymph nodes, and spleen histological sections from test subjects; however it should be noted that subsequent experiments (not described) demonstrated the presence of TiO_2 also in the kidneys. Overall, these experiments highlighted the dependence of the organ distribution of TiO_2 nanoparticles on the route of administration, and the change in aggregation potential in solutions of increasing ionic strength as applicable to a biological model [39, 162].

The administration of TiO_2 nanomaterials by intravenous injection permits the study of systemic effects in the setting of 100% bioavailability, while bypassing any potential route-specific organ pathologies associated with more natural environmental exposure. The biodistribution of injected nanomaterials in male Wistar rats follows the expected clearance by the RES, with TiO_2 found to accumulate primarily in the liver, lymph nodes and spleen at up to 28 days after injection, with a decrease in TiO_2 levels over time [25]. The aggregation of TiO_2 nanoparticles due to changes in the ionic strength of the surrounding milieu may perhaps affect the rate of clearance of TiO_2 once a certain size limit has been reached, although at low doses (up to $560\,mg\,kg^{-1}$ body weight) hematoxylin and eosin staining of histological sections and an analysis of various circulating markers of inflammation and cellular degeneration showed no obvious pathological changes compared to controls [25, 39, 162]. Taken together, the results of these studies indicated a relative lack of toxicity of low-dose TiO_2 nanomaterials in mammals within the setting of intravenous injection, and supported the safety of using TiO_2 nanomaterials in the development of novel diagnostics and treatments [13, 16].

1.6.5 Intraperitoneal Injection

As an alternative to intravenous injection, the intraperitoneal route has been selected as a means of studying the *in vivo* transportation of TiO_2 nanomaterials to various organs, and thus any pathological changes that might arise following the systemic administration of TiO_2 nanomaterials. As the majority of reported applications of TiO_2 have involved concern as to potential adverse environmental effects, toxicology studies have focused on common routes of exposure such as ingestion, inhalation, and dermal absorption [168]. The use of TiO_2 nanomaterials as a diagnostic and therapeutic tool has only recently entered the realm of applications [5, 9, 10, 13–17, 118]. As with most cancer diagnostics and therapeutics, these applications of TiO_2 nanomaterials are likely to involve intravenous administration so as to ensure 100% bioavailability. Whilst, in the laboratory setting, it may be commonplace to use intraperitoneal injections (e.g., the use of streptozotocin to eliminate insulin-producing cells to reproduce a Type 1 diabetes mellitus model [175]), the correlation to environmental or therapeutic exposure in this setting is limited. The effects of intraperitoneal injections can vary, depending on the side at which the material was injected (left or right flank, i.e., liver side or spleen side), or the skill of the individual performing the injection (was the injection consistently accurate, with no damage to the internal organs?). The injected material may adhere to internal organs and viscera as it has complete access to the peritoneal cavity, whereas alternative routes of exposure (e.g., inhalation, ingestion, intravenous injection) would require the material to pass through any combination of layers and types of tissue before gaining access to the serosal surfaces of the organs in the peritoneum. Whilst it has been suggested that particles in the nanoscale regime can readily pass through various types of tissue, it is unlikely that this passive diffusive capacity can approach the availability of an intraperitoneal injection [176]. In addition, the volume of material being injected may have significant effects on the outcome, considering that the introduction of a large volume into the peritoneal cavity does not have any reasonable corollary when considering the effects of environmental exposure. A much more appropriate approach to studying the transportation of TiO_2 nanomaterials *in vivo*, and the possible effects on various organ systems without the concern of a rate-limiting interaction between nanomaterial and organ system of exposure, would be that of intravenous injection. Nevertheless, toxicology studies of TiO_2 nanomaterials administered via intraperitoneal injection have been conducted, albeit with varying results.

One such study of intraperitoneal injection in mice used 80–110 nm anatase TiO_2 nanoparticles, prepared in house using a sol–gel method under acidic conditions. The mice were injected with different dosages (0, 324, 648, 972, 1296, 1944, or 2592 $mg\,kg^{-1}$ body weight) of TiO_2 nanomaterial, and the histopathology was evaluated at 7 and 14 days post-treatment. Blood samples were analyzed for markers of hepatotoxicity and nephrotoxicity at 24 and 48 h post-treatment. The markers included ALT, AST, alkaline phosphatase (ALP), and BUN. It should be noted that, given the body mass of an average human, these dosages translate to upwards of 180–190 g of TiO_2 nanomaterials in a single bolus injection into the peritoneum. From a clinical perspective, it may not be reasonable to interpret this experiment as an adequate representation of potential environmental or even clinically relevant exposure, but rather as an example of acute toxicity at extreme doses. The animals' behavioral patterns and physical characteristics were observed, when it was noted that all mice showed signs of passivity, loss of appetite, tremor and lethargy compared to controls; however, these signs gradually disappeared in the mice receiving lower dosages (324, 648, and 1296 $mg\,kg^{-1}$). The mice receiving high doses (1944 and 2592 $mg\,kg^{-1}$) also showed signs of anorexia, diarrhea, lethargy, tremor, weight loss, and lusterless skin [23]. Unfortunately, no scaled rating system was used to qualify or quantify these findings, and the biochemical assays for serum parameters showed wide variations in ALT and AST levels at different treatment dosages. While an effect may be present, it would be incorrect to consider that a definitive concentration-dependence of AST and ALT could be discerned from these data. The authors noted no significant effects on BUN (kidney function) over time, and asserted that TiO_2 nanoparticles injected into the peritoneum had a greater impact on the liver than on the kidneys [23].

The titanium contents of tissues were evaluated using inductively coupled plasma-mass spectrometry (ICP-MS) in tissues removed at 24, 48, 168, or 336 h post-injection. The spleen, heart, lungs, kidneys, and liver were digested and the titanium content was examined over time. Accumulation in the spleen was seen to be dose-dependent, but not so in other tissues. An examination of the data presented, however, showed a large decrease in titanium content from 24 to 48 h, with little change from 48 h to 7 days, but then another increase between 7 and 14 days. Whilst this discrepancy was unexplained, it is possible that over longer time points the organ distribution of TiO_2 nanoparticles changed. There was also a potential for matter injected into the peritoneum to adhere to the serosal surfaces of the peritoneal organs, and TiO_2 nanoparticles cannot be ruled out as an exception. While studies focusing on alternative routes of administration need not be concerned with this possibility, studies of the intraperitoneal injection of TiO_2 nanomaterials would do well to account for this surface accumulation, which might skew any inferences of the systemic distribution of TiO_2. The liver titanium content was significantly less than that of the spleen at all time points, which contrasted with other studies using intraperitoneal injection [33]. Again, given the propensity for injected material to adhere to the peritoneal surfaces, the side of the peritoneum at which the injection was made could represent a major factor in the location of the greatest TiO_2 accumulation, and this was a clear shortcoming of studies where this method of administration was applied. The total body titanium content was not assessed, and neither was the titanium content of fecal matter and urine collected and analyzed. An evaluation of waste for titanium content would strengthen studies of how mammals clear TiO_2 nanoparticles, although at the time of this writing such studies have yet to be conducted. Mild increases were found in the titanium contents of the lungs and kidneys, but no titanium was detected in the heart [23].

Despite the relatively large splenic content of titanium, the histological examination of target organs showed a greater pathology in the liver, followed in order of decreasing severity by the spleen, kidneys, and lungs. The high-dose groups showed hepatic fibrosis around the central vein, apoptotic bodies, and minor fatty change in the liver, accompanied by swelling in the renal glomerulus and dilatation and an accumulation of protein-rich liquid in the renal tubules. Spleen sections showed massive inflammation and neutrophil infiltration in the high-dose groups, and alveolar septal thickening and mild neutrophil infiltration was found in the lungs. No histological changes were found in the heart. In addition, 8% (5/60) of the injected mice that died were in the high-dose groups. Given that these changes were found only in the highest-dose groups (1944 and 2592 $mg\,kg^{-1}$), these data infer a relatively low toxicity of TiO_2 nanoparticles when injected at low doses. However the dose-dependence of the biochemical parameters and histological changes at low doses cannot be stated accurately from these data [23].

One important point made by Liu *et al.* was that the majority of studies, regardless of the route of administration, tended to focus on a single large bolus injection or the instillation of TiO_2 nanomaterials, whereas ecological exposure would more likely be chronic [33]. In order to study these chronic effects, Liu and coworkers designed a study that included multiple, smaller exposures to 5 nm anatase TiO_2 nanoparticles, and for comparison administered commercially available bulk rutile TiO_2 particles with an average grain size of 15–20 μm [33]. For this study, female ICR mice were injected intraperitoneally daily for 14 days, and then sacrificed. The titanium contents of the liver, kidneys, spleen, lungs, brain and heart were analyzed using ICP-MS and evaluated according to the organ coefficient (the ratio of organ tissue to body weight). A comprehensive list of biochemical parameters was also assessed to evaluate the effects of treatment on organ function and on carbohydrate and lipid metabolism.

The liver, kidneys, and spleen all showed an increase in net weight at doses of 50, 100, and 150 $mg\,kg^{-1}$, but these were equivalent to increases in the organ weights of animals treated with 150 mg kg^{-1} bulk TiO_2. The lungs and brain showed a decrease in net organ weight, indicating possible degenerative damage. All of the organs tested showed a dose-dependent increase in titanium content, although interestingly the bulk TiO_2-treated mice showed titanium contents closer to those of the 100 mg kg^{-1}-treated group, which suggested that 5 nm anatase TiO_2 nanoparticles might enter the organs more easily than bulk TiO_2.

Biochemical parameters evaluated in the liver showed increases at higher doses (notably of ALA and alkaline phosphatase), followed by more moderate increases in leucine acid peptide, pseudocholinesterase, and total protein. Increases were also noted in the livers of bulk TiO_2-treated animals, though to a lesser extent than with equivalent concentrations of nanoparticle-treated animals. Similarly, serum markers of kidney function showed decreases (notably uric acid and BUN), with lesser decreases in bulk TiO_2-treated animals. These data indicate an increased toxicity of nanoscale TiO_2 to the liver and kidneys compared to equivalent concentrations of bulk TiO_2 particles. On evaluating markers of metabolic equilibrium, glucose, total cholesterol, triglycerides and high-density lipoprotein were found to have increased (but with no effect on low-density lipoprotein) at higher doses of TiO_2, indicating metabolic toxicity of TiO_2 nanoparticles. Whilst not evaluated in this study, an accumulation of titanium in the pancreas might suggest the cause of this change in carbohydrate and lipid metabolism. It was concluded that, in the higher doses tested (100 and 150 mg kg^{-1}), repeated exposure to TiO_2 nanoparticles caused toxicity and inflammation in the liver, kidneys and myocardium, in addition to a disruption of carbohydrate and lipid metabolism, compared to bulk TiO_2 particles. Hence, it was suggested that the size, crystal structure (anatase versus rutile) and route of administration might represent significant factors in the toxicity of TiO_2 nanoparticles, though lower concentrations were relatively nontoxic [33].

1.6.6 Subcutaneous Administration

In one (relatively rare) study in pregnant mice, the transfer of subcutaneously injected nanoparticles was examined, and the subsequent effects on the genital and cranial nerve systems were evaluated. (Note: This study is included as an example of TiO_2 nanoparticle administration that does not mimic environmental exposure. Moreover, as the site of injection was not indicated, the role of possible TiO_2 nanoparticle migration and proximity of the site to the target organs are again drawn into question.)

Pregnant Slc:ICR mice were administered 25–70 nm anatase TiO_2 nanoparticles subcutaneously, with each mouse receiving 100 μl of a 1 mg ml^{-1} TiO_2 in saline plus 0.05% Tween 80 surfactant (to prevent nanoparticle aggregation) [39, 162]. The TiO_2 doses were administered at 3, 7, 10, and 14 days post-coitum, and the male offspring sacrificed at 4 days or 6 weeks of age.

Histological sectioning and immunohistochemical staining for caspase-3 (a marker of apoptosis) showed significant toxicity relating to TiO_2 nanoparticle exposure in the testes, epididymes, and seminal vesicles. At both time points, TiO_2 nanoparticles were detected in the Leydig cells, Sertoli cells and spermatids in the testis. There was, in addition, an obvious disorganization and disruption of the normal morphology of the seminiferous tubules, and fewer mature spermatozoa were observed, indicating a severe toxic effect in the male reproductive organs. Sperm motility was notably decreased, and the mitochondria of spermatozoa collected from TiO_2 nanoparticle-treated animals showed significant damage. The mean body weight of TiO_2 nanoparticle-exposed offspring was only 88% of that of their nonexposed counterparts. An inspection of the olfactory bulb and cerebral cortex of the brain of these mice at six weeks showed the presence of crescent-shaped cells (a known feature of apoptosis) and numerous caspase-3-positive cells in the TiO_2 nanoparticle-treated mice. Whilst previously, TiO_2 nanoparticles have been shown not to accumulate in the brain, it was postulated that because the blood–brain barrier was not yet fully developed at the time of injection, the TiO_2 could pass from mother to fetus, so as to exert pathological effects on the brain and interfere with normal fetal central nervous system development. Although, in other studies, the effects on the male reproductive organs were not examined, the blood–testis barrier (much like the blood–brain barrier) may not have been fully developed at the time of treatment [43].

While studies of the intraperitoneal or subcutaneous injection of TiO_2 nanoparticles do not translate readily into comparable situations of natural environmental or therapeutic exposure, they do confirm the results obtained after intravenous injection. At low doses, TiO_2 nanoparticles injected into the peritoneum of mammals proved to be relatively nontoxic and to accumulate predominantly in the RES, with minimal pathological changes to those organs where accumulation had occurred. However, high doses result in significant TiO_2 nanoparticle aggregate accumulation and toxicity to the liver, spleen and kidneys, as was evident from the fluctuations in serum levels of biochemical markers and pathological changes.

1.7 Conclusions

TiO_2 nanoparticles, nanocomposites and nanoconjugates show variable degrees of photocatalytic reactivity, photoresponse and surface reactivity, all of which influence their interactions with biological systems. Whilst it is difficult – and perhaps even impossible – to arrive at a consensus based on currently available information, several conclusions have come to the fore:

1. Anatase TiO_2 nanoparticles – especially those below 20 nm in size, where this crystal phase predominates – have the most pronounced surface reactivity, as evidenced by their interactions with phosphoproteins, cells in culture, and tissues *in vivo*. This surface reactivity in and of itself can cause derangements in cells and organisms, depending on the nanoparticle concentration. According to the most conservative estimates, anatase TiO_2 induces some signs of

cytotoxicity at concentrations above 5–10 μg ml^{-1}, although more relaxed estimates claim that 10-fold higher concentrations still pose no risk of cytotoxicity. Any therapeutic or diagnostic treatment with anatase TiO_2 nanoparticles must take these dose limits into consideration in order to protect healthy tissues from such cytotoxic effects. Consequently, modifications of the nanoparticle surface will be essential in order to prevent nanoparticles from aggregating and behaving as macroparticles, which are subject to different *in vivo* clearance mechanisms than nanoparticles. With aggregates, a greater accumulation can be expected, for example in the Kupffer cells of the liver and in alveolar macrophages, and in turn an extended retention in the liver, lungs and spleen and a relatively slow clearance of aggregates via the kidneys. Generally, nano-anatase TiO_2 nanomaterials will show good renal clearance, provided that their surface is rendered "inert" with respect to any crossreaction with tissues and other nanoparticles.

2. The photoresponse and photocatalytic reactivity of TiO_2 nanomaterials may be harnessed to deliberately induce cell ablation. This capacity depends, as in chemical degradation and microbicidal activity, on the ability to illuminate and activate nanoparticles or nanocomposites. Composites of TiO_2 will likely be better agents than pure TiO_2 for these applications, and the main obstacles during the course of the therapeutic use of such nanomaterials will be the ability to target the nanomaterials to the cells in question and to deliver light of the correct wavelength so as to induce cytotoxicity only in the desired cells and tissue locations.

In the meantime, TiO_2 nanoparticles, nanocomposites and nanoconjugates will continue to be used for different industrial purposes, from water purification to self-sterilizing surgical devices.

References

1 Naicker, P.K., Cummings, P.T. *et al.* (2005) Characterization of titanium dioxide nanoparticles using molecular dynamics simulations. *Journal of Physical Chemistry B*, **109** (32), 15243–9.

2 Fujishima, A. and Honda, K. (1972) Electrochemical photolysis of water at a semiconductor electrode. *Nature*, **238** (5358), 37–8.

3 Chen, X. and Mao, S.S. (2006) Synthesis of titanium dioxide (TiO2) nanomaterials. *Journal of Nanoscience and Nanotechnology*, **6** (4), 906–25.

4 Liang, S.S., Makamba, H. *et al.* (2006) Nano-titanium dioxide composites for the enrichment of phosphopeptides. *Journal of Chromatography A*, **1116** (1–2), 38–45.

5 Brown, E.M.B., Paunesku, T. *et al.* (2008) Methods for assessing DNA hybridization of peptide nucleic acid-titanium dioxide nanoconjugates. *Analytical Biochemistry*, **383** (2), 226–35.

6 Chen, L.X., Rajh, T. *et al.* (1997) *Journal of Physical Chemistry B*, **101**, 10688.

7 Rabatic, B.M., Dimitrijevic, N.M. *et al.* (2006) Spacially confined corner defects induce chemical functionality of TiO_2 nanorods. *Advanced Materials*, **18**, 1033–7.

8 Rajh, T., Chen, L.X. *et al.* (2002) Surface restructuring of nanoparticles: an efficient route for ligand-metal oxide

crosstalk. *Journal of Physical Chemistry B*, **106** (41), 10543–52.

9 Paunesku, T., Rajh, T. *et al.* (2003) Biology of TiO_2-oligonucleotide nanocomposites. *Nature Materials*, **2** (5), 343–6.

10 Thurn, K.T., Paunesku, T. *et al.* (2009) Labeling TiO_2 nanoparticles with dyes for optical fluorescence microscopy and determination of TiO_2-DNA nanoconjugate stability. *Small*, **5** (11), 1318–25.

11 Cozzoli, P.D., Comparelli, R. *et al.* (2004) Photocatalytic synthesis of silver nanoparticles stabilized by TiO_2 nanorods: a semiconductor/metal nanocomposite in homogeneous nonpolar solution. *Journal of the American Chemical Society*, **126** (12), 3868–79.

12 Yao, K.F., Peng, Z. *et al.* (2009) Preparation and photocatalytic property of TiO_2-Fe_3O_4 core-shell nanoparticles. *Journal of Nanoscience and Nanotechnology*, **9** (2), 1458–61.

13 Endres, P.J., Paunesku, T. *et al.* (2007) DNA-TiO_2 nanoconjugates labeled with magnetic resonance contrast agents. *Journal of the American Chemical Society*, **129** (51), 15760–+.

14 Paunesku, T., Stojicevic, N. *et al.* (2003) Intracellular localization of titanium dioxide-biomolecule nanocomposites. *Journal De Physique IV*, **104**, 317–19.

15 Paunesku, T., Vogt, S. *et al.* (2007) Intracellular distribution of TiO_2-DNA oligonucleotide nanoconjugates directed to nucleolus and mitochondria indicates sequence specificity. *Nano Letters*, **7** (3), 596–601.

16 Paunesku, T., Ke, T. *et al.* (2008) Gadolinium-conjugated TiO_2-DNA oligonucleotide nanoconjugates show prolonged intracellular retention period and T_1-weighted contrast enhancement in magnetic resonance images. *Nanomedicine: Nanotechnology, Biology and Medicine*, **4** (3), 201–7.

17 Wu, A., Paunesku, T. *et al.* (2008) Titanium dioxide nanoparticles assembled by DNA molecules hybridization and loading of DNA interacting proteins. *NANO*, **3** (1), 27–36.

18 Gentleman, D.J. and Chan, W.C. (2009) A systematic nomenclature for codifying engineered nanostructures. *Small*, **5** (4), 426–31.

19 Baggs, R.B., Ferin, J. *et al.* (1997) Regression of pulmonary lesions produced by inhaled titanium dioxide in rats. *Veterinary Pathology*, **34** (6), 592–7.

20 Bermudez, E., Mangum, J.B. *et al.* (2004) Pulmonary responses of mice, rats, and hamsters to subchronic inhalation of ultrafine titanium dioxide particles. *Toxicological Sciences*, **77** (2), 347–57.

21 Blaise, C., Gagne, F. *et al.* (2007) Ecotoxicity of selected nano-materials to aquatic organisms. *13th International Symposium on Toxicity Assessment, Toyama, Japan.*

22 Chen, H.W., Su, S.F. *et al.* (2006) Titanium dioxide nanoparticles induce emphysema-like lung injury in mice. *FASEB Journal*, **20** (13), 2393–5.

23 Chen, J.Y., Dong, X. *et al.* (2009) In vivo acute toxicity of titanium dioxide nanoparticles to mice after intraperitoneal injection. *Journal of Applied Toxicology*, **29** (4), 330–7.

24 Drobne, D., Jemec, A. *et al.* (2009) In vivo screening to determine hazards of nanoparticles: nanosized TiO_2. *Environmental Pollution*, **157** (4), 1157–64.

25 Fabian, E., Landsiedel, R. *et al.* (2008) Tissue distribution and toxicity of intravenously administered titanium dioxide nanoparticles in rats. *Archives of Toxicology*, **82** (3), 151–7.

26 Federici, G., Shaw, B.J. *et al.* (2007) Toxicity of titanium dioxide nanoparticles to rainbow trout (*Oncorhynchus mykiss*): gill injury, oxidative stress, and other physiological effects. *Aquatic Toxicology*, **84** (4), 415–30.

27 Grassian, V.H., O'Shaughnessy, P.T. *et al.* (2007) Inhalation exposure study of titanium dioxide nanoparticles with a primary particle size of 2 to 5 nm. *Environmental Health Perspectives*, **115** (3), 397–402.

28 Griffitt, R.J., Hyndman, K. *et al.* (2009) Comparison of molecular and histological changes in zebrafish gills exposed to metallic nanoparticles. *Toxicological Sciences*, **107** (2), 404–15.

29 Heinlaan, M., Ivask, A. *et al.* (2008) Toxicity of nanosized and bulk ZnO, CuO and TiO_2 to bacteria *Vibrio fischeri* and crustaceans *Daphnia magna* and *Thamnocephalus platyurus*. *Chemosphere*, **71** (7), 1308–16.
30 Jemec, A., Drobne, D. *et al.* (2008) Effects of ingested nano-sized titanium dioxide on terrestrial isopods (*Porcellio scaber*). *Environmental Toxicology and Chemistry*, **27** (9), 1904–14.
31 Lee, S.W., Kim, S.M. *et al.* (2009) Genotoxicity and ecotoxicity assays using the freshwater crustacean *Daphnia magna* and the larva of the aquatic midge *Chironomus riparius* to screen the ecological risks of nanoparticle exposure. *Environmental Toxicology and Pharmacology*, **28** (1), 86–91.
32 Li, J.A., Li, Q.N. *et al.* (2007) Comparative study on the acute pulmonary toxicity induced by 3 and 20 nm TiO_2 primary particles in mice. *Environmental Toxicology and Pharmacology*, **24** (3), 239–44.
33 Liu, H.T., Ma, L.L. *et al.* (2009) Biochemical toxicity of nano-anatase TiO_2 particles in mice. *Biological Trace Element Research*, **129** (1–3), 170–80.
34 Lovern, S.B., Strickler, J.R. *et al.* (2007) Behavioral and physiological changes in *Daphnia magna* when exposed to nanoparticle suspensions (titanium dioxide, nano-C-60, and C(60)HxC(70) Hx. *Environmental Science and Technology*, **41** (12), 4465–70.
35 Nemmar, A., Melghit, K. *et al.* (2008) The acute proinflammatory and prothrombotic effects of pulmonary exposure to rutile TiO_2 nanorods in rats. *Experimental Biology and Medicine*, **233** (5), 610–19.
36 Nurkiewicz, T.R., Porter, D.W. *et al.* (2008) Nanoparticle inhalation augments particle-dependent systemic microvascular dysfunction. *Particle and Fibre Toxicology*, **5**, Article no. 1.
37 Oberdorster, G., Ferin, J. *et al.* (1994) Correlation between particle size, in vivo particle persistence, and lung injury. *Environmental Health Perspectives*, **102** (Suppl. 5) (96), 173–9.
38 Park, E.J., Yoon, J. *et al.* (2009) Induction of chronic inflammation in mice treated with titanium dioxide nanoparticles by intratracheal instillation. *Toxicology*, **260** (1–3), 37–46.
39 Patri, A., Umbreit, T. *et al.* (2009) Energy dispersive X-ray analysis of titanium dioxide nanoparticle distribution after intravenous and subcutaneous injection in mice. *Journal of Applied Toxicology*, **22**, 22.
40 Sager, T.M., Kommineni, C. *et al.* (2008) Pulmonary response to intratracheal instillation of ultrafine versus fine titanium dioxide: role of particle surface area. *Particle and Fibre Toxicology*, **5**, Article no. 17.
41 Sager, T.M. and Castranova, V. (2009) Surface area of particle administered versus mass in determining the pulmonary toxicity of ultrafine and fine carbon black: comparison to ultrafine titanium dioxide. *Particle and Fibre Toxicology*, **6**, Article no. 15.
42 Scown, T.M., van Aerle, R. *et al.* (2009) High doses of intravenously administered titanium dioxide nanoparticles accumulate in the kidneys of rainbow trout but with no observable impairment of renal function. *Toxicological Sciences*, **109** (2), 372–80.
43 Takeda, K., Suzuki, K.I. *et al.* (2009) Nanoparticles transferred from pregnant mice to their offspring can damage the genital and cranial nerve systems. *Journal of Health Science*, **55** (1), 95–102.
44 van Ravenzwaay, B., Landsiedel, R. *et al.* (2009) Comparing fate and effects of three particles of different surface properties: nano-TiO_2, pigmentary TiO_2 and quartz. *Toxicology Letters*, **186** (3), 152–9.
45 Wang, J.X., Zhou, G.Q. *et al.* (2007) Acute toxicity and biodistribution of different sized titanium dioxide particles in mice after oral administration. *Toxicology Letters*, **168** (2), 176–85.
46 Wang, J.X., Chen, C.Y. *et al.* (2008) Potential neurological lesion after nasal instillation of TiO_2 nanoparticles in the anatase and rutile crystal phases. *Toxicology Letters*, **183** (1–3), 72–80.
47 Wang, J.X., Liu, Y. *et al.* (2008) Time-dependent translocation and potential impairment on central nervous system by intranasally instilled TiO_2

nanoparticles. *Toxicology*, **254** (1–2), 82–90.

48 Warheit, D.B., Webb, T.R. *et al.* (2006) Pulmonary instillation studies with nanoscale TiO_2 rods and dots in rats: toxicity is not dependent upon particle size and surface area. *Toxicological Sciences*, **91** (1), 227–36.

49 Warheit, D.B., Hoke, R.A. *et al.* (2007) Development of a base set of toxicity tests using ultrafine TiO_2 particles as a component of nanoparticle risk management. *Toxicology Letters*, **171** (3), 99–110.

50 Warheit, D.B., Webb, T.R. *et al.* (2007) Pulmonary toxicity study in rats with three forms of ultrafine-TiO_2 particles: differential responses related to surface properties. *Toxicology*, **230** (1), 90–104.

51 Matsumoto, T., Iyi, N. *et al.* (2007) High visible-light photocatalytic activity of nitrogen-doped titania prepared from layered titania/isostearate nanocomposite. *Catalysis Today*, **120** (2), 226–32.

52 Xue, M., Huang, L. *et al.* (2008) The direct synthesis of mesoporous structured MnO_2/TiO_2 nanocomposite: a novel visible-light active photocatalyst with large pore size. *Nanotechnology*, **19** (18), Article no. 185604.

53 Li, J.H., Yang, X. *et al.* (2009) Rare earth oxide-doped titania nanocomposites with enhanced photocatalytic activity towards the degradation of partially hydrolysis polyacrylamide. *Applied Surface Science*, **255** (6), 3731–8.

54 Wang, Q., Yang, D. *et al.* (2007) Synthesis of anatase titania-carbon nanotubes nanocomposites with enhanced photocatalytic activity through a nanocoating-hydrothermal process. *Journal of Nanoparticle Research*, **9**, 1087–96.

55 Yang, C.S., Wang, Y.J. *et al.* (2009) Photocatalytic performance of alumina-incorporated titania composite nanoparticles: surface area and crystallinity. *Applied Catalysis A: General*, **364** (1–2), 182–90.

56 Yu, J.G., Xiong, J.F. *et al.* (2005) Fabrication and characterization of Ag-TiO_2 multiphase nanocomposite thin films with enhanced photocatalytic activity. *Applied Catalysis B: Environmental*, **60** (3–4), 211–21.

57 Rana, S., Rawat, J. *et al.* (2005) Antimicrobial active composite nanoparticles with magnetic core and photocatalytic shell: TiO_2-$NiFe_2O_4$ biomaterial system. *Acta Biomaterialia*, **1** (6), 691–703.

58 Kun, R., Mogyorosi, K. *et al.* (2006) Synthesis and structural and photocatalytic properties of TiO_2/montmorillonite nanocomposites. *Applied Clay Science*, **32** (1–2), 99–110.

59 Li, S.Y., Chen, M.K. *et al.* (2009) Preparation and characterization of polypyrrole/TiO_2 nanocomposite and its photocatalytic activity under visible light irradiation. *Journal of Materials Research*, **24** (8), 2547–54.

60 Konovalova, T.A. (2004) Generation of superoxide anion and most likely singlet oxygen in irradiated TiO_2 nanoparticles modified by carotenoids. *Journal of Photochemistry and Photobiology*, **162**, 1–8.

61 Zhu, J.H., Yang, D. *et al.* (2008) Synthesis and characterization of bamboo-like CdS/TiO_2 nanotubes composites with enhanced visible-light photocatalytic activity. *Journal of Nanoparticle Research*, **10** (5), 729–36.

62 Kuo, C.S., Tseng, Y.H. *et al.* (2007) Synthesis of a CNT-grafted TiO_2 nanocatalyst and its activity triggered by a DC voltage. *Nanotechnology*, **18**, Article no. 465607.

63 Li, G.H. and Gray, K.A. (2007) The solid-solid interface: explaining the high and unique photocatalytic reactivity of TiO_2-based nanocomposite materials. *Chemical Physics*, **339**, 173–87.

64 Yu, J.G., Yue, L. *et al.* (2009) Hydrothermal preparation and photocatalytic activity of mesoporous Au-TiO_2 nanocomposite microspheres. *Journal of Colloid and Interface Science*, **334** (1), 58–64.

65 Wang, L.S., Xiao, M.W. *et al.* (2009) Synthesis, characterization, and photocatalytic activities of titanate nanotubes surface-decorated by zinc oxide nanoparticles. *Journal of Hazardous Materials*, **161** (1), 49–54.

66 Min, S.X., Wan, F. *et al.* (2009) Preparation and photocatalytic activity of

PANI/AMTES-TiO_2 nanocomposite materials. *Acta Physico-Chimica Sinica*, **25** (7), 1303–10.

67 Cozzoli, P.D., Fanizza, E. *et al.* (2004) Role of metal nanoparticles in TiO_2/Ag nanocomposite-based microheterogeneous photocatalysis. *Journal of Physical Chemistry B*, **108** (28), 9623–30.

68 Fu, G.F., Vary, P.S. *et al.* (2005) Anatase TiO_2 nanocomposites for antimicrobial coatings. *Journal of Physical Chemistry B*, **109** (18), 8889–98.

69 Zhang, H.J. and Chen, G.H. (2009) Potent antibacterial activities of Ag/TiO_2 nanocomposite powders synthesized by a one-pot sol-gel method. *Environmental Science and Technology*, **43** (8), 2905–10.

70 Liu, Y., Wang, X.L. *et al.* (2008) Excellent antimicrobial properties of mesoporous anatase TiO_2 and Ag/TiO_2 composite films. *Microporous and Mesoporous Materials*, **114** (1–3), 431–9.

71 Yao, Y., Ohko, Y. *et al.* (2008) Self-sterilization using silicone catheters coated with Ag and TiO_2 nanocomposite thin film. *Journal of Biomedical Materials Research Part B - Applied Biomaterials*, **85B** (2), 453–60.

72 Akhavan, O. (2009) Lasting antibacterial activities of Ag-TiO_2/Ag/a-TiO_2 nanocomposite thin film photocatalysts under solar light irradiation. *Journal of Colloid and Interface Science*, **336** (1), 117–24.

73 Sayilkan, F., Asilturk, M. *et al.* (2009) Photocatalytic antibacterial performance of Sn(4+)-doped TiO(2) thin films on glass substrate. *Journal of Hazardous Materials*, **162** (2–3), 1309–16.

74 Lee, S.H., Pumprueg, S. *et al.* (2005) Inactivation of bacterial endospores by photocatalytic nanocomposites. *Colloids and Surfaces B: Biointerfaces*, **40** (2), 93–8.

75 Yu, J.C., Ho, W. *et al.* (2005) Efficient visible light induced photocatalytic disinfection on sulfur-doped nanocrystalline titania. *Environmental Science and Technology*, **39**, 1175–9.

76 Venkatasubramanian, R., Srivastava, R.S. *et al.* (2008) Comparative study of antimicrobial and photocatalytic activity in titania encapsulated composite nanoparticles with different dopants. *Materials Science and Technology*, **24** (5), 589–95.

77 Cerrada, M.L., Serrano, C. *et al.* (2008) Self-sterilized EVOH-TiO_2 nanocomposites: interface effects on biocidal properties. *Advanced Functional Materials*, **18** (13), 1949–60.

78 Kubacka, A., Ferrer, M. *et al.* (2009) Boosting TiO_2-anatase antimicrobial activity: polymer-oxide thin films. *Applied Catalysis B: Environmental*, **89** (3–4), 441–7.

79 Chen, W.-J., Tsai, P.-J. *et al.* (2008) Functional Fe_3O_4/TiO_2 core/shell magnetic nanoparticles as photokilling agents for pathogenic bacteria. *Small*, **4** (4), 485–91.

80 Bannat, I., Wessels, K. *et al.* (2009) Improving the photocatalytic performance of mesoporous titania films by modification with gold nanostructures. *Chemistry of Materials*, **21** (8), 1645–53.

81 Li, Y., Xu, X.Q. *et al.* (2008) Novel Fe_3O_4@TiO_2 core-shell microspheres for selective enrichment of phosphopeptides in phosphoproteome analysis. *Journal of Proteome Research*, **7** (6), 2526–38.

82 Webster, T.J., Ergun, C. *et al.* (2000) Specific proteins mediate enhanced osteoblast adhesion on nanophase ceramics. *Journal of Biomedical Materials Research*, **51** (3), 475–83.

83 McManus, A.J., Doremus, R.H. *et al.* (2005) Evaluation of cytocompatibility and bending modulus of nanoceramic/polymer composites. *Journal of Biomedical Materials Research Part A*, **72** (1), 98–106.

84 Kay, S., Thapa, A. *et al.* (2002) Nanostructured polymer/nanophase ceramic composites enhance osteoblast and chondrocyte adhesion. *Tissue Engineering*, **8** (5), 753–61.

85 Webster, T.J. and Smith, T.A. (2005) Increased osteoblast function on PLGA composites containing nanophase titania. *Journal of Biomedical Materials Research Part A*, **74** (4), 677–86.

86 Webster, T.J., Siegel, R.W. *et al.* (1999) Osteoblast adhesion on nanophase ceramics. *Biomaterials*, **20** (13), 1221–7.

87 Mills, A. and Le Hunte, S. (1997) An overview of semiconductor photocatalysis. *Journal of Photochemistry and Photobiology A: Chemistry*, **108**, 1–35.

88 Blake, D.M., Maness, P.C. *et al.* (1999) Application of the photocatalytic chemistry of titanium dioxide to disinfection and the killing of cancer cells. *Separation and Purification Methods*, **28** (1), 1–50.

89 Yang, H.M., Shi, R.R. *et al.* (2005) Synthesis of WO_3/TiO_2 nanocomposites via sol-gel method. *Journal of Alloys and Compounds*, **398** (1–2), 200–2.

90 Gratzel, M. (2004) Conversion of sunlight to electric power by nanocrystalline dye-sensitized solar cells. *Journal of Photochemistry and Photobiology A: Chemistry*, **164**, 3–14.

91 Yu, H.T., Quan, X. *et al.* (2008) TiO_2-carbon nanotube heterojunction arrays with a controllable thickness of TiO_2 layer and their first application in photocatalysis. *Journal of Photochemistry and Photobiology A: Chemistry*, **200** (2–3), 301–6.

92 Neves, M.C., Monteiro, O.C. *et al.* (2008) From single-molecule precursors to coupled Ag_2S/TiO_2 nanocomposites. *European Journal of Inorganic Chemistry*, **2008** (28), 4380–6.

93 Su, H.L., Dong, Q. *et al.* (2008) Biogenic synthesis and photocatalysis of Pd-PdO nanoclusters reinforced hierarchical TiO_2 films with interwoven and tubular conformations. *Biomacromolecules*, **9** (2), 499–504.

94 Xiao, M.W., Wang, L.S. *et al.* (2009) Synthesis and characterization of WO_3/titanate nanotubes nanocomposite with enhanced photocatalytic properties. *Journal of Alloys and Compounds*, **470** (1–2), 486–91.

95 Yu, M.D., Guo, Y. *et al.* (2008) A novel preparation of mesoporous $Cs_xH_{3-x}PW_{12}O_{40}/TiO_2$ nanocomposites with enhanced photocatalytic activity. *Colloids and Surfaces A: Physicochemical and Engineering Aspects*, **316** (1–3), 110–18.

96 Shchukin, D., Poznyak, S. *et al.* (2004) TiO_2-In_2O_3 photocatalysts: preparation, characterisations and activity for 2-chlorophenol degradation in water. *Journal of Photochemistry and Photobiology A: Chemistry*, **162** (2–3), 423–30.

97 Lei, S.M., Gong, W.Q. *et al.* (2006) Preparation of TiO_2/kaolinite nanocomposite and its photocatalytical activity. *Journal of Wuhan University of Technology: Materials Science Edition*, **21** (4), 12–15.

98 Luo, Y.B., Li, W.D. *et al.* (2009) Preparation and properties of nanocomposites based on poly(lactic acid) and functionalized TiO_2. *Acta Materialia*, **57** (11), 3182–91.

99 Matsunaga, T., Tomoda, T. *et al.* (1985) Photoelectrochemical sterilization of microbial cells by semiconductor powder. *FEMS Microbiology Letters*, **29**, 211–16.

100 Maness, P.C., Smolinski, S. *et al.* (1999) Bactericidal activity of photocatalytic TiO(2) reaction: toward an understanding of its killing mechanism. *Applied and Environmental Microbiology*, **65** (9), 4094–8.

101 Mo, A.C., Xu, W. *et al.* (2007) Antibacterial activity of silver-hydroxyapatite/titania nanocomposite coating on titanium against oral bacteria. *Bioceramics*, **19** (Pts 1 and 2) (330–332), 455–8.

102 Liu, L.F., Barford, J. *et al.* (2007) Non-UV based germicidal activity of metal-doped TiO_2 coating on solid surfaces. *Journal of Environmental Sciences: China*, **19**, 745–50.

103 Mahltig, B., Gutmann, E. *et al.* (2007) Solvothermal preparation of metallized titania sols for photocatalytic and antimicrobial coatings. *Journal of Materials Chemistry*, **17** (22), 2367–74.

104 Sheel, D.W., Brook, L.A. *et al.* (2008) Biocidal silver and silver/titania composite films grown by chemical vapour deposition. *International Journal of Photoenergy*, Article no. 168185.

105 Wang, X.C., Yu, J.C. *et al.* (2005) A mesoporous Pt/TiO_2 nanoarchitecture with catalytic and photocatalytic functions. *Chemistry*, **11** (10), 2997–3004.

106 Kangwansupamonkon, W., Lauruengtana, V. *et al.* (2009) Antibacterial effect of apatite-coated titanium dioxide for textiles applications. *Nanomedicine-Nanotechnology Biology and Medicine*, **5** (2), 240–9.

107 Mazanek, M., Mituloviae, G. *et al.* (2007) Titanium dioxide as a chemo-affinity solid phase in offline phosphopeptide chromatography prior to HPLC-MS/MS analysis. *Nature Protocols,* **2** (5), 1059–69.

108 Yu, L.R., Zhu, Z.Y. *et al.* (2007) Improved titanium dioxide enrichment of phosphopeptides from HeLa cells and high confident phosphopeptide identification by cross-validation of MS/MS and MS/MS/MS spectra. *Journal of Proteome Research,* **6** (11), 4150–62.

109 Ashman, K. and Villar, E.L. (2009) Phosphoproteomics and cancer research. *Clinical and Translational Oncology,* **11** (6), 356–62.

110 Paradela, A. and Albar, J.P. (2008) Advances in the analysis of protein phosphorylation. *Journal of Proteome Research,* **7** (5), 1809–18.

111 Karlsson, H.L., Gustafsson, J. *et al.* (2009) Size-dependent toxicity of metal oxide particles – a comparison between nano- and micrometer size. *Toxicology Letters,* **188** (2), 112–18.

112 Geiser, M., Rothen-Rutishauser, B. *et al.* (2005) Ultrafine particles cross cellular membranes by nonphagocytic mechanisms in lungs and in cultured cells. *Environmental Health Perspectives,* **113** (11), 1555–60.

113 Singh, S., Shi, T.M. *et al.* (2007) Endocytosis, oxidative stress and IL-8 expression in human lung epithelial cells upon treatment with fine and ultrafine TiO_2: role of the specific surface area and of surface methylation of the particles. *Toxicology and Applied Pharmacology,* **222** (2), 141–51.

114 Kang, S.J., Kim, B.M. *et al.* (2009) Titanium dioxide nanoparticles induce apoptosis through the JNK/p38-caspase-8-Bid pathway in phytohemagglutinin-stimulated human lymphocytes. *Biochemical and Biophysical Research Communications,* **386** (4), 682–7.

115 Kong, S.J., Kim, B.M. *et al.* (2008) Titanium dioxide nanoparticles trigger p53-mediated damage response in peripheral blood lymphocytes. *Environmental and Molecular Mutagenesis,* **49** (5), 399–405.

116 Park, E.J., Yi, J. *et al.* (2008) Oxidative stress and apoptosis induced by titanium dioxide nanoparticles in cultured BEAS-2B cells. *Toxicology Letters,* **180** (3), 222–9.

117 Wan, R., Mo, Y.Q. *et al.* (2008) Matrix metalloproteinase-2 and-9 are induced differently by metal nanoparticles in human monocytes: the role of oxidative stress and protein tyrosine kinase activation. *Toxicology and Applied Pharmacology,* **233** (2), 276–85.

118 Rozhkova, E.A., Ulasov, I. *et al.* (2009) A high-performance nanobio photocatalyst for targeted brain cancer therapy. *Nano Letters,* **9** (9), 3337–42.

119 Johannes, L. and Lamaze, C. (2002) Clathrin-dependent or not: is it still the question? *Traffic,* **3** (7), 443–51.

120 Kirkham, M. and Parton, R.G. (2005) Clathrin-independent endocytosis: new insights into caveolae and non-caveolar lipid raft carriers. *Biochimica et Biophysica Acta,* **1746** (3), 349–63.

121 Mosesson, Y., Mills, G.B. *et al.* (2008) Derailed endocytosis: an emerging feature of cancer. *Nature Reviews Cancer,* **8** (11), 835–50.

122 Hussain, S., Boland, S. *et al.* (2009) Oxidative stress and proinflammatory effects of carbon black and titanium dioxide nanoparticles: role of particle surface area and internalized amount. *Toxicology,* **260** (1–3), 142–9.

123 Barua, S. and Rege, K. (2009) Cancer-cell-phenotype-dependent differential intracellular trafficking of unconjugated quantum dots. *Small,* **5** (3), 370–6.

124 Koeneman, B.A., Zhang, Y. *et al.* (2009) Toxicity and cellular responses of intestinal cells exposed to titanium dioxide. *Cell Biology and Toxicology,* [Epub ahead of print].

125 Jin, C.Y., Zhu, B.S. *et al.* (2008) Cytotoxicity of titanium dioxide nanoparticles in mouse fibroblast cells. *Chemical Research in Toxicology,* **21** (9), 1871–7.

126 Suh, W.H., Suslick, K.S. *et al.* (2009) Nanotechnology, nanotoxicology, and neuroscience. *Progress in Neurobiology,* **87** (3), 133–70.

127 Hirakawa, K., Mori, M. *et al.* (2004) Photo-irradiated titanium dioxide catalyzes site specific DNA damage via generation of hydrogen peroxide. *Free Radical Research*, **38** (5), 439–47.

128 Serpone, N., Salinaro, A.E. *et al.* (2006) Beneficial effects of photo-inactive titanium dioxide specimens on plasmid DNA, human cells and yeast cells exposed to UVA/UVB simulated sunlight. *Journal of Photochemistry and Photobiology A: Chemistry*, **179** (1–2), 200–12.

129 Wamer, W.G., Yin, J.J. *et al.* (1997) Oxidative damage to nucleic acids photosensitized by titanium dioxide. *Free Radical Biology and Medicine*, **23** (6), 851–8.

130 Cai, R.X., Kubota, Y. *et al.* (1992) Induction of cytotoxicity by photoexcited TiO_2 particles. *Cancer Research*, **52** (8), 2346–8.

131 Chihara, Y., Fujimoto, K. *et al.* (2007) Anti-tumor effects of liposome-encapsulated titanium dioxide in nude mice. *Pathobiology*, **74** (6), 353–8.

132 Kubota, Y., Shuin, T. *et al.* (1994) Photokilling of T-24 human bladder-cancer cells with titanium-dioxide. *British Journal of Cancer*, **70** (6), 1107–11.

133 Liu, J., de la Garza, L. *et al.* (2007) Photocatalytic probing of DNA sequence by using TiO_2/dopamine-DNA triads. *Chemical Physics*, **339**, 154–63.

134 Tachikawa, T., Asanoi, Y. *et al.* (2008) Photocatalytic cleavage of single TiO_2/DNA nanoconjugates. *Chemistry*, **14** (5), 1492–8.

135 Tachikawa, T. and Majima, T. (2009) Single-molecule fluorescence imaging of TiO_2 photocatalytic reactions. *Langmuir*, **25** (14), 7791–802.

136 Branemark, P.I. (1959) Vital microscopy of bone marrow in rabbit. *Scandinavian Journal of Clinical and Laboratory Investigation*, **11** (Suppl. 38), 1–82.

137 Branemark, R., Branemark, P.I. *et al.* (2001) Osseointegration in skeletal reconstruction and rehabilitation: a review. *Journal of Rehabilitation Research and Development*, **38** (2), 175–81.

138 Adell, R., Lekholm, U. *et al.* (1981) A 15-year study of osseointegrated implants in the treatment of the edentulous jaw. *International Journal of Oral Surgery*, **10** (6), 387–416.

139 Franchi, M., Fini, M. *et al.* (2005) Biological fixation of endosseous implants. *Micron*, **36** (7–8), 665–71.

140 Lindgren, M., Astrand, M. *et al.* (2009) Investigation of boundary conditions for biomimetic HA deposition on titanium oxide surfaces. *Journal of Materials Science: Materials in Medicine*, **20** (7), 1401–8.

141 Rho, J.Y., Kuhn-Spearing, L. *et al.* (1998) Mechanical properties and the hierarchical structure of bone. *Medical Engineering and Physics*, **20** (2), 92–102.

142 Webster, T.J., Ergun, C. *et al.* (2000) Enhanced functions of osteoblasts on nanophase ceramics. *Biomaterials*, **21** (17), 1803–10.

143 Misra, R.D., Thein-Han, W.W. *et al.* (2009) Cellular response of preosteoblasts to nanograined/ultrafine-grained structures. *Acta Biomaterialia*, **5** (5), 1455–67.

144 Ramires, P.A., Cosentino, F. *et al.* (2002) In vitro response of primary rat osteoblasts to titania/hydroxyapatite coatings compared with transformed human osteoblast-like cells. *Journal of Materials Science: Materials in Medicine*, **13** (8), 797–801.

145 Sailaja, G.S., Ramesh, P. *et al.* (2006) Human osteosarcoma cell adhesion behaviour on hydroxyapatite integrated chitosan-poly(acrylic acid) polyelectrolyte complex. *Acta Biomaterialia*, **2** (6), 651–7.

146 Zhao, G., Schwartz, Z. *et al.* (2005) High surface energy enhances cell response to titanium substrate microstructure. *Journal of Biomedical Materials Research Part A*, **74** (1), 49–58.

147 Meng, D., Ioannou, J. *et al.* (2009) Bioglass®-based scaffolds with carbon nanotube coating for bone tissue engineering. *Journal of Materials Science: Materials in Medicine*, **20** (10), 2139–44.

148 Jarcho, M., Kay, J.F. *et al.* (1977) Tissue, cellular and subcellular events at a bone-ceramic hydroxylapatite interface. *Journal of Bioengineering*, **1** (2), 79–92.

149 Paderni, S., Terzi, S. *et al.* (2009) Major bone defect treatment with an osteoconductive bone substitute. *Musculoskeletal Surgery*, **93** (2), 89–96.

150 Daculsi, G., LeGeros, R.Z. *et al.* (1990) Formation of carbonate-apatite crystals after implantation of calcium phosphate ceramics. *Calcified Tissue International*, **46** (1), 20–7.

151 Kitsugi, T., Yamamuro, T. *et al.* (1986) Bone bonding behavior of three kinds of apatite containing glass ceramics. *Journal of Biomedical Materials Research*, **20** (9), 1295–307.

152 Valimaki, V.V., Yrjans, J.J. *et al.* (2005) Combined effect of BMP-2 gene transfer and bioactive glass microspheres on enhancement of new bone formation. *Journal of Biomedical Materials Research Part A*, **75** (3), 501–9.

153 Valimaki, V.V. and Aro, H.T. (2006) Molecular basis for action of bioactive glasses as bone graft substitute. *Scandinavian Journal of Surgery*, **95** (2), 95–102.

154 Abou Neel, E.A. and Knowles, J.C. (2008) Physical and biocompatibility studies of novel titanium dioxide doped phosphate-based glasses for bone tissue engineering applications. *Journal of Materials Science: Materials in Medicine*, **19** (1), 377–86.

155 Abou Neel, E.A., Ahmed, I. *et al.* (2005) Effect of iron on the surface, degradation and ion release properties of phosphate-based glass fibres. *Acta Biomaterialia*, **1** (5), 553–63.

156 Abou Neel, E.A., Chrzanowski, W. *et al.* (2008) Effect of increasing titanium dioxide content on bulk and surface properties of phosphate-based glasses. *Acta Biomaterialia*, **4** (3), 523–34.

157 Ahmed, I., Collins, C.A. *et al.* (2004) Processing, characterisation and biocompatibility of iron-phosphate glass fibres for tissue engineering. *Biomaterials*, **25** (16), 3223–32.

158 Neel, E.A., Ahmed, I. *et al.* (2005) Characterisation of antibacterial copper releasing degradable phosphate glass fibres. *Biomaterials*, **26** (15), 2247–54.

159 Prabhakar, R.L., Brocchini, S. *et al.* (2005) Effect of glass composition on the degradation properties and ion release characteristics of phosphate glass–polycaprolactone composites. *Biomaterials*, **26** (15), 2209–18.

160 Zhao, L., Chang, J. *et al.* (2009) Preparation and HL-7702 cell functionality of titania/chitosan composite scaffolds. *Journal of Materials Science: Materials in Medicine*, **20**, 949–57.

161 Dobrovolskaia, M.A., Patri, A.K. *et al.* (2009) Interaction of colloidal gold nanoparticles with human blood: effects on particle size and analysis of plasma protein binding profiles. *Nanomedicine: Nanotechnology, Biology, and Medicine*, **5** (2), 106–17.

162 Patri, A., Umbreit, T. *et al.* (2009) Energy dispersive X-ray analysis of titanium dioxide nanoparticle distribution after intravenous and subcutaneous injection in mice. *Toxicology*, **230** (1), 90–104.

163 Olmedo, D.G., Tasat, D.R. *et al.* (2008) Biological response of tissues with macrophagic activity to titanium dioxide. *Journal of Biomedical Materials Research Part A*, **84** (4), 1087–93.

164 Tran, C.L., Buchanan, D. *et al.* (2000) Inhalation of poorly soluble particles. II. Influence Of particle surface area on inflammation and clearance. *Inhalation Toxicology*, **12** (12), 1113–26.

165 Liu, R., Yin, L.H. *et al.* (2009) Pulmonary toxicity induced by three forms of titanium dioxide nanoparticles via intra-tracheal instillation in rats. *Progress in Natural Science*, **19** (5), 573–9.

166 Oberdorster, G. (2001) Pulmonary effects of inhaled ultrafine particles. *International Archives of Occupational and Environmental Health*, **74** (1), 1–8.

167 Yokohira, M., Kuno, T. *et al.* (2008) Lung toxicity of 16 fine particles on intratracheal instillation in a bioassay model using F344 male rats. *Toxicologic Pathology*, **36** (4), 620–31.

168 Handy, R.D. and Shaw, B.J. (2007) Toxic effects of nanoparticles and nanomaterials: implications for public health, risk assessment and the public perception of nanotechnology. *Health Risk & Society*, **9** (2), 125–44.

169 Crosera, M., Bovenzi, M. *et al.* (2009) Nanoparticle dermal absorption and toxicity: a review of the literature. *International Archives of Occupational and Environmental Health*, **82** (9), 1043–55.

170 Schulz, J., Hohenberg, H. *et al.* (2002) Distribution of sunscreens on skin. *Advanced Drug Delivery Reviews*, **54**, S157–63.

171 Sayes, C.M., Wahi, R. *et al.* (2006) Correlating nanoscale titania structure with toxicity: a cytotoxicity and inflammatory response study with human dermal fibroblasts and human lung epithelial cells. *Toxicological Sciences*, **92** (1), 174–85.

172 Chen, Z., Meng, H. *et al.* (2006) Acute toxicological effects of copper nanoparticles in vivo. *Toxicology Letters*, **163** (2), 109–20.

173 Huggins, C.B., Froehlich, J.P. *et al.* (1966) High concentration of injected titanium dioxide in abdominal lymph nodes. *The Journal of Experimental Medicine*, **124** (6), 1099–106.

174 Longmire, M., Choyke, P.L. *et al.* (2008) Clearance properties of nano-sized particles and molecules as imaging agents: considerations and caveats. *Nanomedicine*, **3** (5), 703–17.

175 Paik, S.G., Fleischer, N. *et al.* (1980) Insulin-dependent diabetes mellitus induced by subdiabetogenic doses of streptozotocin: obligatory role of cell-mediated autoimmune processes. *Proceedings of the National Academy of Sciences of the United States of America*, **77** (10), 6129–33.

176 Garnett, M.C. and Kallinteri, P. (2006) Nanomedicines and nanotoxicology: some physiological principles. *Occupational Medicine*, **56** (5), 307–11.

177 Kikuchi, Y., Sunada, K. *et al.* (1997) Photocatalytic bactericidal effect of TiO_2 thin films: Dynamic view of the active oxygen species responsible for the effect. *Journal of Photochemistry and Photobiology A - Chemistry* **106** (1–3), 51–6.

178 Le Roy, C. and Wrana, J.L. (2005) Clathrin- and non-clathrin-mediated endocytic regulation of cell signalling. *Nature Reviews of Molecular and Cell Biology*, **6** (2), 112–26.

179 Hillaireau, H. and Couvreur, P. (2009) Nanocarriers' entry into the cell: relevance to drug delivery. *Cellular and Molecular Life Sciences*, **66** (17), 2873–96.

180 Huang, S., Chueh, P.J. *et al.* (2009) Disturbed mitotic progression and genome segregation are involved in cell transformation mediated by nano-TiO_2 long-term exposure. *Toxicology and Applied Pharmacology*, **241** (2), 182–94.

181 Cai, R.X., Kubota, Y. *et al.* (1992) Induction of cytotoxicity by photoexcited TiO_2 particles. *Cancer Research*, **52** (8), 2346–8.

182 Zhu, R.R., Wang, S.L. *et al.* (2009) Bio-effects of Nano-TiO_2 on DNA and cellular ultrastructure with different polymorph and size. *Materials Science and Engineering C - Biomimetic and Supramolecular Systems*, **29** (3), 691–6.

183 Seo, J.W., Chung, H. *et al.* (2007) Development of water-soluble single-crystalline TiO_2 nanoparticles for photocatalytic cancer-cell treatment. *Small*, **3** (5), 850–3.

2
Chitin Nanocomposites for Medical Applications

Aji P. Mathew and Kristiina Oksman

2.1
Introduction

Novel biomaterials in the form of composites with nanosized reinforcements may bring unpredictable new characteristics to material, such as mechanical (stronger), physical (lighter and more porous), optical (tunable optical emission), color, chemical reactivity (more active or less corrosive), electronic properties (more electrically conductive), or magnetic properties (super-paramagnetic), in addition to new functionalities that may be unavailable at the micro- or macroscale [1, 2]. During recent years, an extensive interest has arisen in the development of products from bio-based and natural resources, and their extensive use in a variety of applications including biomedical products for wound dressing, artificial skin, sutures, and the controlled release of drugs [3]. It is well known that naturally occurring polymers have a greater biocompatibility and immunogenicity than synthetic polymers when used in biomedical applications, with protein-based polymers such as collagen, elastin, gelatin and silk fibroin having undergone extensive investigation in recent years to identify biomedical applications [4, 5]. Polysaccharide-based polymers such as cellulose, chitin, and chitosan represent another class of polymers used to develop biomaterials for medical applications [6, 7].

Chitin, a natural polysaccharide found widely in crustaceans and insects, is composed of β-(1–4)-linked units of *N*-acetyl-2-amino-2-deoxy-D-glucose. Natural chitin also contains a small proportion of chitosan, in the deacetylated form. Chitosan, the product of extensive deacetylation of chitin, is soluble in dilute acid solutions and can be found in many applications in the fields of biomedicine, cosmetics, and agriculture [7, 8]. Depending on the degree of deacetylation, chitosan contains various proportions of (1–4) linked units of 2-amino-2-deoxy-D-glucose and *N*-acetyl-2-amino-2-deoxy-D-glucose. The chemical structures of chitin and chitosan are shown in Figure 2.1 [2].

Chitin polymers form highly crystalline networks, with the α polymorph being more common and stable than its β polymorph counterpart [8, 9]. Chitin in the form of nanocrystals was successfully isolated from crab shells by Marchessault

Nanomaterials for the Life Sciences Vol.8: Nanocomposites. Edited by Challa S. S. R. Kumar

ISBN: 978-3-527-32168-1

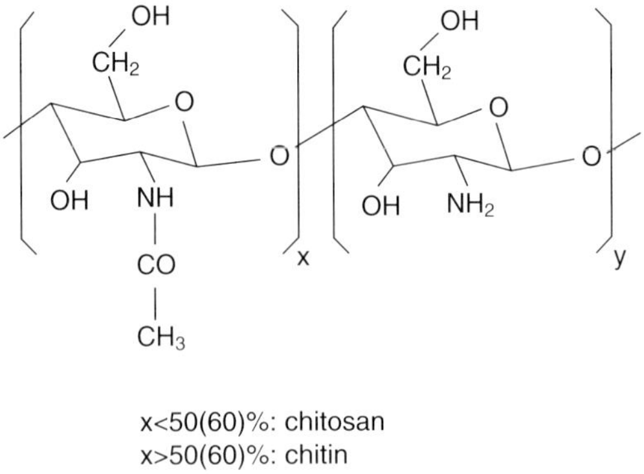

$x < 50(60)\%$: chitosan
$x > 50(60)\%$: chitin

Figure 2.1 Chemical structures of chitin and chitosan.

Table 2.1 Products of chitosan-based materials and suitability for applications.

Application	Suitability/Properties
Artificial skin	Renewable, nontoxic, bioactive
Sutures	Biocompatible, biodegradable
Wound dressing	Antibacterial, antiviral, antifungal, nontoxic
Implants	Biocompatible, biodegradable
Bone reconstruction	Biocompatible, nontoxic, bioactive, film-forming
Corneal contact lenses	Hydrating ability
Controlled drug release	Biocompatible, nontoxic, water permeable

as early as 1959 [10], and used subsequently by Dufresne and coworkers as a reinforcing material in bionanocomposites [11–13]. Thus, a major difference between chitosan and chitin resides in the abundant free amino groups on chitosan compared to chitin, that confer to the former a much greater reactivity. Chitosan is most often amorphous and can be processed into flexible films; it is also nontoxic, biosorbable, antibacterial, bioactive and biocompatible, which makes it an interesting and unique polymer for the production of hydrogels, films, fibers and sponges for use as biomedical products [14–22]. The characteristics of chitosan with respect to some potential biomedical products are listed in Table 2.1 [7].

Although several materials based on chitosan have been described, very few have been found to be commercially viable, due mainly to the material's poor stability because of its hydrophilic nature and pH sensitivity. In contrast, chitin has found very few applications to date, due to its insolubility in most solvents. Consequently, in order to develop useful products based on chitosan and chitin, both physical- and chemical-based treatments and manipulations of the materials are required.

One of the earliest methods used to stabilize chitosan was to crosslink it using reagents such as glutaraldehyde, or to apply gamma irradiation [22–24]. Another approach was to blend the chitosan with different polymers to form stable products with improved chemical resistance and mechanical properties [24–26]. The development of composites, where a reinforcing phase is added to chitosan, also represents a viable method of stabilization and improvement of properties [17–19]. Although, in the past, the biomedical application of chitosan-based products has been relatively extensive, only two areas have been identified to date involving the use of chitosan-based nanocomposites [27]:

- Nanocomposites with chitosan as the matrix phase and nanosized reinforcements such as carbon nanotubes (CNTs), organically modified nanoclays and biobased reinforcements, which results in more stable and stronger materials. In these cases, the percentage of reinforcing phase is normally low (5–10 wt%), such that making proportion of the chitosan component is relatively high. This results in a product which closely resembles the matrix in terms of biocompatibility and bioactivity [17–19].
- Nanocomposites with "chitin whiskers" as nanoreinforcements in a biopolymer or synthetic polymers [11–13]. In this case, the chitin fraction serves mainly as the reinforcing phase, and has a limited effect on material properties such as biocompatibility and bioactivity, which rather depend on the matrix phase being the major component.

Whilst in this chapter attention is focused on these two types of chitosan-based nanocomposite, the development of nanocomposite-based products such as films, hydrogels, fiber meshes and porous scaffolds will also be discussed, as will their potential in biomedical applications.

2.2 Chitosan/Chitin Nanocomposites

2.2.1 Chitosan as the Matrix Phase

2.2.1.1 Carbon Nanotube-Based Chitosan Nanocomposites

Although currently, several reports of chitosan nanocomposites based on CNTs have been made, very few studies of these materials have been relevant to biomedical applications. In general, the chitosan-based CNT composites are used as biosensors (to detect molecules such as glucose, cholesterol or creatine), as biogels for the controlled release of drugs, or as conductive scaffolds for cell growth [28–35].

Biosensors: Tiwari and Dhakate have shown that chitosan-SiO_2-multiwalled carbon nanotube (MWCNT) nanocomposites can be used to immobilize creatine amidinohydrolase (CAH), which is used as a biosensor [28]. The nanocomposite was prepared using a spin-coating technique, after which CAH was linked covalently to the nanocomposite, using glutaraldehyde as the crosslinking agent. The

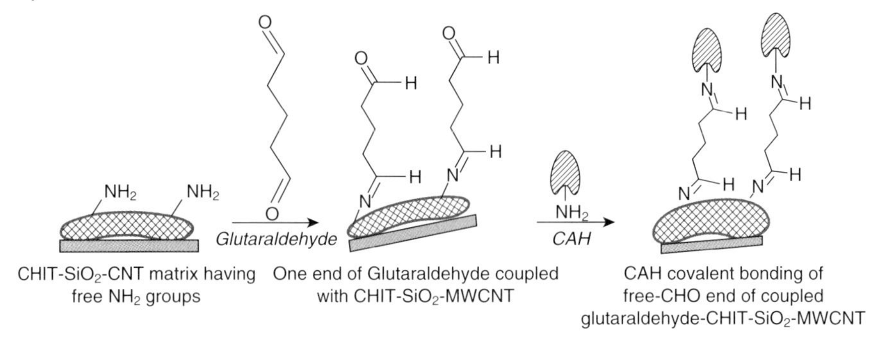

Figure 2.2 Schematic representation of the immobilization of creatine amidinohydrolase (CAH) on the chitosan (CHIT)–SiO_2–MWCNT matrix, using glutaraldehyde as a linker. Reproduced with permission from Ref. [28]; © Elsevier.

functional mechanism of the biosensor is shown schematically in Figure 2.2. It was suggested that this development should lead to the creation of an electrochemical biosensor to detect creatine in the blood and urine.

Recently, Malhotra and coworkers described the immobilization (via glutaraldehyde crosslinking) of cholesterol esterase and cholesterol oxidase on silica/chitosan/MWCNT-based nanocomposite films to provide a biosensor capable of estimating the total cholesterol content of serum samples [29]. The same group also described a similar system in which zinc oxide would act as a glucose sensor [30].

Similarly, Goplan *et al.* developed a cholesterol biosensor using gold particles and MWCNTs in a crosslinked chitosan matrix, using an ionic liquid (IL) at room temperature [31]. In this case, the synergistic influence of the MWCNTs, gold particles, chitosan and IL contributed to an excellent performance of the material as a biosensor for cholesterol, that could be extended to other biosensors. The same group have also developed a similar system that acts as a glucose biosensor [32], and is capable of detecting glucose (electrochemically) over a wide linear concentration range (1–10 mM), with a high sensitivity (4.101 A mM^{-1}) and a low response time (6 s).

Qui *et al.* developed a novel amperometric sensor for glucose by entrapping glucose oxidase (GOD) in a chitosan matrix that had been doped with ferrocene monocarboxylic acid-modified 3-(aminopropyl) triethoxysilane enwrapping MWCNT (FMC-AMWCNT) [33, 34]. In this case, the FMC-AMWCNTs could be effectively employed to fabricate reagentless glucose biosensors, by forming a GOD/FMCAMWNT/CS film on the electrode.

Controlled drug release: Recently, Brianna *et al.* reported chitosan-based biogels with single-walled carbon nanotubes (SWCNTs), prepared via a solution-casting technique [35]. The nanocomposite films in the nonswollen state, and after swelling in a phosphate-buffered solution for 30 min, is shown in Figure 2.3. In this case, the swelling process was shown to be reversible, with the films shrinking to

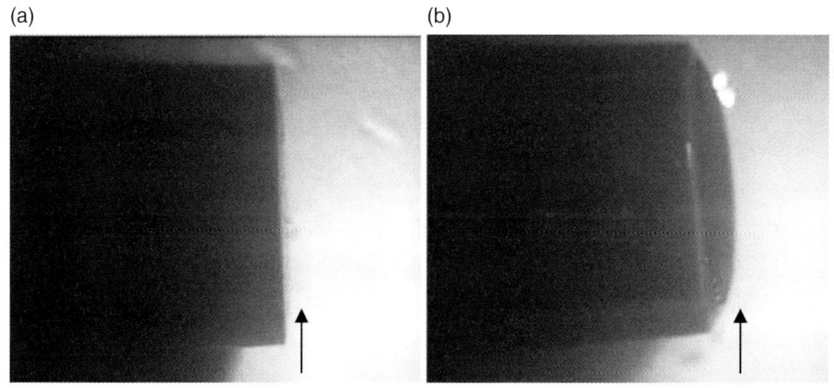

Figure 2.3 Optical image of a SWCNT–chitosan biofilm (a) in the nonswollen state (0 min) and (b) in the fully swollen state (30 min). The SWCNT biofilms were immersed in phosphate-buffered solution (pH 7.2). The vertical arrows indicate the CNT–gel/buffer interface. Reproduced with permission from Ref. [35]; © Elsevier.

their original size when removed from the buffered medium. Subsequently, when a drug (neurotropin; NT-3) was incorporated into the gel, its release could be controlled accurately, simply by applying an electrical stimulation to the gel.

Cell growth scaffolds: The chitosan-based/SWCNT biogels were also found to be noncytotoxic, and to serve as an excellent platform for fibroblast cell growth *in vitro*.

2.2.1.2 Inorganic Nanoreinforcement-Based Chitosan Nanocomposites

Inorganic nanoreinforcements such as hydroxyapatite, clay and metallic nanoparticles have been used to develop chitosan composites for use in medical applications [36–42]. These nanocomposites have typically demonstrated potential as implants, as controlled drug-release agents, as bone substitutes, and also as catheter materials.

Implants: Chitosan nanocomposites with hydroxyapatite represent an interesting material for biomedical application as implants. Kashiwazaki *et al.* [36] developed porous nanocomposites with a high porosity and pore size of 100–500 μm, the main advantage of this material being its mechanical flexibility as well its potential use for implantation. Subsequent studies in rats showed the implanted material to create only minimal inflammation and to generate good blood vessel growth, confirming the biocompatibility of the implant material.

Likewise, Cui *et al.* prepared a magnetite/hydroxyapatite/chitosan nanocomposite, using an *in situ* compositing method, the aim being to use the material as a three-dimensional (3-D) component with potential for bone repair and regeneration [37].

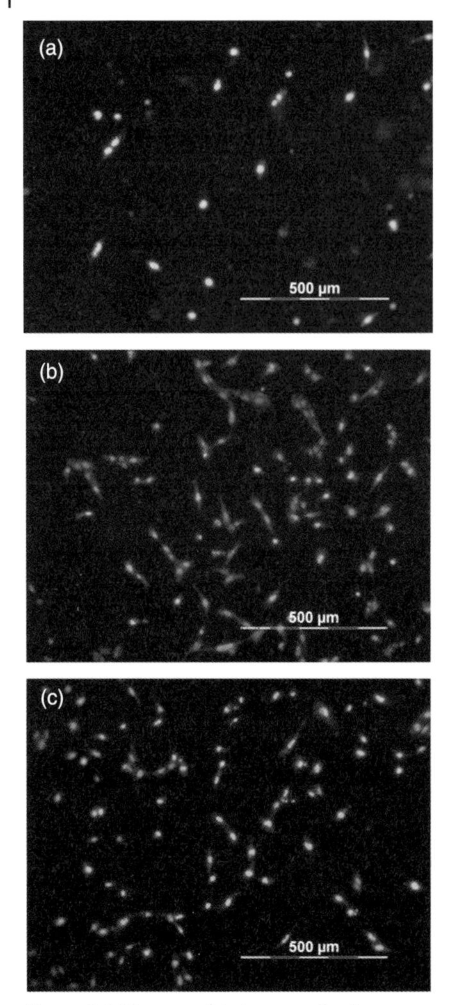

Figure 2.4 Photographic images of cells grown over samples of (a) chitosan/hydroxyapatite; (b) chitosan/MMT; (c) chitosan/MMT/hydroxyapatite after 3 days. Reproduced with permission from Ref. [38]; © IOP Publications.

Bone tissue engineering: A more recent study, conducted by Katti *et al.*, demonstrated the potential biomedical application of chitosan nanocomposites with montmorillonite (MMT) and hydroxyapatite in bone tissue engineering [38]. The study results showed these materials to have good mechanical properties and to cause a better cell proliferation than did the chitosan/hydroxyapatite nanocomposite. Cell growth on the nanocomposite samples was superior in the presence of MMT (Figure 2.4).

Controlled drug release: Wang and coworkers showed that chitosan-based clay nanocomposites had great biomedical potential [39, 40] by developing modified chitosan/montmorillonite (HTCC/MMT) nanocomposites and investigating their application as drug-release agents [40]. For this, intercalated nanoparticles were prepared and loaded with bovine serum albumin (BSA) as a model protein drug incorporated into the nanoparticles, and their controlled drug-release properties then tested *in vitro*. The nanocomposite particles were shown to have drug-loading capability and release properties, while cytotoxicity studies showed them to be biocompatible. Although no cell death was observed at low levels of HTCC/MMT usage, cell growth was seen to be inhibited when larger amount of the material were used.

When nanocomposites of chitosan-g-lactic acid and MMT were investigated in terms of their cell proliferation and controlled release of drugs [41], MMT was seen to have a positive effect not only on the mechanical properties of the composites, but also on the controlled release of ibuprofen in a phosphate-buffered saline medium. Moreover, the drug release occurred faster when nanocomposite (porous) scaffolds were used, rather than films.

Bone repair: Nanohydroxyapatite/chitosan/carboxymethyl cellulose-based nanocomposites were reported to be very useful for bone repair [42, 43]. With the morphology and size of the composite being similar to that of natural bone, subsequent *in vitro* biodegradation and bioactivity studies using a simulated body fluid (SBF) system showed positive results. These materials demonstrated good mechanical properties, and were considered to show great promise for effective bone repair.

Catheters: Chitosan nanocomposite technology has also been used to improve the antibacterial properties of biomedical catheter materials. For example, Shen and coworkers developed silver–chitosan, silver–chitosan/clay and polydimethylsiloxane (PDMS)/silver–chitosan/clay nanocomposites for use as catheter materials, and subsequently monitored their antibacterial activities [44] (see Table 2.2). The prepared PDMS/silver–chitosan/clay material showed excellent antimicrobial and

Table 2.2 Bacteriostatic effects on the mass of predominant urinary bacteria.

Bacteria	Bacteriostatic effect		
	PDMS/ chitosan–Ag	PDMS/ clay–chitosan–Ag	$AgNO_3$
Escherichia coli (ATCC25922)	+++	+++	+
Pseudomonas aeruginosa (ATCC27853)	++	++	+
Staphylococcus aureus (ATCC25923)	++	++	+
Candida albicans (ATCC14053)	+	+	+

+, Degree of bacteriostatic effect.
PDMS, polydimethylsiloxane.
Reprinted with permission from Ref. [44]; © Elsevier.

controlled drug-release characteristics, and was suggested to have (potentially) an effective role in indwelling urinary catheters.

Bone fracture: The creation has been reported of chitosan/hydroxyapatite nanocomposite rods, prepared via an *in situ* hybridization, which should serve as a potential material for the internal fixation of bone fractures [45]. The bending strength and modulus of the composite with 5% hydroxyapatite content, prepared by *in situ* hybridization, were 86 MPa and 3.4 GPa, respectively. In addition, this material was reported to be two- to threefold stronger than other bone replacement materials such as poly(methylmethacrylate) (PMMA) and bone cement.

Hybrid composites: An interesting class of electrically charged hybrid composites based on chitosan and clay was studied by Liu *et al.* [46]. In this study, the incorporation of polysilicate magadiite ($Na_2Si_{14}O_{29} \cdot nH_2O$) induces the formation of crosslinks due to an electrostatic interaction of the negatively charged clay sheets with the positively charged $-NH^{3+}$group of chitosan. In this case, the clay was found to enhance not only the thermal stability and mechanical properties of the chitosan matrix, but also its response to an electrical stimulus. It was suggested that the incorporation of negatively charged clay as a crosslinker would provides a more effective antifatigue property for pure positively charged chitosan under cyclic electrostimulation. The controlled-release profile of several clinically valuable drugs, using this new class of hybrid composite, is currently under investigation.

Biocompatible shells: Using a layer-by-layer (LBL) technique, Theodor *et al.* created hydrogel–magnetic nanoparticles with a magnetic core (Fe_3O_4) encapsulated in biocompatible shells composed of chitosan and hydraunic acid [47]. It was suggested that this nanocomposite might be used for medical and/or biotechnological purposes and, indeed, the hydrogels proved suitable for cellular wall penetration due to their dimensions (180–264 nm when swollen, 40–90 nm when dried), and to their homogeneous distribution and swelling capacity. Subsequent biocompatibility tests indicated that the hydrogel–magnetic nanoparticles were biocompatible and relatively inert to microorganisms, making them suitable for use when loading and delivering active compounds.

2.2.1.3 Biobased Reinforcement-Based Chitosan Nanocomposites

Li and coworkers recently reported the use of cellulose nanowhiskers as a reinforcing phase in a chitosan matrix [48]. The results indicated a good interaction between the matrix and the whisker, which in turn led to improved mechanical properties, thermal stability, and water resistance. Whilst the suggestion was made that these materials might (potentially) be of use for food packaging, no reports have emerged regarding the use of chitosan nanocomposites reinforced with cellulose whiskers or fibrils for medical applications. Chitin nanowhiskers as biobased nanoreinforcements have also been used in the chitosan matrix (see Section 2.2.3).

2.2.2
Chitin as the Reinforcing Phase

Whilst very few studies in this area have referred to biomedical applications of chitin, several investigations have been conducted in which chitin nanofibrils or nanowhiskers have served as the reinforcing phase in either natural or synthetic polymers. For example, when Dufresne and coworkers used chitin nanowhiskers (in the form of nanocrystals) as reinforcements in bionanocomposites, it was suggested that chitin could be used to reinforce biopolymers such as starch, natural rubber and polycaprolactone (PCL) [11–13]. Unfortunately, however, no studies were reported on the biomedical potential of these materials.

Supaphol and coworkers developed nanofibers by electrospinning poly(vinyl alcohol) (PVA) nanocomposite nanofibers that had been reinforced with α-chitin whiskers [49]. Such incorporation of chitin whiskers in the nanocomposite fiber mats caused an increase in the Young's modulus, to a level four- to eightfold that of the undiluted PVA fiber mats.

In 2007, Phongyoing and coworkers described the preparation of a chitosan nanoscaffold directly from chitin whiskers [50]. In this case, the one-pot deacetylation reaction yielded large amounts of the chitosan nanoscaffold, and also produced safer scaffolds because no organic solvents or chemicals were involved. The authors expected that these materials would have potential in biomedical applications.

The use of nanocomposite sponges based on chitin whisker-reinforced silk fibroin (as developed by Wongpanit and coworkers), supported the proposal that the incorporation of chitin whiskers into a silk fibroin matrix would improve both the dimensional stability and mechanical properties of the matrix [51]. Subsequent cell culture studies showed that those nanocomposites with chitin whiskers had a better cell-spreading potential than neat silk fibroin (as shown graphically in Figure 2.5). Surprisingly, the nanocomposite sponges did not demonstrate any cytotoxicity, indicating the potential use of this material as scaffold materials for cell growth *in vitro*.

Neves and colleagues have reported the use of chitosan nanofiber meshes, created by electrospinning, to reinforce a microcomposite of polybutylsuccinate (PBS) and chitosan microparticles [52]. A composite of chitosan microparticles dispersed in polybutylenesuccinate demonstrated good biodegradability and excellent biological performance, both *in vitro* and *in vivo*, for bone and cartilage tissue-engineering applications [53]. The chitosan nanofiber mesh produced by electrospinning was used to reinforce the composite, without detracting from its good biological performance. A schematic representation of the composite structure, with and without chitosan nanofibers, is shown in Figure 2.6.

The tensile modulus of the chitosan nanofiber mesh was improved by 70% at 0.05 wt%, and the water uptake increased by 24%, rendering it more biodegradable. It was suggested by the authors that this composite processing methodology might represent a new strategy for designing materials with properties tailored towards many biomedical applications.

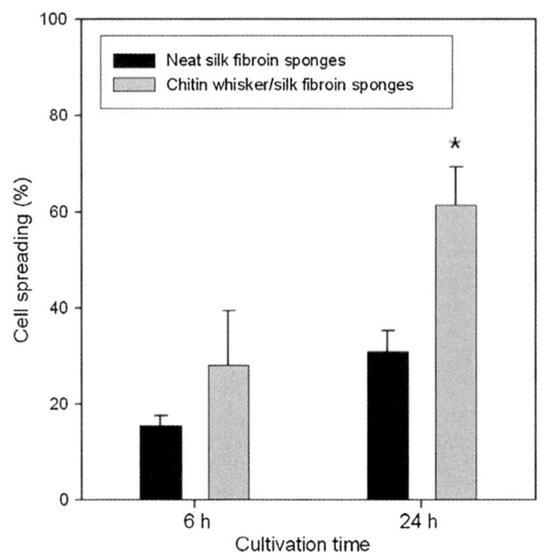

Figure 2.5 Percentage cell spreading of L929 cells on methanol-treated neat silk fibroin sponges and methanol-treated chitin whisker/silk fibroin sponges having a chitin whiskers/silk fibroin (C/S) ratio at 4 : 8, *, $P < 0.05$; significant versus the percentage cell spreading of methanol-treated neat silk fibroin sponges. Reproduced with permission from Ref. [51]; © Elsevier.

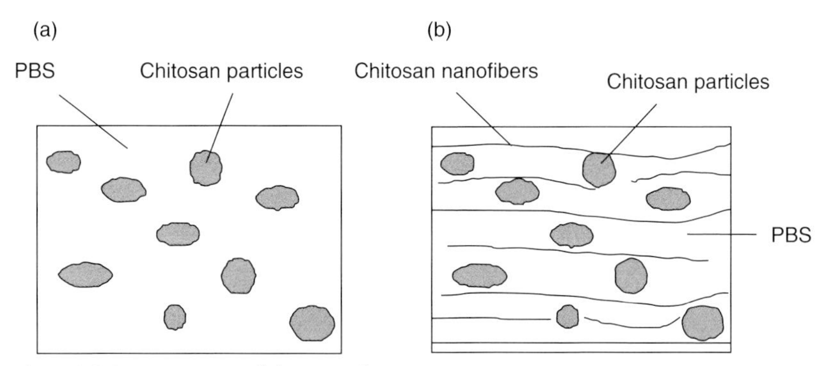

Figure 2.6 Representation of the particulate composites. (a) Polybutylsuccinate (PBS) matrix reinforced with chitosan particles – particulate microfibers; (b) PBS matrix reinforced with chitosan particles and nanofiber meshes – particulate microfibers reinforced with nanofibers. Reproduced with permission from Ref. [52]; © Elsevier.

2.2.3
All-Chitin Nanocomposites

In a study of chitin whisker-reinforced chitosan nanocomposites [54], heat treatment was shown to improve the water resistance of the nanocomposites films; however, no medical applications were suggested for these materials.

Subsequently, Muzzarelli and coworkers reported the details of chitin nanofibril-based chitosan glycolate composites, which had been developed in the form of a spray, a gel and a gauze, to be used for wound medications [55]. Each of the developed products provided satisfactory results in all clinical cases examined; notably, while the chitosan promoted antimicrobial activity, cell growth and film formation, the chitin nanofibrils facilitated the slow release of the active therapeutic agent.

Shelma *et al.* described the development of a chitin whisker-reinforced chitosan, and its potential application in wound-healing [56]. As pure chitosan films will have a poor tensile strength and elasticity, it is important to develop films with not only better mechanical properties and biocompatibility but also a wound-healing capacity suited to wound dressing applications. The study results showed that the tensile strength of the chitosan films could be increased to a significant level by incorporating chitin nanofibers, without any appreciable change in permeability to water vapor.

Mathew and coworkers recently reported a study on chitosan nanocomposites that were reinforced with chitin whiskers, where the matrix phase was crosslinked with glutaraldehyde [57]. The nanocomposites in which chitin whiskers had been used for nanoreinforcement was considered an efficient means of producing chitosan-based biomaterials with enhanced properties, without affecting the materials' biodegradability, transparency, and antibacterial properties. The study results also showed that crosslinking increased the stability of chitosan-based nanocomposite materials towards an acidic pH.

The dynamic mechanical thermal analysis of this composite showed that the storage modulus increased with the addition of chitin whiskers, whereas crosslinking showed no significant impact (Figure 2.7). The material was also clearly stable not only at body temperature (37 °C) but also at temperatures up to 200 °C, which would permit its sterilization before use. The stability of the product over a wide range of pH and temperature conditions, combined with their enhanced mechanical properties and permeation selectivity, confirm the potential role of these materials in biomedical applications such as the controlled release of drugs and wound dressings [57].

2.3
Processing of Chitosan/Chitin Nanocomposites

Details of some commonly used methods for processing chitosan-based nanocomposites for medical applications are provided in the following subsections. It

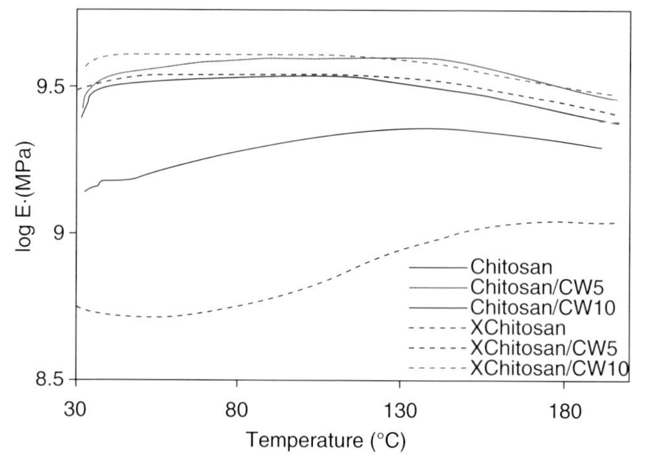

Figure 2.7 Dynamic mechanical thermal analysis curves showing the storage modulus of uncrosslinked and crosslinked chitosan/chitin nanocomposites in the temperature range of 30 to 200 °C.

should be noted that conventional polymer-processing methods such as melt extrusion, compression molding or injection molding have not been reported for chitosan-based nanocomposites intended for medical application. Similarly, in cases where chitin nanocrystals are used as reinforcement in different polymers, no reports have been made of any high-temperature processing methods, except for hot pressing, which was reported for the production of natural rubber (NR)-based chitin nanocomposites [58]. Low-temperature processes such as film-casting, freeze-drying, LBL assembly and electrospinning are frequently used to produce chitosan nanocomposites for medical applications.

2.3.1 Film-Casting

As chitosan has excellent film-forming properties, the most common (and straightforward) method used to prepare chitosan-based nanocomposites, irrespective of the nature of the nanoreinforcement, is that of solution-casting from a dilute acetic acid medium, followed by solvent evaporation [54–57, 59, 60]. In this case, whether the drying is carried out at different temperatures in the laboratory, or in a vacuum oven, it results in films or membranes.

In this procedure, as reported by Xianmaio *et al.*, chitosan and nanohydroxyapatite were mixed together in a 2% acetic acid medium, cast onto glass plates, and dried to form nanocomposite membranes [59]. The membranes were subsequently washed with very dilute NaOH solutions to neutralize them, and then rinsed thoroughly with distilled water. Cell-culture studies on n-HA/CS composite membranes showed not only a lack of any cytotoxic effect but also a good biocom-

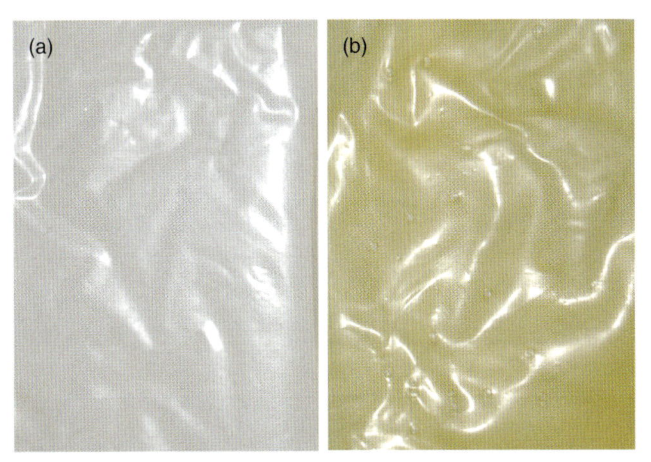

Figure 2.8 Photographic images of (a) uncrosslinked and (b) crosslinked chitosan/chitin nanocomposite films obtained by solution-casting from 2 vol% acetic acid.

patibility. Hence, the prospects for future guided bone regeneration membrane applications should be good.

All chitin nanocomposites created recently have been developed following a similar procedure [57]. In addition, crosslinked nanocomposites have been prepared by treating the solution-cast films with 0.2% glutaraldehyde so as to produce crosslinked chitosan–chitin nanowhisker composites. The chitosan and the nanocomposite thin films each showed a high optical clarity; indeed, the addition of 5 and 10 wt% chitin whiskers had no significant effect on the optical clarity of the films. The crosslinked chitosan and nanocomposites (see Figure 2.2b) films were yellow-brown in color; this was characteristic of chitosan/glutaraldehyde gels, and attributed to the formation of a chromophoric imine group (–N=C–) during crosslinking.

Photographic images of the uncrosslinked and crosslinked nanocomposite films with 5% chitin whiskers are shown in Figure 2.8, for comparison. The initial studies on chitosan-based bionanocomposite films showed that the permeation properties could be tailored by controlling the chitin whisker content and the crosslinking, again indicating a potential application in controlled drug release. The same could be considered for wound-dressing applications, on the basis of reinforcement due to the chitin whiskers, together with an enhanced water resistance, pH stability and antibacterial properties. However, further studies must be conducted to evaluate the biocompatibility of these nanocomposites.

Lui and Huang investigated a novel nanocomposite (GC/Ag) which was based on a genipin crosslinked chitosan (GC) film via solution casting, with different amounts of embedded Ag nanoparticles, for wound-dressing applications [60]. The superior cell spreading on nanocomposites with 50 ppm and 200 ppm of Ag particles, compared to a control sample and the matrix without Ag nanoparticles, is

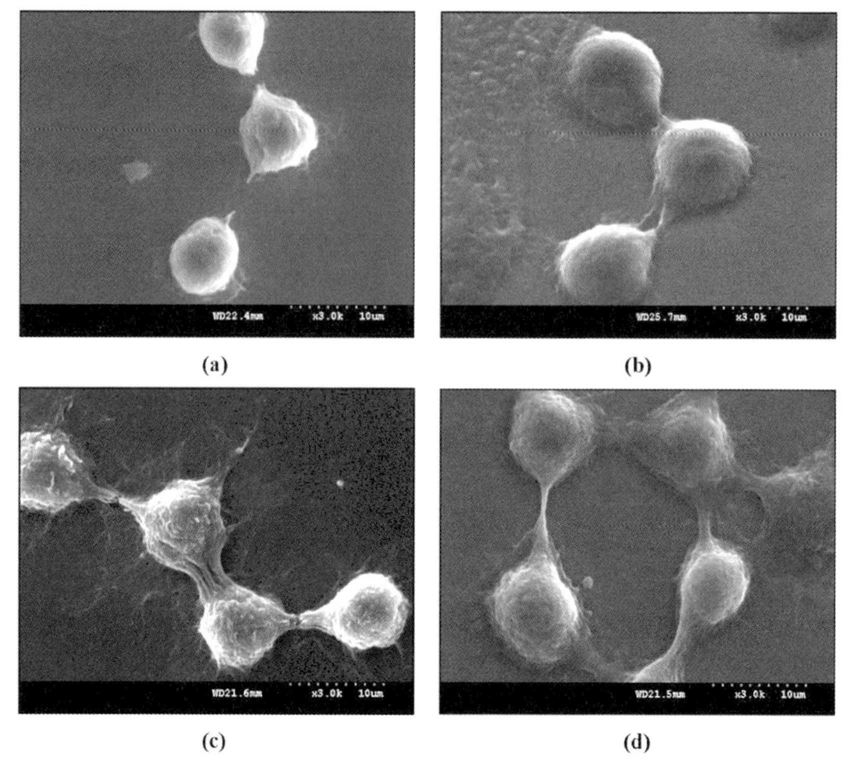

Figure 2.9 Scanning electron microscopy images of L929 cells cultured on surfaces of: (a) control; (b) pure GC film; (c) GC/Ag 50; (d) GC/Ag 200 nanocomposites after a 24 h culture. Reproduced with permission from Ref. [60]; © Wiley-VCH Verlag GmbH & Co. KGaA.

shown in Figure 2.9. The results of the study showed that these materials are biocompatible, and favor cell growth. It was also suggested that the silver ions had a dual function, namely structural reinforcement and the provision of antimicrobial properties to a biocompatible polymer.

2.3.2 Freeze-Drying

The development of porous scaffolds based on chitosan nanocomposites is normally achieved via a freeze-drying technique, whereby the matrix and reinforcements are mixed in a dilute acetic acid medium and then frozen at low temperature, perhaps using liquid N_2. The frozen mixture is then freeze-dried/lyophilized to remove the solvent, thus forming a porous structure.

A recent study on chitosan–nanohydroxyapatite composite scaffolds for bone tissue engineering demonstrated the formation of a well-defined pore structure

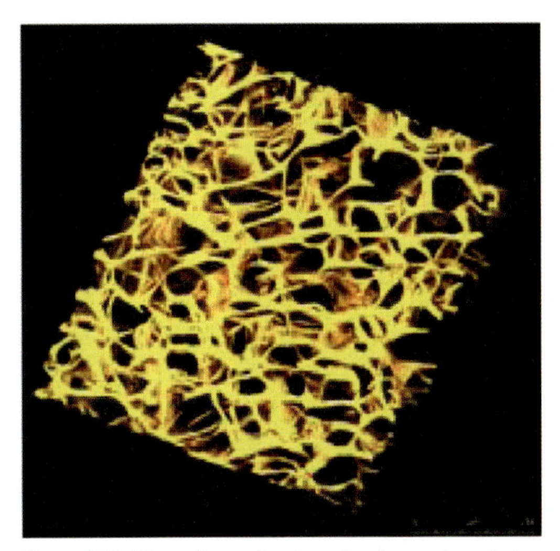

Figure 2.10 Three-dimensional confocal scanning electron microscopy image of a chitosan–nHA (CH1) scaffold. Reproduced with permission from Ref. [61]; © Elsevier.

obtained by freeze drying [61]. A three-dimensional (3-D) confocal micrograph of the foam with open pore structure in different sizes and different directions is shown in Figure 2.10. The nanocomposite foams showed a better cell adhesion, a higher degree of proliferation and a better morphology compared to the chitosan matrix foam, the improvements being attributed to the properties of nanohydroxyapatite.

Freeze-dried chitosan-based SWCNT composites incorporating poly (2-methoxy-aniline-5-sulfonic acid) (PMAS) as the gelation agent were prepared by Sweetman *et al.*, to be used as highly porous conductive scaffolds [62]. These scaffolds had a highly porous structure, which was repeatable, and the porosity was controllable based on the initial freezing temperature. A representative sample showing the uniform pore size obtained by freeze-drying the dispersion is shown in Figure 2.11. The CNTs and PMAS provided conductivity for the scaffolds; notably, whereas the CNTs improved the mechanical performance of the porous scaffold, the PMAS improved its resistance to degradation.

Recently, Wongpanit and coworkers developed nanocomposite sponges of chitin whisker-reinforced silk fibroin, using a freeze-drying technique [51]. In this case, the suspension containing chitin whiskers and silk fibroin was frozen at 40 °C overnight under a controlled cooling rate, followed by freeze-drying under vacuum conditions of <10 Pa for 24 h to obtain the nanocomposite sponges. The incorporation of chitin whiskers into the silk fibroin matrix led to improvements not only in dimensional stability but also to the mechanical properties of the silk fibroin sponges. Taken together, these results highlighted the potential of these nanocomposite sponges for scaffold applications.

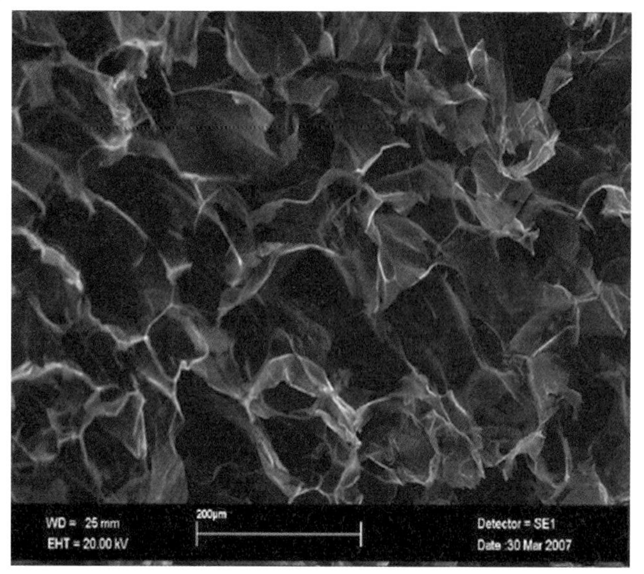

Figure 2.11 Scanning electron microscopy image of a 0.25% SWNT–1.0% chitosan scaffold containing 0.25% (w/v) poly (2-methoxyaniline-5-sulfonic acid) (PMAS), revealing the porous structure. Original magnification, ×500. Reproduced with permission from Ref. [62]; © Royal Society of Chemistry.

2.3.3
Layer-by-Layer Assembly

The LBL assembly technique represents a very useful method for multilayering structures with unique mechanical properties, as well as providing good control over the film's composition and thickness [63–65]. Ou *et al.* have developed an amperometric immunosensor from chitosan-based multilayer nanocomposite films containing MWCNT and a platinum (Pt)–Prussian blue (PB) hybrid by LBL self-assembly [63]. According to these authors, chitosan – as a polymer with positive charges in solution – can disperse the nanotubes so as to form a positively charged MWCNT–chitosan composite, which then can electrostatically immobilize the Pt–PB hybrid nanoparticles. The chitosan can also prevent PB from leaking from the multilayer films, and enhance the stability of the biosensor.

Yuan *et al.* have developed a chitosan-based multifunctional antibacterial coating containing TiO_2 nanoparticles and nanosilver particles by LBL assembly [64]. These materials showed a low bacterial activity in the dark and under low ultraviolet (UV) light conditions. The authors suggested that such multilayer films would have great potential for preclinical sterilization processes, thereby preventing the infection often caused by implanted medical devices.

2.3.4 Electrospinning

Electrospinning represents another very useful technology for producing continuous submicron fibers and/nanofibers. The electrospinning of nanocomposites was shown to offer much promise in the development of biomaterials for tissue engineering applications. In particular, the large surface area-to-volume ratio and the porous structure of nanofibers developed in this way favors cell adhesion, proliferation, migration, and differentiation, all of which are desirable properties when engineering tissues [66].

Zhang *et al.* have developed nanocomposite nanofibers of hydroxyapatite/chitosan for bone tissue engineering by electrospinning (Figure 2.12a) [67]. In this case, the hydroxyapatite-incorporated nanofibrous scaffolds showed a significantly enhanced bone-forming ability, as indicated by the cell proliferation, mineral deposition and morphology observations, when compared to the chitosan matrix. The results also demonstrated the great potential for using hydroxyapatite/chitosan nanocomposite nanofibers in bone tissue engineering applications (Figure 2.12b,c).

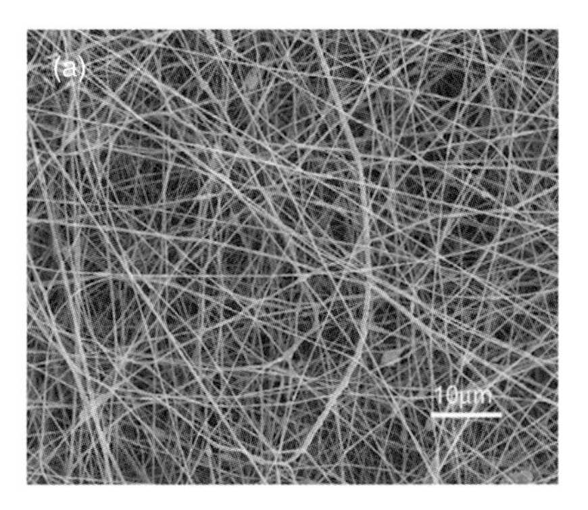

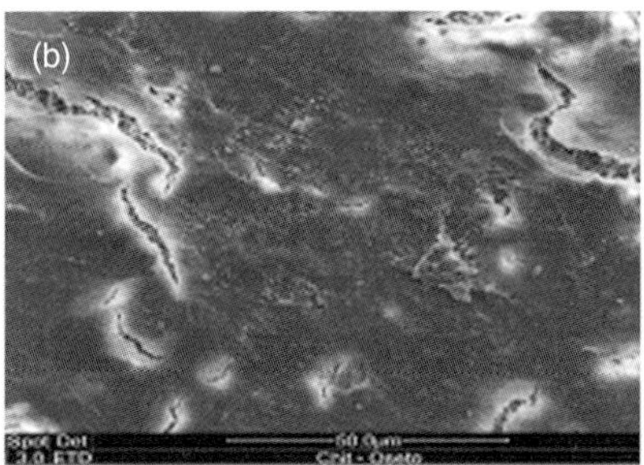

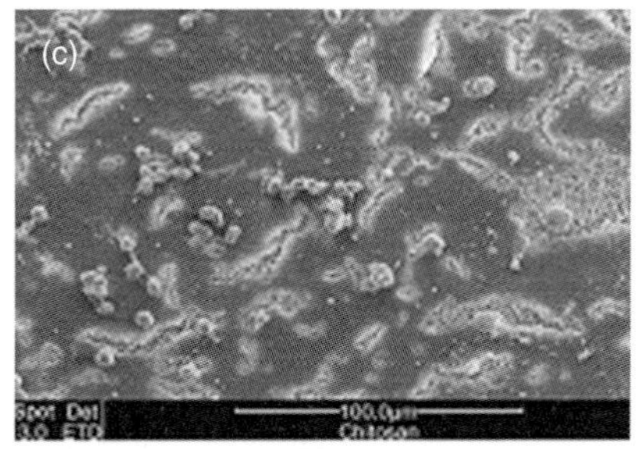

Figure 2.12 Field-emission scanning electron microscopy images. (a) Electrospun hydroxyapatite/chitosan nanocomposite nanofibers, doped with 10 wt% ultrahigh-molecular-weight poly(ethylene oxide) (UHMWPEO); (b,c) Mineral depositions of human fetal osteoblast (hFOB) cells on the electrospun nanofibrous scaffolds of (b) chitosan and (c) hydroxyapatite/chitosan after 10 days. Reproduced with permission from Ref. [67]; © Elsevier.

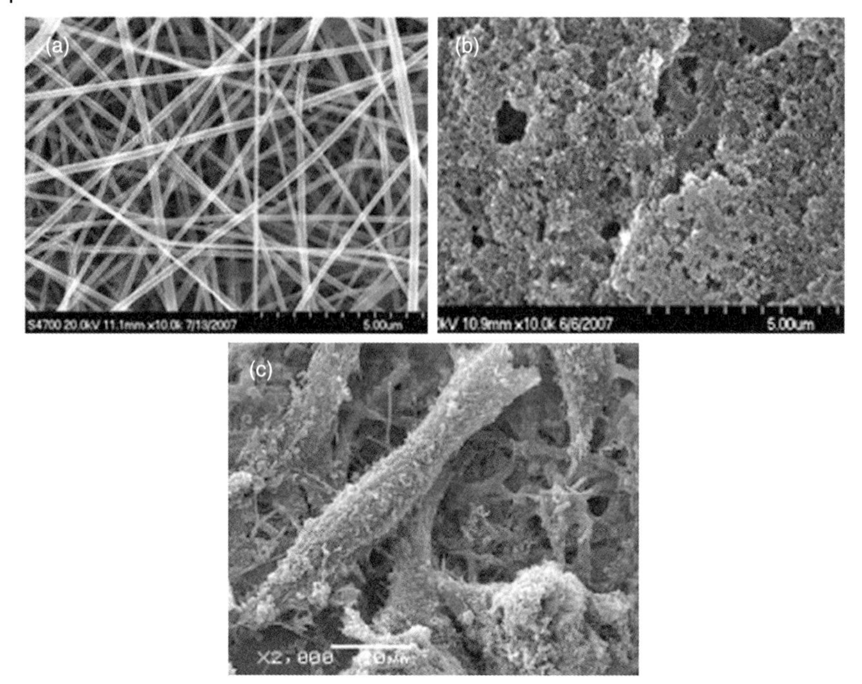

Figure 2.13 (a) Scanning electron microscopy (SEM) image of the electrospun chitosan/PVA nanofibrous scaffold; (b) SEM image of electrospun CECS/PVA scaffold mineralized in incubation solution in the presence of 2.0×10^{-3} wt% PVA for 25 days; (c) L929 cells seeded on fibrous membranes of hydroxyapatite-CECS/PVA after 48 h culture. Reproduced with permission from Ref. [68]; © Wiley-VCH Verlag GmbH & Co. KGaA.

In a slightly different approach, chitosan and water-soluble chitosan (*N*-carboxyethyl chitosan; CECS) derivative-based nanofibers were first electrospun (Figure 2.13a) and then deposited with hydroxyapatite (Figure 2.13b) [68]. The crystal growth of calcium compounds was induced on the polymer membrane surfaces by soaking the electrospun nanofibrous matrix in supersaturated calcium chloride and potassium dihydrogen phosphate solutions. Cell culture scanning electron microscopy (SEM) images of the scaffold showed that mouse fibroblasts (L929) grew on the surface, and that both the cell morphology and viability were well maintained (Figure 2.13c). These results suggested that biocomposites of hydroxyapatite with electrospun nanofibrous CECS/PVA scaffolds represented a promising scaffold biomaterial for bone tissue engineering.

Min *et al.* produced a composite nanofibrous scaffold with chitin nanoparticles that were distributed evenly in a matrix by the co-electrospinning of a poly(lactic acid-*co*-glycolic acid) (PLGA)/1,1,1,3,3,3-hexafluoro-2-propanol (HFP) solution and chitin : formic acid solution at a weight ratio of 80 : 20 [69]. PLGA, the random copolymer of glycolide (G) and lactide (L), is a popular and well-studied system that has been broadly used as an electrospun scaffold for biomedical applications

[70]. Normal human keratinocytes and fibroblasts were used to recognize the potential of this scaffold for tissue engineering. The PLGA/chitin composite scaffolds showed better results than pure PLGA scaffolds for normal human keratinocytes, although PLGA/chitin and PLGA showed similar performances in the case of fibroblasts.

2.4 Biomedical Products from Chitosan/Chitin Nanocomposites

Both, chitosan- and chitin-based nanocomposites have potential biomedical applications, owing to their good biocompatibility, and their nontoxic, biodegradable, and inherent wound-healing and film-forming properties. When nanoreinforcements such as MMT, CNTs, metal nanoparticles and chitin whiskers are added, the mechanical strength and thermal stability of the films, in addition to specific properties such as response to electric simulation, water permeability, drug loading and controlled release, and cell proliferation, are normally enhanced.

2.4.1 Films/Membranes/Coatings

Several reports have indicated the use of chitosan-based nanocomposites in the form of films, membranes or film coatings produced using techniques such as solution-casting, spin-coating, electrochemical deposition, LBL assembly, or electrospinning [11, 12, 39, 48, 50, 54, 56, 57, 59, 71]. The chitosan-based nanocomposite films are used as wound dressings, and/or as drug delivery or biosensing devices to detect molecules such as glucose, cholesterol and creatine in blood and urine samples [28, 32, 39, 41, 55, 56, 63].

Not only cell growth, but also drug delivery for healing and repair purposes, would be expected as possible, as in the case of nanocomposite films of chitosan-g-lactic acid and MMT developed by Singh and coworkers [41].

Chitosan nanocomposite films with gold nanoparticles, prepared by a one-step electrochemical deposition, was reported by Chen and coworkers [71], and found to be effective as glucose sensors.

2.4.2 Mats and Sponges

Nanocomposite mats and sponges usually have a high porosity, which provides structural space for cell accommodation and also facilitates an efficient exchange of nutrients and metabolic waste between a scaffold and the environment. A scaffold used for bone tissue engineering requires a highly porous and interconnected pore structure, so as to ensure that the biological environment is conducive to cell attachment, proliferation, tissue growth and adequate nutrient flow. Whilst these biomaterials resemble natural tissues morphologically, they are characterized chemically by a wide range of pore diameter distribution, a high porosity, good mechanical properties, and specific biochemical properties.

Several nanocomposite materials based on chitosan have been fabricated in the form of porous mats or sponges, and used for both drug delivery and tissue engineering studies [41, 51, 61, 62, 72]. For example, Thein-Han and Misra described the synthesis of a bone-like organic–inorganic biomimetic nanocomposite consisting of chitosan and (nanohydroxyapatite) (nHAp) for potential use as a bone tissue engineering material [61]. The lyophilized chitosan and nanocomposite scaffolds after hydration and neutralization were soft, spongy, flexible, and elastic. The compression modulus of the hydrated scaffolds was significantly increased following the incorporation of nHA, and was increased with in line with increases in the nHA content of the scaffold. Typically, the compression moduli of high-molecular-weight chitosan and the chitosan–nHA (1%) nanocomposite scaffold were 6.0 ± 0.3 kPa and 9.2 ± 0.2 kPa, respectively. In addition to providing mechanical strength, the highly porous nanocomposite scaffold exhibited superior initial cell attachment, spreading, proliferation and morphology when compared to chitosan scaffolds.

Chitosan-g-lactic acid and sodium MMT nanohybrid films and porous scaffolds have been prepared, using solvent-casting and freeze-drying, by Depan *et al.* [41]. The SEM images of porous scaffolds developed in this way are shown in Figure 2.14. A significant difference in drug release from the smooth films and porous scaffolds was observed, with porous scaffolds showing a greater and more rapid drug release that was attributed to the porous morphology of the freeze-dried materials.

Nanocomposite sponges developed by Wongpanit and coworkers based on chitin whisker-reinforced silk fibroin were described in Section 2.3 [51]. The cell culture studies showed that the nanocomposites with chitin whiskers had a better cell-spreading potential than neat silk fibroin, thus highlighting the potential application of this bionanocomposite in the biomedical field.

2.4.3 Hydrogels

Chemical crosslinks (covalent bonds) or physical junctions (e.g., secondary forces, crystallite formation, chain entanglements) provide the hydrogels' unique swelling behavior and 3-D structure [73, 74].

Gariepy and Leroux, in 2004, described the preparation of thermosensitive chitosan–polyol salt hydrogels, while chitosan-based hydrogels were also investigated as potential cell carriers for tissue engineering applications by Cho *et al.* [75, 76]. Although several have been made describing chitosan-based hydrogels for various biomedical applications, the number of reports relating to nanocomposite gels based on chitosan have been limited [77–80].

The incorporation of SWCNTs into chitosan biogels, as reported relatively recently, rendered the chitosan gel electroresponsive [35]. Nonetheless, these biogels were shown to be nontoxic, to serve as an excellent platform for fibroblast

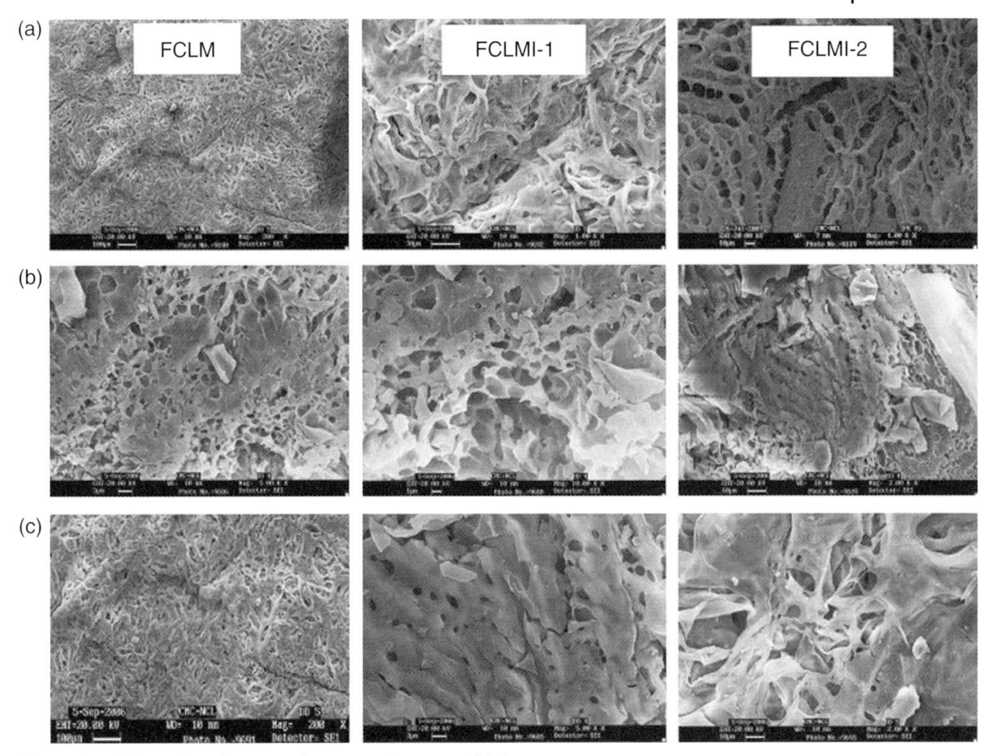

Figure 2.14 Scanning electron microscopy images of (a) top, (b) middle, and (c) bottom portions of freeze-dried porous scaffolds of chitosan-g-lactic acid and sodium MMT. The images show different layers of the freeze-dried foams. FCLM, freeze-dried chitosan-grafted-lactic acid montmorillonite. Reproduced with permission from Ref. [41]; © Elsevier.

cell growth, and demonstrated a capability for the controlled release of drugs (e.g., NT-3).

2.5 The Path Forward

Whilst the field of chitosan-based nanocomposites for biomedical applications has evolved rapidly during recent years, advancements in terms of their medical application may be allocated to three major categories, depending on the functionality and application:

- CNT-based smart chitosan nanocomposites in which high mechanical strength, together with conductivity, is the main focus. Chitosan/CNT-based nanocomposites in medical applications, in which an electrical response is useful, will have great potential [35, 62]. In these cases, the reinforcement is neither biobased nor biodegradable. However, in these cases, the reinforcement content is usually very low (>5%), such that the products are environmentally friendly and interesting.

- Inorganic nanoreinforcement-based nanocomposites in which mechanical strength and nontoxicity are important. It has been shown that, in medical applications where bone substitutes are intended, clay and hydroxyapatite-based nanocomposites have precedence [37, 38, 42, 45]. Here, the reinforcing phase, although biobased, is not biodegradable. However, those nanocomposites based on chitosan and inorganic fillers are interesting owing to their superior mechanical strength, biocompatibility, cell viability and the possibility of obtaining stable porous scaffolds for bone regeneration and tissue engineering, etc.
- Biobased nanoreinforcement-based nanocomposites in which biodegradability and nontoxicity are the main focus. For artificial implants, wound dressing, and controlled drug release, fully biobased nanocomposites will be the best choice. In this context, all chitin nanocomposites and chitosan–cellulose hybrid nanocomposite systems are promising materials for the future [55–57]. The advantages of these materials are their fully biobased and biodegradable nature, as well as their improved mechanical properties, thermal stability and barrier properties. Further, the development of these fully biodegradable materials in the form of films, foams or electrospun fibers will open up new applications and new possibilities in the field of chitin-based nanocomposites for medical applications. To date, very few studies have been reported in this field, and further research is clearly required. The all-chitin nanocomposites are particularly interesting, due to the possibility to obtain biobased materials for medical applications with improved mechanical and barrier properties, without adversely affecting the inherent properties such as nontoxicity, antibacterial activity, biocompatibility, and the biodegradability of chitosan.

Acknowledgments

The authors gratefully acknowledge the European Commission for their financial support, under the contract number NMP4-CT-2006-033277 TEM-PLANT. Dr Marie-Pierre G. Laborie is acknowledged for providing the redrawn Figure 2.1.

References

1 Yang, K.-K., Wang, X.-L. and Wang, Y.-Z. (2007) Progress in nanocomposite of biodegradable polymer. *Journal of Industrial and Engineering Chemistry*, **13** (4), 485–500.

2 Hule, R.A. and Pochan, D.J. (2007) Polymer nanocomposites for biomedical applications. *MRS Bulletin*, **32**, 354–8.

3 Hayashi, T. (1994) Biodegradable polymers for biomedical usages.

Progress in Polymer Science, **9**, 663–702.

4 Li, M., Mondrinos, M.J., Gandhi, M.R., Ko, F.K., Weiss, A.S. and Lelkes, P.I. (2005) Electrospun protein fibers as matrices for tissue engineering. *Biomaterials*, **26**, 5999–6008.

5 Zhang, Y.Z., Venugopal, J., Huang, Z.M., Lim, C.T. and Ramakrishna, S. (2006) Crosslinking of the electrospun gelatin nanofibers. *Polymer*, **47**, 2911–17.

6 Czaja, W.K., Young, D.J., Kawecki, M. and Brown, R.M. Jr (2007) The future prospects of microbial cellulose. *Biomacromolecules*, **8** (1), 1–12.

7 Rinaudo, M. (2006) Chitin and chitosan: properties and applications. *Progress in Polymer Science*, **31**, 603–32.

8 Peniche, C., Arguelles-Monal, W. and Goycoolea, F.M. (2008) Chapter 25: chitin and chitosan: major sources, properties and applications, in *Monomers, Polymers and Composites from Renewable Resources* (eds M.N. Belgacem and A. Gandini), Elsevier, Amsterdam, pp. 517–42.

9 Van de Velde, K. and Kiekens, P. (2004) Structure analysis and degree of substitution of chitin, chitosan and dibutyrylchitin by FT-IR spectroscopy and solid-state ^{13}C NMR. *Carbohydrate Polymers*, **58**, 409–16.

10 Marchessault, R.H., Morehead, R.R. and Walter, N.M. (1959) Liquid crystal systems from fibrillar polysaccharides. *Nature*, **184**, 632–4.

11 Morin, A. and Dufresne, A. (2002) Nanocomposites of chitin whiskers from *Riftia* tubes and poly(caprolactone). *Macromolecules*, **35** (6), 2190–9.

12 Paillet, M. and Dufresne, A. (2001) Chitin whisker reinforced thermoplastic nanocomposites. *Macromolecules*, **34**, 6527–9.

13 Nair, K.G. and Dufresne, A. (2003) Crab shell chitin whisker reinforced natural rubber nanocomposites. 2. Mechanical behavior. *Biomacromolecules*, **4**, 666–74.

14 Madihally, S.V. and Matthew, H.W.T. (1999) Porous scaffolds for tissue engineering. *Biomaterials*, **20**, 1133–11.

15 Wang, M., Chen, L.J., Ni, J., Weng, J. and Yue, C.Y. (2001) Manufacture and evaluation of bioactive and biodegradable materials and scaffolds for tissue engineering. *Journal of Materials Science: Materials in Medicine*, **12** (10), 855–60.

16 Chow, K.S., Khor, E. and Wan, A.C.A. (2001) Porous chitin matrices for tissue engineering: fabrication and in-vitro cytotoxic assessment. *Journal of Polymer Research*, **8**, 27–35.

17 Ito, M., Hidaka, Y., Nakajima, M., Yagasaki, H. and Kafrawy, A.H. (1999) Effect of hydroxyapatite content on physical properties and connective tissue reactions to a chitosan–hydroxyapatite composite membrane. *Journal of Biomedical Materials Research*, **45**, 204–8.

18 Zhang, Y. and Zhang, M. (2002) Three-dimensional macroporous calcium phosphate bioceramics with nested chitosan sponges for load-bearing bone implants. *Journal of Biomedical Materials Research*, **61**, 1–8.

19 Yamaguchi, I., Tokuchi, K., Fukuzaki, H., Koyama, Y., Takakuda, K., Monma, H. and Tanaka, J. (2001) Preparation and microstructure analysis of chitosan/hydroxyapatite nanocomposites. *Journal of Biomedical Materials Research*, **55**, 20–7.

20 Hidaka, Y., Ito, M., Mori, K., Yagasaki, H. and Kafrawy, A.H. (1999) Histopathological and immunohistochemical studies of membranes of deacetylated chitin derivatives implanted over rat calvaria. *Journal of Biomedical Materials Research*, **46**, 418–23.

21 Yokoyama, A., Yamamoto, S., Kawasaki, T., Kohgo, T. and Nakasu, M. (2002) Development of calcium phosphate cement using chitosan and citric acid for bone substitute materials. *Biomaterials*, **23**, 1091–101.

22 Gupta, K.C. and Jabrail, F.H. (2006) Glutaraldehyde and glyoxal crosslinked chitosan microspheres for the controlled delivery of centchroman. *Carbohydrate Research*, **341**, 744–56.

23 Ramani, S.P., Chaudari, C.V., Patil, N.D. and Sabharwal, S. (2004) Synthesis and characterization of crosslinked chitosan formed by γ-irradiation in the presence of carbon tetrachloride as sensitizer. *Journal of Polymer Science Part A: Polymer Chemistry*, **42**, 3897–909.

24 Neto, C.G.T., Dantas, T.N.-C., Foonseca, J.L.C. and Pereira, M.R. (2005) Permeability studies in chitosan membranes. Effect of crosslinking and polyethylene oxide addition. *Carbohydrate Research*, **340**, 2630–6.

25 Malheiro, V.N., Caridade, S.G., Alves, N.M. and Mano, J.F. (2009) New poly(ε-caprolactone)/chitosan blend fibers for tissue engineering applications. *Acta Biomaterialia*, **6** (2), 418–28.

26 Costa-Júnior, E.S., Barbosa-Stancioli, E.F., Mansur, A.A.P., Vasconcelos, W.L. and Mansur, H.S. (2009) Preparation and characterization of chitosan/poly(vinyl alcohol) chemically crosslinked blends for biomedical applications. *Carbohydrate Polymers*, **76** (3), 472–81.

27 Khora, E.E. and Lim, L.Y. (2003) Implantable applications of chitin and chitosan. *Biomaterials*, **24**, 2339–49.

28 Tiwari, A. and Dhakate, S.R. (2009) Chitosan–SiO_2–multiwall carbon nanotubes nanocomposite: a novel matrix for the immobilization of creatine midinohydrolase. *International Journal of Biological Macromolecules*, **44**, 408–12.

29 Solanki, P.R., Kaushik, A., Ansari, A.A., Tiwari, A. and Malhotra, B.D. (2009) Multi-walled carbon nanotubes/sol-gel-derived silica/chitosan nanobiocomposite for total cholesterol sensor. *Sensors and Actuators*, **B 137**, 727–35.

30 Khan, R., Kaushik, A., Solanki, P.R., Ansari, A.A., Pandey, M.K. and Malhotra, B.D. (2008) Zinc oxide nanoparticles-chitosan composite film for cholesterol biosensor. *Analytica Chimica Acta*, **616** (2), 207–13.

31 Gopalan, A.I., Lee, K.-P. and Ragupathy, D. (2009) Development of a stable cholesterol biosensor based on multi-walled carbon nanotubes–gold nanoparticles composite covered with a layer of chitosan–room-temperature ionic liquid network. *Biosensors and Bioelectronics*, **24**, 2211–17.

32 Ragupathy, D., Gopalan, A.I. and Lee, K.-P. (2009) Synergistic contributions of multiwall carbon nanotubes and gold nanoparticles in a chitosan-ionic liquid matrix towards improved performance for a glucose sensor. *Electrochemistry Communications*, **11** (2), 397–401.

33 Qui, J.-D., Zhou, W.-M., Guo, J., Wang, R. and Liang, R.P. (2009) Amperometric sensor based on ferrocene-modified multiwalled carbon nanotube nanocomposites as electron mediator for the determination of glucose. *Analytical Biochemistry*, **385** (2), 264–9.

34 Qui, J.-D., Deng, M.-Q., Liang, R.-P. and Xiong, M. (2008) Ferrocene-modified multiwalled carbon nanotubes as building block for construction of reagent less enzyme-based biosensors. *Sensors and Actuators B: Chemical*, **135** (1), 181–7.

35 Thompson, B.C., Moulton, S.E., Gilmore, K.J., Higgins, M.J., Whitten, P.G. and Wallace, G.G. (2009) Carbon nanotube biogels. *Carbon*, **47**, 1282–91.

36 Kashiwazaki, H., Kishiya, Y., Matsuda, A., Yamaguchi, K., Iizuka, T., Tanaka, J. and Inoue, N. (2009) Fabrication of porous chitosan/hydroxyapatite nanocomposites: their mechanical and biological properties. *Bio-Medical Materials and Engineering*, **19** (2–3), 133–40.

37 Cui, W., Hu, Q., Wu, J., Li, B. and Shen, J. (2008) Preparation and characterization of magnetite/hydroxyapatite/chitosan nanocomposite by in situ compositing method. *Journal of Applied Polymer Science*, **109**, 2081–8.

38 Katti, K.S., Katti, D.R. and Dash, R. (2008) Synthesis and characterization of a novel chitosan/montmorillonite/hydroxyapatite nanocomposite for bone tissue engineering. *Biomedical Materials*, **3**, 034122–34.

39 Wang, X., Du, Y., Luo, J., Lin, B. and Kennedy, J.F. (2007) Chitosan/organic rectorite nanocomposite films: structure, characteristic and drug delivery behaviour. *Carbohydrate Polymers*, **69**, 41–9.

40 Wang, X., Du, Y. and Luo, J. (2008) Biopolymer/montmorillonite nanocomposite: preparation, drug-controlled release property and cytotoxicity. *Nanotechnology*, **19**, 065707–14.

41 Depan, D., Pratheep Kumar, A. and Singh, R.P. (2009) Cell proliferation and controlled drug release studies of nanohybrids based on chitosan-g-lactic acid and montmorillonite. *Acta Biomaterialia*, **5** (1), 93–100.

42 Liuyun, J., Yubao, L., Li, Z. and Jianguo, L. (2008) Preparation and properties of a novel bone repair composite: nano-hydroxyapatite/chitosan/carboxymethyl cellulose. *Journal of Materials Science: Materials in Medicine*, **19**, 981–7.

43 Jiang, L., Li, Y., Zhang, L. and Wang, X. (2009) Preparation and characterization of a novel composite containing carboxymethyl cellulose used for bone repair. *Materials Science and Engineering*, **29** (1), 193–8.

44 Zhou, N.-I., Liu, Y., Li, L., Meng, N., Huang, Y.-X., Zhang, J., Wei, S.-H. and Shen, J. (2007) A new nanocomposite biomedical material of polymer/ Clay–Cts–Ag nanocomposites. *Current Applied Physics*, **7S1**, e58–62.

45 Hu, Q.L., Li, B.Q., Wang, M. and Shen, J.C. (2004) Preparation and characterization of biodegradable chitosan/ hydroxyapatite nanocomposite rods via in situ hybridization: a potential material as internal fixation of bone fracture. *Biomaterials*, **25**, 779–85.

46 Liu, K.-H., Liu, T.-Y., Chen, S.-Y. and Liu, D.-M. (2007) Effect of clay content on electrostimulus deformation and volume recovery behavior of a clay–chitosan hybrid composite. *Acta Biomaterialia*, **3**, 919–26.

47 Theodor, E., Litescu, S.C., Petcu, C., Mihalache, M. and Somoghi, R. (2009) Nanostructured biomaterials with controlled properties synthesis and characterization. *Nanoscale Research Letters*, **4**, 544–9.

48 Li, Q., Zhou, J. and Zhang, L. (2009) Structure and properties of the nanocomposite films of chitosan reinforced with cellulose whiskers. *Journal of Polymer Science Part B: Polymer Physics*, **47**, 1069–77.

49 Junkasem, J., Rujiravanit, R. and Supaphol, P. (2006) Fabrication of α-chitin whisker-reinforced poly(vinyl alcohol) nanocomposite nanofibres by electrospinning. *Nanotechnology*, **17**, 4519–28.

50 Phongying, S., Aiba, S.-I. and Chirachanchai, S. (2007) Direct chitosan scaffolds formation via chitin whiskers. *Polymer*, **48**, 393–400.

51 Wongpanit, P., Sanchavanakit, N., Pavasant, P., Bunaprasert, T., Tabata, T.Y. and Rujiravanit, R. (2007) Preparation and characterization of chitin whisker-reinforced silk fibroin nanocomposite sponges. *European Polymer Journal*, **43**, 4123–35.

52 Pinho, E.D., Martins, A., Araújo, J.V., Reis, R.L. and Neves, N.M. (2009) Degradable particulate composite reinforced with nanofibres for biomedical applications. *Acta Biomaterialia*, **5**, 1104–14.

53 Coutinho, D.F., Pashkuleva, I., Alves, C.M., Marques, A.P., Neves, N.M. and Reis, R.L. (2008) The effect of chitosan on the in vitro biological performance of chitosan-poly(butylene succinate) blends. *Biomacromolecules*, **9**, 1139–45.

54 Sriupayo, J., Supapol, P., Blackwell, J. and Rujiravanit, P. (2005) Preparation and characterisation of A-chitin whisker reinforced chitosan nanocomposite films with and without heat treatment. *Carbohydrate Polymers*, **62**, 130–6.

55 Muzzarelli, R.A.A., Morganti, P., Morganti, G., Palombo, P., Palambo, M., Biagini, G., Belmonte, M.M., Giantomassi, F., Orlandi, F. and Muzzarelli, C. (2007) Chitin nanofibrils/ chitosan glycolate composites as wound medicaments. *Carbohydrate Polymers*, **70**, 274–81.

56 Shelma, R., Paul, W. and Sharma, C.P. (2008) Chitin nanofibre reinforced thin chitosan films for wound healing application. *Trends in Biomaterials and Artificial Organs*, **22** (2), 107–11.

57 Mathew, A.P., Laborie, M.-P.G. and Oksman, K. (2009) Cross-linked chitosan/ chitin crystal nanocomposites with improved permeation selectivity and pH stability. *Biomacromolecules*, **10**, 1627–32.

58 Nair, K.G. and Dufresne, A. (2003) Crab shell chitin whisker reinforced natural rubber nanocomposites. 1. Processing and swelling behaviour. *Biomacromolecules*, **4** (3), 657–65.

59 Xianmiao, C., Yubao, Li., Yi, Z., Li, Z., Jidong, L. and Huanan, W. (2009) Properties and in vitro biological evaluation of nano-hydroxyapatite/ chitosan membranes for bone guided regeneration. *Materials Science and Engineering*, **C 29**, 29–35.

60 Liu, B.S. and Huang, T.B. (2008) Nanocomposites of genipin-crosslinked chitosan/silver nanoparticles – structural reinforcement and antimicrobial properties. *Macromolecular Bioscience*, **8** (10), 932–41.

61 Thein-Han, W.W. and Misra, R.D.K. (2009) Biomimetic chitosan–nanohydroxyapatite composite scaffolds for bone tissue engineering. *Acta Biomaterialia*, **5**, 1182–97.

62 Sweetman, L.J., Moulton, S.E. and Wallace, G.G. (2008) Characterisation of porous freeze dried conducting carbon nanotube–chitosan scaffolds. *Journal of Materials Chemistry*, **18**, 5417–22.

63 Ou, C., Chen, S., Yuan, R., Chai, Y. and Zhong, X. (2008) Layer-by-layer self-assembled multilayer films of multi-walled carbon nanotubes and platinum-prussian blue hybrid nanoparticles for the fabrication of amperometric immunosensor. *Journal of Electroanalytical Chemistry*, **624** (1–2), 287–92.

64 Yuan, W., Ji, J., Fu, J. and Shen, J. (2008) Facile method to construct hybrid multilayered films as a strong and multifunctional antibacterial coating. *Journal of Biomedical Materials Research: Part B Applied Biomaterials*, **85** (2), 556–63.

65 Podsiadlo, P., Tang, Z., Shim, B.S. and Kotov, N.A. (2007) Counterintuitive effect of molecular strength and role of molecular rigidity on mechanical properties of layer-by-layer assembled nanocomposites. *Nano Letters*, **7** (5), 1224–31.

66 Venugopal, J., Low, S., Choon, A.T. and Ramakrishna, S. (2008) Interaction of cells and nanofiber scaffolds in tissue engineering. *Journal of Biomedical Materials Research: Part B Applied Biomaterials*, **84B**, 34–48.

67 Zhang, Y., Venugopal, J.R., El-Turki, A., Ramakrishna, B.S.S. and Lim, C.T. (2008) Electrospun biomimetic nanocomposite nanofibers of hydroxyapatite/chitosan for bone tissue engineering. *Biomaterials*, **29**, 4314–22.

68 Yang, D., Jin, Y., Zhou, Y., Ma, G., Chen, X., Lu, F. and Nie, J. (2008) In situ mineralization of hydroxyapatite on electrospun chitosan-based nanofibrous scaffolds. *Macromolecular Bioscience*, **8**, 239–46.

69 Min, B.-M., You, Y., Kim, J.-M., Lee, S.J. and Park, W.H. (2004) Formation of nanostructured poly(lactic-*co*-glycolic acid)/chitin matrix and its cellular response to normal human keratinocytes and fibroblasts. *Carbohydrate Polymers*, **57**, 285–92.

70 Zong, X., Fang, D., Kim, K., Ran, S., Hsiao, B.S., Chu, B., Brathwaite, C., Li, S. and Chen, E. (2002) Nonwoven nanofiber membranes of poly(lactide) and poly (glycolide-*co*-lactide) via electrospinning and application for antiadhesions. *Polymer Preprints (American Chemical Society, Division of Polymer Chemistry)*, **43**, 659–60.

71 Feng, D., Wang, F. and Chen, Z. (2009) Electrochemical glucose sensor based on one-step construction of gold nanoparticle–chitosan composite film. *Sensors and Actuators*, **B 138**, 539–44.

72 Lau, C. and Cooney, M.J. (2008) Conductive macroporous composite chitosan–carbon nanotube scaffolds. *Langmuir*, **24**, 7004–10.

73 Peppas, N.A. and Mikos, A.G. (1986) Preparation methods and structure of hydrogels, in *Hydrogels in Medicine and Pharmacy*, vol. 1 (ed. N.A. Peppas), CRC Press, Boca Raton, Florida, pp. 1–25.

74 Hoffman, A.S. (2002) Hydrogels for biomedical applications. *Advanced Drug Delivery Reviews*, **43**, 3–12.

75 Gariepy, E.R. and Leroux, J.C. (2004) In situ forming hydrogels; Review of temperature-sensitive systems. *European Journal of Pharmaceutics and Biopharmaceutics*, **58**, 409–26.

76 Cho, J.H., Kim, S.H., Park, K.D., Jung, M.C., Yang, W.I., Han, S.W., Noh, J.Y. and Lee, J.W. (2004) Chondrogenic differentiation of human mesenchymal stem cells using a thermosensitive poly(*N*-isopropylacrylamide) and water-soluble chitosan copolymer. *Biomaterials*, **25**, 5743–51.

77 Muzzarelli, R.A.A. (2009) Genipin-crosslinked chitosan hydrogels as biomedical and pharmaceutical aids. *Carbohydrate Polymers*, **77** (1), 1–9.

78 Ta, H.T., Han, H., Larson, I., Dass, C.R. and Dunstan, D.E. (2009) Chitosan-dibasic orthophosphate hydrogel: a potential drug delivery system.

International Journal of Pharmaceutics, **371** (1–2), 134–41.

79 Schuetz, Y.B., Gurny, R. and Jordan, O. (2008) A novel thermoresponsive hydrogel based on chitosan. *European Journal of Pharmaceutics and Biopharmaceutics*, **68** (1), 19–25.

80 Ji, Q.X., Chen, X.G., Zhao, Q.S., Liu, C.S., Cheng, X.J. and Wang, L.C. (2009) Injectable thermosensitive hydrogel based on chitosan and quaternized chitosan and the biomedical properties. *Journal of Materials Science: Materials in Medicine*, **1**, 1–8.

3
Silver Nanocomposites and Their Biomedical Applications

Andrea Travan, Eleonora Marsich, Ivan Donati, and Sergio Paoletti

3.1
Introduction to Silver Nanocomposites

By definition, nanocomposites are materials that contain domains or inclusions which are of nanometer size scale [1]. Today, nanocomposite materials that exploit the properties of silver at the nanoscale for biomedical applications are of increasing interest (Figure 3.1), and may be found in a variety of commercial products, including wound dressings and antiseptic creams.

The idea behind these novel materials is that silver, at the nanometer scale, displays unique properties that can be used for different purposes, ranging from antimicrobial to optical and catalytic applications [2, 3]. This chapter provides an overview of the current state of the art of silver nanocomposites in the area of biomedicine. Following a brief introduction to "silver nanocomposites," details of the main routes of preparation of silver nanoparticles and the final composite constructs are discussed. The applications of these materials, and the results obtained over the past few years are then presented, with particular attention being paid to silver's antimicrobial properties. Hence, a summary of results is provided, together with a hypothesis of the antimicrobial mechanism. Subsequent sections of the chapter describe the use of silver nanocomposites in applications related to wound healing, inflammation and biological sensing. Details are also provided of the possible biological hazards of these materials, which are rarely taken into account but have proven to be somewhat controversial. Finally, some possible future biomedical applications of the materials are suggested.

Nanomaterials for the Life Sciences Vol.8: Nanocomposites. Edited by Challa S. S. R. Kumar

ISBN: 978-3-527-32168-1

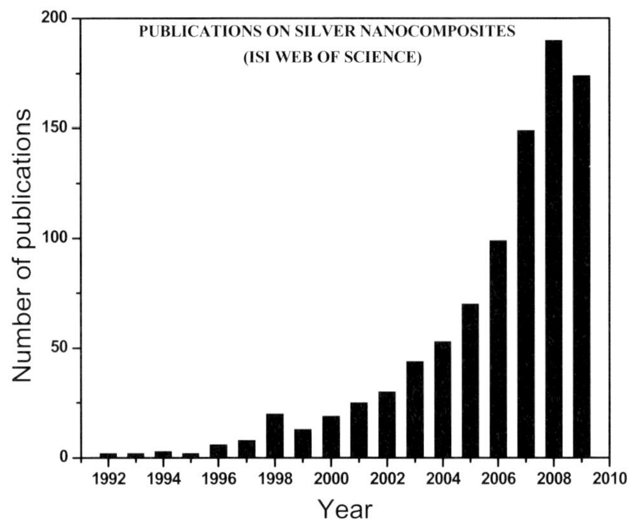

Figure 3.1 Publication counts derived from the Thompson ISI Web of Science database on October 2009, using the key word "silver nanocomposites".

3.2 Preparation and Characterization of Silver Nanocomposites

3.2.1 Preparation Techniques

Since silver nanocomposite constructs are meant to exploit the properties of silver at the nanoscale, much effort has been targeted at identifying efficient and reproducible routes for the production of silver nanoparticles (AgNPs). Today, many potential uses for these novel materials are under investigation, ranging from antimicrobial therapies to molecular imaging.

When preparing homogeneous nanocomposite materials containing AgNPs, one crucial issue is the tendency for nanoparticles to aggregate, which leads to a loss of the peculiar properties associated with the nanoscale. For instance, in the area of antimicrobials, studies conducted by Lok *et al.* [4] revealed that nonstabilized AgNPs prepared via standard chemical methods (the reduction of silver salt solutions) tend to aggregate in culture media and in biological buffers with a high salt content (chloride and phosphate are the most problematic anions). Such aggregation will lead to a reduction in the effective surface of the nanoparticles, or the degree to which they can associate to the bacteria.

The so-called "*ex-situ*" preparation methods consist of the synthesis of nanoparticles as a first step, followed by mixing the preformed nanoparticles with a matrix (typically a polymer). Unfortunately, however, the use of this conventional technique tends to prevent the homogeneous dispersion of the nanoparticles [5]. In contrast, the aim of an "*in situ*" approach is to synthesize the nanoparticles within a suitable matrix, so as to achieve an homogeneous dispersion within the composite materials.

To date, the preparation and stabilization of metal nanoparticles represent an open challenge, with the methods used to prepare nonaggregated AgNPs for biomedical applications divisible into two groups [6]:

- Wet chemical syntheses in the presence of a reducing agent and a stabilizing agent.
- Synthesis through physical processes.

The chemical approach is based on the use of a silver salt in an aqueous environment; the silver ions are then reduced to zeroth-valent silver in the presence of a stabilizing agent, in order to limit the aggregation of the thus-formed nanoparticles. The role of the stabilizing agent is to cap the particles and prevent further growth. The most widely accepted mechanism for the synthesis of a particle involves a two-step process: nucleation, followed by successive growth. In the first step, a proportion of the metal ions in solution is reduced by a suitable reducing agent; the atoms thus produced then act as nucleation centers and catalyze a reduction of the remaining metal ions present in the bulk solution. The atomic coalescence leads to the formation of metal clusters, the dimensions of which can be controlled by ligands, surfactants or polymers. In the absence of a stabilizer, clusters in aqueous solution can undergo further growth, leading ultimately to precipitation of the metal [7].

Polymers and, more generally, many organic molecules can bind to the particle surface and thus play the role of a stabilizer. In general, the stabilization of metal nanoparticles is explained by the electronic interaction of the polymer functional groups with the metal particles. In fact, nucleophilic groups can bind the metal particles by donating electrons [3]. A common approach for the preparation of stable AgNPs involves the use of polymer solutions, and for this a variety of polymers can be used ranging from synthetic to natural forms. Nitrogen-containing stabilizing polymers, such as poly-(ethyleneimine) (PEI) and poly(vinylpyrrolidone) (PVP), act via lone electron pairs. Protective polymers can coordinate metal ions before reduction, forming a polymer–ion complex that can then be reduced under mild conditions; this results in metal particles of smaller dimensions and a narrower size distribution than those obtained without protective polymers [8]. Once the reduction has occurred, the stabilizing effect of these macromolecules can be attributed to the fact that either the particles are attached to the much larger protecting polymers, or the protecting molecules cover or encapsulate the metal particles [9]. This chemical approach is shown schematically in Figure 3.2.

In particular, owing to the presence of many different functional groups, polyelectrolytes have been used successfully in the preparation of stable AgNPs. Polyelectrolytes at low concentrations (e.g., polyphosphate, polyacrylate, poly(vinylsulfate), PEI [2, 3, 10], poly(allyl-amine) [9], and chitosan [11–13]) have all been used, with different outcomes, to stabilize nanoparticles and prevent the growth of aggregates. As noted above, PVP [5, 14] (a neutral polymer) has been

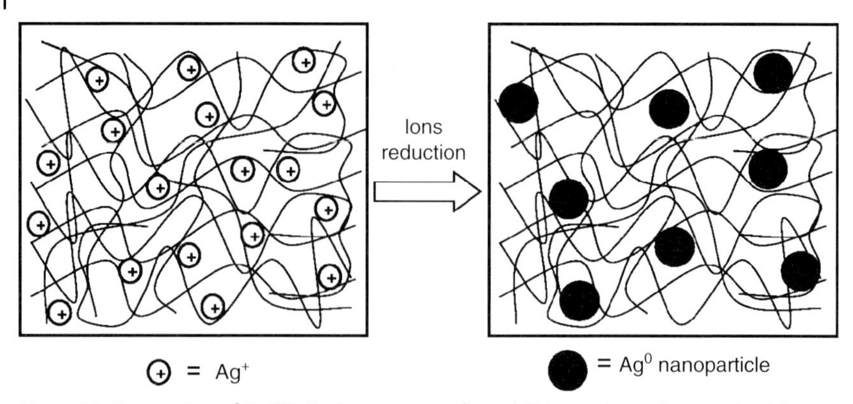

Figure 3.2 Preparation of AgNPs in the presence of a stabilizing polymer (wet synthesis).

widely used for the stabilization of silver nanoparticles. It should be noted that, in order to identify applications in the field of biomaterials, neither the stabilizing agent nor the reducing agent must represent a biological hazard [15].

With regards to the reduction step, a variety of exogenous agents may be used, including ascorbic acid [16, 17], sodium borohydride [18], sodium citrate [19, 20], alcohols [20], hydrogen [21], polyols, hydroxyalkyl radicals, and aldehyde groups of reducing sugars [22–24]. For a "green synthesis" of nanoparticles [15], reducing saccharides can be used as nontoxic and environment-friendly reducing agents. In this case, the synthesis is based on a Tollens reaction that involves the reduction of a silver ammoniacal solution with a reducing sugar (e.g., glucose, maltose, xylose) [25].

By selecting the stabilizer and reducing agent, it is possible to control the growth process and manipulate the shape and size of the metal nanoparticles in the nanocomposites. For instance, a lactose-substituted chitosan (1-deoxylactit-1-yl chitosan) was used successfully to prepare stable silver nanoparticles, using this particular polymer either as stabilizing agent alone or as both a reducing and stabilizing agent simultaneously [26, 27]. In the first case, the nanoparticles were obtained in a polysaccharide solution with the silver ions reduced by ascorbic acid ($C_6H_8O_6$) according to the following stoichiometry [16, 17]:

$$2Ag^+ + C_6H_8O_6 \rightleftharpoons 2Ag^0 + C_6H_6O_6 + 2H^+$$

The nanoparticles thus obtained were mostly round-shaped and well dispersed, with an average diameter of 30 nm as revealed by image analysis based on transmission electron microscopy (TEM) images.

In this case, the polysaccharide acts as an efficient stabilizing ligand for silver ions and AgNPs, based on the presence of amino groups. Esumi *et al.* [8] showed that metal nanoparticles can be protected by the exterior amino groups of dendrimers which act as stabilizers. In the case of the lactose-modified chitosan (chitlac), it was suggested that when the $AgNO_3$ is mixed with the polyelectrolyte solutions,

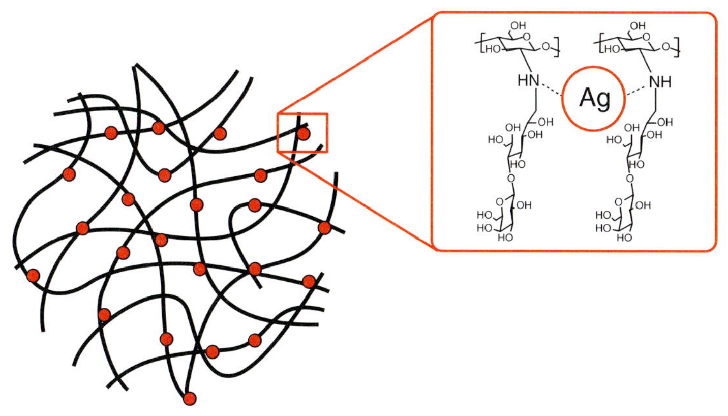

Figure 3.3 Schematic representation of the polymeric chains of lactose-modified chitosan providing the nitrogen atoms for the coordination and stabilization of silver nanoparticles. Reproduced with permission from Ref. [27].

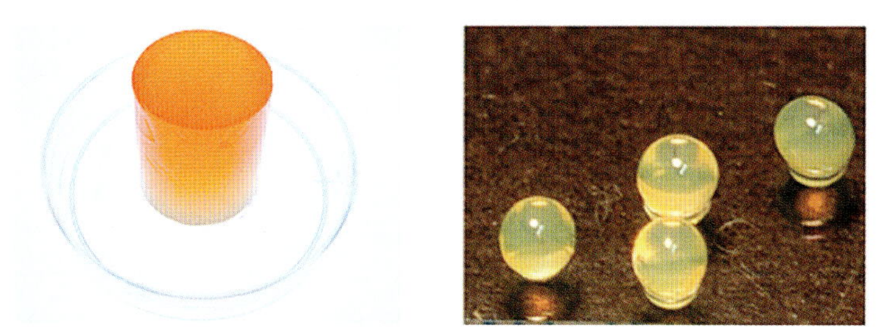

Figure 3.4 Nanocomposite hydrogels based on AgNPs and polysaccharides. Reproduced with permission from Ref. [27].

the Ag^+ ions give rise to a localized binding to the polymer macromolecules via an amino group chelation that persist also with the formed AgNPs. The hydrophilic lactitol side-chains also play a fundamental role in the stabilization by embedding the AgNPs bound in the proximity of the polymer backbone and isolating them from the surrounding species (Figure 3.3). Furthermore, the complete miscibility of this colloidal solution with the anionic polysaccharide alginate can lead to the preparation of homogeneous nanocomposite hydrogels that entrap highly stabilized silver nanoparticles, with the yellow coloration providing evidence of the good dispersion of AgNPs within the gel matrix (Figure 3.4).

Huang *et al.* [12] discussed the preparation of silver–chitosan nanocomposites obtained by chemical synthesis from a chitosan solution in the presence of $AgNO_3$ reduced by $NaBH_4$; the evaporation of the solvent in this colloidal solution leads to the formation of a nanocomposite film. Morphologically, the film shows a rod-like structure, the formation of which is attributed to the presence of silver clusters that act as an accelerant to the precipitation of polymer crystallites.

Figure 3.5 Transmission electron microscopy image of Ag/PVP nanocomposite electrospun fiber. Reproduced with permission from Ref. [5].

Silver nanocomposites can be prepared in the form of nanofibers by means of an *electrospinning* technique, a process by which a suspended droplet of polymer solution is charged to high voltage to produce fibers with diameters typically less 500 nm. If the solution contains AgNPs stabilized by the polymer, it is possible to obtain nanocomposite fibers. Wang *et al.* [5] successfully prepared AgNPs through the reduction of $AgNO_3$ by ethanol in PVP/ethanol solution, so as to obtain nanocomposite fibers with AgNPs less than 10 nm in size (Figure 3.5).

The need to obtain antimicrobial coatings and films can be fulfilled by incorporating AgNPs into polymeric matrixes by means of the so-called "layer-by-layer" (LBL) technique. This is a simple and straightforward approach based on the alternate deposition on a substrate of layers of polyelectrolytes with opposite charges through a dipping process (Figure 3.6).

The dipping process is repeated until a desired number of layers is obtained, each layer having a thickness in the range of a few nanometers. With this method it is possible to incorporate one or more layers bioactive agents such as silver. By using this approach, Grunlan *et al.* [10] prepared polyelectrolyte multilayers with silver nitrate and/or cetrimide (an antiseptic agent) as the antimicrobial agents. The films were prepared by alternately dipping a poly(ethylene terephthalate) (PET) substrate into solutions of poly(acrylic acid) (PAA) and PEI in a mixture with the bioactive agents. Dai *et al.* [2] also followed the LBL approach to obtain multilayered polyelectrolyte films that incorporated homogeneously dispersed AgNPs; in this case, a PEI solution was used to prepare stabilized silver ions from $AgNO_3$. The subsequent alternate deposition of the PEI–silver layer and of a PAA solution onto a substrate, followed by reduction with $NaBH_4$, led to the formation of a multilayered film based on the two oppositely charged polysaccharides containing finely dispersed AgNPs (Figure 3.7).

Multilayers can be fabricated by means of the LBL technique not only on planar substrates, but also on three-dimensional (3-D) templates. In the case of a spheri-

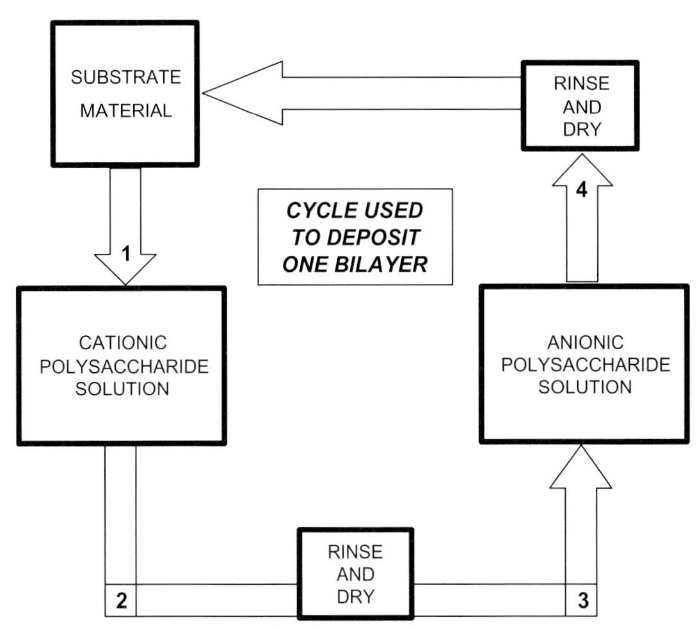

Figure 3.6 Schematics of the layer-by-layer self-assembly procedure for creating antimicrobial thin films. Adapted from Ref. [10].

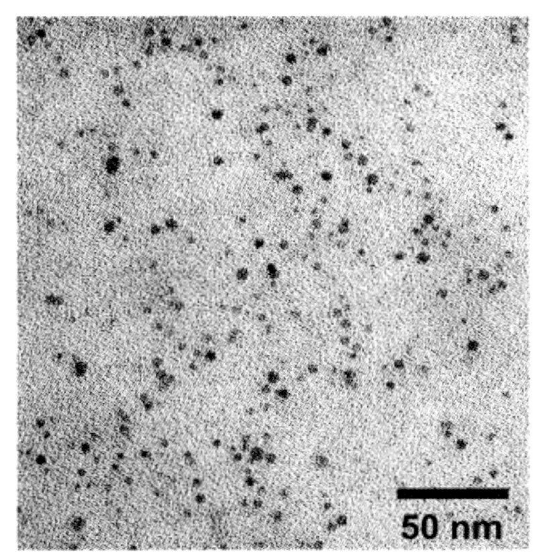

Figure 3.7 TEM image of a silver–PEI–PAA film, showing the dispersion of spherical AgNPs with a mean diameter of 4 nm. Reproduced with permission from Ref. [2].

cal template, its eventual dissolution after the film formation allows multilayered capsules to be obtained that can embed bioactive agents such as AgNPs. As an example, Choi *et al.* [28] synthesized polyelectrolyte capsules in which two types of nanoparticle were embedded: (i) AgNPs for the antimicrobial activity; and (ii) goethite nanoparticles which endowed the capsules with ability to be mobilized by the application of an external magnetic field. The capsules were prepared by means of copolymerized polyelectrolytes composed of poly(styrene sulfonate) (PSS) and PAA, in addition to poly(allylamine hydrochloride) (PAH). The metal ions were then loaded from aqueous solution into the capsules and finally allowed to react so as to create the two different types of nanoparticle.

Ho *et al.* [29] developed a nanocomposite film in which a polymer network based on PEI derivatized with double bonds was copolymerized with 2-hydroxyethyl acrylate; following an ultraviolet (UV)-initiated polymerization, the film was loaded with silver ions that were subsequently reduced to form AgNPs. After immersing the film in $AgNO_3$ solution, a chemical reduction of the Ag^+ complexed within the polymer network was carried out by soaking the film in an ascorbic acid solution until it turned yellowish-brown, which was suggestive of the formation of well-dispersed nanoparticles.

Polyols may be used as reducing agents for silver ions, in which case the silver-reduction mechanism typically involves a heat treatment. Wiley *et al.* [23, 30] prepared AgNPs of different shapes (quasi-spherical particles, single-crystal cubes, tetrahedrons, rods, triangular nanoplates, nanowires with pentagonal cross-sections, twinned structures) by reducing $AgNO_3$ with ethylene glycol at high temperature in the presence of PVP.

The results of these studies showed how the crystallinity of seeds could be determined by the molar ratio between the capping agent (PVP) and $AgNO_3$, as well as by the strength of the chemical interaction between the polymer and various crystallographic planes of silver (Figure 3.8).

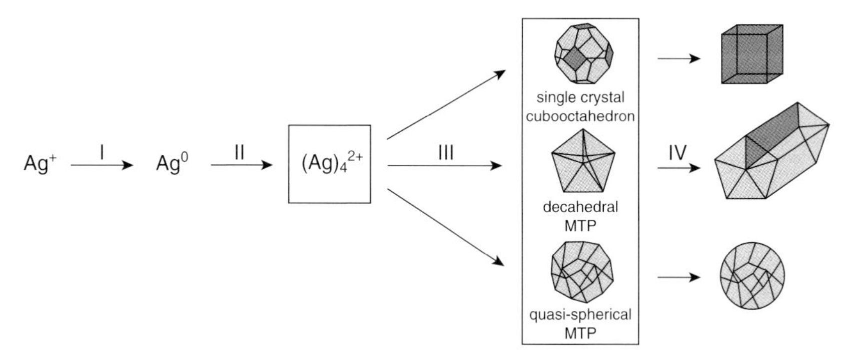

Figure 3.8 Schematic showing the reduction of silver ions by ethylene glycol (I); the formation of silver clusters (II); the nucleation of seeds (III); and the growth of seeds into nanocubes, nanorods or nanowires, and nanospheres (IV). Reproduced with permission from Ref. [30].

Silver nanocomposites may be prepared using polymeric dendrimers as the stabilizing agent. *Dendrimers* are branched macromolecules that possess architectural and ligand sites which allow the preorganization of metal ions and an effective stabilization of silver, based on the formation of stable complexes at atomic/molecular level dispersion. Balogh *et al.* [31] discussed the preparation of silver complexes within poly(amidoamine) dendrimers for antimicrobial applications, while Kuo *et al.* [9] reported the formation of AgNPs stabilized in pseudo-dendritic poly(allylamine) derivates.

Following the same strategy, other organic macromolecules can be used to stabilize AgNPs: for example, Lok *et al.* [4] described the preparation of antimicrobial AgNPs stabilized by bovine serum albumin (BSA).

Inorganic stabilizers have also been investigated for the preparation of silver nanocomposites. For example, Su *et al.* [6] prepared AgNPs immobilized within inorganic phyllosilicate clays in order to obtain nanohybrids that could exploit an ion exchange between Ag^+ (provided by $AgNO_3$) and Na^+ from the clay, followed by the *in situ* reduction of silver ions by methanol at 80 °C. In this case, the nanoparticles formed were free from polymeric surfactants and had a narrow size distribution (average diameter ca. 30 nm) (Figure 3.9).

Silver nanocomposites can be obtained in the form of bioceramics in order to endow osteoconductive materials with antimicrobial properties. Rameshbabu *et al.* [32] prepared silver-substituted hydroxyapatite nanocrystals by microwave processing.

Although, the conventional chemical synthesis represents to date the most popular approach for the preparation of silver nanocomposites, other techniques are currently under investigation which are based on physical processes. One such promising technique is a radiolytic method, used to create AgNPs in solution, whereby radiolytically generated species, solvated electrons and secondary radicals each exhibit strong reducing potentials towards metal ions. Krkljes *et al.* [33] reported the preparation of Ag–PVA nanocomposites via a radiolytic procedure, using steady-state gamma irradiation.

Alternatively, the *ablation* of a metal surface immersed in a liquid can lead to the production of nanoparticles of that metal in the liquid. In laser ablation, the

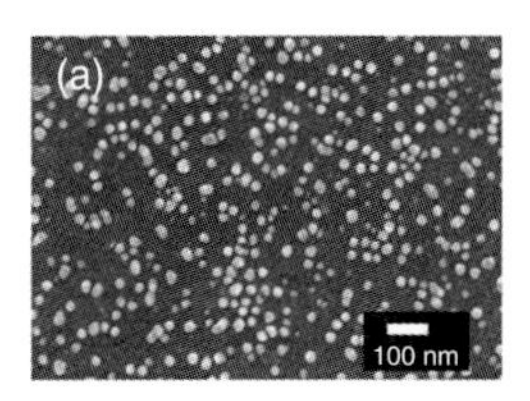

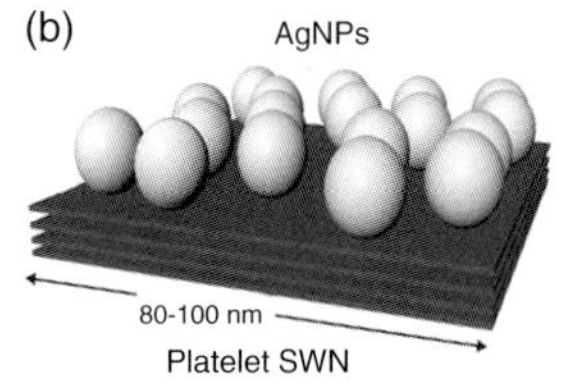

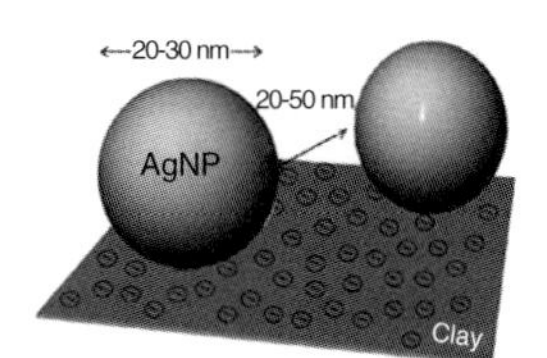

Figure 3.9 Silver nanoparticles dispersed on silicate clay. (a) The surface structure of the AgNP/clay is characterized using field-emission SEM; (b) Representations of the nanocomposite structure. Reproduced with permission from Ref. [6].

metal atoms and small metal clusters are ablated from a metal rod by laser irradiation; any subsequent self-aggregation of the nanoparticles within the liquid can be prevented using a surfactant to hinder any direct contact between the nanoparticles. An example of this procedure was reported by Mafunè *et al.* [34], who prepared AgNPs by laser ablation against a silver plate in an aqueous solution of sodium dodecyl sulfate.

Photochemical methods can be used to prepare polymer-stabilized AgNPs [35]. Mallick *et al.* [7] prepared colloidal silver via the UV photoirradiation of silver nitrate solution in the presence of methoxy poly(ethylene glycol) (PEG), which served as both a reducing agent and stabilizing agent.

Further interesting routes for the preparation of silver nanocomposites include sonochemical treatments [36], potentiostatic and galvanostatic methods [37], vapor deposition [38], microwave irradiation [39], electron-beam irradiation [40], synthesis templated by DNA [41], Langmuir–Blodgett-based techniques [42], ion implantation, sputtering, vapor-phase co-deposition, and vacuum co-condensation [43].

3.2.2
Characterization Techniques

The characterization of silver nanocomposites can be achieved through a variety of techniques, the most important of which are spectroscopic in nature (UV-Visible, infrared and X-ray photoelectron spectroscopy) and X-ray diffraction (XRD). In addition, various forms of microscopy may be applied, including TEM, scanning electron microscopy (SEM), dark-field illumination, atomic force microscopy (AFM), and near-field scanning optical microscopy (NSOM).

In the chemical synthesis, colloidal solutions of AgNPs are obtained when Ag^+ ions are reduced to metal atoms, the consecutive coalescence of which yields larger particles. The transition from atom to metal nanoparticle can be studied using pulse radiolysis techniques [3, 20]. In this case, when a sufficient number of atoms have coalesced, the particles begin to display the property of surface plasmon resonance (SPR), which results from the collective excitation of all free electrons present in the particles [3]. Under the influence of an electric field due to the incoming light, such movement of the electrons leads to a dipole excitation across the particle sphere that causes the electrons to oscillate (Figure 3.10). The result is that the electron density within a surface layer (which is a few angstroms thick) will oscillate, while the density in the interior of the particle will remain constant. When the condition of resonance is reached, the UV-Visible spectrum displays an intense absorption band termed the "plasmon resonance band" (Figure 3.11).

A typical example is provided by the formation of a plasmon resonance peak at about 400 nm during the chemical synthesis of AgNPs in a chitosan-derivative (chitlac) solution [27]. In this case, the absorption peak increases with time following addition of the reducing agent, which indicates a coalescence of the reduced ions to form metallic nanoparticles (Figure 3.12).

A symmetrical shape of the plasmon band suggests that the nanoparticles are well dispersed and spherical [9]. At variance, the aggregation of nanoparticles leads

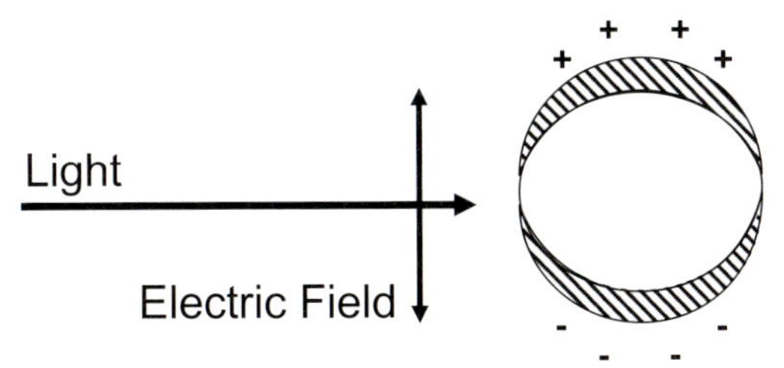

Figure 3.10 Polarization of a spherical metal particle by the electrical field vector of the incoming light. Adapted from Ref. [3].

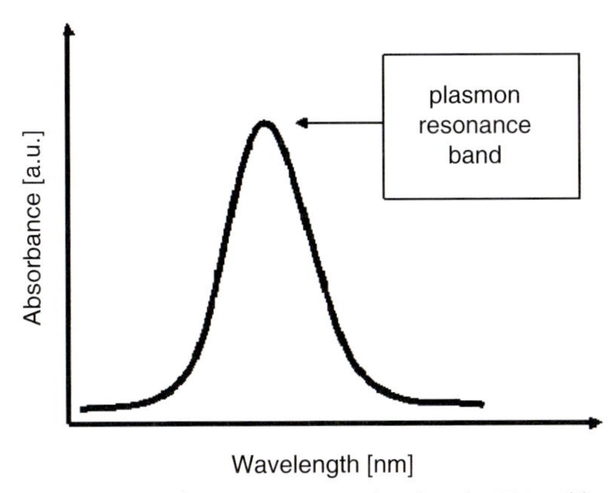

Figure 3.11 A plasmon resonance band in the UV-visible spectrum accounts for the presence of metallic nanoparticles.

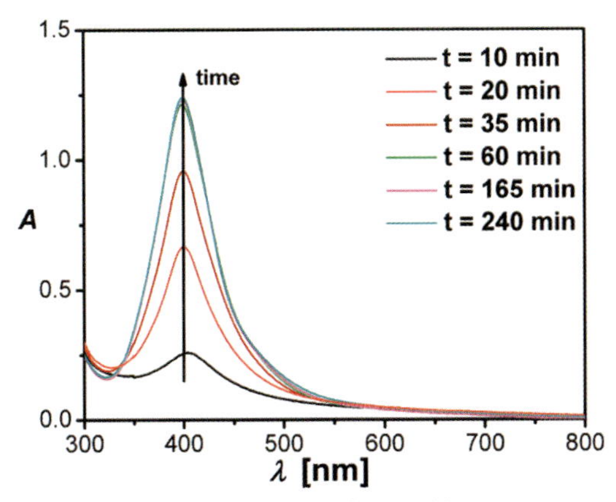

Figure 3.12 Time-dependence of UV-visible spectra variations of polymer-stabilized silver nanoparticles after the addition of the reducing agent. Reproduced with permission from Ref. [27].

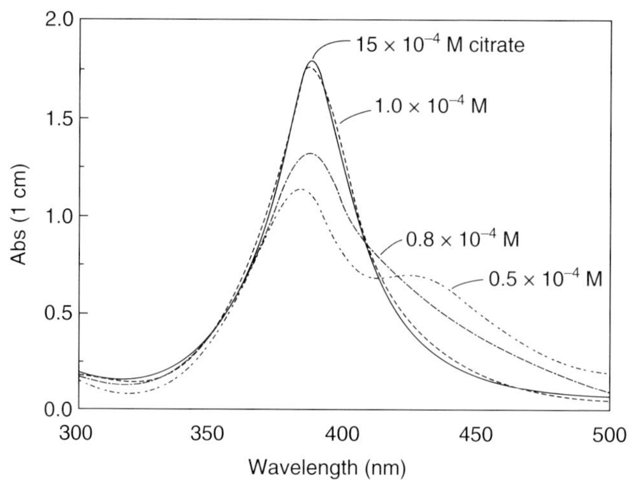

Figure 3.13 Absorption spectra of silver colloids obtained at various citrate concentrations. Reproduced with permission from Ref. [45].

to a broader plasmon band, with a red-shifted maximum [44]. A systematic study on the formation of silver nanoparticles by absorption spectroscopy was conducted by Henglein [45], who reported the preparation of AgNPs from $AgClO_4$ solutions at a fixed concentration and studied the variations of the UV-Visible spectra as a function of the concentration of sodium citrate used as both reducing and capping agent.

The UV-Visible spectra in Figure 3.13 suggest that the presence of sodium citrate has a drastic effect on the formation of AgNPs with different sizes; in fact, the citrate concentration can determine whether the formation of well-stabilized AgNPs will occur, or a coalescence of poorly stabilized polycrystallites.

Excess charge carriers may influence the wavelength of the plasmon resonance band; for example, a blue shift of the plasmon peak will occur upon electron donation to the particles, whereas a red shift will occur upon the injection of positive holes into the particles. Furthermore, any chemical modification of the surfaces of the particles, such as interactions with organic molecules, will have a strong effect on the plasmon absorption band.

Those anions capable of forming complexes or insoluble salts with silver ions will be strongly adsorbed onto silver particles. The Fermi level floats upon chemisorption, depending on whether the adsorbed molecule is nucleophilic or electrophilic. A surface atom carrying an adsorbed nucleophile molecule will acquire a slightly positive charge ("preoxidation state"), and the excess electron density will be transferred simultaneously into the metal particle. Thus, the chemisorption of a nucleophile will be accompanied by a shift of the Fermi potential to a more negative value. Since AgNPs are highly sensitive to oxygen, this will result in the formation of partially oxidized AgNPs with chemisorbed Ag^+ onto the surface. In the absorption spectrum, this partial oxidation will lead to a red-shift of the SPR band, according to the following equation [44]:

$$\lambda = \lambda_0\left(1+[Ag^+]/[Ag]^{1/2}\right)$$

where λ_0 is the wavelength of the plasmon peak before oxidation, and λ after oxidation.

Lok *et al.* [4] reported that borohydride-reduced (zeroth-valent) AgNPs in the presence of citrate exhibited a SPR peak at 375 nm, while subsequent exposure to oxygen led to a rapid shift of the absorption peak to 398 nm, with a broadening of the band width and lowering of the maximum absorption. These changes in the absorption spectrum can be attributed to the partial oxidation of the nanoparticles with the formation of chemisorbed Ag^+ on their surface, indicating the sensitivity of AgNPs to oxygen. Any further addition of borohydride induced a shift in the plasmon resonance band to the intensity and shape of nonoxidized, zeroth-valent nanoparticles (Figure 3.14).

It is important to note that the presence of oxidized (or partially oxidized) atoms on the surface of the nanoparticle affects its biological properties (as discussed in Section 3.3.1.4).

Infrared spectroscopy (IR) can provide information regarding the interaction (chemical bonding) between the nanoparticles and the matrix in a silver nanocomposite structure. Krkljes *et al.* [33] performed a series of IR analyses to identify the interaction between AgNPs and PVA chains via the OH groups. By means of Fourier transform infrared (FT-IR) analysis, Dai *et al.* [2] characterized PAA/PEI–silver nanocomposites in order to evaluate the influence of silver reduction on the

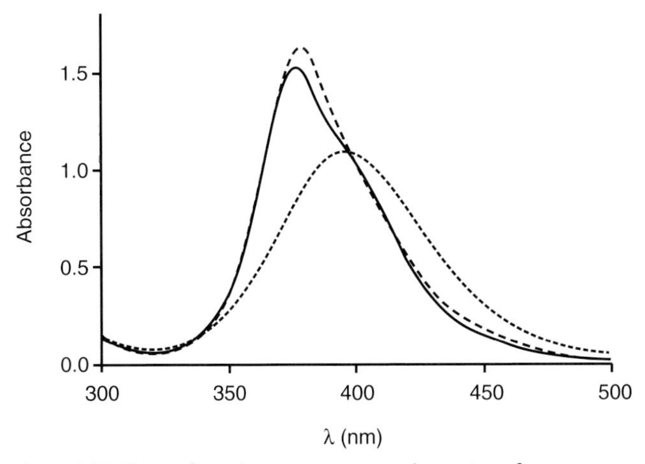

Figure 3.14 The surface plasmon resonance absorption of reduced AgNPs (solid line), oxidized AgNPs (dotted line), and oxidized AgNPs re-reduced by $NaBH_4$ (dashed line). Reproduced with permission from Ref. [4].

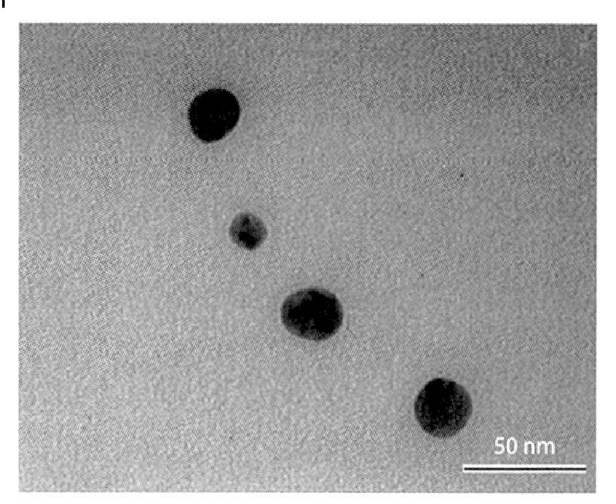

Figure 3.15 TEM image of silver nanoparticles stabilized by a lactose-modified chitosan. Reproduced with permission from Ref. [27].

polymer matrix, while Zhang *et al.* [46] verified the formation of coordination bonds between the amino/amide groups of a poly(amidoamine) derivative and AgNPs.

Currently, TEM is widely used to investigate the size, structure (crystallography, existence of twin planes, stacking faults) and dimensional distribution of AgNPs. Typically, AgNPs prepared via chemical reduction in the presence of chitlac [27] will appear as spheroidal particles with average dimensions between 20 and 30 nm (Figure 3.15).

Wiley *et al.* [30, 47] carried out a series of TEM investigations to study a polymer-mediated polyol process that allows for the preparation of silver nanostructures with a number of different morphologies (e.g., cubes, rods, wires, spheres), as shown in Figure 3.16.

Wang *et al.* [48] prepared polyelectrolyte multilayer films containing AgNPs, and evaluated the distribution of particles in the polymer matrix by means of cross-sectional TEM imaging. The subsequent images showed the presence of spherical particles, distributed randomly throughout the film (Figure 3.17).

Metal nanocomposites have been investigated using a specially adapted SEM system, termed field-emission SEM (FE-SEM), that allows the sputtering step of conventional SEM (which might cover or affect the nanoparticles on the surface of the nanocomposite material) to be avoided [28, 49]. When Wiley *et al.* [23] carried out a polyol synthesis of AgNPs, the structure of which was analyzed using FE-SEM, the images confirmed the formation of single-crystal, truncated cubes and tetrahedrons (Figure 3.18).

Among other techniques, AFM represents a useful means of characterizing the morphology of nanocomposite materials that contain a dispersion of nanoparticles [50, 51]. By using AFM, Deshmukh *et al.* [43] explored the surface and bulk morphology of poly(methyl methacrylate) (PMMA) nanocomposite films containing

Figure 3.16 TEM characterization of AgNPs with different morphologies obtained by polyol synthesis. (a) Silver nanocubes. The inset shows a typical electron diffraction pattern taken from any individual nanocube by directing the electron beam perpendicular to one of its square faces; (b) Silver nanobelts. The inset at lower left shows that the nanobelts were single crystals, and grew along the [101] direction. The inset at upper right shows a typical diffraction pattern taken from an individual nanobelt; (c) Triangular nanoplates. The inset shows a typical electron microdiffraction pattern taken from an individual nanoplate supported on the TEM grid against one of its triangular faces; (d) Pentagonal cross-sections of silver nanowires. The inset shows that each wire has a fivefold twinned structure, characterized by five single-crystal subunits. Reproduced with permission from Ref. [30].

AgNPs. The presence of the nanoparticles on the surface of the polymer film was confirmed using AFM (Figure 3.19).

Ho *et al.* [29] compared AFM and TEM techniques in the characterization of nanocomposite films based on PEI and AgNPs. The AFM images in phase-contrast mode showed the dispersion of nanoparticles that appeared as bright spots on the surface of the film, whereas the corresponding TEM image of a cross-section of the film indicated the presence of non-aggregated nanoparticles with sizes ranging from 4 to 50 nm (Figure 3.20).

The crystallographic structure of AgNPs formed within a matrix may also be studied using XRD. Indeed, when Wang *et al.* [5] used XRD to characterize Ag/PVP nanocomposite films, the pattern revealed the presence of face-centered cubic (fcc) nanocrystals formed within the polymer matrix (Figure 3.21).

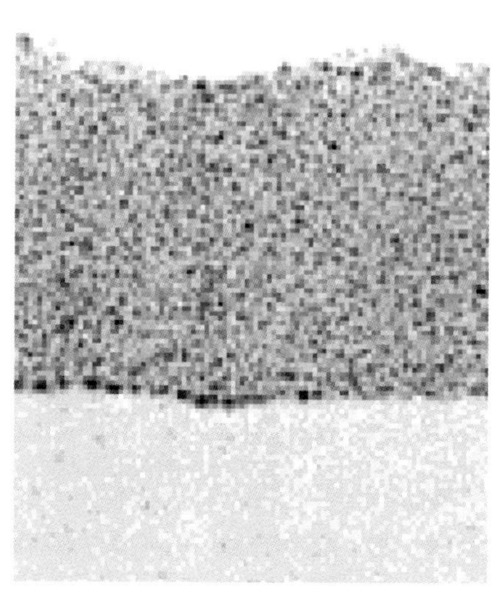

Figure 3.17 Cross-sectional TEM image of a polyelectrolyte multilayer containing AgNPs. Reproduced with permission from Ref. [48].

As a surface chemical analytical technique, X-ray photoelectron spectroscopy (XPS) can be used to analyze silver–nanocomposite materials. When XPS was used by Zeng *et al.* [52] to characterize polymer films (poly(styrene) and acrylonitrile–styrene copolymer) filled with silver nanoparticles, the results pointed to the existence of a charge-transfer interaction between the AgNPs and the acrylonitrile segments, although no obvious interaction was apparent between the silver and styrene segments. Stofik *et al.* [53] prepared silver–dendrimer nanocomposites for use as immunosensors, and confirmed with XPS the synthesis of AgNP. Subsequent XPS studies on PVP–silver nanocomposite fibers demonstrated an interaction between silver and the carbonyl oxygen; the strong Ag:O coordination proved to be capable of preventing the AgNPs from aggregating within the polymer matrix [5] (Figure 3.22).

The extremely bright nature of silver nanoparticles means that they can be observed directly by using dark-field single-nanoparticle optical microscopy and spectroscopy. By using this technique, Lee *et al.* [54] were able to characterize AgNPs embedded in zebrafish embryos to study their transport, biocompatibility, and toxicity in real time. In another study, Lu *et al.* [42] produced dark-field images of a high-density nanocomposite film obtained from poly (*N*-isopropylacrylamide) and AgNP (Figure 3.23).

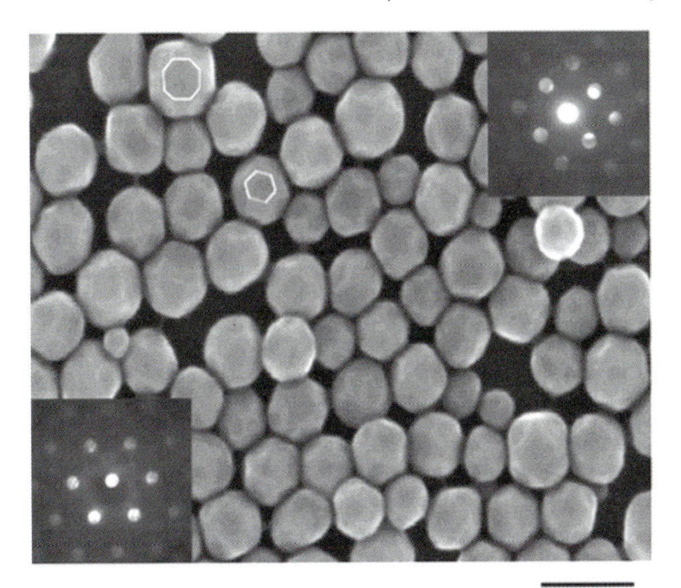

Figure 3.18 FE-SEM image of AgNPs obtained by polyol synthesis. The main image shows truncated cubes (indicated by an octagon) and truncated tetrahedrons (indicated by a hexagon). The insets show the convergent beam electron diffraction patterns showing that these particles are single crystals. Scale bar = 100 nm. Reproduced with permission from Ref. [23].

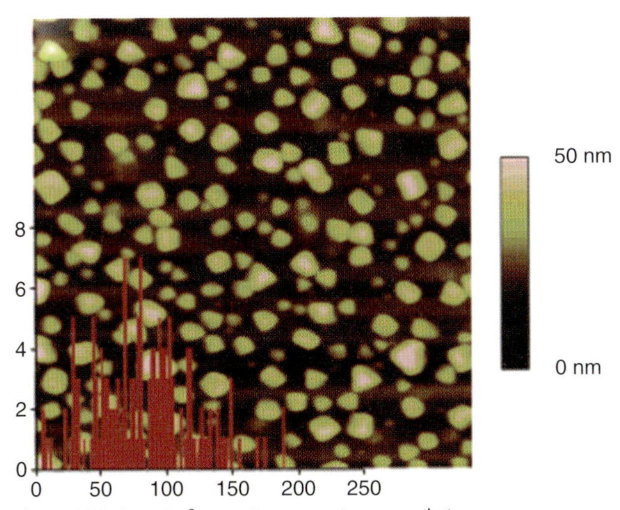

Figure 3.19 Atomic force microscopy image and size distribution of silver nanoparticles dispersed in a PMMA film. The x-axis corresponds to the size of the protruding nanoparticles covered by the PMMA layer. Reproduced with permission from Ref. [43].

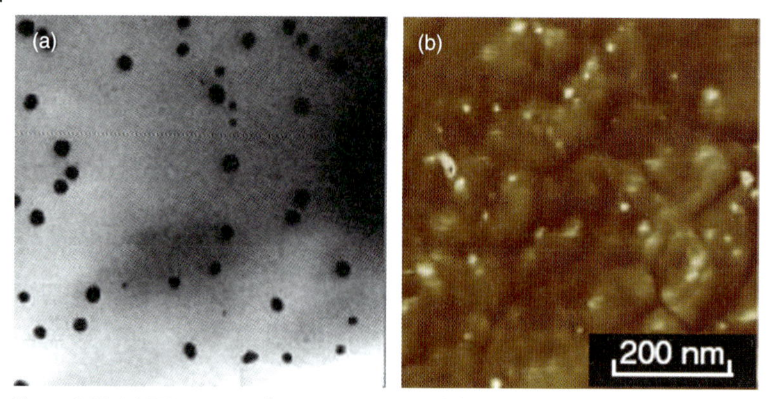

Figure 3.20 (a) TEM image of a cross-section and (b) AFM (phase mode) image of the surface of poly(ethylene imine)-based films containing silver nanoparticles. Reproduced with permission from Ref. [29].

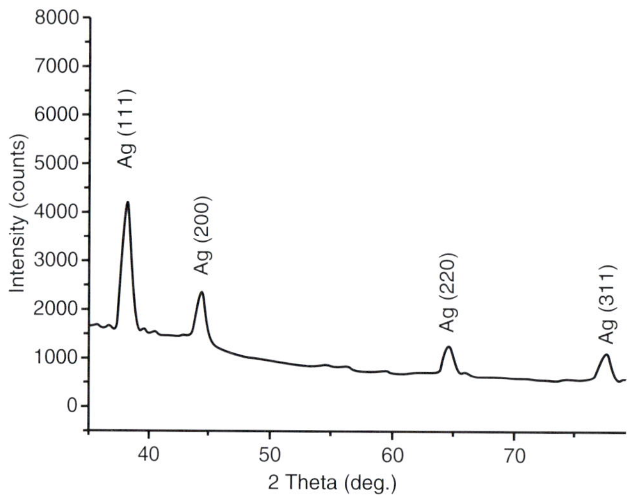

Figure 3.21 X-ray diffraction pattern of Ag/PVP nanocomposite film. Reproduced with permission from Ref. [5].

Silver nanocomposites have also been characterized using NSOM, a microscopic technique that permits surface inspections to be made with high spatial, spectral and temporal resolution, thus overcoming the far-field resolution limit by exploiting the properties of evanescent waves. The NSOM technique was subsequently used by Zhou *et al.* [55] to study the dispersion of AgNPs in polymer films containing azo groups (Figure 3.24).

Whilst all of the above-mentioned techniques are used widely to characterize biomedical silver nanocomposites, other techniques may also provide valuable

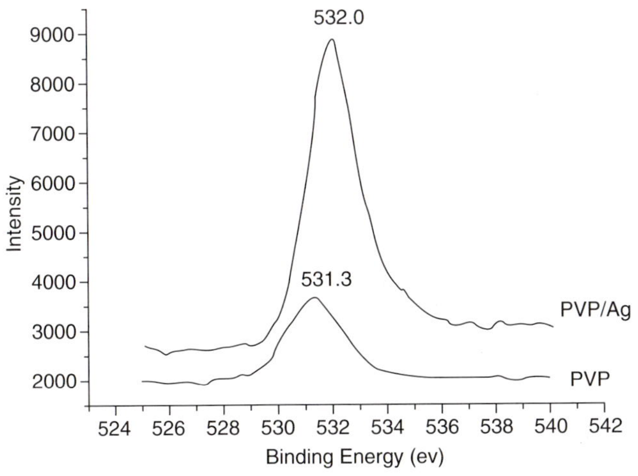

Figure 3.22 X-ray photoelectron spectra of O 1s of pure PVP and PVP/Ag nanocomposite. Reproduced with permission from Ref. [5].

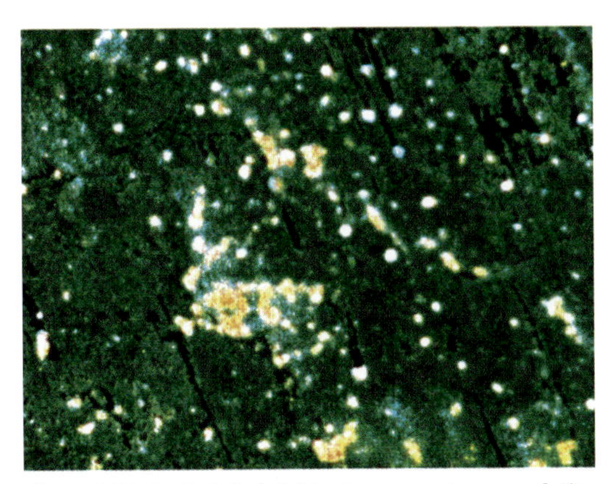

Figure 3.23 Optical dark-field microscope images of silver nanoparticles on a poly(*N*-isopropylacrylamide) film. Reproduced with permission from Ref. [42].

information relating to this new class of materials. Techniques such as differential scanning calorimetry (DSC) [33], thermogravimetric analysis (TGA) [56], dynamic light scattering (DLS) [57, 58] and zeta potential measurements [58] may each provide deeper insights into the specific properties of the constructs, notably with regards to thermodynamic properties and thermal stability, size, surface charge, and diffusion.

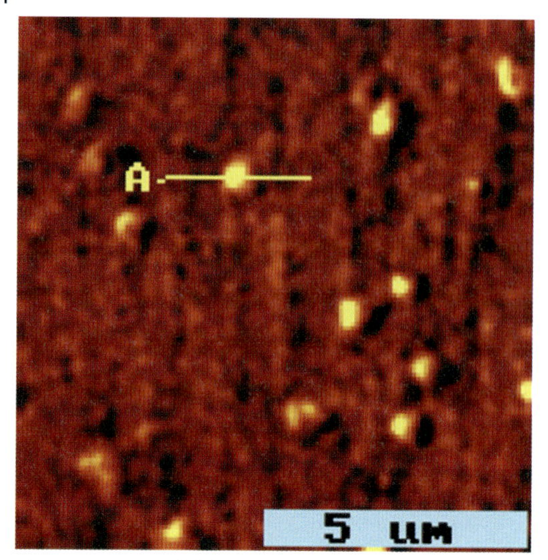

Figure 3.24 Near-field scanning optical image of azo film sample doped with silver nanoparticles. The line "A" represents a NSOM line scan which referred to optical profile studies (not reported in the figure).Reproduced with permission from Ref. [55].

3.3
Biomedical Applications

3.3.1
Silver Nanocomposites for Biocidal Applications (Antimicrobial, Antiviral, Antifungal)

3.3.1.1 General Considerations

In recent years, nanoscale materials have emerged as novel bioactive agents on the basis of their unique physical-chemical properties and their high surface area-to-volume ratios. Yet, the major interest in silver-based nanocomposites has been due to silver's biocidal properties [59, 60]. Historically, silver has been used extensively to control infections, especially for the treatment of burns and chronic wounds. Although, during the 1940s penicillin was introduced and the use of silver to treat bacterial infections diminished, its use was revitalized in 1968 when silver nitrate was combined with sulfonamide to produce a silver sulfadiazine cream that could be used to treat burns [59]. Moreover, because antibiotic-resistant bacterial strains have today become a major issue in public healthcare, silver-based nanocomposites in a variety of forms (e.g., wound dressings, coated medical devices, hydrogels) have made a major comeback in terms of anti-infective applications [27]. Currently available silver-based medical products, ranging from topical ointments and bandages for wound healing to coated stents, have been proven to be effective in retarding and preventing bacterial infections [61].

Improvements in the development of novel silver nanoparticle-containing products are continuously sought. In particular, there is an increasing interest towards the exploitation of silver nanoparticle technology in the development of new bioactive biomaterials, the aim being to combine the unique antibacterial properties of the metal at the nanoscale with the performance of the biomaterial [9, 12, 16, 31, 62]. Silver-containing nanomaterials represent a promising strategy to combat those infections related to indwelling medical devices such as catheters, stents and bone prostheses. Somewhat surprisingly, these types of infection (as a group) represent the fifth-highest cause of in-hospital death in the United States [63].

Although silver is known and used primarily for its antibacterial properties, silver nanoparticles have been shown to exhibit also promising antiviral and antifungal properties. AgNPs exert cytoprotective activities towards HIV-infected T cells by inhibiting the *in vitro* production of extracellular virions. It is hypothesized that the direct interaction between the nanoparticles and the double-stranded DNA of hepatitis B virus (HBV) particles is responsible for their antiviral mechanism [64]; however, the effects of silver nanoparticles towards other types of virus remain largely unexplored.

Although resistance to antifungal drugs seems to be less of a problem than resistance to antibacterial drugs, the long-term concern is that the number of fundamentally different types of antifungal agent available for treatment is extremely limited. Yet, there is an inevitable and urgent medical need for drugs with novel antifungal mechanisms. During recent years, attention has focused on the potential use of silver as an antifungal agent, with experimental evidence that AgNPs are capable of exhibiting potent antifungal effects, most likely by destroying the membrane integrity of fungal cells [65–67].

3.3.1.2 **Overview of in vitro Results**

During the past few years, a number of promising applications for silver nanocomposites as biocidal systems have been developed. Currently, silver-based materials are considered to be good candidates for coating medical devices and, indeed, many reports have recently been made concerning the preparation of nanocomposite-coatings based on polymers and silver nanoparticles. As discussed in Section 3.3.1.4, the mechanism by which silver-based materials exert their biocidal activity is only partially understood, and this often leads to different interpretations of experimental results. For this reason, the results and discussions summarized in the following paragraphs may, in some cases, appear conflicting.

For central venous catheter (CVC) applications, Stevens *et al.* [63] reported the use of various hydrophilic polymer coatings loaded with silver nanoparticles in order to assess both the antimicrobial efficacy and the impact of silver on the coagulation of contacting blood. The roll-plate assay [68] showed that bacterial inhibition begins when the concentration of silver ions released from the nanoparticles into the suspension exceeded 100 nM, with no bacteria surviving Ag^+ concentrations >10 μM. On the other hand, thrombin generation and platelet activation was shown to start at Ag^+ concentrations higher than 100 μM. Interestingly, thrombin generation can occur also upon the activation of blood platelets through collision (direct contact) with silver particles exposed on the surface; hence, it was suggested that the use of silver nanocomposite coatings for CVCs might enhance thrombus formation. Su *et al.* [6] successfully prepared AgNPs/clay nanocomposites and, by using inductively coupled plasma mass spectrometry (ICP-MS) analysis, showed that immersion of the nanocomposite in water at 0.1 wt% after centrifugation caused a release of silver ions of only 139 ppb. A nanocomposite concentration of 0.05 wt% was sufficient to completely inhibit the growth of *Staphylococcus aureus, Streptococcus pyogenes, Pseudomonas aeruginosa,* and *Escherichia coli*. It must be noted that the solution obtained from a 0.5 wt% dispersion after centrifugation of the slurry was not effective for the inhibition of *S. aureus*. Hence, in this case it was the AgNPs, and not the released Ag^+, that appeared to be involved in the antibacterial mechanism. In the case of multilayered nanocomposites, Dai *et al.* [2] showed that a film based on the alternated deposition of PEI and PAA, including AgNPs, was effective in inhibiting *E. coli* growth. Remarkably, the effect was the same when the film contained silver ions or when it contained AgNPs; however, the latter case may be more desirable because the amount of Ag^+ absorbed in the body should be minimized. Grunlan *et al.* [10] prepared a multilayered film based on PAA and PEI containing silver ions and cetrimide (an organic quaternary ammonium molecule) as antimicrobial agents. The biocidal activity was monitored using a zone of inhibition (ZOI) test with *E. coli* and *S. aureus* (Figure 3.25).

These materials were also shown to be effective in preventing bacterial growth, and the antimicrobial efficacy of the silver-containing films was further enhanced by the use of cetrimide. Choi *et al.* [28] tested nanocomposite capsules based on polyelectrolytes with AgNPs with a suspension of *E. coli*; the optical density meas-

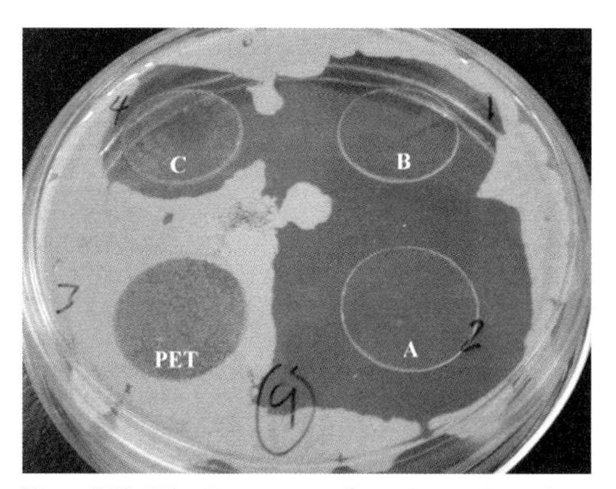

Figure 3.25 Kirby–Bauer test performed to evaluate the zone of inhibition (ZOI) after 24 h of *S. aureus* incubation. The letters (A-B-C) associated with the test films refer to different amounts of silver and cetrimide. The film in the lower left corner of the plate is a PET film with no antimicrobial coating. Reproduced with permission from Ref. [10].

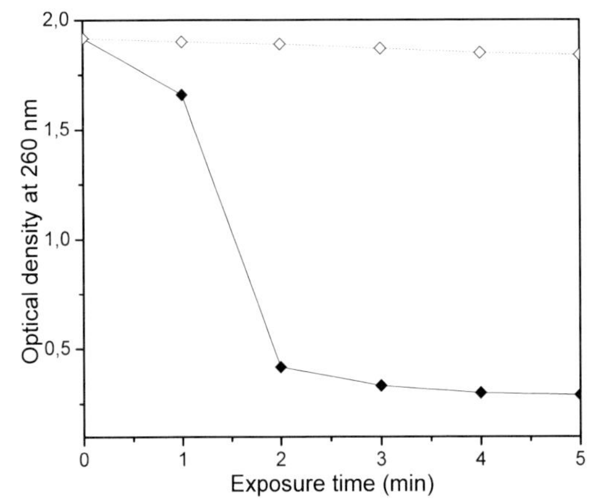

Figure 3.26 Optical density of a suspension of *E. coli* with silver-embedded (♦) or without silver (◊) capsules as a function of exposure time. Redrawn from Ref. [28].

urements at 260 nm (which are proportional to the number of bacteria) showed the number of bacterial cells to have decreased after only 1 min exposure to silver-embedded capsules. As a control, in the absence of silver the optical density of the *E. coli* suspension remained almost unchanged (Figure 3.26).

Antimicrobial studies were carried out on a nanocomposite systems based on lactose-modified chitosan (chitlac) and AgNPs, either as a colloidal solution or as hydrogels obtained in association with alginate [27]. In both cases, the materials displayed a remarkable bactericidal effect on four bacterial strains: *Staphylococcus epidermidis, E. coli, S. aureus,* and *P. aeruginosa*. In the case of *S. epidermidis* (see Figure 3.27a), the number of viable cells decreased drastically (a drop of 3 log units in colony-forming units (CFU) ml^{-1}) after only a 30 min incubation period with the colloidal system, and a complete inactivation of bacterial cells was found after 2 h of treatment. The activity was preserved also in the hydrogel state; bacteria smeared onto the surface of hydrogels without silver were able to grow (control gel), but were completely absent from AgNPs-containing gels (Figure 3.27b).

When the release of silver ions from hydrogels into a saline solution was monitored using ICP-MS, the concentration after four weeks was quite low (58 $\mu g\,l^{-1}$). Subsequent MTT [3-(4,5-dimethylthiazol-2-yl)-2,5-diphenyltetrazolium bromide] assays confirmed this concentration to be nontoxic to three eukaryotic cell lines (fibroblasts NIH-3T3, osteoblasts MG63, and hepatocytes HepG2). Interestingly, lactate dehydrogenase (LDH) cytotoxicity assays revealed that the polysaccharide–AgNPs colloidal system in solution was toxic to these eukaryotic cell lines, whereas the hydrogel system obtained from its combination with alginate was not. This suggests that the 3-D system might prevent nanoparticles from being available for eukaryotic cellular uptake whilst, at the same time, preserving its antimicrobial activity and allowing a direct interaction between the nanoparticles and the proteins localized on the bacterial surface. Ho *et al.* [29] described the preparation of

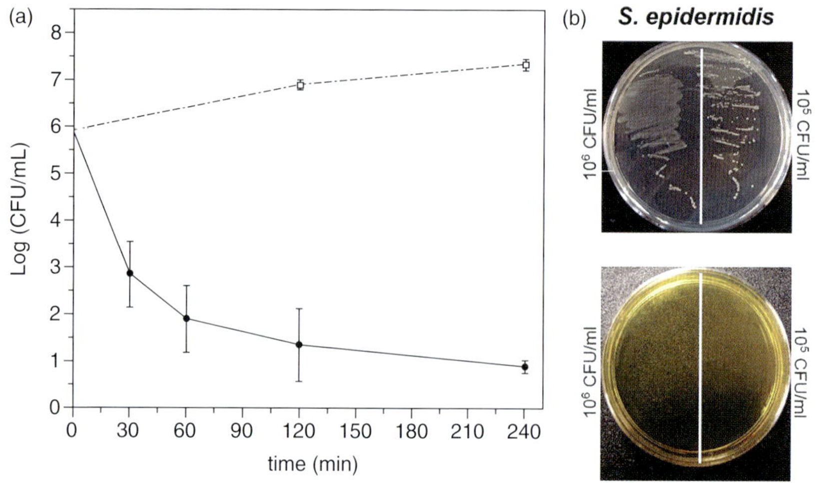

Figure 3.27 (a) Killing kinetics of polymer-stabilized silver colloidal solution (solid line) against *S. epidermidis*; the dashed line indicates control runs in the absence of silver; (b) Growth of *S. epidermidis* on polysaccharide-based hydrogels without (upper Petri dish) and with (lower Petri dish) silver nanoparticles. Reproduced with permission from Ref. [27].

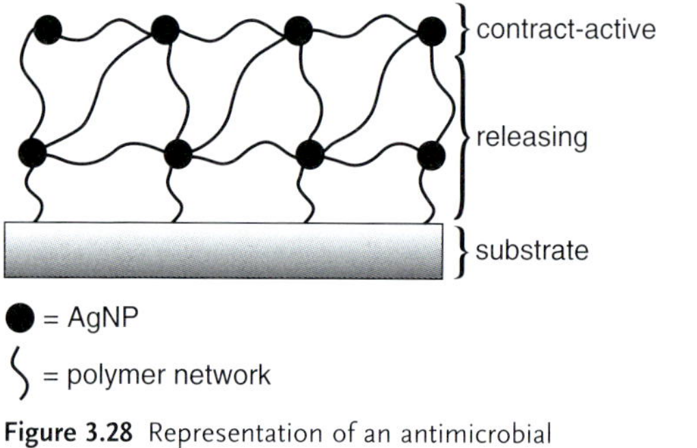

Figure 3.28 Representation of an antimicrobial nanocomposite coating that exploits both the release of silver ions and the direct contact of silver nanoparticles within a polymer matrix. Adapted from Ref. [29].

antimicrobial nanocomposite films based on PEI and AgNPs that exploited both the release of silver ions and the direct contact between bacteria and nanoparticles (Figure 3.28).

When *S. aureus* was allowed first to adhere to the film surface and was then cultivated in agar to evaluate the viable cells, the film was shown to inhibit bacterial growth for only 12 h. However, when the silver content of the film reached $10\,\mu g\,cm^{-2}$, no bacterial growth occurred even after two weeks, indicating a prolonged bactericidal effect of the nanocomposite material.

3.3.1.3 Effects of Nanoparticles Properties and Role of the Matrix

As many factors may influence the biocidal activity associated with AgNPs-based constructs, this can complicate the present understanding of antibacterial mechanisms (as discussed in Section 3.1.4). Consequently, it is difficult to evaluate one property at a time, as this may lead to conflicting results. Moreover, the biological effects of AgNPs may also be affected by the dispersing agent, or by the matrix of each particular nanocomposite system.

In general, it can be stated that nanoparticle size, shape, surface properties, dispersion and stability are important issues for tailoring biological performance. In particular, the effect of particle dimension has often been taken into account while characterizing antibacterial activity, with various reports having been made that the size of the AgNPs affects biocidal efficacy [46, 69]. The antibacterial activity of AgNPs can be related to their size, as the activity of smaller particles will be higher because of their much greater surface area, when compared on the basis of equivalent silver mass content. For such speculations, it may be useful to evaluate the number of silver atoms in a nanoparticle (n) from the following relationship [45]:

$$n = \frac{0.5 \cdot \pi \cdot N_A \cdot d^3}{3 \cdot V_m}$$

where d is the particle diameter (in nanometers), N_A is Avogadro's number, and V_m is the molar volume of silver ($ml\,mol^{-1}$). When investigating the effect of AgNPs dimensions on *E. coli* growth, Lok *et al.* [4] showed that AgNPs of average diameter 9.2 nm were ninefold more active than those of average diameter 62 nm. Morones *et al.* [70] reported that the bactericidal properties of the carbon-stabilized AgNPs were size-dependent in four types of Gram-negative bacteria, as the only nanoparticles that displayed any direct interaction with bacteria preferentially had a diameter of 1–10 nm. In contrast to this, when Su *et al.* [6] prepared silver nanocomposites of different sizes (45.7 nm and 25.9 nm) within a silicate clay, they showed that the degree of bacterial growth inhibition did not depend on particle size, but rather on the quantity of silver present (Figure 3.29).

The crystallographic structure and shape of the nanoparticles are considered to be important properties that affect antimicrobial behavior. Notably, the results of recent studies have shown reactivity to be favored by high atom density facets, such as {111} [70]. According to Pal *et al.* [71], truncated triangular nanoparticles display a higher antibacterial activity as compared to spherical nanoparticles and ionic silver (Figure 3.30).

When comparing the antibacterial activities of metallic and partially oxidized AgNPs, Lok *et al.* [4] suggested that only partially oxidized particles would exhibit antibacterial activities. Moreover, since smaller AgNPs have a higher surface area-to-mass ratio, they would provide a higher relative concentration of chemisorbed silver ions.

The matrix in which nanoparticles are dispersed is of primary importance to determine the performance of the material; in fact, in the final nanocomposite

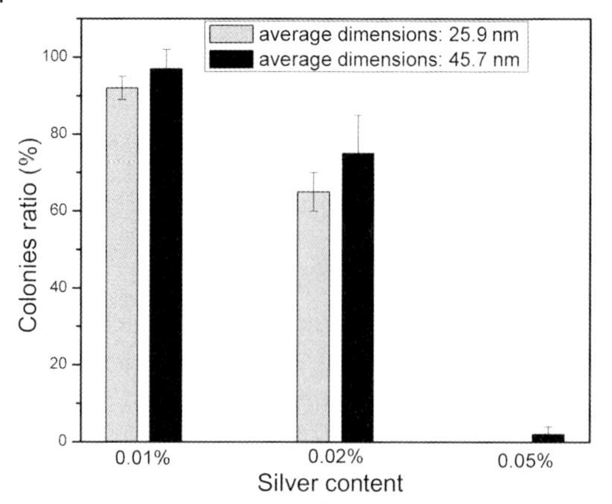

Figure 3.29 Antimicrobial study on AgNP/clay nanocomposites. In this system, the biocidal effect depends on the silver content and not on the particles size. Redrawn from Ref. [6].

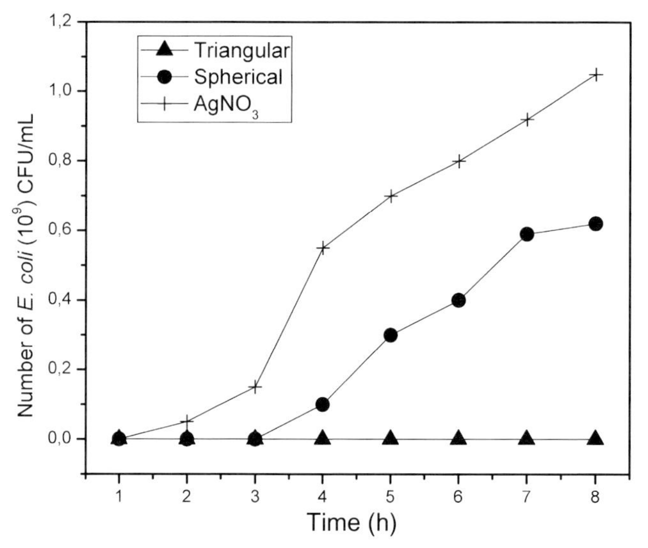

Figure 3.30 Comparative dynamics of *E. coli* growth in the presence of spherical silver nanoparticles, triangular nanoparticles, and ionic silver ($AgNO_3$). Redrawn from Ref. [71].

constructs, stabilizers play a fundamental role in controlling not only the formation of nanoparticles but also their dispersion stability. In the case of polymer solutions, the concentration of the stabilizer will function as a controller of nucleation, affecting the size of the final nanoparticles (as monitored by UV-Visible spectra and TEM observations). Although polymeric dispersants or capping agents are generally used to stabilize AgNPs, it is possible that they might also deactivate the nanoparticles' functions because the organic wrapping on a metal's surface may limit or prevent its surface reactivity [6].

3.3.1.4 Antimicrobial Mechanism

Three main strategies can be pursued to produce materials with antimicrobial properties [29], by selecting: (i) their anti-adhesiveness; (ii) their biocide-release activity; and (iii) their activity by contact.

Silver-based systems may be developed as both biocide-releasing systems (Ag^+) and contact-active materials. Although the toxic effects of silver on bacteria have been investigated for more than 60 years, the bactericidal mechanism of silver remains only partially understood [59, 70]. Nonetheless, several possible mechanisms have been suggested that involve the interaction of silver ions with biological macromolecules.

Within the microbial cell, most of the Ag^+-sensitive sites are likely to be proteins which, if altered, will lead to cell disruption as the result of structural and/or severe metabolic damage (Figure 3.31). Silver ions inhibit several enzymatic activities, by reacting with electron donor groups that contain sulfur, oxygen, or nitrogen, such as carboxylates, phosphates, hydroxyl, amines, imidazoles, indoles and, especially, sulfhydryl groups [72–75]. In the bacterial cell wall, the free sulfhydryl groups localized on the transmembrane and outer-membrane proteins (including proteins of the electron-transport chain) protrude into the extracellular portion of the membrane, where they represent a highly accessible interaction site for silver ions [70, 73, 76–78].

The Na^+-translocating NADH:ubiquinone oxidoreductase (NQR), which has been recognized as a primary target for Ag^+ ions, is a component of the respiratory chain of various bacteria, and generates a redox-driven transmembrane electrochemical Na^+ potential. In two independent studies, submicromolar concentrations of Ag^+ ions were shown to inhibit energy-dependent Na^+ transport in membrane vesicles of the NQR-possessing *Bacillus* sp. strain [79], and also to

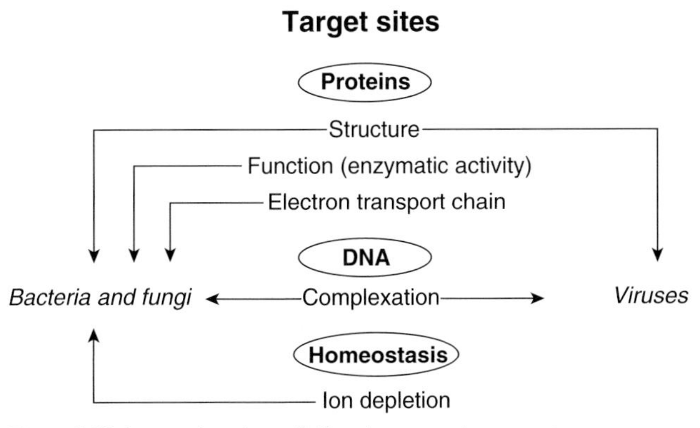

Figure 3.31 Interaction sites of silver ions on microorganisms.

inhibit the purified NQR of *V. alginolyticus* [80]. Likewise, Dibrov *et al.* [81] showed that low (submicromolar) concentrations of Ag^+ induced a massive proton leakage through the *Vibrio cholerae* membrane, which resulted in a complete alteration and elimination of the transmembrane proton gradient, followed by de-energization and cell death.

In a study conducted with *E. coli* as a bacterial model, it was suggested that the bactericidal action of silver ions was correlated with the interaction with ribosomal subunit proteins, and also with a suppression of the enzymes and proteins necessary for ATP production [82].

Ag^+ also forms complexes with the DNA bases, inducing DNA condensation. It is known that the replication of DNA molecules is effectively conducted only when DNA molecules are in a relaxed state since, when in a condensed form, the DNA molecules lose their replicating abilities [59, 70, 73].

Feng *et al.* [73] have conducted a morphological and structural study in which they monitored the changes that occurred on bacteria when treated with silver ions. Thus, a detachment of the cytoplasm membrane from the cell, and the presence of electron-dense granules around the cell wall and in the cytoplasm, were observed. The authors suggested that the electron-dense granules, which most likely were formed by a combination of silver and proteins, were prevented from permeating through the membrane, thus denying electron transport. It was proposed, therefore, that the silver ions might cause the formation of a low-molecular-weight region in the center of the bacterium, that would serve as a defense mechanism by which the bacterium would conglomerate its DNA so as to protect it against toxic compounds when it sensed a disturbance of the membrane (Figure 3.32).

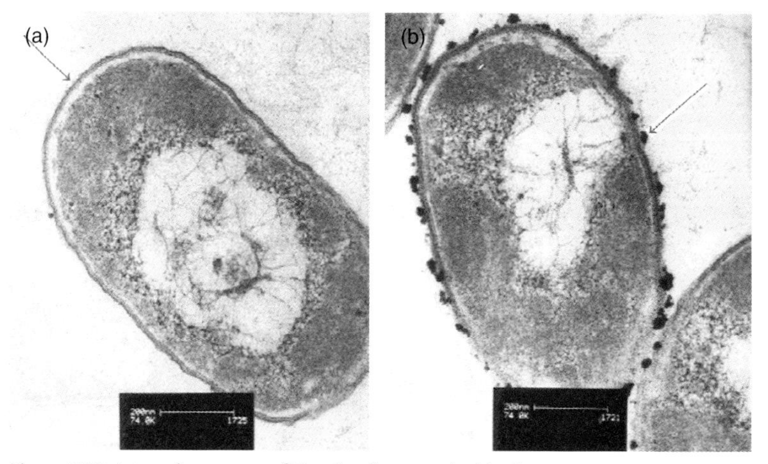

Figure 3.32 Internal structure of *E. coli* cells treated with silver ions. (a) Detachment of the cytoplasmic membrane from the cell wall (arrow); (b) Electron-dense granules around the cell wall (arrow). Reproduced with permission from Ref. [73].

In the case of silver nanocomposite materials, it remains unclear whether the biocidal mechanism of AgNPs involves only silver ions, or whether it also follows different routes [6]. A variety of mechanisms has been suggested according to the morphological and structural changes observed in bacteria. In general, silver is believed to interact with the bacterial membrane, either by direct contact between the nanoparticle and the membrane causing a direct transfer (solvent-free) of Ag^+ ions, or by the release of silver ions into the medium. A combination of these two mechanisms is also possible, however. Stevens *et al.* [63] suggested that, when bacteria are brought into direct contact with silver nanoparticles in the medium, they encounter (locally) high concentrations of silver ions, and that this results in cell death. In fact, the high surface-to-volume ratio of the nanoparticles accounts for a sustained local supply of silver ions at the material–bacterium interface, thus preventing bacterial adhesion and biofilm formation (Figure 3.33). It should also be noted that, if the plasma proteins are adsorbed onto the biomaterial surface, the release of Ag^+ might also be hampered as the direct contact between nanoparticles and bacteria can be prevented.

Catellano *et al.* [83] have suggested that, when metallic silver reacts with the moisture in wounded skin, it would be ionized and bind to the bacterial membrane proteins, DNA and RNA, leading to bacterial cell death. Lok *et al.* [4] suggested that the antibacterial activity of AgNPs was dependent on surface oxidation; in contrast to zeroth-valent nanoparticles, only partially oxidized AgNPs would exhibit antibacterial activities, which suggested that partially oxidized AgNPs might serve as carriers of chemisorbed Ag^+ in quantities sufficient to cause bacterial damage (Figure 3.34). One possible method of delivering Ag^+ from oxidized AgNPs might involve a direct interaction with the bacterial membrane.

When Su *et al.* [6] studied the antimicrobial mechanism of AgNPs/silicate clay nanocomposites, the constructs appeared to exert a biocidal effect by means of a direct contact with the nanoparticles, and not by means of the silver ions released.

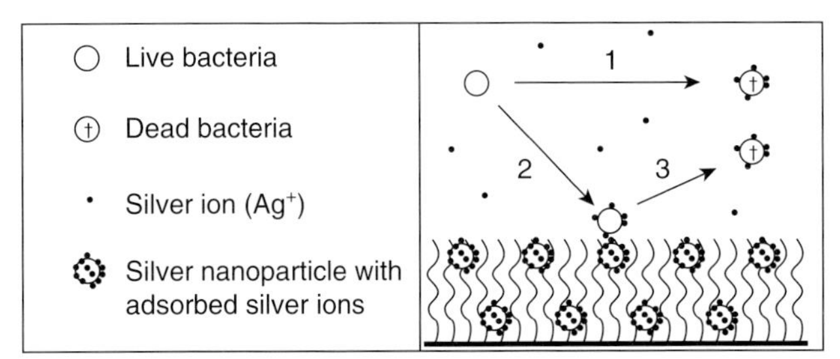

Figure 3.33 Schematic representation of the possible antimicrobial mechanisms by ion elution or contact-dependent transfer of silver ions following collision with the silver nanoparticle-containing surface. Key: 1, concentrations of Ag^+ ions built up by elution from the silver nanoparticle coating sufficiently high to kill bacteria; 2 and 3, possible killing of bacteria by contact-dependent transfer of Ag^+ ions following collision with the silver nanoparticle-containing surface. Reproduced with permission from Ref. [63].

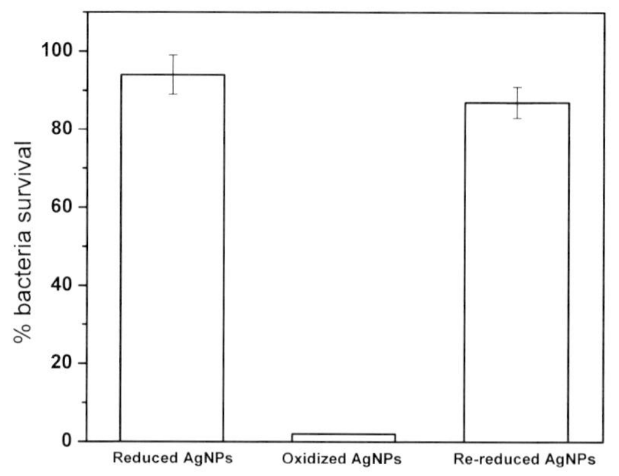

Figure 3.34 Antibacterial activity of reduced, oxidized and re-reduced silver nanoparticles. Redrawn from Ref. [4].

This indicated that, in this system, a simple contact with AgNPs was sufficient to trigger membrane leakage and cell death.

One of the main target sites for silver nanoparticles is the bacterial membrane, where the induction of deep morphological changes will lead to significant increased in membrane permeability and to changes in transport mechanisms through the plasma membrane [27, 84]. As reported previously [27], a colloidal system based on a lactose-modified chitosan and AgNPs had a notable effect on membrane potential and permeability. Cell membrane depolarization has been assessed by the addition of the fluorescent probe bis-(1,3-dibarbituric acid)-trimethine oxanol) (DiBAC4(3)), which is able to selectively enter and label those bacteria in which the membrane potential has collapsed. In this case, the fluorescence intensity of AgNPs-treated bacteria was increased compared to untreated cells after only a 10 min incubation with the silver-based colloidal solution. This revealed a remarkable depolarization of the cell population (97.5% of cells) that rose to 99.6% after 30 min, thus confirming a strong interaction between silver and the bacterial membrane. As the antimicrobial mechanism is thought to involve damage to the bacterial membrane, membrane integrity was also evaluated using the fluorescent probe propidium iodide (PI); this revealed a remarkable permeabilizing effect as a function of incubation time (Figure 3.35).

When Morones *et al.* [70] investigated the effect of AgNPs on four types of Gram-negative bacteria, namely *E. coli*, *V. cholera*, *P. aeruginosa*, and *Salmonella typhus*, AgNPs were seen to be attached to the cell membrane and also present in the bacterial cytoplasm (Figure 3.36). Whilst the mechanism by which the nanoparticles penetrate the bacteria is not fully understood, their presence on the cell surface and inside the bacteria is fundamental to understanding their bactericidal mechanism. In analogy with the mechanism suggested for silver ions, the AgNPs tend to react with sulfur-containing proteins, as well as with phosphorus-containing compounds such as DNA, so as to induce irreversible cellular damage.

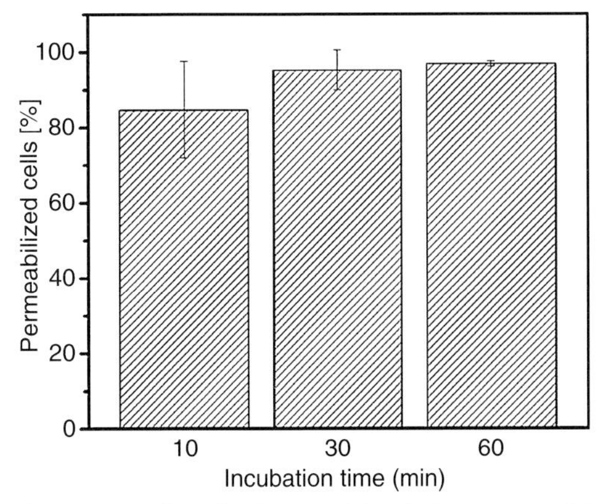

Figure 3.35 Effect of polymer-stabilized silver nanoparticles on the membrane integrity of *S. epidermidis*. Reproduced with permission from Ref. [27].

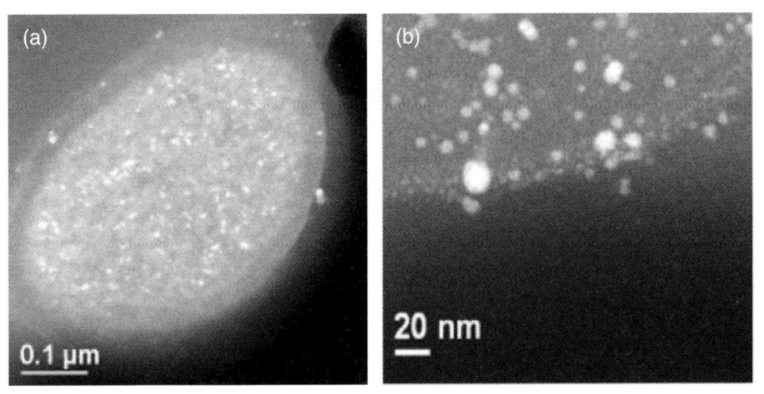

Figure 3.36 (a) Silver nanoparticles on the membrane and inside the *E. coli* bacterium; (b) Magnification of the *E. coli* cell membrane, where the presence of silver nanoparticles is clearly observed. Reproduced with permission from Ref. [70].

Proteomic data have revealed that a short exposure of *E. coli* cells to antibacterial concentrations of AgNPs resulted in an accumulation of envelope protein precursors, indicative of the dissipation of proton motive force. Consistent with these proteomic findings, nano-Ag was shown to destabilize the outer membrane, to collapse the plasma membrane potential, and to deplete the levels of intracellular ATP [85].

As an increase in free-radical content has been considered a possible explanation of the antimicrobial mechanism [86, 87], Su *et al.* [6] studied the free-radical burst and creation of reactive oxygen species (ROS) in AgNPs/clay-treated bacteria, using 2,7-dichlorofluorescin-diacetate (DCFH-DA) as an indicator or intracellular

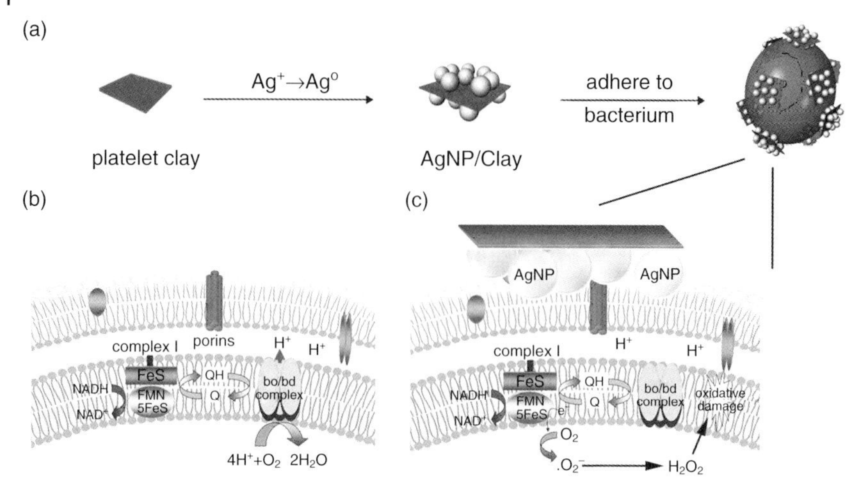

Figure 3.37 Possible mechanisms of AgNP/clay-mediated cytotoxicity. (a) AgNPs on synthetic platelet clay are attracted electrostatically to the bacterium, and form AgNP clusters on the bacterial cell wall; (b) Electronic transport through the respiratory chain on the plasma membrane of *E. coli* through complex I, ubiquinone oxidoreductase (Q) and cytochrome bo/bd ubiquinone oxidase (bo/bd complex); (c) AgNP/clay on the cell wall interacts with transmembrane proteins and consequently interferes with the proton pool in the intermembrane space or the electronic flow through the respiratory chain. Accumulated electrons due to a disturbance of complex I are transferred to oxygen to form O_2 and H_2O_2, contributing to oxidative damage and membrane leakage in the bacterium. Reproduced with permission from Ref. [6].

ROS levels (Figure 3.37). The fact that 40.3% of the AgNPs/clay-treated bacteria became DCF^+ indicated that ROS had been generated and had played a role in the killing mechanism. That the bacteria had also lost their mobility when treated with the nanocomposites suggested that the motor function of the cytoskeleton had been hampered, with a consequent prevention of cytokinesis.

A nanocrystalline, silver-supported activated carbon matrix [88] demonstrated kinetics of bacterial inactivation, in the presence of hydroxyl radical scavengers and superoxide anion radical inducers, that suggested a contribution of ROS to the antibacterial effect. However, the fact that ROS scavengers had no inhibitory effect on the bactericidal activity after about 1 h suggested that the generated ROS were responsible for *E. coli* inactivation during only the first hour of incubation. Never the less, the antibacterial process was seen to be greatly increased at higher temperatures, this being ascribed to an enhanced ROS formation and Ag^+ elution.

3.3.2 Silver Nanocomposites in Wound Healing

Nonhealing chronic wounds, such as diabetic ulcers, pressure, venous and arterial ulcers, and burn wounds, represent a serious problem in the healthcare system worldwide. In addition to causing great pain and suffering to patients, and adversely affecting their quality of life, such wounds impose a serious financial burden on healthcare systems.

A wound may become infected when the wound bioburden exceeds a host-manageable level, with chronic wounds being easily contaminated by several species, and the progression to local infection occurring in stages, often leading to critical colonization [89]. Chronic infections are clearly detrimental to wound healing, and represent one of the main factors contributing to the formation of a nonhealing wound [90]. When such infections do occur, and the host response is suppressed, the normal healing process may be blocked due to a prolonged inflammatory response, in addition to molecular and cellular abnormalities in the wound bed and granulation tissue disruption. The result is a clear deterioration in the condition of the wound [91]. An appropriate management of infected and critically colonized wounds is, therefore, essential to encourage wound-healing progression [92]. Unfortunately, however, it is not only critical colonization and/or infection that cause problems in diagnosis; rather, traditional topical antimicrobials may also be toxic towards granulation tissues, or they may raise the potential for the development of resistant organisms. Silver, whether in its ionic form or as nanoparticles, is particularly attractive as an antibacterial agent for infected wound treatments, mainly because it can be readily incorporated into wound-dressing materials or included in topically applied antimicrobial ointments.

Topically applied antimicrobial agents, such as silver-based formulations, are often used to prepare the wound for healing. Silver ions are effective against a broad range of microorganisms such as yeasts, molds, and bacteria, including MRSA (methicillin-resistant *Staphylococcus aureus*) and VRE (vancomycin-resistant enterococci) when provided at appropriate concentrations. Silver-coated dressings, which represent the most commonly used form, are more effective and kill a broader range of bacteria than the cream-based silver applications; moreover, they are also less-irritant than silver nitrate solutions and better tolerated [93]. Chronic wounds respond to AgNPs-based devices on the basis of their antimicrobial activity, and also of the biological properties displayed by metal nanoparticles. Hence, the opportunity for wound healing is improved by creating conditions that are unfavorable to microorganisms, but favorable to the host's repair mechanisms.

One of the major contributors to delayed wound healing is a prolonged inflammatory response in the wound [94]. Normally, an inflammatory response will occur immediately after wounding, and begin to induce phagocytosis and the removal of bacteria and tissue debris. It also initiates the release of factors that cause the migration and division of cells involved in the proliferative phase, and in the deposition of new tissue components. Unfortunately, however, a prolonged inflammatory response can result in the destruction of tissue, via the same processes that normally would have protective and restorative functions. Currently, evidence obtained from *in vivo* and *in vitro* studies has indicated that AgNPs promote wound healing on the basis of their potent anti-inflammatory activity [95–98]. This activity appears to be correlated to an increased expression of anti-inflammatory molecular signals such as vascular endothelial growth factor (VEGF) and interleukin (IL)-10, as well as to a reduction in pro-inflammatory cytokines such as IL-6 and interferon-γ. VEGF promotes wound healing by inducing vascular permeability and endothelial cell proliferation, whereas IL-10 is a vital mediator of the anti-inflammatory cascade and is produced not only by keratinocytes but also by inflammatory cells involved in the healing process, including T and B lymphocytes and macrophages. A unique actions of IL-10 is its ability to inhibit the synthesis of pro-inflammatory cytokines, including IL-6. The latter is secreted by polymorphonuclear cells and fibroblasts, and has been recognized as an initiator of events in the physiological alterations of inflammation following thermal injury [99]. Interleukin-6 also promotes inflammation through monocyte and macrophage chemotaxis and activation [100]. Decreased levels of IL-6 may result in fewer neutrophils and macrophages being recruited to the wound, and fewer cytokines being released into the wound, which leads in turn to a lower paracrine-mediated stimulation of cellular proliferation, fibroblast and keratinocyte migration, and extracellular matrix (ECM) production. This lack of amplification of the inflammatory cytokine cascade may be important in providing a permissive environment in which a scarless wound repair may proceed. Moreover, silver-induced neutrophil apoptosis, decreased matrix metalloproteinase (MMP) activity, and the inhibition of free radical formation (ROS/reactive nitrogen species) also contribute to the overall decrease in inflammatory response and, as a consequence, an increased rate of wound healing [97, 101–104]. MMPs represent a group of proteinases that includes collagenases, elastases and gelatinases, all of which may be either endogenous (cellular) or exogenous (bacterial) in origin and present in chronic ulcers at abnormally high levels (as compared to acute wounds), and may contribute to the chronicity of the wounds [105]. MMPs also play a role in the controlled degradation of the ECM, by removing damaged components and allowing cell migration and angiogenesis to occur [106]. It has also been proposed that elevated levels of these enzymes might contribute to an excessive matrix destruction and, therefore, to a delay in wound repair.

In the field of biomedicine, a wide selection of wound dressings containing silver in different chemical forms, such as silver salts, silver oxides and metallic silver, are currently available. With the advent of nanotechnology, however, interest has focused primarily on silver salts since, in comparison to silver ions and metallic silver, they exhibit both improved bactericidal and fungicidal effectiveness due to their nanoscale dimensions and good silver release kinetics based on their minimal solubility. The details of some commercially available silver-based dressings of different types and compositions are listed in Table 3.1.

Today, all newly developed commercially available wound care products have, typically, a multilayer structure. Starting from the top, these include an outer acrylic adhesive layer that protects the wound from the external environment, a middle layer to absorb any exudates and, finally, a wound-contact layer in which the silver particles are incorporated. The selection of a suitable silver substrate depends on many factors, including the design of the wound care product, the

Table 3.1 Commercially available silver-based dressings.

Silver formulation	Product name	Manufacturer
SILCRYST Nanocrystalline (Patented Silver technology)	Acticoat®	Smith & Nephew
Nanocrystals of metallic silver	Acticoat 7®	Smith & Nephew
Ionic silver	Aquagel Ag®	Convatec
Metallic silver	Actisorb Silver 220®	Johnson & Johnson
Silver particles	Polymem Silver®	Ferris Mfg. Corp.
Ionic silver	Suprasorb A + Ag®	Activa Healthcare
Silver sulfadiazine	Urgotul SSD®	Urgo Medical

type of wound, and the fabric strength and thickness. The substrate materials may include cotton, viscose, silk, polyamide fibers, hydrogels (alginate and agarose), and a hydrophilic polyurethane foam.

3.3.3 Silver Nanocomposites and Inflammation

Besides the anti-inflammatory activity of AgNPs in relation to wound healing, the results of many studies have suggested that nanocrystalline silver might possess a general anti-inflammatory effect, although the molecular and cellular mechanisms of action of this have not yet been fully elucidated [107–111].

In an allergic contact dermatitis model, the anti-inflammatory activity of topical nanocrystalline silver cream has been shown comparable with that of topical steroids and currently available immunosuppressants [107–109, 112]. It has also been shown that nanocrystalline silver treatments uniquely decrease any induced erythema and edema, increase apoptosis in inflammatory cells, decrease MMP activity, and inhibit the expression of proinflammatory cytokines such as tumor necrosis factor-alpha (TNF-α), transforming growth factor-beta (TGF-β), and IL-12 and IL-8.

The reduction of pro-inflammatory factors and MMP activity has been ascribed both to a specific inhibition of gene expression [108] and to the reduction of inflammatory cells via apoptosis [109]. Programmed cell death (apoptosis), which occurs in various physiological and pathological conditions, involves a characteristic mechanism of intercellular sequential reactions. During inflammation events, apoptosis contributes to the elimination of inflammatory cells from the inflamed area. When cells (such as neutrophils or T lymphocytes) die as a result of apoptosis, they initially maintain their plasma membrane integrity, whereas granulocytes lose their ability to secrete granular contents that consist of numerous cytotoxic and pro-inflammatory factors [113, 114]. The apoptotic cells are subsequently recognized and phagocytosed by neighboring macrophages, further minimizing any local inflammation and tissue injury [114].

In contrast, when inflammatory cells die by necrosis, they burst and release numerous cytotoxic compounds, including proteases, oxygen radicals, and various acids, all of which further amplify any local inflammation [115–118]. In this respect, cell death by apoptosis represents an injury-limiting clearance mechanism, and plays an important role in the resolution of local inflammation.

Interestingly, anti-inflammatory effects were observed only or mostly for silver nanoparticles, and not for comparable concentrations of silver salts [102, 109]. Moreover, silver salts such as $AgNO_3$ may even have an opposite effect, being pro-inflammatory with a subsequent delay in healing [119, 120]. This apparent discrepancy may be resolved by the recently proposed hypothesis of Nadworny *et al.* [109], who showed in a porcine model of contact dermatitis, that dressings containing nanostructured silver would induce apoptotic processes in a discriminatory manner towards dermal cells, but would not target keratinocytes. Otherwise, apoptotic activity was induced indiscriminately by $AgNO_3$-based dressings in all cell types at the tissue surface, including keratinocytes. The authors assumed, therefore, that the anti-inflammatory activity shown by nanocrystalline silver was not due to the Ag^+ form, but rather may be related to the Ag[0] form. This species may have anti-inflammatory properties, similar to other noble metals; colloidal gold, for example, has been used successfully in the treatment of rheumatoid arthritis (RA) [121, 122]. Various gold-containing compounds, when used as anti-inflammatory agents to treat RA, have been shown to induce apoptosis in cells (including T cells and macrophages) through multiple mechanisms. Moreover, Au[0] nanoparticles appear to suppress the activity of IL-6 and TNF-α, thus relieving the symptoms of RA [123, 124]. The results of both *in vitro* and *in vivo* studies have indicated that, within a biological environment, gold in its monovalent state (Au(I))–when not tightly bound to ligands such as cyanide, phosphine or molecules containing sulfur(II)–will dismutate spontaneously to form finely divided metallic gold (Au[0]) and auric (Au(III)) complexes. As the crystal structures (fcc) and Pauling covalent radii for silver and gold are equal, it was concluded that these metals could replace each other in crystal lattices. Hence, it was suggested that Au[0] and Ag[0] clusters should be physically identical, and therefore have similar biological activities.

3.3.4
Silver Nanocomposites for Applications in Biological Sensing and Nanoscale Photonics

Today, metal nanocomposites are undergoing investigation with respect to nanoparticle plasmonics, an emerging area of research that is related to the optical properties of noble metal nanoparticles of various sizes, shapes, and structures. In particular, the scattering of plasmon-resonant nanoparticles and their colors in dark-field microscopy allow for many applications in biomedical imaging. Depending on the type of nanoparticle, and on the nature of the stabilizing shell on their surface, plasmon resonant nanoparticles can emit bright resonance light-scattering of various wavelengths [125]. These nanoparticles can be addressed to well-defined biological targets by means of surface functionalization with the correct molecular signals. For this, selected biomacromolecules can be attached to the metal nanoparticles, and used as optical labels to target specific receptors. Based on this rationale, noble metal nanoparticles (especially gold and silver) are currently being studied for *in vivo* imaging applications, to overcome the limitations associated with traditional fluorescence probes (e.g., photobleaching, autofluorescence of living cells). Among the noble metal nanoparticles, AgNPs offer the highest quantum yield of Rayleigh scattering, and for this reason colloidal silver may be used efficiently as an optical probe for *in vivo* imaging in real time, with sub-100 nm spatial resolution and millisecond time resolution [54]. Schrand *et al.* [126] showed that low concentrations of AgNPs ($<25\,\mu g\,ml^{-1}$) would bind to the plasma membranes of living cells, and cause an intense scattering of light when imaged at submicron resolution with high-illumination light microscopy, demonstrating their potential as biological labels. Lee *et al.* [54] reported the use of silver colloids for the study of nanoparticles transport, biocompatibility, and toxicity in the early development of zebrafish embryos in real time, using dark-field single-nanoparticle optical microscopy and spectroscopy. Huang *et al.* [57] developed functionalized silver nanoparticle biosensors to quantify the binding kinetics and affinity of single protein molecules on single living cells for an extended period of time (hours), using single-nanoparticle optical microscopy and spectroscopy (Figure 3.38). These silver nanocomposites offer the possibility to monitor cascades of biochemical reactions in real time, through quantitative molecular imaging.

Hu *et al.* [127] have developed biocompatible polyelectrolyte-coated AgNPs for the targeted *in vitro* labeling of pancreatic cancer cells. In this case, the particles could be conjugated with monoclonal antibodies and proteins, and serve as Plasmon-enhanced scattering probes for dark-field multiplex and TEM imaging of pancreatic cancer cells.

Nanoparticle optics can also be used as a tool for real-time monitoring of the effect of antibiotics in living bacterial cells, with millisecond temporal and nanometer-sized resolution. for example, Kyriacou *et al.* [128] studied the modes of action and pharmacokinetics of antibiotics in *P. aeruginosa* by treating the bacteria with silver colloidal solutions and evaluating membrane permeability and disruption of the cell wall, by using optical darkfield microscopy and spectroscopy.

Lesniak *et al.* [58] synthesized fluorescent and biocompatible silver/poly(amidoamine) dendrimers carrying various surface functionalities for applications as cell labeling agents, as demonstrated by *in vitro* assays with four different cell lines.

In addition to bioimaging, the strong plasmon resonance of AgNPs has been used to identify other applications in the field of nanophotonics. One particular area of interest of silver nanocomposite constructs for biomedical purposes is in the field of Raman spectroscopy, as a significant enhancement of the Raman signal can be achieved in the presence of plasmon-resonant nanoparticles. This

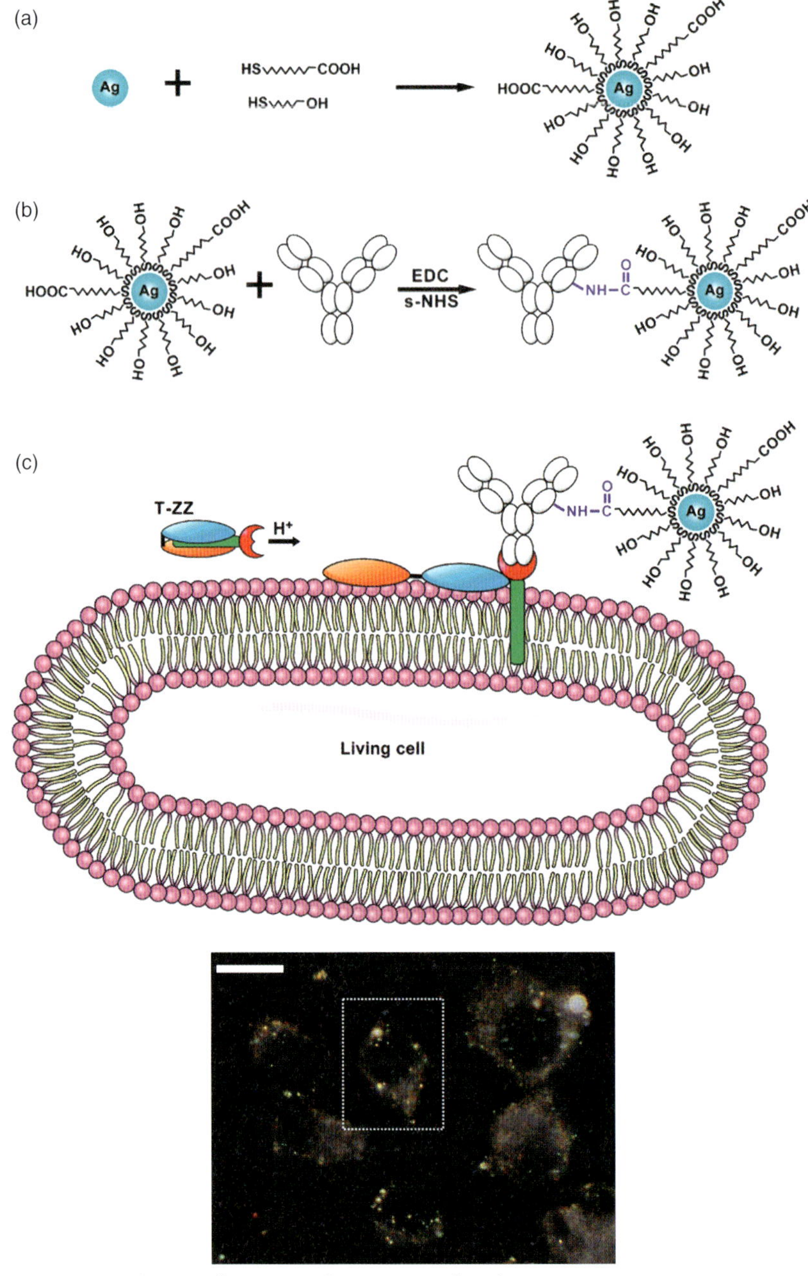

Figure 3.38 Schematic illustration of preparation of single-nanoparticle biosensors to image and detect single T-ZZ receptor molecules on single living cells. (a) Functionalizing AgNPs; (b) Covalently linking with immunoglobulin G; (c) Attaching T-ZZ onto living cells to detect single T-ZZ molecules on single living cells. Reproduced with permission from Ref. [57].

phenomenon, which is referred to as surface-enhanced Raman scattering (SERS), provides chemical information for molecules in the proximity of metal nanostructures. Moreover, as the enhancement factor may be up to 10^{14}–10^{15}, this technique can be used to detect very diluted compounds, and even single molecules [129]. Both, the size and arrangement of the nanoparticles can strongly affect the enhancement factor, such that the structural properties of the nanocomposite system will determine its efficient use in SERS applications. Jia *et al.* [130] have developed a "green synthesis" preparation of silver nanocomposites based on the formation of AgNPs within a cuttlebone-derived organic matrix as a natural macroporous dispersing material. In this case, the resultant nanocomposite films could be successfully employed for trace analysis in SERS applications. In the past, much effort has been targeted at developing effective ways in which to organize nanoscale particles into functional structures and devices. In particular, efficient routes for the fabrication of films characterized by a high density of nanoparticles with controllable spacing on solid substrates have been sought. For example, Lu *et al.* [42] developed a silver nanocomposite film with temperature-controllable interparticle spacing for a tunable SERS substrate. In such constructs, the scattering signal enhancement factor can be dynamically tuned by thermal activation. In fact, the spacing between the nanoparticles can be controlled by adjusting the temperature in order to approach the strongest coupling between adjacent particles, and to match the plasmon resonance wavelength to the laser excitation wavelength (Figure 3.39). This nanocomposite system may find use in label-free biomolecular detections, environmental monitoring, and in the sensing of biological warfare agents.

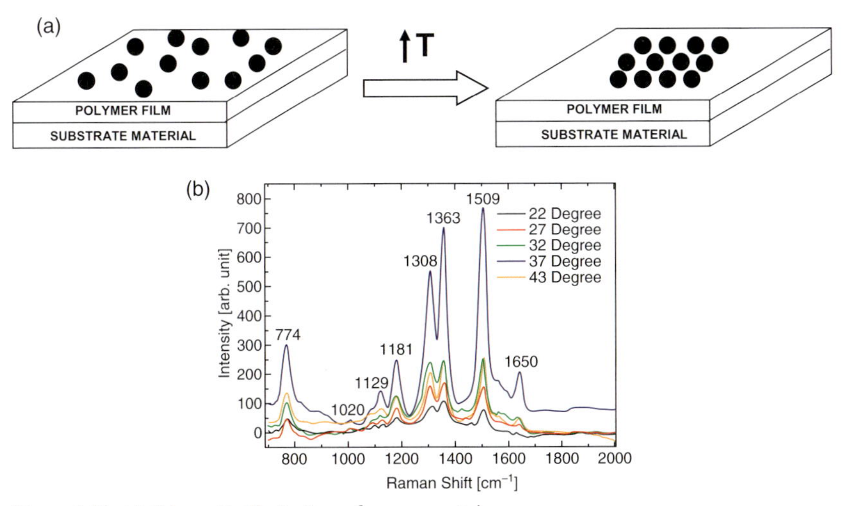

Figure 3.39 (a) Schematic illustration of a nanoparticle monolayer on a thermoresponsive polymer film; (b) Surface-enhanced Raman scattering spectra of Rhodamine 6G on the nanocomposite film at different temperatures. Reproduced with permission from Ref. [42].

Biocompatible SERS substrates based on chitin (from the cicada wing) and silver were prepared by Stoddart *et al.* [131], who proposed a molecular detection limit as low as the nanomolar level.

3.4 Biological Hazards of Silver Nanocomposites

Nanosized materials are currently being used in medicine, biotechnology, and in both energy and environmental technologies. Given the wide variety and growing number of applications available commercially, there is a serious lack of information concerning the effects of silver nanoparticles on general human health and the environment. Today, silver-based products are widely available, primarily as formulations for topical application in the area of medicine (as wound dressings or as surgical instruments that have been coated and/or embedded with AgNPs), and in daily life in the form of silver-containing detergents and soaps, room sprays, water purificants, textiles, personal care products, handles, and furniture for public places. Silver-based systems are also used in some areas of the food processing industry, where pipelines may be susceptible to biofilm formation. AgNPs have also more recently been investigated for systemic applications, and for the preparation of internal prostheses and devices (e.g., bone cement, catheters). Although, currently, many hundreds of silver-based products are available commercially, these carry no specific reporting requirements, and neither are any risk assessments or official government indications required in order for these products to be marketed.

Nonetheless, the widespread use of AgNPs and silver ion-based products is partly justified by the fact that, until a few years ago (and even today), ionic silver was associated with a low toxicity in the human body, with minimal risk expected following clinical exposure via inhalation, ingestion or dermal application, nor through urological or hematogeneous routes. When administered systemically, silver is not recognized as a toxic agent in humans, except at extreme doses. The most evident secondary effect derived from the chronic ingestion or inhalation of silver preparations (especially colloidal silver) is the deposition of particles in the skin (argyria), eye (argyrosis), and other organs. While not generally considered to be life-threatening, these conditions are simply cosmetically undesirable. Following its absorption in humans, ionic silver enters the systemic circulation as a protein complex, to be eliminated via the liver and kidneys with a metabolism that is modulated by induction and binding to metallothioneins. These protein complexes may mitigate the cellular toxicity of silver, and also contribute to tissue repair [132]. Whilst, to date, no cases have been reported of people being harmed specifically by the use of, or exposure to, AgNPs, this situation might be attributed to a lack of knowledge regarding what effects might be expected. Recently, Larese *et al.* demonstrated the absorption of silver nanoparticles through intact and damaged skin, detecting the presence of nanoparticles in the stratum corneum and the outermost surface of the epidermis [133].

In fact, in sharp contrast to the attention paid to new applications of silver nanocomposites, very few studies have provided any insight into the possible interaction of AgNPs with the human body following absorption via routes other than through the skin. Although, until now, the systemic distribution and translocation, organ accumulation, degradation, possible adverse effects and toxicity of AgNPs in human tissues have been recognized only slowly, this has led to major questions being asked in relation to the increased medical use of these materials. One important–but often disregarded–aspect that should be considered when investigating the biological effects of silver nanocomposites is that nanoparticles demonstrate an impressive array of unusual physical-chemical properties that confer upon them a greater and often unpredictable bioactivity compared to the identical bulk materials. For example, based on size alone and not on any physical or chemical properties, nanomaterials have the ability (which normally is attributed to microorganisms such as viruses) not only to penetrate the circulatory system, but also to reach and translocate in all living organs, including the blood–brain barrier.

During the past five years, most *in vitro* studies of the exposure of cell lines to silver nanoparticles have witnessed the presence of a potential cytotoxic mechanism that is strictly dependent on the particle size and concentration. Yet, despite the limited number of cell lines tested, all experimental data have agreed that the mitochondria represent the main intracellular target of silver nanoparticles. AgNPs mediate their toxicity through an oxidative stress with increased ROS levels following activation of the apoptotic mitochondrial pathway [134–138]. This mechanism was shown to be shared by all of the cell lines tested [139–145], whereby apoptosis and apoptosis-like changes of morphology would occur when the AgNPs are applied. For example, AgNPs were shown to induce mitochondrial membrane perturbation, to generate ROS, to deplete levels of the antioxidant glutathione (GSH), and to reduce mitochondrial function in BRL 3A rat liver cells [143], in rat alveolar macrophages [140], and in human THP-1 monocytes [141]. GSH depletion in human skin carcinoma and fibrosarcoma cell lines was also shown to be associated with the expression of apoptotic markers [146] and, in human hepatoma cells (HepG2), with the overexpression of oxidative stress-related genes such as catalase and superoxide dismutase [144].

In a study where the cytotoxicity of four commercially available silver dressings towards keratinocyte HaCaT and fibroblast 142BR cells [147] was monitored, all of the dressings induced (albeit to different degrees) apoptotic cell death in a manner that was dependent on the cell line and type of dressing. Similar results were reported by Burd *et al.* [148].

Although the molecular mechanism of silver nanoparticle (potential) toxicity has yet to be completely elucidated, a recent study conducted by Hsin *et al.* [142] has managed to shed some light on the problem. The mitochondria-dependent apoptotic mechanisms identified in eukaryotic cells constitute intrinsic and extrinsic pathways, as characterized by the activation of pro-apoptotic proteins such as Bid, Bad, and Bak, and the inactivation of anti-apoptotic proteins such as Bcl-2 and Bcl-Xl. The activation of Bad and Bak results in their translocation and oligomerization on the mitochondrial membrane, to form ion channels that allow the exit of cytochrome C and other apoptotic factors. The tumor suppressor protein p53 is capable of exerting a pro-apoptotic effect, by a direct transcriptional activation of pro-apoptotic genes such as BAK, by inhibiting the expression of anti-apoptotic proteins such as Bcl-2 and Bcl-X, and by its direct binding to pro-apoptotic (Bax and Bak) and anti-apoptotic protein effectors (Bcl-2, BcXL) [149–151]. In NIH-3T3 (mouse fibroblast) cells, pro-apoptotic internalized AgNPs induced intracellular ROS generation, leading to an activation of JNK and p53 genes and, subsequently, to Bax translocation, cytochrome c release and PARP (poly ADP ribose polymerase) cleavage. Moreover, the elevated expression of the anti-apoptotic Bcl-2 protein induced in HCT116 (human colon cancer) cells appeared to shield these cells from silver-mediated apoptosis. In the same manner, a study [152] comparing the biological effects of uncoated and polysaccharide-coated AgNPs in mouse embryonic stem cells (mES) and in mouse embryonic fibroblasts (MEF) has demonstrated p53 upregulation in both cell lines, and p53 phosphorylation in mES cells.

Based on experimental data acquired to date, the prevailing hypothesis is that AgNPs interact with the thiol groups of proteins and enzymes after passing through the eukaryotic cell membrane. Most of those proteins can be involved in the antioxidant defense mechanism, such as reduced GSH, catalase and superoxide dismutase, and prevent tissue damage under normal conditions by neutralizing the ROS produced via aerobic energy metabolism [153, 154]. With the over-accumulation of ROS, an inflammatory response is induced such that irreversible mitochondrial membrane damage and permeabilization occurs, with the subsequent release of cytochrome c and other apoptogenic factors [155]. Besides mitochondrial pathway activation, cell membrane destruction and lipoperoxidation may also take place, which together represent another aspect of AgNPs toxicity (Figure 3.40).

Silver nanoparticles may be internalized through the membrane by different mechanisms, including passive diffusion, receptor-mediated endocytosis, and clathrin- or caveolae-mediated endocytosis [140, 144, 152, 156] (Figure 3.41).

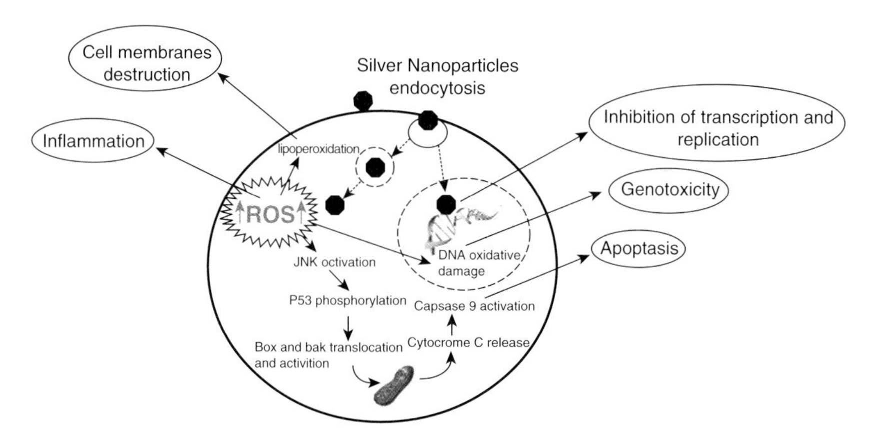

Figure 3.40 Cytotoxic effect of silver nanoparticles on eukaryote cells. See text for details.

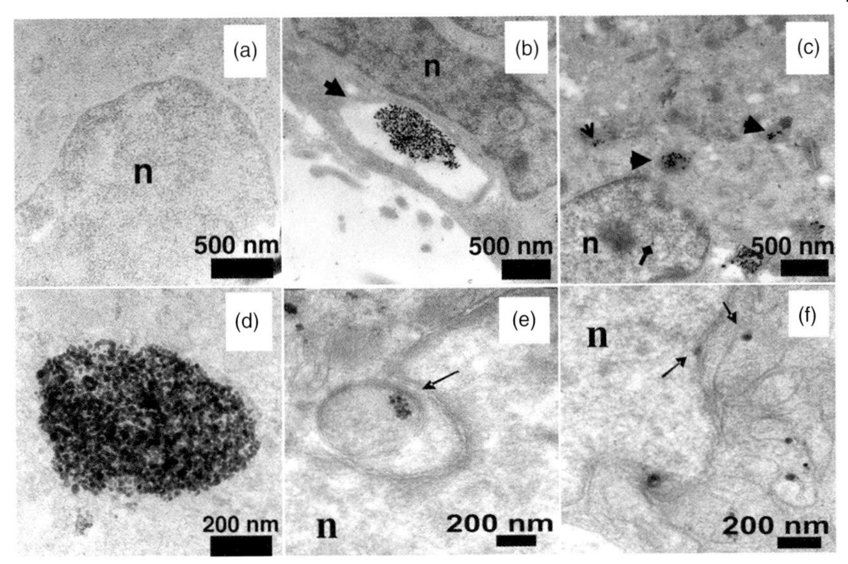

Figure 3.41 TEM images of ultrathin sections of cells. (a) Untreated cells showed no abnormalities; (b) Cells treated with AgNPs showed large endosomes near the cell membrane containing many nanoparticles; (c) Lysosomes containing nanoparticles are visible (arrowheads) and scattered in the cytoplasm (arrow). The diamond-headed arrow shows the presence of the nanoparticle in the nucleus; (d) Magnified images of nanogroups, showing the cluster to be composed of individual nanoparticles rather than clumps; (e) The endosomes in the cytosol are lodged in the nuclear membrane invaginations; (f) The presence of nanoparticles in the mitochondria and on the nuclear membrane. Reproduced with permission from Ref. [181].

The above-described mechanism of action of AgNPs appears to be shared by silver ions [157], and supports the hypothesis that part of the toxic potential of AgNPs is due to Ag^+ release from the particles.

An examination of recent reports shows clearly that, over the past few years, the majority of investigations have been conducted to assess the adverse biological effects of AgNPs by their administration at doses and concentrations that would lead to cell death, or cause irreversible cellular damage. Yet, these studies have failed to consider that concentrations of nanoparticles not causing cellular death might in fact cause sublethal cellular alterations, with serious consequences to human health. The most important examples of this are DNA damage and the induction of inflammation processes. The mutation of DNA, induced by genotoxic materials, leads to carcinogenesis and has a deep impact on the biology of the reproductive cells. As reviewed recently by Singh *et al.* [145], metal nanoparticles may create DNA damage indirectly, by inducing oxidative stress and inflammation responses. For example, ROS – the increased levels of which are associated with oxidative stress and a subsequent redox imbalance – react with many biological macromolecules, including DNA, enzymes, and lipids. In particular, the ROS may cause oxidative damage to DNA in the form of breaks in both single and double strands, as well as purine, pyrimidine and deoxyribose modifications, abasic sites and DNA–DNA coupling [158]. At the same time, if sufficiently small, the nanoparticles can gain direct access to the nucleus, either though passive diffusion or via transport across the nuclear pore complexes, where they may trigger the aggregation of nuclear proteins to cause a subsequent inhibition of transcription, replication, and cell proliferation [159–162].

Although substantial experimental evidence exists regarding the genotoxic potential of many metal-based nanomaterials, the data accumulated on AgNPs point only very marginally to genotoxicity. The detection of γ-H2AX phosphorylation, which is indicative of DNA double-strand breakage, was described in HepG2 cells treated with $2\,\mu g\,ml^{-1}$ of AgNPs [144], while mouse embryonic stem cells and fibroblasts exhibited a severe DNA damage response. This was suggested by an increased expression of Rad51, a key double-strand break repair protein, and phosphorylation at Ser139 of the histone H2AX [152].

Whilst the initial toxicity screening of nanomaterials must always be accomplished *in vitro*, supportive *in vivo* testing is essential, as the *in vitro–in vivo* gap associated with the complexity of biological interactions in higher-order organisms is not always reproducible in an *in vitro* system. *In vivo* testing requires the use of animal models to evaluate markers of inflammation, oxidant stress, and cell proliferation, both at the portal-of-entry and in selected remote organs and tissues. Also required are details of the deposition and translocation of the material, of its degradation products, toxicokinetics and biopersistence, the effects of multiple exposures and, finally, of the potential effects on the reproductive system, placenta, and fetus. In the case of silver nanocomposite materials, the scarcity of toxicity data available from cellular models runs in parallel with the lack of exhaustive studies performed using *in vivo* models. In zebrafish embryos, for example, AgNPs administered over a concentration range between 250 and $0.25\,\mu M$, caused toxic lethal and sublethal (morphological malformations) effects, albeit in a size-dependent manner, for certain concentrations and time points [163]. Silver nanoparticles administered by intraperitoneal injection to adult mice were able to translocate to the circulatory system and reach the brain, where they caused a neurotoxic response by inducing free-radical oxidative stress, and by altering gene expression and producing apoptosis [164]. A study performed to compare adverse effects on the liver caused by feeding mice with nanosized and microsized silver particles, revealed an induction of liver inflammation in both cases [156]. Both, microscale and nanoscale silver particles, when implanted into the back muscle of rats, revealed an effective biological effect on days 7 and 14, but caused inflammation by day 30 which was more serious in rats treated with AgNPs than with microscale silver particles [165].

Notably, all of the studies conducted were to monitor acute toxicity; no information was obtained from more informative, chronic testing.

The major point which underlines the risk of using AgNPs is that, based on the concerns of many research groups, many reports have been made on the development and characterization of materials employing silver nanocomposites for eukaryotic systems. For example, Fu *et al.* [16] described the realization of antibacterial multilayer films containing AgNPs via a LBL assembly of heparin and chitosan, but which lacked any toxic effect on osteoblasts. Nanocomposite materials based on acrylic resins and AgNPs tend to stimulate fibroblast and osteoblast aggregation and growth, without displaying any toxic effect [166, 167]. Similarly, stainless steel orthopedic materials (many of which are already used widely in biomedicine) that had been coated with silver were able to efficiently sustain the growth of osteoblasts, and did not show any genotoxicity [168]. Recently, Travan *et al.* [27] used silver nanoparticles that had been formed and stabilized by a bioactive chitosan-derived nanoparticle to develop alginate-based hydrogel structures that displayed antimicrobial properties, but which were noncytotoxic towards different eukaryotic cell lines. The same nanocomposite systems, based on chitosan-derived and on silver nanoparticles, have been exploited to realize a noncytotoxic coating for acrylic resins (A. Tavan *et al.*, unpublished results).

In addition to the potential toxicity of silver administered either topically or systemically, general concerns have been expressed regarding the hazardous effects of nanoparticles as they accumulate in the environment. As has often been pointed out, there is a major lack of information concerning the impact of AgNPs on human health, and even less is known about the environmental pathways and effects of these materials. At present, silver is classified by the US Environmental Protection Agency (EPA) as an environmental hazard because, under certain circumstances, it is toxic, persistent, and bioaccumulative. Moreover, as nanotechnology has not been well researched or regulated, the environmental impact and risks of silver nanoparticles remain unknown, at least for the present.

Although nanoparticles may accumulate in the air and soil, it is primarily in water where most materials demonstrate their most significant effects, mainly because in an aqueous environment materials can be more easily degraded, transformed, and accumulated in a number of ways. When silver is present in a bioavailable chemical form [169, 170], it is more toxic towards aquatic organisms than any other metal except for mercury [171]; however, no comparable information is available for AgNPs. The environmental hazard of silver can be mitigated by its complexation with other compounds (especially with sulfide), so as to reduce its bioavailability. The inhibition of nitrification by silver nanoparticles has been demonstrated in nitrifying bacteria isolated from wastewater treatment plants. However, the results obtained suggested that AgNPs had a similar behavior of surface complexation as silver ions, and that the inhibition by nanosilver in wastewater treatment might be nullified by reacting the AgNPs with soluble sulfide species [87, 172]. The exposure of bacteria, algae, invertebrates and fish to AgNPs confirmed that different models of trophic levels and feeding strategies (zebrafish, daphnia and algae) showed differential susceptibilities to AgNPs, with filter-feeding invertebrates being markedly more susceptible compared to larger organisms (e.g., zebrafish) [173].

Whilst, today, the scientific community is in agreement that it might be premature to formulate any definitive risk assessment for silver-based materials, an increasing–and contradictory–public opinion is emerging to encourage the generation of experimental evidence on the possible impact of these nanomaterials on health [174, 175]. One central point to be considered is that silver nanocomposites should not be regarded as a uniform group of materials, as their toxicity may be influenced by the particle concentration, size distribution, agglomeration state, shape, the chemical and physical nature of the matrix, the physical status of the composite and, finally, the site and time of exposure. The ultimate – and main – goal in the field of silver nanotechnology remains the development and choice of products with a superior profile of functionality – such as high infection control – associated with a reduced host cell cytotoxicity and a moderate environmental risk, so as to fully exploit the potential benefits of this technology while limiting any unnecessary risks.

Currently, the rapid proliferation of many differently engineered nanomaterials presents a dilemma to regulators with regards to hazard identification. Recently, The International Life Sciences Institute Research Foundation/Risk Science Institute [176] convened an expert working group to develop a screening strategy for the hazard identification of engineered nanomaterials. Based on the evaluation of the limited data currently available, the report presents only a broad data-gathering strategy applicable to this very early stage in the development of a risk assessment process for nanomaterials. Oral, dermal and inhalation routes, as well as various injected routes of exposure, must be considered recognizing that, depending on the patterns of use, the exposure to nanomaterials may occur by any of these routes. In particular, three key elements of the toxicity screening strategy have been identified: (i) physico-chemical characteristics; (ii) *in vitro* assays (cellular and noncellular); and (iii) *in vivo* assays. It is common opinion that, in the near future, every new nanomaterial entering the market will need to be screened for toxicity and biopersistence, using low-cost, high-throughput – but scientifically rigorous and standardized – tests.

3.5 Perspectives

Novel potential applications of silver nanocomposites are continuously sought, especially in the field of biomedicine. Metal nanoparticles-based systems are today being considered as a means of minimally invasive diagnosis for the early detection of diseases, to facilitate targeted drug delivery, and to enhance the effectiveness of selected therapies [177, 178]. More specifically, by combining the scattering brightness and biocompatibility of silver nanocomposites, these systems could be employed as a new generation of contrast agents for the early diagnosis of cancer in humans [127].

Liong *et al.* [179] have recently studied multifunctional silica nanoparticles that incorporate iron oxide, gold or silver for applications in drug delivery, magnetic resonance and fluorescence imaging, magnetic manipulation, and cell targeting. These versatile multifunctional nanoparticles could potentially be used for simultaneous imaging and therapy towards cancer cells.

In a different approach, Balogh *et al.* [180] synthesized, by reactive encapsulation, fluorescent poly(amidoamine) dendrimer/silver nanocomposite particles for the labeling and selective destruction of cancer cells. To explore this biochemical targeting, B16 melanoma cancer cells and KB cells were incubated with the nanocomposite particles; subsequently, a laser-induced optical breakdown was able to selectively destroy targeted cells, without affecting the viability of other cells. By using this technique, a range of effects can be monitored by using simultaneous real-time acoustic and optical microscopy.

Huang *et al.* [57] suggested the use of single-nanoparticle biosensors to quantitatively analyze single-ligand and single-receptor molecules on single living cells for the effective characterization of anticancer vaccines, and for a deeper understanding of the biological functions of these materials.

The "tailorability" of AgNPs through surface engineering allows for a great variety of *in vivo* applications [126]; therefore, to take advantage of the various desirable properties of AgNPs, while protecting the body against any harmful side effects, is a worthwhile long-term goal. The surface modification of AgNPs appears to be one of the key factors for the design of nanocomposites that can fulfill these expectations.

References

1 Prasad, P.N. (2004) *Nanophotonics*, John Wiley & Sons, Inc.

2 Dai, J.H. and Bruening, M.L. (2002) Catalytic nanoparticles formed by reduction of metal ions in multilayered polyelectrolyte films. *Nano Letters*, **2** (5), 497–501.

3 Henglein, A. (1993) Physicochemical properties of small metal particles in solution: "microelectrode" reactions, chemisorption, composite metal particles, and the atom-to-metal transition. *Journal of Physical Chemistry*, **97** (21), 5457–71.

4 Lok, C.N., Ho, C.M., Chen, R., He, Q.Y., Yu, W.Y., Sun, H., Tam, P., Chiu, J.F. and Che, C.M. (2007) Silver nanoparticles: partial oxidation and antibacterial activities. *Journal of Biological Inorganic Chemistry*, **12** (4), 527–34.

5 Wang, Y.Z., Li, Y.X., Yang, S.T., Zhang, G.L., An, D.M., Wang, C., Yang, Q.B., Chen, X.S., Jing, X.B. and Wei, Y. (2006) A convenient route to polyvinyl pyrrolidone/silver nanocomposite by electrospinning. *Nanotechnology*, **17** (13), 3304–7.

6 Su, H.L., Chou, C.C., Hung, D.J., Lin, S.H., Pao, I.C., Lin, J.H., Huang, F.L., Dong, R.X. and Lin, J.J. (2009)The disruption of bacterial membrane integrity through ROS generation induced by nanohybrids of silver and clay. *Biomaterials*, **30** (30), 5979–87.

7 Mallick, K., Witcomb, M.J. and Scurrell, M.S. (2004) Polymer stabilized silver nanoparticles: a photochemical synthesis route. *Journal of Materials Science*, **39** (14), 4459–63.

8 Esumi, K., Suzuki, A., Aihara, N., Usui, K. and Torigoe, K. (1998) Preparation of gold colloids with UV irradiation using dendrimers as stabilizer. *Langmuir*, **14** (12), 3157–9.

9 Kuo, P.L. and Chen, W.F. (2003) Formation of silver nanoparticles under structured amino groups in pseudo-dendritic poly(allylamine) derivatives. *Journal of Physical Chemistry B*, **107** (41), 11267–72.

10 Grunlan, J.C., Choi, J.K. and Lin, A. (2005) Antimicrobial behavior of polyelectrolyte multilayer films containing cetrimide and silver. *Biomacromolecules*, **6** (2), 1149–53.

11 dos Santos, D.S., Goulet, P.J.G., Pieczonka, N.P.W., Oliveira, O.N. and Aroca, R.F. (2004) Gold nanoparticle embedded, self-sustained chitosan films

as substrates for surface-enhanced Raman scattering. *Langmuir*, **20** (23), 10273–7.

12 Huang, H., Yuan, Q. and Yang, X. (2004) Preparation and characterization of metal-chitosan nanocomposites. *Colloids and Surfaces B: Biointerfaces*, **39** (1–2), 31–7.

13 Yi, Y., Wang, Y. and Liu, H. (2003) Preparation of new crosslinked chitosan with crown ether and their adsorption for silver ion for antibacterial activities. *Carbohydrate Polymers*, **53** (4), 425–30.

14 Yu, H., Xu, X., Chen, X., Lu, T., Zhang, P. and Jing, X. (2006) Preparation and antibacterial effects of PVA-PVP hydrogels containing silver nanoparticles. *Journal of Applied Polymer Science*, **103**, 125–33.

15 Huang, H. and Yang, X. (2004) Synthesis of polysaccharide-stabilized gold and silver nanoparticles: a green method. *Carbohydrate Research*, **339** (15), 2627–31.

16 Fu, J., Ji, J., Fan, D. and Shen, J. (2006) Construction of antibacterial multilayer films containing nanosilver via layer-by-layer assembly of heparin and chitosan-silver ions complex. *Journal of Biomedical Materials Research Part A*, **79** (3), 665–74.

17 Sondi, I., Goia, D.V. and Matijevic, E. (2003) Preparation of highly concentrated stable dispersions of uniform silver nanoparticles. *Journal of Colloid and Interface Science*, **260** (1), 75–81.

18 Huang, H., Yuan, Q. and Yang, X. (2005) Morphology study of gold-chitosan nanocomposites. *Journal of Colloid and Interface Science*, **282** (1), 26–31.

19 Bonifacio, A., van der Sneppen, L., Gooijer, C. and van der Zwan, G. (2004) Citrate-reduced silver hydrosol modified with ω-mercaptoalkanoic acids self-assembled monolayers as a substrate for surface-enhanced resonance Raman scattering. A study with cytochrome c. *Langmuir*, **20** (14), 5858–64.

20 Pillai, Z.S. and Kamat, P.V. (2003) What factors control the size and shape of silver nanoparticles in the citrate ion reduction method? *Journal of Physical Chemistry B*, **108** (3), 945–51.

21 Liu, Z., Wang, X., Wu, H. and Li, C. (2005) Silver nanocomposite layer-by-layer films based on assembled polyelectrolyte/dendrimer. *Journal of Colloid and Interface Science*, **287** (2), 604–11.

22 Sun, Y., Mayers, B., Herricks, T. and Xia, Y. (2003) Polyol synthesis of uniform silver nanowires: a plausible growth mechanism and the supporting evidence. *Nano Letters*, **3**, 955–60.

23 Wiley, B., Herricks, T., Sun, Y. and Xia, Y. (2004) Polyol synthesis of silver nanoparticles: use of chloride and oxygen to promote the formation of single-crystal, truncated cubes and tetrahedrons. *Nano Letters*, **4** (9), 1733–9.

24 Yin, Y., Lu, Y., Sun, Y. and Xia, Y. (2002) Silver nanowires can be directly coated with amorphous silica to generate well-controlled coaxial nanocables of silver/silica. *Nano Letters*, **2** (4), 427–30.

25 Panacek, A., Kvitek, L., Prucek, R., Kolar, M., Vecerova, R., Pizurova, N., Sharma, V.K., Nevecna, T. and Zboril, R. (2006) Silver colloid nanoparticles: synthesis, characterization, and their antibacterial activity. *Journal of Physical Chemistry B*, **110** (33), 16248–53.

26 Donati, I., Travan, A., Pelillo, C., Scarpa, T., Coslovi, A., Bonifacio, A., Sergo, V. and Paoletti, S. (2009) Polyol synthesis of silver nanoparticles: mechanism of reduction by alditol bearing polysaccharides. *Biomacromolecules*, **10** (2), 210–13.

27 Travan, A., Pelillo, C., Donati, I., Marsich, E., Benincasa, M., Scarpa, T., Semeraro, S., Turco, G., Gennaro, R. and Paoletti, S. (2009) Non-cytotoxic silver nanoparticle-polysaccharide nanocomposites with antimicrobial activity. *Biomacromolecules*, **10** (6), 1429.

28 Choi, W.S., Koo, H.Y., Park, J.H. and Kim, D.Y. (2005) Synthesis of two types of nanoparticles in polyelectrolyte capsule nanoreactors and their dual functionality. *Journal of the American Chemical Society*, **127** (46), 16136–42.

29 Ho, C.H., Tobis, J., Sprich, C., Thomann, R. and Tiller, J.C. (2004) Nanoseparated polymeric network with multiple antimicrobial properties. *Advanced Materials*, **16** (12), 957–61.

30 Wiley, B., Sun, Y., Mayers, B. and Xia, Y. (2005) Shape-controlled synthesis of metal nanostructures: the case of silver. *Chemistry: A European Journal*, **11** (2), 454–63.

31 Balogh, L., Swanson, D.R., Tomalia, D.A., Hagnauer, G.L. and McManus, A.T. (2001) Dendrimer-silver complexes and nanocomposites as antimicrobial agents. *Nano Letters*, **1** (1), 18–21.

32 Rameshbabu, N., Sampath Kumar, T.S., Prabhakar, T.G., Sastry, V.S., Murty, K.V. and Prasad, R.K. (2007) Antibacterial nanosized silver substituted hydroxyapatite: synthesis and characterization. *Journal of Biomedical Materials Research Part A*, **80** (3), 581–91.

33 Krkljes, A.N., Marinovic-Cincovic, M.T., Kacarevic-Popovic, Z.M. and Nedeljkovic, J.M. (2007) Radiolytic synthesis and characterization of Ag-PVA nanocomposites. *European Polymer Journal*, **43** (6), 2171–6.

34 Mafune, F., Kohno, J.Y., Takeda, Y., Kondow, T. and Sawabe, H. (2000) Formation and size control of silver nanoparticles by laser ablation in aqueous solution. *Journal of Physical Chemistry B*, **104** (39), 9111–17.

35 Kundu, S., Mandal, M., Ghosh, S.K. and Pal, T. (2004) Photochemical deposition of SERS active silver nanoparticles on silica gel and their application as catalysts for the reduction of aromatic nitro compounds. *Journal of Colloid and Interface Science*, **272** (1), 134–44.

36 Zhu, J., Liu, S., Palchik, O., Koltypin, Y. and Gedanken, A. (2000) Shape-controlled synthesis of silver nanoparticles by pulse sonoelectrochemical methods. *Langmuir*, **16** (16), 6396–9.

37 Starowicz, M., Stypula, B. and Banas, J. (2006) Electrochemical synthesis of silver nanoparticles. *Electrochemistry Communications*, **8** (2), 227–30.

38 Yan, X.M., Ni, J., Robbins, M., Park, H.J., Zhao, W. and White, J.M. (2002) Silver nanoparticles synthesized by vapor deposition onto an ice matrix. *Journal of Nanoparticle Research*, **4** (6), 525–33.

39 Yin, H., Yamamoto, T., Wada, Y. and Yanagida, S. (2004) Large-scale and size-controlled synthesis of silver nanoparticles under microwave irradiation. *Materials Chemistry and Physics*, **83** (1), 66–70.

40 Li, K. and Zhang, F.S. (2009) A novel approach for preparing silver nanoparticles under electron beam irradiation. *Journal of Nanoparticle Research*, **12**, 1423–8 doi: 10.1007/s11051-009-9690-2.

41 Braun, E., Eichen, Y., Sivan, U. and Ben-Yoseph, G. (1998) DNA-templated assembly and electrode attachment of a conducting silver wire. *Nature*, **391** (6669), 775–8.

42 Lu, Y., Liu, G.L. and Lee, L.P. (2005) High-density silver nanoparticle film with temperature-controllable interparticle spacing for a tunable surface enhanced Raman scattering substrate. *Nano Letters*, **5** (1), 5–9.

43 Deshmukh, R.D. and Composto, R.J. (2007) Segregation and formation of silver nanoparticles created in situ in poly(methyl methacrylate) films. *Chemistry of Materials*, **19** (4), 745–54.

44 Henglein, A. (1998) Colloidal silver nanoparticles: photochemical preparation and interaction with O_2, CCl_4, and some metal ions. *Chemistry of Materials*, **10** (1), 444–50.

45 Henglein, A. and Giersig, M. (1999) Formation of colloidal silver nanoparticles: capping action of citrate. *Journal of Physical Chemistry B*, **103** (44), 9533–9.

46 Zhang, Y., Peng, H., Huang, W., Zhou, Y. and Yan, D. (2008) Facile preparation and characterization of highly antimicrobial colloid Ag or Au nanoparticles. *Journal of Colloid and Interface Science*, **325** (2), 371–6.

47 Wiley, B., Sun, Y. and Xia, Y. (2007) Synthesis of silver nanostructures with controlled shapes and properties. *Accounts of Chemical Research*, **40** (10), 1067–76.

48 Wang, T.C., Rubner, M.F. and Cohen, R.E. (2002) Polyelectrolyte multilayer nanoreactors for preparing silver nanoparticle composites: controlling metal concentration and nanoparticle size. *Langmuir*, **18** (8), 3370–5.

49 Dotzauer, D.M., Dai, J., Sun, L. and Bruening, M.L. (2006) Catalytic membranes prepared using layer-by-layer adsorption of polyelectrolyte/metal nanoparticle films in porous supports. *Nano Letters*, **6** (10), 2268–72.

50 Malikova, N., Pastoriza-Santos, I., Schierhorn, M., Kotov, N.A. and Liz-Marzan, L.M. (2002) Layer-by-layer assembled mixed spherical and planar gold nanoparticles: control of interparticle interactions. *Langmuir*, **18** (9), 3694–7.

51 Csaki, A., Garwe, F., Steinbruck, A., Maubach, G., Festag, G., Weise, A., Riemann, I., Konig, K. and Fritzsche, W. (2007) A parallel approach for subwavelength molecular surgery using gene-specific positioned metal nanoparticles as laser light antennas. *Nano Letters*, **7** (2), 247–53.

52 Zeng, R., Rong, M.Z., Zhang, M.Q., Liang, H.C. and Zeng, H.M. (2002) Laser ablation of polymer-based silver nanocomposites. *Applied Surface Science*, **187** (3-4), 239–47.

53 Stofik, M., Strýhal, Z. and Malý, J. (2009) Dendrimer-encapsulated silver nanoparticles as a novel electrochemical label for sensitive immunosensors. *Biosensors and Bioelectronics*, **24** (7), 1918–23.

54 Lee, K.J., Nallathamby, P.D., Browning, L.M., Osgood, C.J. and Xu, X.H.N. (2007) In vivo imaging of transport and biocompatibility of single silver nanoparticles in early development of zebrafish embryos. *ACS Nano*, **1** (2), 133–43.

55 Zhou, J., Yang, J., Sun, Y., Zhang, D., Shen, J., Zhang, Q. and Wang, K. (2007) Effect of silver nanoparticles on photo-induced reorientation of azo groups in polymer films. *Thin Solid Films*, **515** (18), 7242–6.

56 Murthy, P.S.K., Murali Mohan, Y., Varaprasad, K., Sreedhar, B. and Mohana Raju, K. (2008) First successful design of semi-IPN hydrogel-silver nanocomposites: a facile approach for antibacterial application. *Journal of Colloid and Interface Science*, **318** (2), 217–24.

57 Huang, T., Nallathamby, P.D., Gillet, D. and Xu, X.H.N. (2007) Design and synthesis of single-nanoparticle optical biosensors for imaging and characterization of single receptor molecules on single living cells. *Analytical Chemistry*, **79** (20), 7708–18.

58 Lesniak, W., Bielinska, A.U., Sun, K., Janczak, K.W., Shi, X., Baker, J.R. and Balogh, L.P. (2005) Silver/dendrimer nanocomposites as biomarkers: fabrication, characterization, in vitro toxicity, and intracellular detection. *Nano Letters*, **5** (11), 2123–30.

59 Rai, M., Yadav, A. and Gade, A. (2001) Silver nanoparticles as a new generation of antimicrobials. *Biotechnology Advances*, **27** (1), 76–83.

60 Grishchenko, L., Medvedeva, S., Aleksandrova, G., Feoktistova, L., Sapozhnikov, A., Sukhov, B. and Trofimov, B. (2006) Redox reactions of arabinogalactan with silver ions and formation of nanocomposites. *Russian Journal of General Chemistry*, **76** (7), 1111–16.

61 Chen, J.P. (2007) Late angiographic stent thrombosis (LAST): the cloud behind the drug-eluting stent silver lining? *Journal of Invasive Cardiology*, **19** (9), 395–400.

62 Sanpui, P., Murugadoss, A., Prasad, P.V.D., Ghosh, S.S. and Chattopadhyay, A. (2008) The antibacterial properties of a novel chitosan-Ag-nanoparticle composite. *International Journal of Food Microbiology*, **124** (2), 142–6.

63 Stevens, K.N.J., Crespo-Biel, O., van den Bosch, E.E.M., Dias, A.A., Knetsch, M.L.W., Aldenhoff, Y.B.J., van der Veen, F.H., Maessen, J.G., Stobberingh, E.E. and Koole, L.H. (2009) The relationship between the antimicrobial effect of catheter coatings containing silver nanoparticles and the coagulation of contacting blood. *Biomaterials*, **30** (22), 3682–90.

64 Lu, L., Sun, R.W., Chen, R., Hui, C.K., Ho, C.M., Luk, J.M., Lau, G.K. and Che, C.M. (2008) Silver nanoparticles inhibit hepatitis B virus replication. *Antiviral Therapy*, **13** (2), 253–62.

65 Kim, K.J., Sung, W.S., Suh, B.K., Moon, S.K., Choi, J.S., Kim, J.G. and Lee, D.G. (2009) Antifungal activity and mode of action of silver nano-particles on *Candida albicans*. *Biometals*, **22** (2), 235–42.

66 Esteban-Tejeda, L., Malpartida, F., Esteban-Cubillo, A., Pecharroman, C. and Moya, J.S. (2009) The antibacterial and antifungal activity of a soda-lime glass containing silver nanoparticles. *Nanotechnology*, **20** (8), 85103.

67 Gajbhiye, M.B., Kesharwani, J.G., Ingle, A.P., Gade, A.K. and Rai, M.K. (2009) Fungus-mediated synthesis of silver nanoparticles and their activity against pathogenic fungi in combination with fluconazole. *Nanomedicine*, **5**, 382–6.

68 Maki, D.G., Weise, C.E. and Sarafin, H.W. (1977) A semiquantitative culture method for identifying intravenous-catheter-related infection. *New England Journal of Medicine*, **296** (23), 1305–9.

69 Sondi, I. and Salopek-Sondi, B. (2004) Silver nanoparticles as antimicrobial agent: a case study on *E. coli* as a model for Gram-negative bacteria. *Journal of Colloid and Interface Science*, **275** (1), 177–82.

70 Morones, J.R., Elechiguerra, J.L., Camacho, A., Holt, K., Kouri, J.B., Ramirez, J.T. and Yacaman, M.J. (2005) The bactericidal effect of silver nanoparticles. *Nanotechnology*, **16** (10), 2346–53.

71 Pal, S., Tak, Y.K. and Song, J.M. (2007) Does the antibacterial activity of silver nanoparticles depend on the shape of the nanoparticle? A study of the gram-negative bacterium *Escherichia coli*. *Applied and Environmental Microbiology*, **73** (6), 1712–20.

72 Bragg, P.D. and Rainnie, D.J. (1974) The effect of silver ions on the respiratory chain of *Escherichia coli*. *Canadian Journal of Microbiology*, **20** (6), 883–9.

73 Feng, Q.L., Wu, J., Chen, G.Q., Cui, F.Z., Kim, T.N. and Kim, J.O. (2000) A mechanistic study of the antibacterial effect of silver ions on *Escherichia coli* and *Staphylococcus aureus*. *Journal of Biomedical Materials Research*, **52** (4), 662–8.

74 Furr, J.R., Russell, A.D., Turner, T.D. and Andrews, A. (1994) Antibacterial activity of Actisorb Plus, Actisorb and silver nitrate. *Journal of Hospital Infection*, **27** (3), 201–8.

75 Gupta, A., Matsui, K., Lo, J.F. and Silver, S. (1999) Molecular basis for resistance to silver cations in *Salmonella*. *Nature Medicine*, **5** (2), 183–8.

76 Clement, J.L. and Jarrett, P.S. (1994) Antibacterial silver. *Metal-Based Drugs*, **1** (5–6), 467–82.

77 Elechiguerra, J., Burt, J., Morones, J., Camacho-Bragado, A., Gao, X., Lara, H. and Yacaman, M. (2005) Interaction of silver nanoparticles with HIV-1. *Journal of Nanobiotechnology*, **3** (1), 6.

78 Nel, A. (2005) Air pollution-related illness: effects of particles. *Science*, **308** (5723), 804–6.

79 Semeykina, A.L. and Skulachev, V.P. Submicromolar Ag^+ increases passive Na^+ permeability and inhibits the respiration-supported formation of Na^+ gradient in *Bacillus* FTU vesicles. *FEBS Letters*, 1990, **269** (1), 69–72.

80 Hayashi, M., Miyoshi, T., Sato, M. and Unemoto, T. (1992) Properties of respiratory chain-linked Na(+)-independent NADH-quinone reductase in a marine *Vibrio alginolyticus*. *Biochimica et Biophysica Acta*, **1099** (2), 145–51.

81 Dibrov, P., Dzioba, J., Gosink, K.K. and Hase, C.C. (2002) Chemiosmotic mechanism of antimicrobial activity of Ag(+) in *Vibrio cholerae*. *Antimicrobial Agents and Chemotherapy*, **46** (8), 2668–70.

82 Yamanaka, M., Hara, K. and Kudo, J. (2005) Bactericidal actions of a silver ion solution on *Escherichia coli*, studied by energy-filtering transmission electron microscopy and proteomic analysis. *Applied and Environmental Microbiology*, **71** (11), 7589–93.

83 Castellano, J.J., Shafii, S.M., Ko, F., Donate, G., Wright, T.E., Mannari, R.J., Payne, W.G., Smith, D.J. and Robson, M.C. (2007) Comparative evaluation of silver-containing antimicrobial dressings and drugs. *International Wound Journal*, **4** (2), 114–22.

84 Sondi, I. and Salopek-Sondi, B. (2004) Silver nanoparticles as antimicrobial agent: a case study on *E. coli* as a model for Gram-negative bacteria. *Journal of Colloid and Interface Science*, **275** (1), 177–82.

85 Lok, C.N., Ho, C.M., Chen, R., He, Q.Y., Yu, W.Y., Sun, H., Tam, P.K.H., Chiu,

J.F. and Che, C.M. (2006) Proteomic analysis of the mode of antibacterial action of silver nanoparticles. *Journal of Proteome Research*, **5** (4), 916–24.

86 Kim, J.S., Kuk, E., Yu, K.N., Kim, J.H., Park, S.J., Lee, H.J., Kim, S.H., Park, Y.K., Park, Y.H., Hwang, C.Y., Kim, Y.K., Lee, Y.S., Jeong, D.H. and Cho, M.H. (2007) Antimicrobial effects of silver nanoparticles. *Nanomedicine: Nanotechnology, Biology and Medicine*, **3** (1), 95–101.

87 Choi, O. and Hu, Z. (2008) Size dependent and reactive oxygen species related nanosilver toxicity to nitrifying bacteria. *Environmental Science and Technology*, **42** (12), 4583–8.

88 Pal, S., Tak, Y.K., Joardar, J., Kim, W., Lee, J.E., Han, M.S. and Song, J.M. (2009) Nanocrystalline silver supported on activated carbon matrix from hydrosol: antibacterial mechanism under prolonged incubation conditions. *Journal of Nanoscience and Nanotechnology*, **9** (3), 2092–103.

89 Ayello, E.A. and Cuddigan, J.E. (2004) Conquer chronic wounds with wound bed preparation. *The Nurse Practitioner*, **29** (3), 8–25.

90 Tomaselli, N. (2006) The role of topical silver preparations in wound healing. *Journal of Wound, Ostomy, and Continence Nursing*, **33** (4), 367–78.

91 Gray, M. and Ratliff, C.R. (2006) Is hyperbaric oxygen therapy effective for the management of chronic wounds? *Journal of Wound, Ostomy, and Continence Nursing*, **33** (1), 21–5.

92 Bowler, P.G. (2002) Wound pathophysiology, infection and therapeutic options. *Annals of Medicine*, **34** (6), 419–27.

93 Ip, M., Lui, S.L., Poon, V.K., Lung, I. and Burd, A. (2006) Antimicrobial activities of silver dressings: an in vitro comparison. *Journal of Medical Microbiology*, **55** (Pt 1), 59–63.

94 Ehrlich, H.P. (1998) The physiology of wound healing. A summary of normal and abnormal wound healing processes. *Advances in Wound Care*, **11** (7), 326–8.

95 Tian, J., Wong, K.K., Ho, C.M., Lok, C.N., Yu, W.Y., Che, C.M., Chiu, J.F. and Tam, P.K. (2007) Topical delivery of silver nanoparticles promotes wound healing. *ChemMedChem*, **2** (1), 129–36.

96 Olson, M.E., Wright, J.B., Lam, K. and Burrell, R.E. (2000) Healing of porcine donor sites covered with silver-coated dressings. *European Journal of Surgery*, **166** (6), 486–9.

97 Wiegand, C., Heinze, T. and Hipler, U.C. (2009) Comparative in vitro study on cytotoxicity, antimicrobial activity, and binding capacity for pathophysiological factors in chronic wounds of alginate and silver-containing alginate. *Wound Repair and Regeneration*, **17** (4), 511–21.

98 Sibbald, R.G., Contreras-Ruiz, J., Coutts, P., Fierheller, M., Rothman, A. and Woo, K. (2007) Bacteriology, inflammation, and healing: a study of nanocrystalline silver dressings in chronic venous leg ulcers. *Advances in Skin and Wound Care*, **20** (10), 549–58.

99 Paquet, P. and Pierard, G.E. (1996) Invasive atypical fibroxanthoma and eruptive actinic keratoses in a heart transplant patient. *Dermatology*, **192** (4), 411–13.

100 Biswas, P., Delfanti, F., Bernasconi, S., Mengozzi, M., Cota, M., Polentarutti, N., Mantovani, A., Lazzarin, A., Sozzani, S. and Poli, G. (1998) Interleukin-6 induces monocyte chemotactic protein-1 in peripheral blood mononuclear cells and in the U937 cell line. *Blood*, **91** (1), 258–65.

101 Wright, J.B., Lam, K. and Burrell, R.E. (1998) Wound management in an era of increasing bacterial antibiotic resistance: a role for topical silver treatment. *American Journal of Infection Control*, **26** (6), 572–7.

102 Wright, J.B., Lam, K., Buret, A.G., Olson, M.E. and Burrell, R.E. (2002) Early healing events in a porcine model of contaminated wounds: effects of nanocrystalline silver on matrix metalloproteinases, cell apoptosis, and healing. *Wound Repair and Regeneration*, **10** (3), 141–51.

103 Vachon, D.J. and Yager, D.R. (2006) Novel sulfonated hydrogel composite with the ability to inhibit proteases and bacterial growth. *Journal of Biomedical Materials Research Part A*, **76** (1), 35–43.

104 Walker, M., Bowler, P.G. and Cochrane, C.A. (2007) In vitro studies to show sequestration of matrix metalloproteinases by silver-containing wound care products. *Ostomy/Wound Management*, **53** (9), 18–25.

105 Rogers, A.A., Burnett, S., Moore, J.C., Shakespeare, P.G. and Chen, W.Y. (1995) Involvement of proteolytic enzymes – plasminogen activators and matrix metalloproteinases – in the pathophysiology of pressure ulcers. *Wound Repair and Regeneration*, **3** (3), 273–83.

106 Kahari, V.M. and Saarialho-Kere, U. (1999) Matrix metalloproteinases and their inhibitors in tumour growth and invasion. *Annals of Medicine*, **31** (1), 34–45.

107 Bhol, K.C., Alroy, J. and Schechter, P.J. (2004) Anti-inflammatory effect of topical nanocrystalline silver cream on allergic contact dermatitis in a guinea pig model. *Clinical and Experimental Dermatology*, **29** (3), 282–7.

108 Bhol, K.C. and Schechter, P.J. (2005) Topical nanocrystalline silver cream suppresses inflammatory cytokines and induces apoptosis of inflammatory cells in a murine model of allergic contact dermatitis. *British Journal of Dermatology*, **152** (6), 1235–42.

109 Nadworny, P.L., Wang, J., Tredget, E.E. and Burrell, R.E. (2008) Anti-inflammatory activity of nanocrystalline silver in a porcine contact dermatitis model. *Nanomedicine*, **4** (3), 241–51.

110 Suska, F., Svensson, S., Johansson, A., Emanuelsson, L., Karlholm, H., Ohrlander, M. and Thomsen, P. (2009) In vivo evaluation of noble metal coatings. *Journal of Biomedical Materials Research Part B: Applied Biomaterials*, **92**, 86–94.

111 Wong, K.K., Cheung, S.O., Huang, L., Niu, J., Tao, C., Ho, C.M., Che, C.M. and Tam, P.K. (2009) Further evidence of the anti-inflammatory effects of silver nanoparticles. *ChemMedChem*, **4** (7), 1129–35.

112 Bhol, K.C. and Schechter, P.J. (2007) Effects of nanocrystalline silver (NPI 32101) in a rat model of ulcerative colitis. *Digestive Diseases and Sciences*, **52** (10), 2732–42.

113 Squier, M.K., Sehnert, A.J. and Cohen, J.J. (1995) Apoptosis in leukocytes. *Journal of Leukocyte Biology*, **57** (1), 2–10.

114 Savill, J., Dransfield, I., Hogg, N. and Haslett, C. (1990) Vitronectin receptor-mediated phagocytosis of cells undergoing apoptosis. *Nature*, **343** (6254), 170–3.

115 Henson, P.M. and Johnston, R.B. (1987) Tissue injury in inflammation. Oxidants, proteinases, and cationic proteins. *Journal of Clinical Investigation*, **79** (3), 669–74.

116 Anderson, B.O., Brown, J.M. and Harken, A.H. (1991) Mechanisms of neutrophil-mediated tissue injury. *Journal of Surgical Research*, **51** (2), 170–9.

117 Dive, C., Gregory, C.D., Phipps, D.J., Evans, D.L., Milner, A.E. and Wyllie, A.H. (1992) Analysis and discrimination of necrosis and apoptosis (programmed cell death) by multiparameter flow cytometry. *Biochimica et Biophysica Acta*, **1133** (3), 275–85.

118 Squier, M.K., Sehnert, A.J. and Cohen, J.J. (1995) Apoptosis in leukocytes. *Journal of Leukocyte Biology*, **57** (1), 2–10.

119 Tautenhahn, J., Meyer, F., Buerger, T., Schmidt, U., Lippert, H., Koenig, W. and Koenig, B. (2008) Interactions of neutrophils with silver-coated vascular polyester grafts. *Langenbeck's Archives of Surgery*, **395**, 143–9.

120 Marchi, E., Vargas, F.S., Teixeira, L.R., Acencio, M.M., Antonangelo, L. and Light, R.W. (2005) Intrapleural low-dose silver nitrate elicits more pleural inflammation and less systemic inflammation than low-dose talc. *Chest*, **128** (3), 1798–804.

121 Gumpel, J.M. (1973) Radioactive colloids in the treatment of arthritis. Review of published and personal results. Criteria for selection of patients. *Annals of the Rheumatic Diseases*, **32** (Suppl. 6), 33.

122 Visuthikosol, V. and Kumpolpunth, S. (1981) Intra-articular radioactive colloidal gold (198 Au) in the treatment of rheumatoid arthritis. *Journal of the Medical Association of Thailand*, **64** (9), 419–27.

123 Eisler, R. (2003) Chrysotherapy: a synoptic review. *Inflammation Research*, **52** (12), 487–501.

124 Jessop, J.D. (1979) Gold in the treatment of rheumatoid arthritis–why, when and how? *Journal of Rheumatology Suppl.*, **5**, 12–17.

125 Khlebtsov, N.G. and Dykman, L.A. (2010) Optical properties and biomedical applications of plasmonic nanoparticles. *Journal of Quantitative Spectroscopy and Radiative Transfer*, **111**, 1–35.

126 Schrand, A.M., Braydich-Stolle, L.K., Schlager, J.J., Dai, L. and Hussain, S.M. (2008) Can silver nanoparticles be useful as potential biological labels? *Nanotechnology*, **19** (23), 235104.

127 Hu, R., Yong, K.T., Roy, I., Ding, H., He, S. and Prasad, P.N. (2009) Metallic nanostructures as localized plasmon resonance enhanced scattering probes for multiplex dark-field targeted imaging of cancer cells. *Journal of Physical Chemistry C*, **113** (7), 2676–84.

128 Kyriacou, S.V., Brownlow, W.J. and Xu, X.H.N. (2004) Using nanoparticle optics assay for direct observation of the function of antimicrobial agents in single live bacterial cells. *Biochemistry*, **43** (1), 140–7.

129 Qian, X.M. and Nie, S.M. (2008) Single-molecule and single-nanoparticle SERS: from fundamental mechanisms to biomedical applications. *Chemical Society Reviews*, **37** (5), 912–20.

130 Jia, X., Qian, W., Wu, D., Wei, D., Xu, G. and Liu, X. (2009) Cuttlebone-derived organic matrix as a scaffold for assembly of silver nanoparticles and application of the composite films in surface-enhanced Raman scattering. *Colloids and Surfaces B: Biointerfaces*, **68** (2), 231–7.

131 Stoddart, P.R., Cadusch, P.J., Boyce, T.M., Erasmus, R.M. and Comins, J.D. (2006) Optical properties of chitin: surface-enhanced Raman scattering substrates based on antireflection structures on cicada wings. *Nanotechnology*, **17** (3), 680–6.

132 Lansdown, A.B. (2006) Silver in health care: antimicrobial effects and safety in use. *Current Problems in Dermatology*, **33**, 17–34.

133 Larese, F.F., D'Agostin, F., Crosera, M., Adami, G., Renzi, N., Bovenzi, M. and Maina, G. (2009) Human skin penetration of silver nanoparticles through intact and damaged skin. *Toxicology*, **255** (1–2), 33–7.

134 Um, H.D., Orenstein, J.M. and Wahl, S.M. (1996) Fas mediates apoptosis in human monocytes by a reactive oxygen intermediate dependent pathway. *Journal of Immunology*, **156** (9), 3469–77.

135 Talley, A.K., Dewhurst, S., Perry, S.W., Dollard, S.C., Gummuluru, S., Fine, S.M., New, D., Epstein, L.G., Gendelman, H.E. and Gelbard, H.A. (1995) Tumor necrosis factor alpha-induced apoptosis in human neuronal cells: protection by the antioxidant N-acetylcysteine and the genes bcl-2 and crmA. *Molecular and Cellular Biology*, **15** (5), 2359–66.

136 Zamzami, N., Marchetti, P., Castedo, M., Decaudin, D., Macho, A., Hirsch, T., Susin, S.A., Petit, P.X., Mignotte, B. and Kroemer, G. (1995) Sequential reduction of mitochondrial transmembrane potential and generation of reactive oxygen species in early programmed cell death. *Journal of Experimental Medicine*, **182** (2), 367–77.

137 Lin, M.T. and Beal, M.F. (2006) Mitochondrial dysfunction and oxidative stress in neurodegenerative diseases. *Nature*, **443** (7113), 787–95.

138 Ott, M., Gogvadze, V., Orrenius, S. and Zhivotovsky, B. (2007) Mitochondria, oxidative stress and cell death. *Apoptosis*, **12** (5), 913–22.

139 Braydich-Stolle, L., Hussain, S., Schlager, J.J. and Hofmann, M.C. (2005) In vitro cytotoxicity of nanoparticles in mammalian germline stem cells. *Toxicological Sciences*, **88** (2), 412–19.

140 Carlson, C., Hussain, S.M., Schrand, A.M., Braydich-Stolle, L.K., Hess, K.L., Jones, R.L. and Schlager, J.J. (2008) Unique cellular interaction of silver nanoparticles: size-dependent generation of reactive oxygen species. *Journal of Physical Chemistry B*, **112** (43), 13608–19.

141 Foldbjerg, R., Olesen, P., Hougaard, M., Dang, D.A., Hoffmann, H.J. and Autrup, H. (2009) PVP-coated silver nanoparticles and silver ions induce reactive oxygen species, apoptosis and necrosis in THP-1 monocytes. *Toxicology Letters*, **190**, 156–62.

142 Hsin, Y.H., Chen, C.F., Huang, S., Shih, T.S., Lai, P.S. and Chueh, P.J. (2008) The apoptotic effect of nanosilver is mediated by a ROS- and JNK-dependent mechanism involving the mitochondrial pathway in NIH3T3 cells. *Toxicology Letters*, **179** (3), 130–9.

143 Hussain, S.M., Hess, K.L., Gearhart, J.M., Geiss, K.T. and Schlager, J.J. (2005) In vitro toxicity of nanoparticles in BRL 3A rat liver cells. *Toxicology In Vitro*, **19** (7), 975–83.

144 Kim, S., Choi, J.E., Choi, J., Chung, K.H., Park, K., Yi, J. and Ryu, D.Y. (2009) Oxidative stress-dependent toxicity of silver nanoparticles in human hepatoma cells. *Toxicology In Vitro*, **23** (6), 1076–84.

145 Singh, N., Manshian, B., Jenkins, G.J.S., Griffiths, S.M., Williams, P.M., Maffeis, T.G.G., Wright, C.J. and Doak, S.H. (2009) NanoGenotoxicology: the DNA damaging potential of engineered nanomaterials. *Biomaterials*, **30** (23–24), 3891–914.

146 Arora, S., Jain, J., Rajwade, J.M. and Paknikar, K.M. (2008) Cellular responses induced by silver nanoparticles: in vitro studies. *Toxicology Letters*, **179** (2), 93–100.

147 Van Den, P.D., De, S.K., Lens, D. and Sollie, P. (2008) Differential cell death programmes induced by silver dressings in vitro. *European Journal of Dermatology*, **18** (4), 416–21.

148 Burd, A., Kwok, C.H., Hung, S.C., Chan, H.S., Gu, H., Lam, W.K. and Huang, L. (2007) A comparative study of the cytotoxicity of silver-based dressings in monolayer cell, tissue explant, and animal models. *Wound Repair and Regeneration*, **15** (1), 94–104.

149 Sionov, R.V. and Haupt, Y. (1999) The cellular response to p53: the decision between life and death. *Oncogene*, **18** (45), 6145–57.

150 Polster, B.M. and Fiskum, G. (2004) Mitochondrial mechanisms of neural cell apoptosis. *Journal of Neurochemistry*, **90** (6), 1281–9.

151 Mihara, M., Erster, S., Zaika, A., Petrenko, O., Chittenden, T., Pancoska, P. and Moll, U.M. (2003) p53 has a direct apoptogenic role at the mitochondria. *Molecular Cell*, **11** (3), 577–90.

152 Ahamed, M., Karns, M., Goodson, M., Rowe, J., Hussain, S.M., Schlager, J.J. and Hong, Y. (2008) DNA damage response to different surface chemistry of silver nanoparticles in mammalian cells. *Toxicology and Applied Pharmacology*, **233** (3), 404–10.

153 Djordjevic, V.B. (2004) Free radicals in cell biology. *International Review of Cytology*, **237**, 57–89.

154 Gilca, M., Stoian, I., Atanasiu, V. and Virgolici, B. (2007) The oxidative hypothesis of senescence. *Journal of Postgraduate Medicine*, **53** (3), 207–13.

155 Simon, H.U., Haj-Yehia, A. and Levi-Schaffer, F. (2000) Role of reactive oxygen species (ROS) in apoptosis induction. *Apoptosis*, **5** (5), 415–18.

156 Cha, K., Hong, H.W., Choi, Y.G., Lee, M.J., Park, J.H., Chae, H.K., Ryu, G. and Myung, H. (2008) Comparison of acute responses of mice livers to short-term exposure to nano-sized or micro-sized silver particles. *Biotechnology Letters*, **30** (11), 1893–9.

157 Lubick, N. (2008) Nanosilver toxicity: ions, nanoparticles – or both? *Environmental Science and Technology*, **42** (23), 8617.

158 Paz-Elizur, T., Sevilya, Z., Leitner-Dagan, Y., Elinger, D., Roisman, L.C. and Livneh, Z. (2008) DNA repair of oxidative DNA damage in human carcinogenesis: potential application for cancer risk assessment and prevention. *Cancer Letters*, **266** (1), 60–72.

159 Chen, M. and von Mikecz, M.A. (2005) Formation of nucleoplasmic protein aggregates impairs nuclear function in response to SiO_2 nanoparticles. *Experimental Cell Research*, **305** (1), 51–62.

160 Geiser, M., Rothen-Rutishauser, B., Kapp, N., Schurch, S., Kreyling, W., Schulz, H., Semmler, M., Im, H.V., Heyder, J. and Gehr, P. (2005) Ultrafine particles cross cellular membranes by nonphagocytic mechanisms in lungs and in cultured cells. *Environmental Health Perspectives*, **113** (11), 1555–60.

161 Nabiev, I., Mitchell, S., Davies, A., Williams, Y., Kelleher, D., Moore, R.,

Gun'ko, Y.K., Byrne, S., Rakovich, Y.P., Donegan, J.F., Sukhanova, A., Conroy, J., Cottell, D., Gaponik, N., Rogach, A. and Volkov, Y. (2007) Nonfunctionalized nanocrystals can exploit a cell's active transport machinery delivering them to specific nuclear and cytoplasmic compartments. *Nano Letters*, **7** (11), 3452–61.

162 Yang, W., Shen, C., Ji, Q., An, H., Wang, J., Liu, Q. and Zhang, Z. (2009) Food storage material silver nanoparticles interfere with DNA replication fidelity and bind with DNA. *Nanotechnology*, **20** (8), 85102.

163 Bar-Ilan, O., Albrecht, R.M., Fako, V.E. and Furgeson, D.Y. (2009) Toxicity assessments of multisized gold and silver nanoparticles in zebrafish embryos. *Small*, **5** (16), 1897–910.

164 Rahman, M.F., Wang, J., Patterson, T.A., Saini, U.T., Robinson, B.L., Newport, G.D., Murdock, R.C., Schlager, J.J., Hussain, S.M. and Ali, S.F. (2009) Expression of genes related to oxidative stress in the mouse brain after exposure to silver-25 nanoparticles. *Toxicology Letters*, **187** (1), 15–21.

165 Chen, D., Xi, T. and Bai, J. (2007) Biological effects induced by nanosilver particles: in vivo study. *Biomedical Materials*, **2** (3), S126–8.

166 Alt, V., Bechert, T., Steinrucke, P., Wagener, M., Seidel, P., Dingeldein, E., Domann, E. and Schnettler, R. (2004) An in vitro assessment of the antibacterial properties and cytotoxicity of nanoparticulate silver bone cement9. *Biomaterials*, **25** (18), 4383–91.

167 Wen, H.C., Lin, Y.N., Jian, S.R., Tseng, S.C., Weng, M.X., Liu, Y.P., Lee, P.T., Chen, P.Y., Hsu, R.Q., Wu, W.F. and Chou, C.P. (2007) Observation of growth of human fibroblasts on silver nanoparticles. *Journal of Physics: Conference Series*, **61**, 445–9.

168 Bosetti, M., Masse, A., Tobin, E. and Cannas, M. (2002) Silver coated materials for external fixation devices: in vitro biocompatibility and genotoxicity. *Biomaterials*, **23** (3), 887–92.

169 Bianchini, A., Bowles, K.C., Brauner, C.J., Gorsuch, J.W., Kramer, J.R. and Wood, C.M. (2002) Evaluation of the effect of reactive sulfide on the acute toxicity of silver(I) to *Daphnia magna*. Part 2: toxicity results. *Environmental Toxicology and Chemistry*, **21** (6), 1294–300.

170 Wood, C.M., McDonald, M.D., Walker, P., Grosell, M., Barimo, J.F., Playle, R.C. and Walsh, P.J. (2004) Bioavailability of silver and its relationship to ionoregulation and silver speciation across a range of salinities in the gulf toadfish (*Opsanus beta*). *Aquatic Toxicology*, **70** (2), 137–57.

171 Wang, W.X. (2001) Comparison of metal uptake rate and absorption efficiency in marine bivalves. *Environmental Toxicology and Chemistry*, **20** (6), 1367–73.

172 Choi, O.K. and Hu, Z.Q. (2009) Nitrification inhibition by silver nanoparticles. *Water Science and Technology*, **59** (9), 1699–702.

173 Griffitt, R.J., Luo, J., Gao, J., Bonzongo, J.C. and Barber, D.S. (2008) Effects of particle composition and species on toxicity of metallic nanomaterials in aquatic organisms. *Environmental Toxicology and Chemistry*, **27** (9), 1972–8.

174 Colvin, V.L. (2003) The potential environmental impact of engineered nanomaterials. *Nature Biotechnology*, **21** (10), 1166–70.

175 Hoet, P.H., Nemmar, A. and Nemery, B. (2004) Health impact of nanomaterials? *Nature Biotechnology*, **22** (1), 19.

176 Oberdorster, G., Maynard, A., Donaldson, K., Castranova, V., Fitzpatrick, J., Ausman, K., Carter, J., Karn, B., Kreyling, W., Lai, D., Olin, S., Monteiro-Riviere, N., Warheit, D. and Yang, H. (2005) Principles for characterizing the potential human health effects from exposure to nanomaterials: elements of a screening strategy. *Particle and Fibre Toxicology*, **2**, 8.

177 Wieder, M.E., Hone, D.C., Cook, M.J., Handsley, M.M., Gavrilovic, J. and Russell, D.A. Intracellular photodynamic therapy with photosensitizer-nanoparticle conjugates: cancer therapy using a "Trojan horse". *Photochemical and Photobiological Sciences*, **5** (8), 727–34.

178 Zharov, V.P., Mercer, K.E., Galitovskaya, E.N. and Smeltzer, M.S. (2005) Photothermal nanotherapeutics and nanodiagnostics for selective killing of bacteria targeted with gold nanoparticles. *Biophysical Journal*, **90**, 619–27.

179 Liong, M., Lu, J., Kovochich, M., Xia, T., Ruehm, S.G., Nel, A.E., Tamanoi, F. and Zink, J.I. (2008) Multifunctional inorganic nanoparticles for imaging, targeting, and drug delivery. *ACS Nano*, **2** (5), 889–96.

180 Balogh, L.P., Tse, C., Lesniak, W., Ye, J., O'Donnell, M. and Khan, M.K. (2007) Photomechanical therapy: destruction of nanocomposite labeled cells by laser induced optical breakdown. *Nanomedicine: Nanotechnology, Biology and Medicine*, **3** (4), 350.

181 AshaRani, P.V., Low Kah, M.G., Hande, M.P. and Valiyaveettil, S. (2009) Cytotoxicity and genotoxicity of silver nanoparticles in human cells. *ACS Nano*, **3** (2), 279–90.

4
Gold Nanocomposite Biosensors

María Isabel Pividori and Salvador Alegret

4.1
Introduction

During recent years, nanostructured materials including gold nanoparticles have undergone intensive investigations on the basis of their unique size effects and enhanced chemical and physical properties. Nanocomposites can be fabricated not only with different nanostructured materials, but also with biomolecules and polymers such that they possess unique hybrid and synergistic properties.

Among the different nanostructured materials, the ability of gold nanoparticles to provide a suitable environment for the immobilization of biomolecules, while retaining biological activity, has led to their intensive use in the construction of biosensors with improved performance. Moreover, electron transfer that is easily promoted by gold nanoparticles towards conducting polymers, carbon nanotubes (CNTs) and redox biomolecules represents a promising feature for the construction of a broad range of novel nanocomposite materials to be used as transducers in electrochemical biosensors.

In this chapter, the main features of nanostructured gold composites as transducer materials for electrochemical biosensing are described, and details provided of the synthesis, main properties and modification of gold nanoparticles, as the basis of gold nanocomposites. The chemistry of biomolecule immobilization onto gold-based transducers is also outlined, and novel approaches based on rigid carbon–polymer composite materials modified with gold nanoparticles for the improved electrochemical biosensing of DNA, are discussed.

4.2
Electrochemical Biosensing and Transducing Features

Biosensing offers an exciting alternative to the more traditional sensing methods, allowing rapid "real-time" and decentralized multiple analyses that are essential

Nanomaterials for the Life Sciences Vol.8: Nanocomposites. Edited by Challa S. S. R. Kumar

ISBN: 978-3-527-32168-1

for many areas such as clinical testing, medical research and diagnosis, environmental monitoring, biotechnology, forensic investigations, and food safety.

Biosensors employ the combination of a biological receptor (mainly antibodies, enzymes, nucleic acids, whole cells) and a physical or physico-chemical transducer so as to produce, in most cases, "real-time" observations of a specific biological event (e.g., antibody–antigen interaction) and allow the detection of a broad spectrum of analytes in complex sample matrices [1, 2].

Biosensors are classified into four basic groups – optical, mass, electrochemical, and thermal – depending on the signal transduction method employed [3]. Optical transducers are particularly attractive as they can allow a direct "label-free" and "real-time" detection of biomolecules, although they lack sensitivity. The phenomenon of surface plasmon resonance (SPR) has demonstrated a good biosensing potential, and many commercial SPR systems are now available (e.g., BIAcore™). Electrochemical-based transduction devices are more robust, easy to use, portable, and inexpensive analytical systems [4]. Electrochemical biosensors can also function in turbid media, while offering an enhanced sensitivity.

The development of new transducing materials is a key issue in the current research strategy for electrochemical biosensors. Whilst the immobilization of the bioreceptor and detection of the biological event are important features, the choice of a suitable electrochemical transducer is also of major importance in the overall performance of electrochemical-based biosensors. *Carbonaceous materials* such as carbon paste [5], glassy carbon [6] and pyrolitic graphite [7] represent the most popular choices of electrode used for biosensing devices. However, the use of platinum, gold [8], indium-tin oxide [9], copper solid amalgam [10], mercury [11], and other continuous-conducting metal substrates has also been reported [12]. Both, *conducting polymers* (such as polypyrrole and polyaniline [13]) and *conducting composites* based on a combination of a nonconducting polymer with a conductive filler [4], have undergone continuous investigation during the past few decades. Finally, nanostructured materials such as CNTs [14] and metal nanoparticles [15] have also been reported as base materials or fillers for conducting composites, or as surface-modifiers for many types of electrochemical transducer, in order to improve their electrochemical properties. Other nanostructured materials – including gold nanoparticles – have also been investigated as possible components of electrochemical transducers [16]. *Nanocomposites* may be fabricated not only with nanostructured materials, but also with biomolecules and redox polymers, such that they possess unique hybrid and synergistic properties. It is to be expected that a combination of nanoengineered "smart" polymers with novel biocompatible nanostructured fillers – such as nanoparticles and CNTs – may lead to the creation of composites with new and interesting properties. This would, in turn, provide a higher sensitivity and stability of the immobilized molecules, and constitute the basis for improved electrochemical biosensing [13].

The immobilization of a biorecognition element (which specifically recognizes a target) onto the transducer is also a key issue when constructing biosensing devices. The choice of immobilization method depends mainly on the biomolecule to be immobilized, the nature of the solid surface, and the transducing mechanism [17]. Besides sensitivity, the ability of an electrochemical transducer to provide a stable immobilization environment, while retaining bioactivity, must be also considered. Indeed, one current problem in the immobilization of biomolecules is the resulting lack of stability and activity in the solid transducer, which is usually overwhelmed by the use either of an *in vivo*-like environment or spacer arms.

Among possible immobilization strategies, the two most successful are:

- *Multisite attachment*, which may be effected either by electrochemical adsorption (via by the application of a potential to the solid support) or by physical adsorption.
- *Single-point attachment*, which occurs mainly via covalent immobilization, affinity linkage (such as streptavidin–biotin binding), and chemisorption based on self-assembled monolayers (SAMs) [12].

Multisite adsorption is the simplest and most easily automated procedure, as it avoids the use of any pretreatment procedures based on the previous activation/modification of the surface transducer and subsequent immobilization. Such pretreatment steps are known to be tedious, expensive, and time-consuming. The binding forces involving physical adsorption include hydrogen bonding, electrostatic interaction, van der Waals forces, and hydrophobic interactions (if water molecules are excluded by dryness) [19]. *Wet adsorption* creates a weak binding that leads to an easy desorption of the biomolecule from the surface, with eventual leaching to the sample solution during measurements. In contrast, *dry adsorption* promotes hydrophobic bonding and more stable adsorbed layers on solid surfaces [19]. Classical strategies such as physical entrapment in membranes and crosslinking by bifunctional reagents (e.g., glutaraldehyde) also represent multisite attachment methods to retain the bioreceptor in close contact with the transducer.

Single-point attachment is beneficial for the kinetics of the biological reaction, especially if a spacer arm is used. Single-point *covalent immobilization* can be performed on different surface-modified electrochemical transducers, such as glassy carbon [20], carbon paste [21], gold [22] or platinum [23], and more recently CNTs [24], through the linkage of a –COOH with a $-NH_2$ group by the use of the carbodiimide chemistry. Single-point *affinity linkage* also provides an interesting strategy for the oriented and stable immobilization of biotinylated biomolecules to solid transducers, through biotin/streptavidin binding [18]. Finally, chemisorption based on *SAMs* has been used extensively for single-point attachment on gold-based transducers.

The different strategies used to attach biomolecules (i.e., enzymes, antigens, antibodies, DNA, receptors) onto gold-based transducers (such as continuous gold transducers, gold nanoparticles, or gold nanocomposites) are described in the following section.

4.3 Modification Strategies of Gold Surfaces and Gold Nanoparticles with Biomolecules

4.3.1 Physisorption

Gold has been a widely used as a support for the multisite attachment of biomolecules through weak bonds – electrostatic, van der Waals, hydrogen or hydrophobic interactions. This nonspecific, noncovalent, multisite attachment is often referred to as physical adsorption or *physisorption*, and may be performed with nano-, micro-, or macroscopic particles, as well as with gold surfaces and gold-based composites and nanocomposites.

4.3.1.1 Physisorption of Proteins

It is well known that proteins may be adsorbed physically to the vast majority of inorganic or organic surfaces, including gold continuous surfaces and nanoparticles [25]. The main advantage of the direct adsorption of proteins onto solid supports is its simplicity and applicability to almost any type of macromolecule and solid support. Such adsorption has been used widely in the modification of colloidal gold (from 1 to 20 nm in diameter) with antibodies or other types of biological molecule, for the localization of cellular antigens, in thin sections, or in homogenized suspensions of cells or tissues, all of which have been monitored using the classical technique of transmission electron microscopy (TEM) [26, 27].

The formation and enzymatic activity of gold nanoparticles complexed with redox enzymes such as horseradish peroxidase (HRP) [28], xanthine oxidase [29], glucose oxidase and carbonic anhydrase molecules [30], has also been investigated. One salient feature of these studies has been the demonstration that the enzyme molecules are bound tightly to gold colloidal particles and retain significant biocatalytic activity whilst in the conjugated form, but are denatured when they are adsorbed onto the planar surfaces of metallic gold [31].

Since the physical adsorption of proteins onto gold nanoparticles does not stabilize the nanoparticles, such stability constitutes an additional issue [32]. Notably, the physical adsorption of proteins onto gold nanoparticles is nonreproducible, time-consuming and tedious; it is also very expensive, as the system must be optimized for each new protein that is coupled, which in turn requires large amounts of protein that is costly to produce.

4.3.1.2 Physisorption of Oligonucleotides and DNA

In the case of oligonucleotides and DNA, competitive physisorption experiments showed a relative adsorption affinity of the DNA bases $A > C \geq G > T$, with the adsorption of oligo(dA) strongly dominant over the other oligonucleotides [33]. The high adsorption affinity of oligo(dA) allows it to compete effectively against the chemisorption of thiolated oligo(dT), and causes hybridized oligo(dA)–oligo(dT) duplexes to denature in the presence of Au. This remarkable affinity has clear practical implications, however, as both the probe and target oligonucleotides may be affected in systems where gold surfaces are involved.

Although physisorption provides an easy means of attaching proteins and DNA to gold surfaces and nanoparticles (as no reagent or reactive functions are required), it has one major disadvantage in that the biomolecule is not oriented, because it is bonded to the surface through multiple sites. Consequently, these multiple limitations have led to the development of alternative approaches to couple biological molecules to gold surfaces and nanoparticles in a controlled and oriented manner. In this way it should be possible to ensure exposure not only to the aqueous environment, but also to the complementary sites of the target molecule.

4.3.2
Chemisorption Based on Self-Assembled Monolayers (SAMs)

Chemisorption requires the presence of a nucleophilic group in the surface of the biomolecule that is capable of linking covalently to the gold surface; typical examples include –SH or other gold-reactive groups, such as –NH2 [31]. SAMs are created by the chemisorption of hydrophilic group onto a substrate, while the hydrophobic tail groups assemble far from the substrate. SAMs are basically interfacial layers between a metal surface and a solution; the best-described are those based on the strong adsorption of functionalized alkane sulfides, disulfides and thiols onto gold surfaces. These were first described by Nuzzo and Allara, who showed that SAMs of alkanethiolates on gold could be prepared by the adsorption of di-*n*-alkyl disulfides from dilute solutions [34].

Both, sulfur and selenium compounds have a strong affinity to transition metal surfaces [35], coordinating very strongly not only to gold but also to silver, platinum, or copper. Of these metals gold is the most favored, however, because it is reasonably inert.

In the case of alkanethiole, the reaction may be considered formally as an oxidative addition of the S–H bond to the gold surface, followed by a reductive elimination of hydrogen. When a clean gold surface is used, the proton most likely results in formation of the H_2 molecule. The assumed reaction is [35]:

$$R-S-H+Au_n^0 \Rightarrow R-S^-Au^+\bullet Au_n^0+1/2H_2 \quad (4.1)$$

The number of reported surface active organosulfur compounds capable of forming monolayers on gold has increased during recent years (Figure 4.1). Currently, the list includes (among others) di-*n*-alkyl sulfide, di-*n*-alkyl disulfides, thiophenols, mercaptopyridines, mercaptoanilines, thiophenes, cysteines, xanthates, thiocarbaminates, thiocarbamates, thioureas, mercaptoimidazoles, and alkaneselenols. However, the most studied – and probably most understood SAM – is that of alkanethiolates on Au(111) surfaces.

Although dense monolayers may be assembled in less than 1 h, well-ordered monolayers may take days to be formed [36]. A clean gold surface is important, but not essential, for the chemisorptions as the high affinity of the thiol moiety

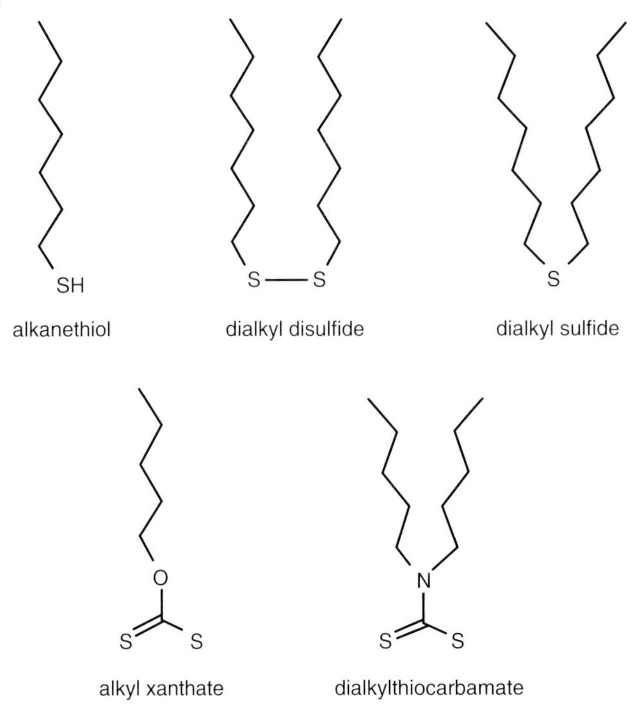

Figure 4.1 Surface-active organosulfur compounds that form monolayers on gold. Reprinted with permission from Ref. [35]; © 1996, American Chemical Society.

for gold will even displace contaminants [37]. It was also suggested that, as gold does not have a stable surface oxide [35], its surface can be cleaned simply by removing the physically and chemically adsorbed contaminants, usually by mechanical polishing with alumina, followed by sonication and etching in either acidic or alkaline solutions. Such treatment is also important for producing a hydrophilic gold surface. Clean gold shows a characteristic anodic peak current close to +1.1 V (versus standard calomel electrode; SCE) and a single cathodic peak close to +0.9 V in dilute sulfuric acid solutions [38].

When used for electrochemical biosensors, the main drawback of this technique is that it produces a tightly packed structure (Figure 4.2), that may limit the diffusion of electroactive species towards the surface, in comparison with bare electrodes. However, by decreasing the electron-transfer rate, SAMs have been shown suitable for kinetic studies; they also greatly reduce nonfaradaic currents and electrode passivation due to the accumulation of unwanted species [39]. Finally, SAMs with long alkanethiols cause a drastic reduction in electrode/electrolyte capacitance compared to bare gold [38]. For this reason, capacitance measurements are considered more suitable than voltammetry, despite the higher sensitivity of the latter.

On the other hand, SAMs may promote steric and electrostatic repulsions between the immobilized probe and the target, especially in the case of the poly-

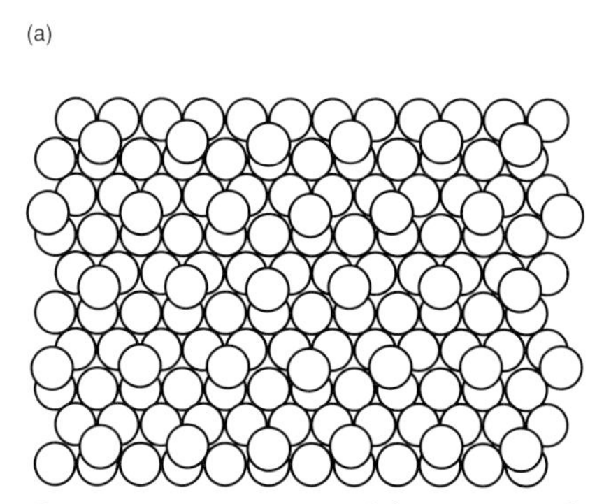

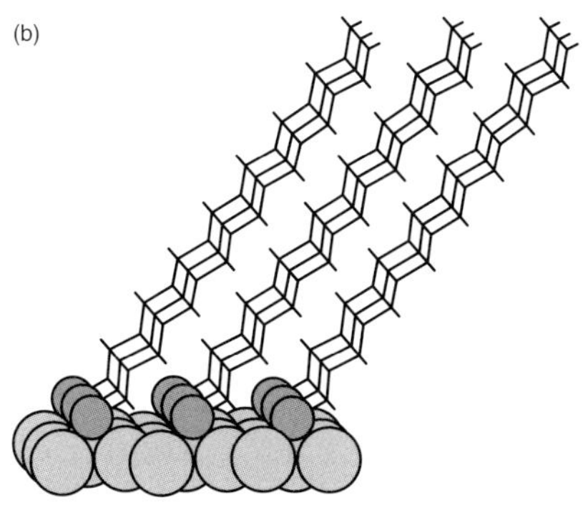

Figure 4.2 Representation of the structure of a fully covered self-assembled monolayer of alkanethiolates on gold. (a) Hexagonal coverage scheme for alkanethiolates on Au(111). The open circles represent gold atoms, and the shaded circles sulfur atoms; (b) A view of the tilt of a $SC_{16}H_{33}$ monolayer on Au(111). The gold atoms are shown in light gray, while sulfur atoms are shown in dark gray and are not drawn to scale. Adapted with permission from Ref. [35]; © 1996, American Chemical Society.

anionic DNA molecule or bulky proteins. For DNA, the standard method to thiolate an oligonucleotide is to attach a $HS(CH_2)_6$-linker molecule to either the 3′- or the 5′-end phosphate group. Although the linker keeps the DNA away from the gold surface and makes the binding event sterically more favorable, it also acts as an electrical insulator between the conducting surface and the DNA molecule [40]. Whilst the van der Waals attraction drives the assembly and ordering in typical SAMs, DNA immobilization is subject to a strong electrostatic repulsion due to the polyanionic nature of DNA [41]. As a result, in hybridization experiments, despite the use of a spacer arm, tightly packed and negatively charged SAMs are obtained, which impedes hybridization with the cDNA probe, due to both steric and electrostatic effects [34].

As such, a stringent control of the surface coverage of DNA is an important factor for maximizing hybridization efficiency, which can be performed by using auxiliary reagents such as lateral spacer thiols and mixed monolayers to obtain bioactive gaps.

As the molar ratio of a mixture of thiols in solution results in the same ratio at the self-assembled surface, the components do not phase-segregate into islands [42], but instead form mixed SAMs. This interesting feature can be exploited to immobilize biomolecules, thus avoiding steric hindrance between the molecules and their binding partners. However, a mixed monolayer of a thiolated probe and a lateral spacer thiol, mercaptohexanol (MCH), in a two-step method, showed a more precise control over coverage of the gold surface [34, 36, 43, 44] (Figure 4.3). One advantage of using this two-step process to form the HS–ssDNA–MCH mixed monolayer is that nonspecifically adsorbed DNA is largely removed from the surface. Before exposure to MCH (Figure 4.3a), the $HS–(CH_2)_6$–ssDNA molecules interact with the gold surface through both the nitrogen-containing nucleotide bases and the sulfur atom of the thiol group. After exposure to MCH, the new thiol groups compete with the nucleotide bases to interact with the gold surface

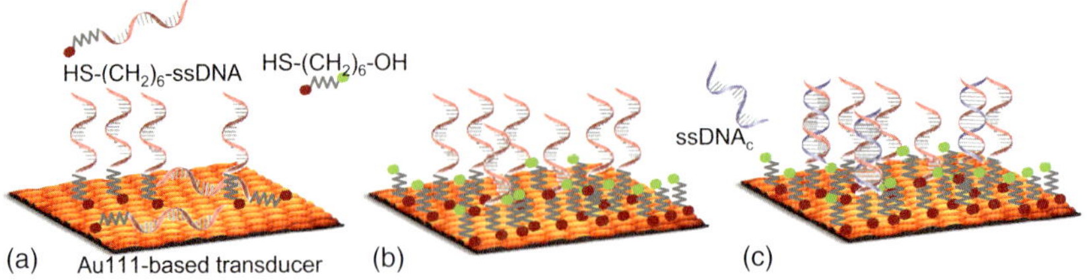

Figure 4.3 Representation of a mixed monolayer of (HS–$(CH_6)_2$-ssDNA) and HS–$(CH_6)_2$–OH performed in a two-step procedure. (a) Chemisorption of thiolated ssDNA (HS–$(CH_6)_2$-ssDNA) to a gold-based transducer through the thiol end as well as through nonspecific adsorption contacts. A multitude of adsorption states exists; (b) Exposure to HS–$(CH_6)_2$–OH (MCH; 6-mercapto-1-hexanol) to minimize nonspecific adsorption of HS–$(CH_6)_2$-ssDNA; (c) Hybridization of the well-oriented HS–$(CH_6)_2$-ssDNA. Adapted with permission from Ref. [43]; © 1997, American Chemical Society.

(Figure 4.3b). Thus, the majority of surface-bound probes are accessible for specific hybridization with complementary DNA (Figure 4.3c) [12].

Despite the above-described problems, one remarkable advantage of SAMs is that they provide a nonrandom orientation of the biomolecules. Specific advantages in analytical applications include the shielding of biomolecules from the sensor surface (thus preventing denaturation), the possibility to tailor the monolayer with functional terminal groups for immobilization purposes, the high stability of the assembly, and the uniformity on the immobilization surface [13]. SAMs constitute a friendly microenvironment for the biocatalytic activity of enzymes, by simulating the optimal conditions found in real biological systems. Moreover, SAMs exhibit a high stability for a wide range of electrochemical conditions, and also allow an efficient electron transfer from the enzyme to the electrode surface. The sequential deposition of monolayers generates highly ordered structures, where a more accurate structure–function relationship of the modified surface may be inferred. This allows the creation of a more homogeneous behavior within the electrode surface and, therefore, more sensitive and reproducible sensors [13]. SAMs of alkanethiols also allow a drastic reduction of nonspecific protein adsorption on macroscopic gold surfaces [45].

4.3.2.1 **Chemisorption of Proteins**

Proteins are able to attach directly to the gold surface if they present one accessible thiol (–SH) group of a cysteine at their surface, as in the case of Fab′-SH antibody fragments. This approach has limited application, however, because only a few proteins present with accessible sulfhydryl groups, even after disulfide reduction. In addition, the production of Fab′-SH proteins has a low yield and is not easy to predict. Rather than use natural –SH groups, however, a second strategy relies on the introduction of this type of moiety. As an example, it is possible to transform amine groups into sulfhydryl groups, with use of Traut's reagent (2-iminothiolane) [46, 47]. Besides chemisorption of the –SH group in cysteine residues, covalent attachment of the $-NH_2$ group of lysine residues of the enzyme pepsin on gold nanoparticles has been also reported [31].

4.3.2.2 Chemisorption of DNA Probes

In order for DNA probes to undergo chemisorptions, they must first be chemically modified; the standard method of thiolating a strand of DNA is to attach a $HS(CH_2)_6$-linker molecule to either the 3′- or the 5′-end phosphate group. This approach has been used extensively for the immobilization of oligonucleotides in gold surfaces [12, 22, 48], for gold- nanoparticles-modified transducers [49–51], and for gold nanocomposites [52]. Moreover, the successful attachment of DNA to gold nanoparticles for detection purposes, and also as elemental building blocks in materials synthesis schemes, has been also achieved [53].

4.3.3 Covalent Immobilization of Self-Assembled Monolayers (SAMs)

The introduction of ligands into SAMs that are complementary to specific binding sites in either native proteins or genetically engineered analogs, represents an attractive avenue for their alignment and orientation [45]. A careful selection of the terminal functionalities of the SAMs and the proper surface chemistry allows tremendous flexibility in their design (Figure 4.4). Three different types of thiol (hydroxyl-, amino-, and carboxyl-terminated) have been investigated for the optimal covalent immobilization of DNA [54]. Both, hydroxyl- and amino-terminated SAMs

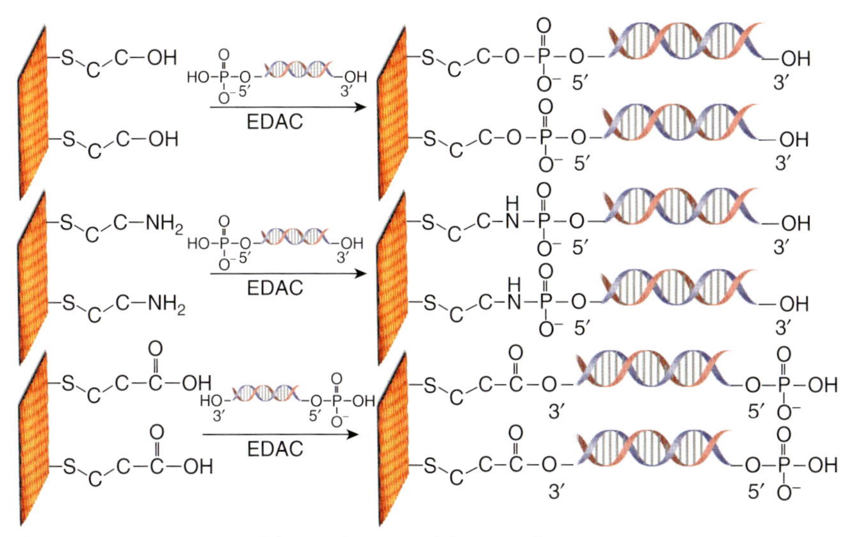

Figure 4.4 Representation of the covalent immobilization of double-stranded (ds) DNA on SAMs with different terminal groups. Adapted with permission from Ref. [54]; © 1999, Elsevier Science B.V.

were coupled to the 5′-phosphate-end of DNA, while carboxyl-terminated SAMs were coupled to the 3′-hydroxy end of the nucleic acid, with the assistance of a water-soluble carbodiimide (EDAC; 1-ethyl-3-(3-dimethylaminopropyl) carbodiimide hydrochloride) in 2-(*N*-morpholino)ethanesulfonic acid (MES) buffer.

The SAMs with different terminal groups were prepared with three ethylene-containing thiols, 2-mercaptoethanol, 2-mercaptoethylamine [22, 54–56], and 3-mercaptopropionic acid, the terminal groups of which were respectively hydroxyl, amino, and carboxyl. The same functionalization strategies can easily be adapted to the immobilization of proteins (enzymes, antibodies, receptors) through the –COOH and $-NH_2$ moieties.

4.3.4
Spacer Arms

The use of spacer arms to separate the metal surface from the biological moiety, so as to avoid their possible denaturation, also represents a very practical approach. Most often, the spacers are covalently linked at one end to the gold surface via thiol chemistry, while being linked at the other end to amine groups exposed at the protein surface, for example, via the use of *N*-hydroxysuccinimide (NHS)-containing ligands. As every protein presents several accessible amine groups, this coupling approach results in a random and multiple orientation of proteins at the surface of gold. In addition, amine groups often participate at active sites or ligand-binding sites, and their modification may lead to a loss of recognition properties. The spacer arms consists of one or several (preferably two) covalently linked spacers selected from the group comprising homo- or hetero-bifunctional polyethylene oxides. In the case of SAMs, some improved strategies have been designed (Figure 4.5) [45, 57–59]. For example, gold clusters may be synthesized in an organic phase and transferred to water using lipoic acid; this results in stable, water-dispersible nanoparticles. Then, by using the thiol-capped clusters as a building platform, a chemical modification of the ligand shell can result in a nitrilotriacetic acid termination, which is suitable for complexation with transition metal ions. Recombinant proteins carrying histidine tags can then be readily attached to this transition metal complex.

4.4
Synthesis and Properties of Gold Nanoparticles

Colloidal Au, as a nanomaterial, possesses a huge specific surface area, good biocompatibility [26, 60, 61], and electrically activity to shuttle electrons from the surface into the conduction band. Moreover, nanoparticles are key nanostructured materials for bridging the gap between “bottom-up” synthetic methods and “top-down” fabrications [62]. Gold nanoparticles can be used as regular shape- and size-building blocks for the creation of larger nanocomposite structures [53]. A combination of the synthetic design with a directed assembly of nanoparticles into

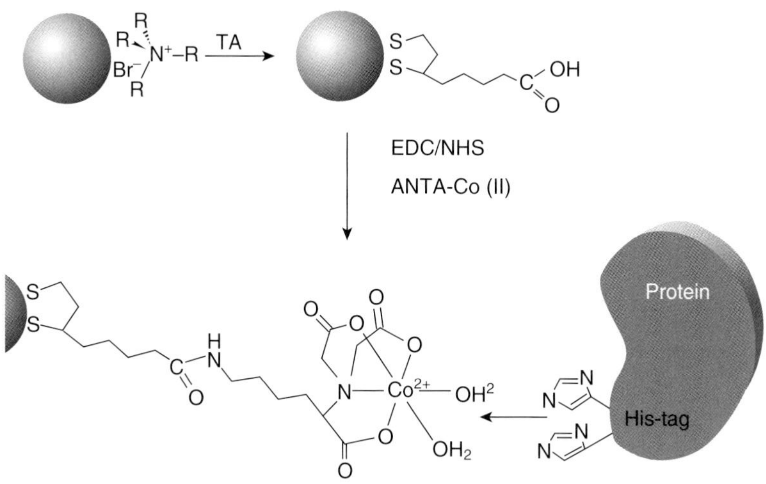

Figure 4.5 Reaction strategy showing the successive steps for the construction of NTA-terminated nanoparticles for specific immobilizationofhistidine-taggedproteins.ANTA-Co^{2+}=aminonitrilotriacetic–Co(II) complex; NTA = nitrilotriacetic acid; TA = 6,8-dithioctic acid; Reprinted with permission from Ref. [45]; © 2005, American Chemical Society.

ensembles provides direct control of the structure, from the molecular to the macroscopic level. Au nanoparticles also exhibit unique optical properties [63, 64]; for example, the color change upon aggregation and surface plasmon resonance (SPR) enhancement have been exploited in biosensing devices [65, 66]. The electrochemical properties of gold nanoparticles have also been used as labels in electrochemical biosensors which, currently, are undergoing investigation [67–69]. On the basis of these somewhat amazing properties, nanosized gold particles have attracted much interest due to their potential applications in the fields of physics, chemistry, biology, medicine, and materials science [70].

Since the original synthesis of gold nanoparticles, as described by Faraday [71] some 150 years ago, much effort has been expended on the preparation of citrate-stabilized gold nanoparticles [72].

The present-day methods used to prepare gold nanoparticles offer not only a reproducible control over nanoparticle size and solubility, but also the ability to produce a variety of nanoparticle shapes, such as rods, cubes, and multi-arm structures [73–80]. Briefly, the preparation of gold nanoparticles generally involves the chemical reduction of gold salt in aqueous phase, in organic phase, or in two phases. Unfortunately, the high surface energy of gold nanoparticles makes them extremely reactive, and without protection or passivation of their surfaces most systems will undergo aggregation. Thus, special precautions must be taken to avoid both aggregation and/or precipitation. Typically, gold nanoparticles are prepared by chemical reduction of the corresponding transition metal salts in the presence of a stabilizer; the latter binds to the particle surface so as to impart a high stability and rich linking chemistry, and provide the desired charge and solubility properties [70]. The most common methods for surface passivation include protection by SAMs, the most popular of which are thiol-functionalized organics [35]. The reduction of gold salts in the presence of alkyl thiols produces gold nanoparticles in the 2–3 nm diameter range, with limited solubility in organic solvents due to the presence of physisorbed molecules or chemisorbed monolayer of alkyl thiols on the nanoparticle surface [81]. Alternative stabilization methods include encapsulation in the H_2O pools of reverse microemulsions, and dispersion in polymeric matrices.

The design of a protecting agent for the synthesis of gold nanoparticles represents a key role, not only to stabilize the gold nanoparticles but also to provide functionalization for further applications, including bioassay, bioimaging, and biosensor devices. A wide range of protecting agents has been reported [70], including peptides, lipids, polysaccharides, polymers, and dendrimers. Microorganisms, such as bacteria and viruses, may also be employed as a constrained template in order to direct nanoparticle synthesis.

Although, in recent years, progress in the synthesis of gold nanoparticles has been extensive, the control of nanoparticle size, dispersion, morphology, and surface chemistry remain a challenge.

4.5 Composites and Nanocomposite Materials in Biosensors

When different materials are combined to form a heterogeneous structure, the properties of the resultant composite will depend on the properties of the constituent materials, the size scale, and the chemical and morphological details of the dispersion. Hence, each individual component will maintain its original characteristics while providing the composite distinctive with chemical, mechanical, physical, and/or biological qualities [82]. Moreover, such global features will be different from and synergistic with those demonstrated by the individual elements of the composite [4, 83].

4.5.1 Biological and Engineering Composites

The primary classification of the composite is that of biological or engineering composites. A typical *biological composite* would be wood, which is composed of fibrous chains of cellulose in a matrix of lignin; likewise, bone is composed of hard inorganic crystals (hydroxyapatite) embedded in a tough organic matrix (collagen). In contrast, a high-performance *engineering composite* would typically consist of a matrix (metal, polymer, ceramic) with an embedded reinforcement (filament, whiskers, particles).

4.5.2
Nanostructured Composite and Nanocomposites

A nanostructured composite or nanocomposite results when the characteristic length scales of at least one of the components is in the nanometer range. Nanometer-sized filler materials, with their inherently large surface area-to-volume ratios, are particularly interesting as they facilitate increases in the efficiency of a given property. Moreover, due to the small size of the filler, some properties may be modified while others remain unaffected [84]. In the nanotechnology era, novel nanostructured composite materials would be expected to be designed that demonstrate improved properties due to their nanostructural characteristics.

Notably, composite materials may be also classified, according to their physical properties, as either *soft* or *rigid* composites.

4.5.3
Conducting Composites

A conducting composite will result if at least one of the phases is an electrical conductor. The overall electrical properties of a conducting composite will be determined by the nature, the relative quantities, and the distribution of each phase. Recent developments in the field of conducting composites, when applied to electrochemistry, have opened a new range of possibilities for the construction of electrochemical sensors and biosensors. The main features of these materials have been described elsewhere [85, 86].

4.5.4
Polymer Composites

A polymer composite will result if at least one of the components is a polymeric matrix, which can be either a *conducting* or *non-conducting* polymer. As such, a conducting polymer composite may be obtained with a conducting polymer matrix or, instead, by using a nonconducting polymer matrix but a conducting filler (such as platinum, gold, carbon, CNTs, metal nanoparticles, etc).

4.5.4.1 Conducting Composites Based on Conducting Polymers

Conducting polymers are basically organic conjugated polymers, the unusual electrochemical characteristics (e.g., low ionization potential, high electrical conductivity, high electronic affinity) of which are due to the conjugated π-electron backbones in their chemical composition. For this reason, conducting polymers are often referred to as "synthetic metals." Their organic chains with single- and double-bonded sp^2-hybridized atoms generate a wide charge delocalization that is responsible for the metal-like semiconductive properties of these polymers [13]. The electrical and optical properties of conducting polymers are similar to those of metals and inorganic semiconductors. Moreover, biomolecules can be immobilized onto conducting polymers without any loss in activity. The mechanical and electronic properties of conducting polymers can be tailored by chemical modeling and synthesis [87]. The attractive feature for biosensor applications results essentially from the rapid electron transfer that these materials provide in electrode surfaces as a consequence of the biological event. Conducting polymers can be synthesized either by chemical or electrochemical oxidation. The electrochemical method is based on the oxidation of monomers, and leads to the formation of cation radicals that repeatedly bind to the growing polymer [88]. Today, both polypyrrole and polyaniline are considered the most promising conducting polymers for the development of biosensor devices, owing to their good biocompatibility, conductivity, and stability. The combination of nanoengineered "smart" conducting polymers with biomolecules and nanostructures, such as metal nanoparticles and CNTs, may generate conducting composites with new and interesting properties, providing higher sensitivity and stability of the immobilized biomolecules.

4.5.4.2 Conducting Composites Based on Nonconducting Polymers

Nonconducting polymers are polymeric binders (epoxy, methacrylate, silicone, araldite) which confer to the conducting composite a certain physical, chemical, or biological stability. The electrical conductivity of these materials is provided by the conducting filler (micro- or nanoparticles of platinum, gold, graphite, CNTs, etc.).

Conducting composites based on *nonconducting polymers* are classified by the nature of the conducting material, and the arrangement of its particles – that is, whether the conducting particles are dispersed in the polymer matrix, or if they are grouped randomly in clearly defined conducting zones and insulating zones.

The inherent electrical properties of conducting composites depend on the nature of each of the components, their relative quantities, and their distribution. The electrical resistance is determined by the connectivity of the conducting particles inside the nonconducting matrix; therefore, the relative amount of each component must be assessed to achieve an optimal composition. A percolation curve [89] (see Figure 4.6) is a representation of the logarithmic variation of the electrical resistance of a composite as a function of its conducting phase content. The construction of a percolation allows the determination of the minimum conductor content required to achieve a certain conductivity; this point is termed "percolation threshold."

The composite acquires particular electrochemical features from the nature of the conductive filler in the bulk. The extensive range of unique properties inherent to metal nanoparticles, including electrical conduction, makes them very attractive candidates for integration into polymers as nanoparticle–polymer composites. The embedding of nanoparticles into a host polymer provides a means of introducing a variety of properties to the polymer-based composite materials, including conductivity in the case of gold nanoparticles, magnetic properties when using cobalt or iron oxide nanoparticles, and mechanical properties when using nanoparticle fillers such as clay [81]. Whatever the particular nanoparticle composition or shape,

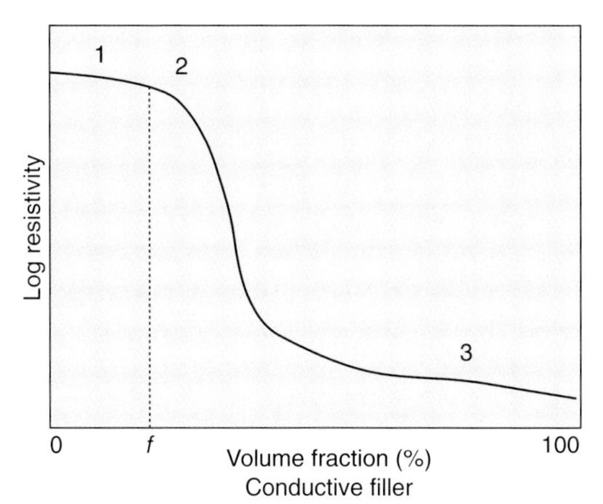

Figure 4.6 Percolation curve of a conducting composite based on a nonconducting polymeric binder with a conductive filler. Theoretical dependence of composite resistivity on conductive filler content. In zone 1, the electrical resistance of the composite is similar to that of the polymer. In zone 2, the percolation fraction *f* represents a critical conductive filler content that permits formation of the first conducting filament consisting of particle-to-particle contacts. In zone 3, the electrical resistance of the composite is similar to that of pure conductive filler.

their blending with most polymers tends towards a phase separation that results in particle clustering or aggregation within the host polymer. However, this problematic issue can be addressed very effectively by appropriate nanoparticle surface modification.

A number of studies have involved gold nanoparticle–polymer composites where thiol-terminated polymers have been introduced at some stage of the nanoparticle growth process [90], or subsequent to the initial nanoparticle synthesis [91, 92]. The covering of metal surfaces with electronically active polymers provides a close proximity of the two materials that facilitates their electronic communication.

4.5.4.3 Carbon Conducting Composites Based on Nonconducting Polymers

In addition to metal micro- and nanoparticles, carbon represents an ideal conductive filler for conducting composites, due to its high chemical inertness, wide range of working potentials, low electrical resistance (ca. $10^{-4}\,\Omega\cdot$cm), and low residual currents [83].

Carbon conducting composites based on nonconducting polymers demonstrate an improved electrochemical performance, similar to an array of carbon fibers separated by an insulating matrix and connected in parallel. The signal produced by this macroelectrode formed by a carbon fiber ensemble is the sum of the signals of the individual microelectrodes. Hence, this type of composite electrode shows a higher signal-to-noise ratio (SNR), and thus an improved (lower) detection limit compared to the corresponding pure conductors [83].

Rigid composites are obtained by mixing graphite powder with a nonconducting polymeric matrix, to obtain a soft paste that becomes rigid after a curing step [83, 85, 86]. Rigid conducting graphite-epoxy composites (GECs) and graphite-epoxy biocomposites (GEB) have been used extensively at the present authors' laboratories, and shown to be a suitable material for electrochemical (bio)sensing due to their unique physical and electrochemical properties [4, 18, 85, 86]. In particular, GEC has been prepared by mixing a nonconducting epoxy resin with graphite powder (particle size $<50\,\mu m$). An ideal material for electrochemical biosensing should allow an effective immobilization of the probe on its surface, a robust reaction of the target with the probe, a negligible nonspecific adsorption of the label, and a sensitive detection of the biological event–all of which are fulfilled by GECs. Moreover, the GECs also provide numerous advantages over more traditional carbon-based materials, including greater sensitivity, robustness, and rigidity. The surface of a GEC can also be regenerated by using a simple polishing procedure. Unlike carbon paste, the rigidity of the GEC permits the design of different configurations, all of which are compatible with nonaqueous solvents.

In order to explore the novel capabilities of rigid conducting GECs, gold nanoparticles were added as a new filler to the composite, so as to create gold nanoparticle-GECs (nanoAu-GECs). The use of these materials for electrochemical biosensing is discussed in the following section.

4.6 Rigid Conducting nanoAu-GECs for Improved Immobilization of the Bioreceptor in Genosensing Devices

As noted above, chemisorption based on SAMs is a single-point immobilization strategy that allows the oriented attachment of a wide range of biomolecules on the gold-based transducer surfaces. However, a major drawback of using SAMs for the immobilization of bioreceptors in electrochemical biosensing is that a compact layer is achieved, and this leads to a dramatic reduction in the diffusion of electroactive species towards the transducer surface. Moreover, the tightly packed layer may also cause steric hindrance that, as a consequence, will reduce the rate of reaction between the probe and the target. Consequently, although a stringent control of the surface coverage of the gold-based transducer is important, this can be achieved by using auxiliary reagents such as lateral spacer thiols and mixed monolayers so as to obtain bioactive gaps.

In order to avoid the stringent control of surface coverage parameters during the immobilization of thiolated oligonucleotides on continuous gold surface films (see Figure 4.3), the use of gold nanoparticles in a GEC (nanoAu-GEC) has been proposed [52]. In this novel transductor, islands of chemisorbing material (gold nanoparticles) surrounded by rigid, nonchemisorbing, conducting GECs, are obtained (see Figure 4.7).

With this arrangement in the electrochemical transducer, the resultant less-packed surface provides improved hybridization features, with a complementary

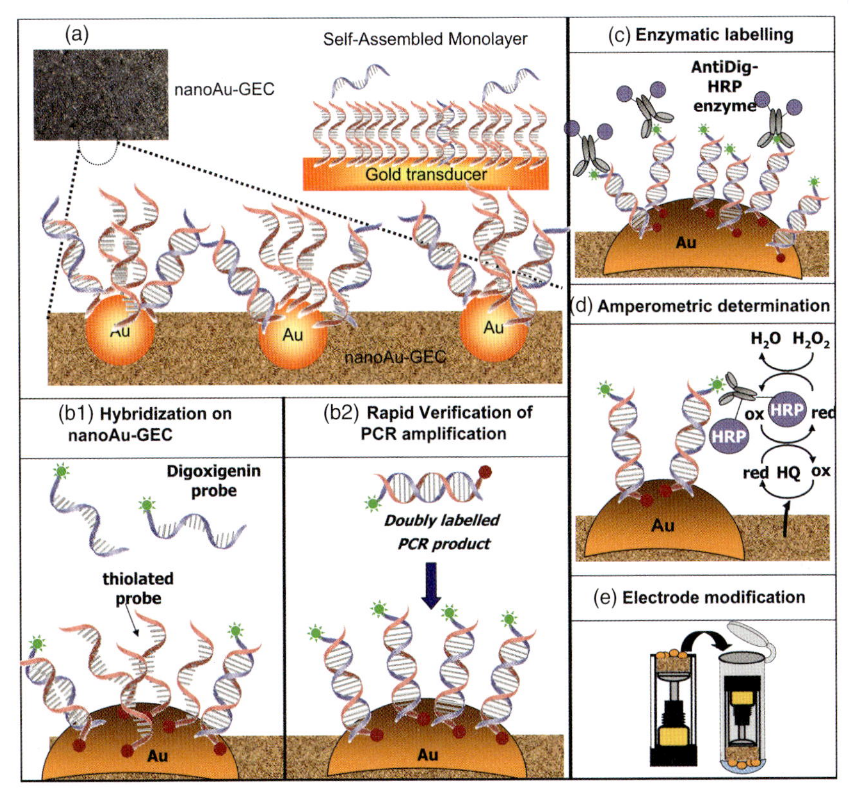

Figure 4.7 Schematic representation of (a) nanoAu-GEC material, showing isolated gold nanoparticles able to produce "bioactive chemisorbing islands" instead of SAMs on a continuous layer of gold; (b1) Hybridization assay on the nanoAu-GEC electrode; (b2) Rapid electrochemical verification of thiolated and double-tagged amplicons on the nanoAu-GEC electrode; (c–e) Common steps (electrode modification, enzymatic labeling, and amperometric determination) for both parts b1 and b2. Reprinted with permission from Ref. [52]; © 2009, American Chemical Society.

probe minimizing any steric and electrostatic repulsion. The spatial resolution of the immobilized thiolated DNA was easily controlled, simply by varying the percentage of gold nanoparticles in the composition of the composite (Figure 4.8).

For the GEC electrodes, graphite powder (particle size <50 μm) and epoxy resin (Epo-Tek; Epoxy Technology, Billerica, MA, USA) in a 1 : 4 (w/w) ratio were thoroughly hand-mixed to ensure a uniform dispersion of the graphite powder throughout the polymer. For nanoAu-GEC electrodes, the following ratios of gold nanoparticles/graphite powder/epoxy resin were prepared: 0.075 : 0.925 : 4 (w/w) for nanoAu(7.5%)-GEC; 0.25 : 0.75 : 4 (w/w) for nanoAu(25%)-GEC; 0.5 : 0.5 : 4 (w/w) for nanoAu(50%)-GEC, and finally, 1 : 0 : 4 (w/w) for nanoAu(100%)-EC. The soft paste became rigid after a curing step of 80 °C, during a one-week period. Hence, the designation of the electrodes was based on the ratio of gold nanoparticles towards graphite particles (the conductive fillers).

0.0%
7.5%
25%
50%
100%

Figure 4.8 Microscopic characterization of nanoAu-GEC electrodes. Microimages showing the distribution of gold nanoparticles on the surface of nanoAu-GEC electrodes, while increasing the amount of gold nanoparticles from 0 to 100% of the conductive phase. Left column: low-resolution (100 μm) SEM with an EDX detector to identify the gold element. Acceleration voltage, 20 kV. Right column: fluorescence stereomicroscopy at low resolution, showing the fluorescence pattern of the different nanoAu-GEC electrodes after the immobilization of 200 pmol of double-tagged oligonucleotides with thiol and fluorescein. Center column: Stereomicroscopy without the fluorescence filter.

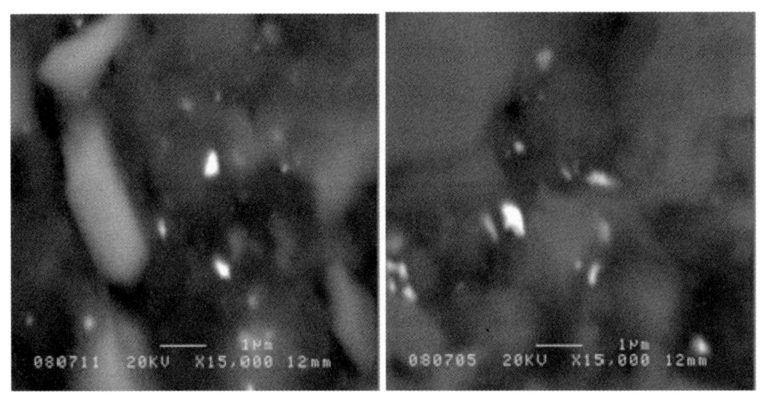

Figure 4.9 High-resolution (1 μm) SEM images showing the isolated gold nanoparticles on the surface of nanoAu-GEC sensors. Acceleration voltage, 20 kV.

The location and spatial pattern of the gold nanoparticles on the surface of the sensor was observed using scanning electron microscopy (SEM) with an energy-dispersive X-ray (EDX) detector. In Figure 4.8, the first column shows, as bright spots, the aggregates of gold nanoparticles for nanoAu(7.5%)-GEC, which appear to be at a higher frequency when the percentage of gold nanoparticles is increased, up to nanoAu(100%)-EC. However, high-resolution SEM images of nanoAu(7.5%)-GEC (Figure 4.9) showed clearly the presence of isolated gold nanoparticles of about 100 nm within the composite, as confirmed by the EDX detector providing the characteristic X-ray spectrum for gold.

The availability of gold nanoparticles in the composite for the immobilization of thiolated oligonucleotides was also studied using fluorescence stereomicroscopy. In this case, 200 pmol of double-tagged oligonucleotides with both a thiolated 5′-end and the fluorescein 3′-end was immobilized on electrodes with different compositions. As can be seen in Figure 4.8 (right column), an increasing amount of fluorescence was obtained with a higher proportion of gold nanoparticles in the composite. The fluorescence showed a discontinuous pattern as fluorescence dots of chemisorbing material surrounded by nonreactive GEC, except in the case of nanoAu(100%)-GEC, in which a continuous fluorescence pattern was clearly observed. Moreover, it should be pointed out that the fluorescence could be related to the isolated gold nanoparticles pattern because it is not located in the aggregate zones, when compared to the same images recorded using the stereomicroscope without the fluorescence filter (Figure 4.8, center column). Thus, the nanometer scale of gold nanoparticles appeared also to play a role in the chemisorbing ability of the gold nanocomposites material, especially in nanoAu(7.5%)-GEC, as the fluorescence pattern was related to the isolated gold nanoparticles instead of the gold aggregates when the amount of gold nanoparticles was increased from 7.5% to 100%.

From an electrochemical evaluation of the electrodes, it was concluded that gold was more available for the electrochemical oxidation in the nanoAu(7.5%)-GEC electrode, was present mostly as nanoparticles rather than aggregates, and showed the characteristic anodic peak current near +1.1 V (versus Ag/AgCl) [38]. The voltammetry for a variety of redox molecules at DNA-modified electrodes can provide additional qualitative information concerning the system's organization on the surface. The voltammetric reversibility of highly charged redox ions is markedly influenced by the attractive/repulsive interactions with the polyanionic DNA layer that the ions must penetrate in order to reach the electrode surface, in the case of highly packed DNA monolayers [44]. The voltammetry for the ferrocyanide ($3^-/4^-$) redox markers at the DNA-modified nanoAu-GEC was slightly affected by the electrostatic interactions with the polyanionic layer, in contrast to previous reported results for SAMs of DNA in continuous gold electrodes [44], thus confirming the laxity of the DNA layer created on the nanoAu(7.5%)-GEC electrode. These data suggest an architecture that is composed of a disperse layer of oligonucleotides immobilized on the isolated gold nanoparticles, and confirms the microscopic pattern achieved with both SEM and fluorescence microscopy [52].

Instead of SAMs on continuous layers of gold, isolated gold nanoparticles are capable of producing "bioactive chemisorbing islands" for the immobilization of thiolated biomolecules, and avoiding stringent conditions for surface coverage as well as the use of auxiliary reagents such as lateral spacer thiols. Less compact layers are thus achieved which favor the biological reaction on biosensing devices. The hybridization efficiency was expected to be higher on the edging of the gold nanoparticles surrounded by nonreactive GEC. The chemisorbing ability of gold nanoparticles in the nano-AuGEC was demonstrated with an excellent limit of detection (9 fmol of ssDNA) in hybridization studies. With regards to other electrochemical transducers reported previously, such as an avidin-GEB (Av-GEB) or protein A graphite-epoxy biocomposite (ProtA-GEB) [18], the main advantages of the inorganic nanoAu-GEC electrode compared to the biocomposite proved to be a lack of any loss of activity, and that the latter required the temperature to be maintained at 4 °C due to the biological nature of the modifier, avidin.

To summarize, the nanoAu-GEC material demonstrates interesting properties for electrochemical genosensing in hybridization experiments, and very promising features for the electrochemical biosensing of a wide range of biomolecules, including dsDNA, PCR products, affinity proteins, antibodies, and/or enzymes.

4.7
Other Approaches Based on Gold Nanocomposites for Electrochemical Biosensing

The integration of redox enzymes with an electrode support, and the formation of an electrical contact between the biocatalysts and electrode, is the fundamental area of bioelectronics and optobioelectronics [93]. Biomolecules – and enzymes in particular – lack any direct electrical communication with electrodes, as the active centers are surrounded by considerably bulky insulating protein shells, such that electron transfer between the electrodes and the active centers is considerably impeded. In this context, gold nanoparticles may permit a direct electron transfer between redox proteins and bulk electrode materials, allowing electrochemical sensing to be carried out without the use of electron-transfer mediators [94].

The incorporation of nanomaterials into *composite electrode matrices* constitutes a useful strategy for the preparation of enzyme biosensors with improved analytical performance. Whilst these devices exhibit the characteristics of the involved nanomaterials, they also bear the advantages of composite electrodes. This great versatility creates the possibility of incorporating different components into the bulk of the electrode matrix, in addition to an easy surface regeneration.

One of the simplest composite approaches for electrochemical biosensors is based on soft *carbon pastes* [95]. Such pastes are created by mixing an inert conductor (e.g., graphite powder) with a nonconducting liquid (e.g., paraffin oil, silicone, Nujol). As the insulating liquid has a specific viscosity, the paste will have a certain consistency. Although these pastes are easy to prepare and inexpensive, they have a limited mechanical and physical stability, especially in flow systems. The pastes may also be dissolved by some nonpolar electrolytic solvents, leading to a deterioration of the signal. Unfortunately, the general and rapid degradation of these materials has limited their use to the research laboratory [4]. Today, many biosensor devices combine the advantages of colloidal gold and carbon paste electrodes: examples include a reagentless glucose biosensor based on the direct electron transfer of the redox enzyme glucose oxidase [96], a hydrogen peroxide biosensor based on the direct electron transfer of HRP [97], and a phenol biosensor based on a tyrosinase–colloidal gold-modified carbon paste electrode [98]. In all cases, the enzyme biosensors were constructed by immobilizing the corresponding enzymes on the electrodes, and prepared by mixing colloidal gold with the carbon paste components. Similarly, a reagentless nitrite biosensor based on a hemoglobin–colloidal gold-modified carbon paste electrode was reported [99]. In this case, the immobilized hemoglobin displayed a direct electron transfer and an excellent response to the reduction of NO_2^-.

In a similar approach, a composite electrode was prepared by modifying glassy carbon microparticles with gold nanoparticles and xanthine oxidase for the detection of xanthine and hypoxanthine [100].

Enzyme biosensors using composite graphite–Teflon electrodes, in which the enzyme(s) and colloidal gold nanoparticles have been physically incorporated, have been also described. Based on this methodology, a tyrosinase biosensor was fabricated [101]. In this case, the presence of colloidal gold in the composite matrix enhanced the kinetics of both the enzyme reaction and the electrochemical reduction of the corresponding *o*-quinones at the electrode, thus providing a high sensitivity.

The conjugation of gold nanoparticles with other nanomaterials and biomolecules represents an attractive area of research within nanobiotechnology [93]. In this context, CNTs are considered an interesting target on the basis of their unique properties. Consequently, a colloidal gold–CNT composite electrode using Teflon as the nonconducting binding material demonstrated significantly improved responses to H_2O_2. The incorporation of glucose oxidase into the new composite matrix allowed the preparation of a glucose biosensor with a remarkably higher sensitivity, without the need for mediators [102].

The covering of metal nanoparticles with electronically active polymers provides a close proximity of the two materials that facilitates their electronic communication. Based on this effect, a biosensing device was fabricated by the covalent incorporation of CNTs and gold nanoparticles onto a GCE (glassy carbon electrode) that had been modified with an electropolymerized poly(thionine). The synergistic effects of the composite nanomaterials, together with the excellent mediating ability of the redox polymer, allowed a rapid electron transfer and a high efficiency of enzyme immobilization [103]. Gold polypyrrole (Au–PPy) nanocomposites were deposited onto a GCE surface by the polymerization of pyrrole in the presence of $AuCl_4^-$, followed by enzyme immobilization. In this case, three enzyme systems were used as models, namely cytochrome c, glucose oxidase, and polyphenol oxidase [104]. Polyaniline, another conducting polymer, was used successfully to design gold nanocomposites. For this, a glucose biosensor based on a mixture of gold nanoparticles and conductive polyaniline with glucose oxidase and Nafion on the surface of the nanocomposite was developed [105]. Interestingly, ultrasound radiation can be also used to prepare PPy–Au and polyaniline–Au nanocomposites [106, 107].

For various reasons, and especially with regards to problems of environmental impact or toxicity, *biopolymers* are today considered increasingly as interesting substitutes for synthetic polymers. Among these, glycosaminoglycans such as chitin and chitosan (which belong to the family of $\beta(1\rightarrow4)$-linked polysaccharides) have emerged as highly promising, as their unique structure gives rise to very good biological and mechanical properties [108]. *Chitosan*, a natural polymer which has one amino group and two hydroxyl groups in the repeating hexosamide residue, demonstrates – besides biocompatibility – an excellent adhesion ability, low cost, and chemical flexibility for modifications. The adsorption of colloidal gold nanoparticles onto a chitosan membrane provides an assembly of multilayers of gold nanoparticles, and a suitable microenvironment that is similar to the native environment of biomolecules. Based on this natural polymer, a biocomposite material made from chitosan hydrogel, gold nanoparticles and a redox enzyme (e.g., glucose oxidase [109] and peroxidase [110]) was reported for the detection of glucose and H_2O_2, respectively. A more hydrophilic and water-soluble carboxymethyl chitosan (CMCS) was also described in a novel CMCS/gold nanoparticles nanocomposite to construct a H_2O_2 biosensor based on the enzyme peroxidase [111].

Nafion is an additional engineering polymer used in the creation of enzyme-based sensors. In this case, a L-cysteine/Au-nanoparticle/Nafion composite membrane was used to prepare a mediator-free H_2O_2 biosensor, in which the Nafion membrane was used to prevent aggregation of the L-cysteine–Au nanocomposite. The fabricated electrode surface had a low background current, and accelerated the electron transfer with an improved electrocatalytic activity [112].

Sol–gel hybrid materials, prepared by physically encapsulating gold nanoparticles (AuNPs) into porous sol–gel networks, were used for the fabrication of biosensors. First, an HRP biosensor was developed by self-assembling AuNPs to a thiol-containing sol–gel network. Subsequently, a cleaned gold electrode was immersed in a hydrolyzed (3-mercaptopropyl)-trimethoxysilane (MPS) sol–gel solution to assemble a three-dimensional (3-D) silica gel, after which the AuNPS were chemisorbed onto the thiol groups of the sol–gel network. Finally, HRP was adsorbed onto the surface of the AuNPS [113]. A sol–gel-derived silicate network assembling gold nanoparticles (AuNPs-SiSG) provided a biocompatible microenvironment around the enzyme molecule so as to stabilize its biological activity and prevented the enzymes from leaking out of the interface [114]. A similar approach was used to create an acetylcholinesterase biosensor to detect monocrotophos, methyl parathion and carbaryl pesticides. For this, an electrochemical biosensor based on the integrated assembly of dehydrogenase enzymes and AuNPs was devised, in which the AuNPs were self-assembled on a thiol-terminated, sol–gel-derived 3-D silicate network that was enlarged by hydroxylamine seeding, and accordingly efficiently catalyzed the oxidation of NADH [115]. Subsequently, a one-step method to fabricate an HRP biosensor was developed by simultaneously embedding AuNPs and HRP in a silica sol–gel network in the presence of cysteine [116].

Electrochemical immunosensing may also benefit from nanocomposite materials based on AuNPs, not only with regards to enhancing the electrochemical signal but also increasing the amount of immunoreagents to be immobilized in stable mode. An electrochemical immunosensor to detect the hepatitis B surface antigen (HBsAg) was constructed by self-assembling AuNPs to a thiol-containing sol–gel network. For this, a cleaned gold electrode was first immersed in a hydrolyzed MPS sol–gel solution to assemble a 3-D silica gel, after which the AuNPs were adsorbed onto thiol groups of the sol–gel network. In the final stage the hepatitis B surface antibody was assembled onto the surface of the AuNPs [117].

An immunosensor to detect α-fetoprotein (AFP) was prepared by entrapping thionine into Nafion, so as to form a composite Thi/Nf membrane which would yield an interface containing amine groups to assemble an AuNPs layer for the immobilization of anti-AFP antibodies. Following incubation of the immunosensor with AFP, the cyclic voltammetry (CV) current was decreased linearly in relation to the concentration range of AFP [118].

A carbon paste–gold nanocomposite was also used for the electrochemical immunosensing of carcinoma antigen 125 (CA125). For this, the immunosensor was designed by immobilizing anti-CA125 on a thionine- and AuNPs-modified carbon paste interface [119]. A second example of this approach involved a rapid determination of the tumor marker CA19-9 in human serum. In this case, the carbon paste was modified with colloidal AuNP and the HRP–anti CA19-9 antibody. Formation of the antigen–antibody complex caused a blockade of the electron transfer of HRP towards the electrode substrate, resulting in a significant decrease in current [120].

A highly hydrophilic, nontoxic and conductive colloidal AuNP/titania sol–gel composite membrane was prepared by the encapsulation of HRP-labeled human chorionic gonadotropin (hCG) antibody (HRP–anti-hCG) in the composite architecture for the detection of hCG as a model antigen. The presence of AuNPs provided a congenial microenvironment for adsorbed biomolecules and decreased the electron transfer impedance, leading to a direct electrochemical behavior of the immobilized HRP. Formation of the immunoconjugate by a simple, one-step immunoreaction between hCG in solution and the immobilized HRP-anti-hCG introduced a barrier of direct electrical communication between the immobilized HRP and the electrode surface. This composite membrane could be used efficiently for the entrapment of different biomarkers and clinical applications [121].

A *composite electrode* based on colloidal AuNPs and carbon paste was used to immobilize living pancreatic adenocarcinoma cells derived from ascites, and to investigate the cytotoxic effect of the anti-tumor drug, adriamycin. The method was based on the measurement of an irreversible voltammetric response from cells in relation to the oxidation of guanine, the peak current of which was decreased in the presence of the drug. The AuNPs proved to be efficient for preserving the activity of the immobilized living cells, and preventing their leakage from the electrode surface [122].

4.8 Final Remarks

Today, the emergence of nanotechnology continues to open new horizons for the development of electrochemical biosensors. A key contribution made by nanotechnology in the area of electrochemical biosensing relies on the design of novel transducers, such as gold nanocomposites. The unique ability of AuNPs to immobilize biomolecules while retaining their biological activity, and serving as efficient conducting interfaces with electrocatalytic ability, reveal AuNPs as a powerful tool in the construction of novel gold nanocomposites. Notably, if combined with polymers and other nanostructured fillers, then novel and improved properties will surely be achieved. There is no doubt that, in the very near future, gold nanocomposite-based biosensors will have a major impact on clinical diagnostics, environmental monitoring, and food safety.

References

1 Patel, P.D. (2002) (Bio)sensors for measurements of analytes implicated in food safety: a review. *Trends in Analytical Chemistry*, **21**, 96–115.

2 Mello, L.D. and Kubota, L.T. (2002) Review of the use of biosensors as analytical tools in the food and drink industries. *Food Chemistry*, **77**, 237–56.
3 Terry, L.A., White, S.F. and Tigwell, L.J. (2005) The application of biosensors to fresh produce and the wider food industry. *Journal of Agricultural and Food Chemistry*, **53**, 1309–16.
4 Alegret, S. (1996) Rigid carbon–polymer biocomposites for electrochemical sensing. A review. *Analyst*, **121**, 1751–8.
5 Wang, J., Cai, X., Jonsson, C. and Balakrishnan, M. (1996) Adsorptive stripping potentiometry of DNA at electrochemically pretreated carbon paste electrodes. *Electroanalysis*, **8**, 20–4.
6 Oliveira Brett, A.M., Serrano, S.H.P., Gutz, I., La-Scalea, M.A. and Cruz, M.L. (1997) Voltammetric behavior of nitroimidazoles at a DNA-biosensor. *Electroanalysis*, **9**, 1132–7.
7 Hashimoto, K., Ito, K. and Ishimori, Y. (1994) Novel DNA sensor for electrochemical gene detection. *Analytica Chimica Acta*, **286**, 219–24.
8 Pang, D.W. and Abruña, H.D. (1998) Micromethod for the investigation of the interactions between DNA and redox-active molecules. *Analytical Chemistry*, **70**, 3162–9.
9 Armistead, P.M. and Thorp, H.H. (2000) Modification of indium tin oxide electrodes with nucleic acids: detection of attomole quantities of immobilized DNA by electrocatalysis. *Analytical Chemistry*, **72**, 3764–70.
10 Jelen, F., Yosypchuk, B., Kourilová, A., Novotný, L. and Paleček, E. (2002) Label-free determination of picogram quantities of DNA by stripping voltammetry with solid copper amalgam mercury in the presence of copper. *Analytical Chemistry*, **74**, 4788–93.
11 Fojta, M. and Paleček, E. (1997) Supercoiled DNA modified mercury electrode: a highly sensitive tool for the detection of DNA damage. *Analytica Chimica Acta*, **342**, 1–12.
12 Pividori, M.I., Merkoçi, A. and Alegret, S. (2000) Electrochemical genosensor design: immobilization of oligonucleotides onto transducer surfaces and detection methods. *Biosensors and Bioelectronics*, **15**, 291–303.
13 Teles, F.R.R. and Fonseca, L.P. (2008) Applications of polymers for biomolecule immobilization in electrochemical biosensors. *Materials Science and Engineering C*, **28**, 1530–43.
14 Wang, J. (2005) Carbon-nanotube based electrochemical biosensors: a review. *Electroanalysis*, **17**, 7–14.
15 Rajesh, Ahuja, T. and Kumar, D. (2009) Recent progress in the development of nano-structured conducting polymers/ nanocomposites for sensor applications. *Sensors and Actuators B*, **136**, 275–86.
16 Pingarrón, J.M., Yáñez-Sedeño, P. and González-Cortés, A. (2008) Gold nanoparticle-based electrochemical biosensors. *Electrochimica Acta*, **53**, 5848–66.
17 Cassidy, J.F., Doherty, A.P. and Vos, J.G. (1998) *Principles of Chemical and Biological Sensors* (ed. D. Diamond), John Wiley & Sons Canada Ltd, Toronto, pp. 73–132.
18 Zacco, E., Pividori, M.I. and Alegret, S. (2006) Electrochemical biosensing based on universal affinity biocomposite platforms. *Biosensors and Bioelectronics*, **21**, 1291–301.
19 Pividori, M.I. and Alegret, S. (2005) Electrochemical genosensing based on rigid carbon composites. A review. *Analytical Letters*, **38**, 2541–65.
20 Millan, K.M. and Mikkelsen, S.K. (1993) Sequence-selective biosensor for DNA based on electroactive hybridization indicators. *Analytical Chemistry*, **65**, 2317–23.
21 Millan, K.M., Saraullo, A. and Mikkelsen, S.K. (1994) Voltammetric DNA biosensor for cystic fibrosis based on a modified carbon paste electrode. *Analytical Chemistry*, **66**, 2943–8.
22 Sun, X., He, P., Liu, S., Ye, L. and Fang, Y. (1998) Immobilization of single stranded deoxyribonucleic acid on gold electrode with self-assembled aminoethanethiol monolayer for DNA electrochemical sensor applications. *Talanta*, **47**, 487–95.
23 Moser, I., Schalkhammer, T., Pittner, F. and Urban, G. (1997) Surface techniques

for an electrochemical DNA biosensor. *Biosensors and Bioelectronics*, **12**, 729–37.

24 Wang, J. and Lin, Y.H. (2008) Functionalized carbon nanotubes and nanofibers for biosensing applications. *Trends in Analytical Chemistry*, **27**, 619–26.

25 Lösche, M. (1997) Protein monolayers at interfaces. *Current Opinion in Solid State and Materials Sciences*, **2**, 546–56.

26 Beesley, J.E. (1989) *Colloidal Gold: Principles, Methods and Applications*, vol. **1** (ed. M.A. Hayat), Academic Press, New York, pp. 421–5.

27 Hayat, M.A. (ed.) (1995) *Immunogold-Silver Staining. Principles, Methods and Applications*, CRC Press, Inc.

28 Stonehuerner, J.G., Zhao, J., O'Daly, J.P., Crumbliss, A.L. and Henkens, R.W. (1992) Comparison of colloidal gold electrode fabrication methods: the preparation of a horseradish peroxidase enzyme electrode. *Biosensors and Bioelectronics*, **7**, 421–8.

29 Zhao, J., O'Daly, J.P., Henkens, R.W., Stonehuerner, J. and Crumbliss, A.L. (1996) A xanthine oxidase/colloidal gold enzyme electrode for amperometric biosensor applications. *Biosensors and Bioelectronics*, **11**, 493–502.

30 Crumbliss, A.L., Perine, S.C., Stonehuerner, J., Tubergen, K.R., Zhao, J. and O'Daly, J.P. (1992) Colloidal gold as a biocompatible immobilization matrix suitable for the fabrication of enzyme electrodes by electrodeposition. *Biotechnology and Bioengineering*, **40**, 483–90.

31 Gole, A., Dash, C., Ramakrishnan, V., Sainkar, S.R., Mandale, A.B., Rao, M. and Sastry, M. (2001) Pepsin-gold colloid conjugates: preparation, characterization, and enzymatic activity. *Langmuir*, **17**, 1674–9.

32 Abad, J.M., Pita, M. and Fernández, V.M. (2006) *Biotechnology: Immobilization of Enzymes and Cells*, vol. **22**, 2nd edn (ed. J.M. Guisan), Humana Press Inc., Totowa, pp. 229–38.

33 Kimura-Suda, H., Petrovykh, D.Y., Tarlov, M.J. and Whitman, L.J. (2003) Base-dependent competitive adsorption of single-stranded DNA on gold. *Journal of the American Chemical Society*, **125**, 9014–15.

34 Nuzzo, R.G. and Allara, D.L. (1983) Adsorption of bifunctional organic disulfides on gold surfaces. *Journal of the American Chemical Society*, **105**, 4481–3.

35 Ulman, A. (1996) Formation and structure of self-assembled monolayers. *Chemical Reviews*, **96**, 1533–54.

36 Bain, C.D., Troughton, E.B., Tao, Y.-T., Evall, J., Whitesides, G.M. and Nuzzo, R.G. (1989) Formation of monolayer films by the spontaneous assembly of organic thiols from solution onto gold. *Journal of the American Chemical Society*, **111**, 321–35.

37 Wink, T.H., van Zuilen, S.J., Bult, A. and van Bennkom, W.P. (1997) Self-assembled monolayers for biosensors. *Analyst*, **122**, 43R–50R.

38 Finklea, H.O., Avery, S., Lynch, M. and Furtsch, T. (1987) Blocking oriented monolayers of alkyl mercaptans on gold electrodes. *Langmuir*, **3**, 409–13.

39 Freire, R.S., Pessoa, C.A. and Kubota, L.T. (2003) Self-assembled monolayers applications for the development of electrochemical sensors. *Quimica Nova*, **26**, 381–9.

40 Wirtz, R., Wälti, C., Tosch, P. and Pepper, M. (2004) Influence of the thiol position on the attachment and subsequent hybridization of thiolated DNA on gold surfaces. *Langmuir*, **20**, 1527–30.

41 Petrovykh, D.Y., Pérez-Dieste, V., Opdahl, A., Kimura-Suda, H., Sullivan, J.M., Tarlov, M.J., Himpsel, F.J. and Whitman, L.J. (2006) Nucleobase orientation and ordering in films of single-stranded DNA on gold. *Journal of the American Chemical Society*, **128**, 2–3.

42 Bain, C.D. and Whitesides, G.M. (1989) Formation of monolayers by coadsorption of thiols on gold: variation in the length of the alkyl chain. *Journal of the American Chemical Society*, **111**, 7164–75.

43 Herne, T.M. and Tarlov, M.J. (1997) Characterization of DNA probes immobilized on gold surfaces. *Journal of the American Chemical Society*, **119**, 8916–20.

44 Steel, A.R., Herne, T.M. and Tarlov, M.J. (1998) Electrochemical quantification of

DNA immobilized on gold. *Analytical Chemistry*, **70**, 4670–7.

45 Abad, J.M., Mertens, S.F.L., Pita, M., Fernández, V.M. and Schiffrin, D.J. (2005) Functionalization of thioctic acid-capped gold nanoparticles for specific immobilization of histidine-tagged proteins. *Journal of the American Chemical Society*, **127**, 5689–94.

46 Traut, R.R., Bollen, A., Sun, T.T., Hershey, J.W., Sundberg, J. and Pierce, L.R. (1973) Methyl 4-mercaptobutyrimidate as a cleavable cross-linking reagent and its application to the *Escherichia coli* 30S ribosome. *Biochemistry*, **12**, 3266–73.

47 Jue, R., Lambert, J.M., Pierce, L.R. and Traut, R.R. (1978) Addition of sulfhydryl groups to *Escherichia coli* ribosomes by protein modification with 2-iminothiolane (methyl 4-mercaptobutyrimidate). *Biochemistry*, **17**, 5399–405.

48 Carpini, G., Lucarelli, F., Marrazza, G. and Mascini, M. (2004) Oligonucleotide-modified screen-printed gold electrodes for enzyme-amplified sensing of nucleic acids. *Biosensors and Bioelectronics*, **20**, 167–75.

49 Cai, H., Xu, C., He, P. and Fang, Y. (2001) Colloid Au-enhanced DNA immobilization for the electrochemical detection of sequence-specific DNA. *Journal of Electroanalytical Chemistry*, **510**, 78–85.

50 Wang, M., Sun, C., Wang, L., Ji, X., Bai, Y., Li, T. and Li, J. (2003) Electrochemical detection of DNA immobilized on gold colloid particles modified self-assembled monolayer electrode with silver nanoparticle label. *Journal of Pharmaceutical and Biomedical Analysis*, **33**, 1117–25.

51 Fu, Y., Yuan, R., Xu, L., Chai, Y., Liu, Y., Tang, D. and Zhang, Y. (2005) Electrochemical impedance behavior of DNA biosensor based on colloidal Ag and bilayer two-dimensional sol–gel as matrices. *Journal of Biochemical and Biophysical Methods*, **62**, 163–74.

52 Oliveira Marques, P.R.B., Lermo, A., Campoy, S., Yamanaka, H., Barbé, J., Alegret, S. and Pividori, M.I. (2009) Double-tagging polymerase chain reaction with a thiolated primer and electrochemical genosensing based on gold nanocomposite sensor for food safety. *Analytical Chemistry*, **81**, 1332–9.

53 Storhoff, J.J. and Mirkin, C.A. (1999) Programmed materials synthesis with DNA. *Chemical Reviews*, **99**, 1849–62.

54 Zhao, Y.-D., Pang, D.-W., Hu, S., Wang, Z.-L., Cheng, J.-K. and Dai, H.-P. (1999) DNA-modified electrodes. 4. Optimisation of covalent immobilisation of DNA on self-assembled monolayers. *Talanta*, **49**, 751–6.

55 Berggren, C., Stalhandske, P., Brundell, J. and Johansson, G. (1999) A feasibility study of a capacitive biosensor for direct detection of DNA hybridization. *Electroanalysis*, **11**, 156–60.

56 Bardea, A., Dagan, A. and Willner, I. (1999) Amplified electronic transduction of oligonucleotide interactions: novel routes for Tay-Sachs biosensors. *Analytica Chimica Acta*, **385**, 33–43.

57 Madoz, J., Kuznetsov, B.A., Medrano, F.J., García, J.L. and Fernández, V.M. (1997) Functionalization of gold surfaces for specific and reversible attachment of a fused β-galactosidase and choline-receptor protein. *Journal of the American Chemical Society*, **119**, 1043–51.

58 Madoz-Gúrpide, J., Abad, J.M., Fernández-Recio, J., Vélez, M., Vázquez, L., Gómez-Moreno, C. and Fernández, V.M. (2000) Modulation of electroenzymatic NADPH oxidation through oriented immobilization of ferredoxin: NADP+ reductase onto modified gold electrodes. *Journal of the American Chemical Society*, **122**, 9808–17.

59 Abad, J.M., Vélez, M., Santamaría, C., Guisán, J.M., Matheus, P.R., Vázquez, L., Gazaryan, I., Gorton, L., Gibson, T. and Fernández, V.M. (2002) Immobilization of peroxidase glycoprotein on gold electrodes modified with mixed epoxy-boronic acid monolayers. *Journal of the American Chemical Society*, **124**, 12845–53.

60 Mann, S., Shenton, W., Li, M., Connolly, S. and Fitzmaurice, D. (2000) Biologically programmed nanoparticle assembly. *Advanced Materials*, **12**, 147–50.

61 Mrksich, M. (2000) A surface chemistry approach to studying cell adhesion. *Chemical Society Reviews*, **29**, 267–73.

62 Shenhar, R. and Rotello, V.M. (2003) Nanoparticles: scaffolds and building blocks. *Accounts of Chemical Research*, **36**, 549–61.

63 Elghanian, R., Storhoff, J.J., Mucic, R.C., Letsinger, R.L. and Mirkin, C.A. (1997) Selective colorimetric detection of polynucleotides based on the distance dependent optical properties of gold nanoparticles. *Science*, **277**, 1078–81.

64 Mulvaney, P. (1996) Surface plasmon spectroscopy of nanosized metal particles. *Langmuir*, **12**, 788–800.

65 Taton, T.A., Mirkin, C.A. and Letsinger, R.L. (2000) Scanometric DNA array detection with nanoparticle probes. *Science*, **289**, 1757–60.

66 Nath, N. and Chilkoti, A. (2002) A colorimetric gold nanoparticle sensor to interrogate biomolecular interactions in real time on a surface. *Analytical Chemistry*, **74**, 504–9.

67 González-García, M.B., Fernández-Sánchez, C. and Costa-García, A. (2000) Colloidal gold as an electrochemical label of streptavidin–biotin interaction. *Biosensors and Bioelectronics*, **15**, 315–21.

68 Dequaire, M., Degrand, C. and Limoges, B. (2000) An electrochemical metalloimmunoassay based on a colloidal gold label. *Analytical Chemistry*, **72**, 5521–8.

69 Wang, J., Polsky, R. and Xu, D. (2001) Silver-enhanced colloidal gold electrochemical stripping detection of DNA hybridization. *Langmuir*, **17**, 5739–41.

70 Guo, S. and Wang, E. (2007) Synthesis and electrochemical applications of gold nanoparticles. *Analytica Chimica Acta*, **598**, 181–92.

71 Faraday, M. (1857) Experimental relations of gold (and other metals) to light. *Philosophical Transactions of the Royal Society of London*, **147**, 145–81.

72 Turkevich, J., Stevenson, P.C. and Hillier, J. (1951) A study of the nucleation and growth processes in the synthesis of colloidal gold. *Discussions of the Faraday Society*, **11**, 55–75.

73 Schmid, G. (1992) Large clusters and colloids. Metals in the embryonic state. *Chemical Reviews*, **92**, 1709–27.

74 Brust, M., Walker, M., Bethell, D., Schriffrin, D.J. and Whyman, R.J. (1994) Synthesis of thiol-derivatized gold nanoparticles in a 2-phase liquid-liquid system. *Journal of the Chemical Society, Chemical Communications*, 801–2.

75 Brust, M., Fink, J., Bethell, D., Schiffrin, D.J. and Kiely, C.J. (1995) Synthesis and reactions of functionalised gold nanoparticles. *Journal of the Chemical Society, Chemical Communications*, 1655–6.

76 Schmid, G. and Chi, L.F. (1998) Metal clusters and colloids. *Advanced Materials*, **10**, 515–26.

77 Schmid, G. (2001) *Nanoscale Materials in Chemistry* (ed. K.J. Klabunde), John Wiley & Sons, Inc., New York, pp. 15–59.

78 Schmid, G., Bäumle, M., Geerkens, M., Heim, I., Osemann, C. and Sawitowski, T. (1999) Current and future applications of nanoclusters. *Chemical Society Reviews*, **28**, 179–85.

79 Feldheim, D.L. and Colby, A.F. Jr. (eds) (2002) *Metal Nanoparticles – Synthesis, Characterization and Applications*, Marcel Dekker, New York.

80 Daniel, M.C. and Astruc, D. (2004) Gold nanoparticles: assembly, supramolecular chemistry, quantum-size-related properties, and applications toward biology, catalysis, and nanotechnology. *Chemical Reviews*, **104**, 293–346.

81 Sudeep, P.K. and Emrick, T. (2007) Polymer-nanoparticle composites: preparative methods and electronically active materials. *Polymer Reviews*, **47**, 155–63.

82 Ruschau, G.R., Newnham, R.E., Runt, J. and Smith, E. (1989) 0–3 ceramic polymer composite chemical sensors. *Sensors and Actuators*, **20**, 269–75.

83 Alegret, S., Merkoçi, A., Pividori, M.I. and Del Valle, M. (2005) *Encyclopedia of Sensors*, vol. **3** (eds C.A. Grimes, E.C. Dickey and M.V. Pishko), American Scientific Publishers, pp. 23–44.

84 Bockstaller, M.R., Mickievitch, R.A. and Thomas, E.L. (2005) Block copolymer nanocomposites: perspectives for tailored functional materials. *Advanced Materials*, **17**, 1331–49.

85 Céspedes, F., Fàbregas, E. and Alegret, S. (1996) New materials for electrochemical sensing I. Rigid conducting composites. *Trends in Analytical Chemistry*, **15**, 296–304.

86 Céspedes, F. and Alegret, S. (2000) New materials for electrochemical sensing II. Rigid carbon–polymer biocomposites trends. *Analytical Chemistry*, **19**, 276–85.

87 Borole, D.D., Kapadi, U.R., Mahulikar, P.P. and Hundiwale, D.G. (2006) Conducting polymers: an emerging field of biosensors. *Designed Monomers and Polymers*, **9**, 1–11.

88 Gerard, M., Chaubey, A. and Malhotra, B.D. (2002) Application of conducting polymers to biosensors. *Biosensors and Bioelectronics*, **17**, 345–59.

89 Godovski, D.Y., Koltypin, E.A., Volkov, A.V. and Moskvina, M.A. (1993) Sensor properties of filled polymer composites. *Analyst*, **118**, 997–9.

90 Lowe, A.B., Sumerlin, B.S., Donovan, M.S. and McCormick, C.L. (2002) Facile preparation of transition metal nanoparticles stabilized by well-defined (Co)polymers synthesized via aqueous reversible addition-fragmentation chain transfer polymerization. *Journal of the American Chemical Society*, **124**, 11562–3.

91 Watson, K.J., Zhu, J., Nguyen, S.T. and Mirkin, C.A. (1999) Hybrid nanoparticles with block copolymer shell structures. *Journal of the American Chemical Society*, **121**, 462–3.

92 Hong, R., Fischer, R., Verma, R., Goodman, C.M., Emrick, T. and Rotello, V.M. (2004) Control of protein structure and function through surface recognition by tailored nanoparticle scaffolds. *Journal of the American Chemical Society*, **126**, 739–43.

93 Willner, I. and Katz, E. (2000) Integration of layered redox proteins and conductive supports for bioelectronic applications. *Angewandte Chemie, International Edition*, **39**, 1180–218.

94 Pandey, P., Datta, M. and Malhotra, B.D. (2008) Prospects of nanomaterials in biosensors. *Analytical Letters*, **41**, 159–209.

95 Adams, R.N. (1958) Carbon paste electrodes. *Analytical Chemistry*, **30**, 1576.

96 Liu, S.Q. and Ju, H.X. (2003) Reagentless glucose biosensor based on direct electron transfer of glucose oxidase immobilized on colloidal gold modified carbon paste electrode. *Biosensors and Bioelectronics*, **19**, 177–83.

97 Liu, S.Q. and Ju, H.X. (2002) Renewable reagentless hydrogen peroxide sensor based on direct electron transfer of horseradish peroxidase immobilized on colloidal gold-modified electrode. *Analytical Biochemistry*, **307**, 110–16.

98 Liu, S., Yu, J. and Ju, H. (2003) Renewable phenol biosensor based on a tyrosinase colloidal gold modified carbon paste electrode. *Journal of Electroanalytical Chemistry*, **540**, 61–7.

99 Liu, S. and Ju, H. (2003) Nitrite reduction and detection at a carbon paste electrode containing hemoglobin and colloidal gold. *Analyst*, **128**, 1420–4.

100 Çubukçu, M., Timur, S. and Anik, Ü. (2007) Examination of performance of glassy carbon paste electrode modified with gold nanoparticle and xanthine oxidase for xanthine and hypoxanthine detection. *Talanta*, **74**, 434–9.

101 Carralero, C., Mena, M.L., González-Cortés, A., Yáñez-Sedeño, P. and Pingarrón, J.M. (2006) Development of a high analytical performance-tyrosinase biosensor based on a composite graphite-Teflon electrode modified with gold nanoparticles. *Biosensors and Bioelectronics*, **22**, 730–6.

102 Manso, J., Mena, M.L., Yáñez-Sedeño, P. and Pingarrón, J.M. (2007) Electrochemical biosensors based on colloidal gold–carbon nanotubes composite electrodes. *Journal of Electroanalytical Chemistry*, **603**, 1–7.

103 Feng, H., Wang, H., Zhang, Y., Yan, B., Shen, G. and Yu, R. (2007) A direct electrochemical biosensing platform constructed by incorporating carbon nanotubes and gold nanoparticles onto redox poly(thionine) film. *Analytical Sciences*, **23**, 235–9.

104 Njagi, J. and Andreescu, S. (2007) Stable enzyme biosensors based on chemically synthesized Au-polypyrrole nanocomposites. *Biosensors and Bioelectronics*, **23**, 168–75.

105 Xian, Y., Hu, Y., Liu, F., Xian, Y., Wang, H. and Jin, L. (2006) Glucose biosensor based on Au nanoparticles-conductive polyaniline nanocomposites. *Biosensors and Bioelectronics*, **21**, 1996–2000.

106 Park, J.E., Atobe, M. and Fuchigami, T. (2005) Sonochemical synthesis of inorganic organic hybrid nanocomposite based on gold nanoparticles and polypyrrole. *Chemistry Letters*, **34**, 96–7.

107 Park, J.E., Atobe, M. and Fuchigami, T. (2005) Sonochemical synthesis of conducing polymer–metal nanoparticles nanocomposite. *Electrochimica Acta*, **51**, 849–54.

108 Sorlier, P., Denuzière, A., Viton, C. and Domard, A. (2001) Relation between the degree of acetylation and the electrostatic properties of chitin and chitosan. *Biomacromolecules*, **2**, 765–72.

109 Luo, X.L., Xu, J.J., Du, Y. and Chen, H.Y. (2004) A glucose biosensor based o chitosan–glucose oxidase–gold nanoparticles biocomposite formed by one-step electrodeposition. *Analytical Biochemistry*, **334**, 284–9.

110 Luo, X.L., Xu, J.J., Zhang, Q., Yang, G.J. and Chen, H.Y. (2005) Electrochemically deposited chitosan hydrogel for horseradish peroxidase immobilization through gold nanoparticles self-assembly. *Biosensors and Bioelectronics*, **21**, 190–6.

111 Xu, Q., Mao, C., Liu, N.N., Zhu, J.J. and Shen, J. (2006) Direct electrochemistry of horseradish peroxidase based on biocompatible carboxymethyl chitosan–gold nanoparticle nanocomposite. *Biosensors and Bioelectronics*, **22**, 768–73.

112 Li, X., Wu, J., Gao, N., Shen, G. and Yu, R. (2006) Electrochemical performance of l-cysteine–gold particle nanocomposite electrode interface as applied to preparation of mediator-free enzymatic biosensors. *Sensors and Actuators B*, **117**, 35–42.

113 Jia, J., Wang, B., Wu, A., Cheng, G., Li, Z. and Dong, S. (2002) A method to construct a third-generation horseradish peroxidase biosensor: self-assembling gold nanoparticles to three-dimensional sol-gel network. *Analytical Chemistry*, **74**, 2217–23.

114 Du, D., Chen, S., Cai, J. and Zhang, A. (2008) Electrochemical pesticide sensitivity test using acetylcholinesterase biosensor based on colloidal gold nanoparticles modified sol-gel interface. *Talanta*, **74**, 766–72.

115 Jena, B.K. and Raj, C.R. (2006) Electrochemical biosensor based on integrated assembly of dehydrogenase enzymes and gold nanoparticles. *Analytical Chemistry*, **78**, 6332–9.

116 Di, J., Shen, C., Peng, S., Tu, Y. and Li, S. (2005) A one-step method to construct a third-generation biosensor based on horseradish peroxidase and gold nanoparticles embedded in silica sol–gel network on gold modified electrode. *Analytica Chimica Acta*, **553**, 196–200.

117 Tang, D., Yuan, R., Chai, Y., Zhong, X., Liu, Y. and Dai, J. (2006) Electrochemical detection of hepatitis B surface antigen using colloidal gold nanoparticles modified by a sol-gel network interface. *Clinical Biochemistry*, **39**, 309–14.

118 Zhuo, Y., Yuan, R., Chai, Y., Tang, D., Zhang, Y., Wang, N., Li, X. and Zhu, Q. (2005) A reagentless amperometric immunosensor based on gold nanoparticles/thionine/Nafion-membrane-modified gold electrode for determination of α-1-fetoprotein. *Electrochemistry Communications*, **7**, 355–60.

119 Tang, D., Yuan, R. and Chai, Y. (2006) Electrochemical immuno-bioanalysis for carcinoma antigen 125 based on thionine and gold nanoparticles-modified carbon paste interface. *Analytica Chimica Acta*, **564**, 158–65.

120 Du, D., Xu, X., Wang, S. and Zhang, A. (2007) Reagentless amperometric carbohydrate antigen 19-9 immunosensor based on direct electrochemistry of immobilized horseradish peroxidase. *Talanta*, **71**, 1257–62.

121 Chen, J., Tang, J., Yan, F. and Ju, H. (2006) A gold nanoparticles/sol-gel composite architecture for encapsulation of immunoconjugate for reagentless electrochemical immunoassay. *Biomaterials*, **27**, 2313–21.

122 Du, D., Liu, S., Chen, J., Ju, H., Lian, H. and Li, J. (2005) Colloidal gold nanoparticle modified carbon paste interface for studies of tumor cell adhesion and viability. *Biomaterials*, **26**, 6487–95.

5
Quantum Dot Nanocomposites

Jianxiu Wang, Yunfei Long, and Feimeng Zhou

5.1
Introduction

Quantum dots (QDs) are group II–VI, III–V, and IV–VI semiconductors, with dimensions ranging from 0.2 to 100 nm. The large surface-to-volume ratios and discrete electronic energy states provide the QDs with unique electronic, optical, magnetic, and mechanical properties. Over recent years, QDs have attracted tremendous interest because their properties not only differ remarkably from those of bulk semiconductors, but they can also be tailored by a judicious control of particle composition, size, and surface [1–7]. The size-dependent photoluminescence (PL) color, absorption and PL spectra of CdSe and (CdSe)ZnS QDs are depicted in Figure 5.1 [8].

Quantum dots, with sizes comparable to those of biomacromolecules, have been used successfully as fluorescent biosensors because they can reasonably resist photobleaching, denaturation of biomolecules, and alterations in pH and temperature [9–11]. Proteins, nucleic acids and other biomolecules have been immobilized on QDs, thus enabling the detection of specific biological processes as diverse as DNA hybridization, antigen–antibody recognition, DNA–protein interaction, and biotin–avidin complexation [12, 13].

5.2
Synthesis and Characterization of QDs

In order to form QD nanocomposites by tethering other species onto the surface of the QDs, the preparation and characterization of QDs represent a crucial step. Typically, QDs can be synthesized in organic media, in aqueous solutions, or in two-phase systems [14–23].

Nanomaterials for the Life Sciences Vol.8: Nanocomposites. Edited by Challa S. S. R. Kumar

ISBN: 978-3-527-32168-1

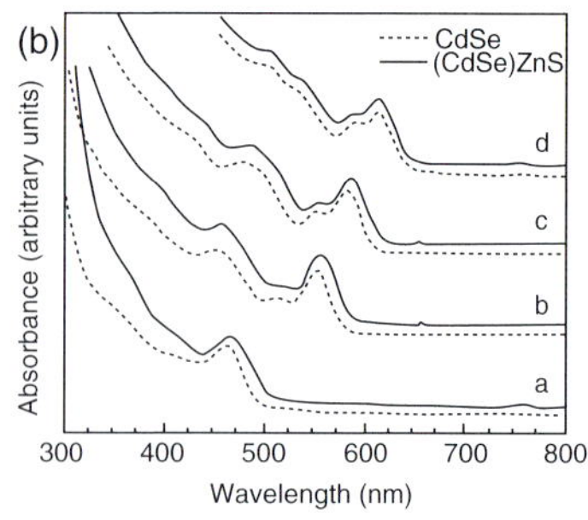

Figure 5.1 (a) Color photograph showing the wide spectral range of bright fluorescence from different size samples of (CdSe)ZnS. The PL peaks occur at (left to right) 470, 480, 520, 560, 594, and 620 nm; (b) Absorption spectra for bare (dashed lines) and CdSe dots coated with one to two monolayers of ZnS (solid lines) with diameters measuring (a) 23, (b) 42, (c) 48, and (d) 55 Å. The absorption spectra for the (CdSe)ZnS dots are broader and slightly red-shifted from the respective bare dot spectra. Reproduced with permission from Ref. [8]; © 1997, American Chemical Society.

5.2.1
Synthesis of QDs in Organic Solvents

Generally, the synthesis of QDs in organic solvents can be conducted via the hot-injection method, though the heating-up method has also been used if a large-scale preparation of monodisperse QDs is desired [15]. Although, during the 1980s, CdSe QDs were prepared using "top-down" techniques, such as lithography [7], problems with size variations, crystal defects, poor reproducibility, and poor optical properties of the resultant QDs made them unsuitable for further applications. Later, Bawendi and coworkers successfully synthesized CdSe by using a "bottom-up" technique that allowed a controlled growth of macroscopic quantities of nanocrystallites [24]. The semiconductor nanocrystallites are nearly monodispersive in a variety of solvents, and exhibit high optical quality, such as sharp absorption features and strong "band-edge" emission.

The synthesis of QDs in organic solvents has proven to be advantageous, because the reaction temperature can be tuned over a wide range, and the specific reactants have been exhaustively explored in this environment [15]. Both, the temperature and solvent composition exert strong influences on the growth kinetics and shape of the nanocrystals [24–27]. For example, CdTe nanocrystals synthesized in organic solvents at higher temperatures generally possess a hexagonal wurtzite phase [24], yet careful control of the reaction conditions can result in a cubic zinc blende phase [27]. Typically, the solvent plays two roles: (i) to solubilize and disperse the nanocrystals and reactants involved in the growth process; and (ii) to control the rate of the reaction. The solvent may include a mixture of different species, such as pure solvents and pure surfactants.

Most often, the synthesis of QDs in organic media involves the use of amphiphilic compounds. For example, in the presence of trioctylphosphine and trioctylphosphine oxide, QDs can be fabricated with a high degree of monodispersity, and the emission quantum yield will be close to 100% [24, 28, 29]. The use of other amphiphilic compounds in this role, such as amines and carboxylic acids, has also been demonstrated [30–32].

5.2.2 Synthesis of QDs in Aqueous Media

The key to using QDs in biological systems is to ensure their water solubility, biocompatibility, and photostability. The surface functionality of QDs enables their efficient coupling to biological species, and provides the possibility to utilize their superior photoluminescence properties in biological studies [33]. To date, a great number of methods for the synthesis of water-soluble QDs have been investigated, including the direct synthesis of size-controlled and highly luminescent CdSe QDs in aqueous phases [20–23], and the synthesis of water-soluble QDs via surface modification or ligand exchange of QDs obtained in organic media [34, 35]. Both, citrate- and thiol-stabilized CdSe QDs have been prepared from sodium hydroselenide and *N,N*-dimethylselenourea by Rogach and coworkers [23]. The preparation of water-soluble QDs by ligand exchange, followed by encapsulation in silica shells and micelles, has been summarized by Colvin *et al.* [36]. Water-soluble QDs have also been prepared by replacing oleic acid (OA) attached to the surface of QDs with mercaptopropionic acid [35].

Kumar *et al.* have reported the synthesis of novel nanostructures of crystalline Zn^{2+}-passivated quantized (Q)-PbS (Zn/PbS) by using RNA as a water-soluble template [37]. The characterization of the above nanomaterials was carried out using field-emission scanning electron microscopy (FE-SEM), coupled with energy-dispersive X-ray (EDX), transmission electron microscopy (TEM), and both infrared (IR) and nuclear magnetic resonance (NMR) spectroscopies. Interestingly, the transformation in morphology from QDs to nanofibers via the formation of nanowires, and an efficient visible light-emitting PbS grown on an RNA template were obtained. Electron microscopy images of the PbS nanoparticles on the RNA matrix containing varied amounts of zinc; typical images of the nanocrystals, aging for different times to form nanowires and nanofibers, are shown in Figure 5.2.

Patel *et al.* have reported the synthesis of PbS QDs using DNA as a surface capping agent [38], whereby the average size of the DNA-capped PbS nanocrystals was 3–8 nm. However, by controlling the surface with surfactant molecules so as to influence the nucleation and growth processes, it was possible to obtain nanoparticles with different sizes and shapes.

Phytochelatins (PCs), as a class of cysteine-containing peptides, play an important role in the detoxification of Cd^{2+} in plants and yeasts [39, 40]. The PCs possess a unique structure of $(\gamma\text{-Glu-Cys})_n$-Gly ($n = 2$–11) that contains at least two cysteine residues capable of coordinating Cd^{2+} through the thiolate ligands. The PC-bound cadmium ions are first transported to vesicles and then combined with physiologically generated sulfide ions to form CdS nanocrystals. The adsorption of PCs onto

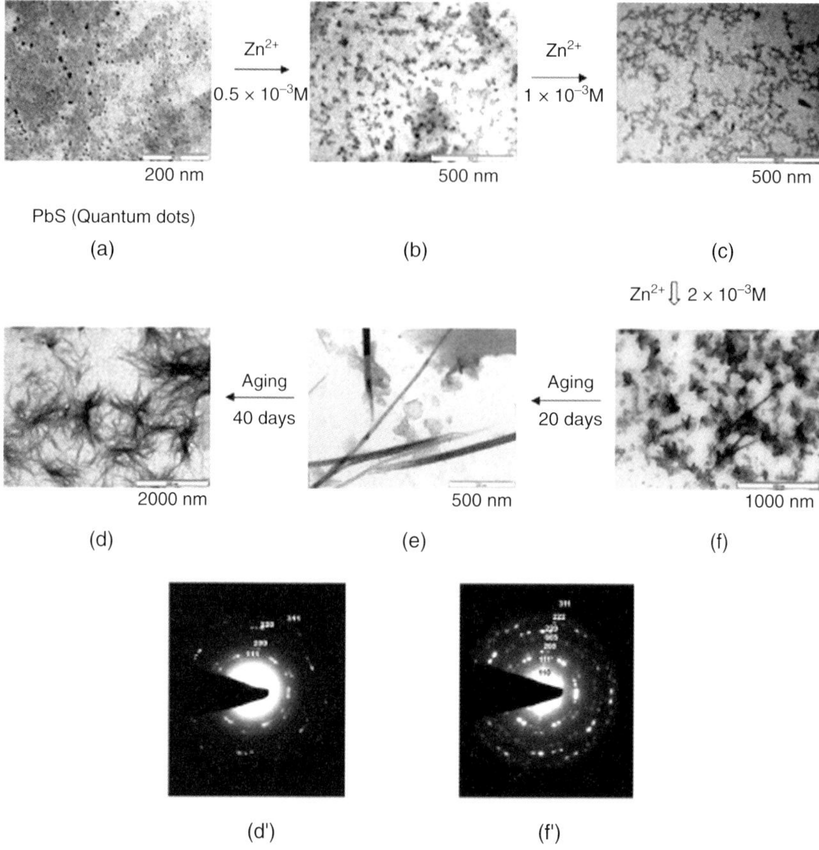

Figure 5.2 Transmission electron microscopy images showing the change in morphology of PbS particles as a function of added [Zn^{2+}] ($\times 10^{-3}$ mol dm^{-3}): 0 (a), 0.5 (b), 1.0 (c), 2.0 (d), and upon aging them for 20 days (e) and 40 days (f). Panels (d′) and (f′) show selected area electron diffraction patterns of the nanostructures shown in images (d) and (f). Reproduced with permission from Ref. [37]; © 2009, American Chemical Society.

the surface of the growing CdS QDs thus inhibits the influx of ions and arrests particle growth. In addition to PCs, glutathione (GSH) has also been found to be effective in limiting the size of the formed CdS particles [41, 42]. The CdS particles produced in the presence of PCs exhibited a high monodispersity with diameters of 2.4 nm, whereas the GSH-capped CdS crystallites were more polydisperse, with diameters ranging from 2.2 to 3.5 nm [41]. However, peptides lacking cysteine residues failed to control the growth of CdS, producing nonfluorescent particles with average diameters of about 5.9 nm [43]. The results of further studies have also indicated that peptides containing two or more cysteine residues are more effective for capping CdS than are those with a single cysteine residue [43]. The binding of cysteine residues to CdS is dynamic, and peptides comprising multiple cysteine residues might remain anchored on the CdS nanoparticles, even if one of these residues is disassociated from the surface [43].

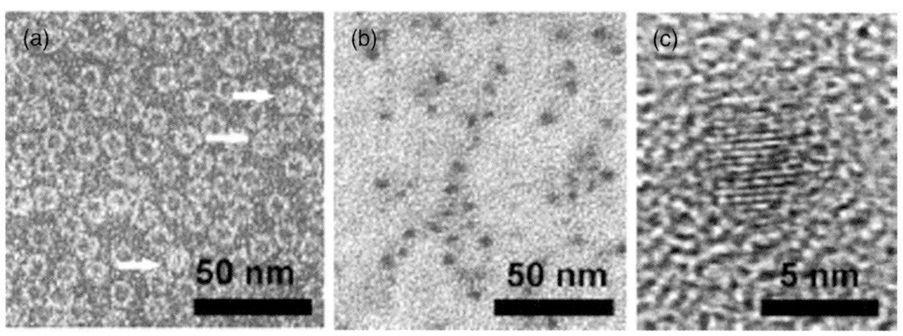

Figure 5.3 Transmission electron microscopy (TEM) images of ferritin and apoferritin synthesized by the newly designed chemical reaction system. (a) The synthesized ZnSe ferritin molecule stained with 1% aurothioglucose. Cores are surrounded by white rings, which are negatively stained protein shells. Almost all molecules accommodate cores, except those indicated by arrowheads, which are apoferritin. The core formation ratio (CFR) was calculated by dividing the number of core-formed apoferritins by all apoferritins in the TEM image; here, the CFR is clearly more than 90%; (b) TEM image of the synthesized ZnSe ferritin molecule without staining. The cores are homogeneous; (c) High-magnification core image and the lattice of the ZnSe core crystal are clearly observed. Reproduced with permission from Ref. [45]; © 2005, American Chemical Society.

The aqueous dispersive stability and constrained internal reaction cavity of *apoferritin* make it an excellent candidate for the synthesis of water-soluble QDs. The use of apoferritin in this role was first reported by Wong *et al.* in 1996 [44] when, due to the strong binding ability of apoferritin to large amounts of Cd^{2+}, the formation of CdS occurred exclusively within the inner cavity of the protein, but not on the outer surface. The as-prepared ferritin/CdS nanocomposites were stable in aqueous medium for a long time [44]. Furthermore, ZnSe nanoparticles have been synthesized in the cavity of the cage-shaped apoferritin protein, via a slow chemical reaction system [45]. Typical TEM images of the synthesized ZnSe ferritin molecules are shown in Figure 5.3.

5.2.3
Synthesis of QDs in the Two-Phase System

By far, a large variety of semiconductor QDs have been synthesized either through one-phase [46–52] or two-phase methods [16–19, 53]. The former approach involves the reaction of molecular monomers in a homogeneous solution at higher temperatures [51, 52], where the nucleation and growth steps are clearly separated to ensure a narrow size distribution [46, 48]. In contrast, the latter approach involves the reaction of molecular precursors in a heterogeneous solvent medium comprising water and organic solvents [16, 19, 53]. Compared to the one-phase method, the two-phase approach experiences an overlapped nucleation process with the growth step, and the as-prepared QDs exhibit extremely narrow size distributions [53]. Pan *et al.* have attempted to elucidate the nucleation and growth processes of CdS QDs prepared with a two-phase method [53]. The effects of monomer

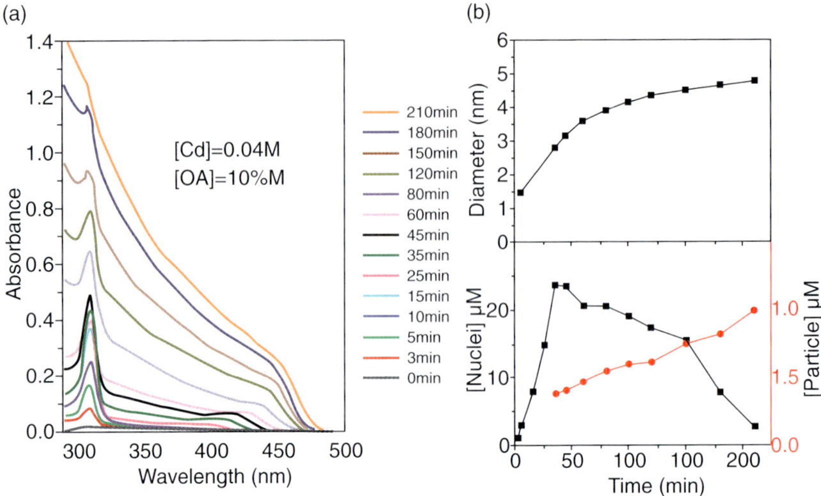

Figure 5.4 (a) Temporal evolution of UV/visible absorption spectra of OA-capped CdS QDs synthesized with a cadmium precursor concentration of 0.04 M in toluene-water media; (b) Calculated size of the QDs (upper graph) and concentration of the nuclei and QDs (lower graph) as a function of time. Reproduced with permission from Ref. [53]; © 2008, American Chemical Society.

concentration, capping-agent concentration, and solvent polarity on the nucleation and growth kinetics have been investigated systematically, using ultraviolet-visible (UV-vis) spectrophotometry. The variation with reaction time of the UV/visible absorption spectra of the critical nuclei and QDs, at a fixed cadmium precursor concentration in a toluene/water reaction medium, is shown in Figure 5.4. The absorbance of the critical nuclei at 311 nm is gradually increased with the reaction time, reaching a maximum after 15 min and disappearing after 210 min. It was of interest to note that the absorption peak at the longer wavelength corresponded to the QDs formed during the nucleation process. Moreover, the growing size of the nanocrystal led to a red shift of the band-edge absorption. Such spectral evolution provides information concerning the dynamic processes involving the generation, consumption, and growth of the nuclei into QDs.

Solvent polarity has a profound influence on the size distribution and PL quantum yield of the nanocrystals [53]. Those nanocrystals synthesized in solvents of relatively high polarity (e.g., toluene, chlorobenzene) exhibited first excitonic absorption peaks and narrower PL bands in comparison with those synthesized in solvents of lower polarity (e.g., octane, 1-octadecene) (Figure 5.5). The size distributions of the nanocrystals synthesized in toluene and chlorobenzene (Figure 5.5b, upper panel) were approximately 5–15%, while those synthesized in 1-octadecene and octane were 15–25% (Figure 5.5b, lower panel). The PL quantum yields for the nanocrystals synthesized in toluene or chlorobenzene (10–20%) were much higher than those synthesized in octane or 1-octadecene (1–2%). Thus, the higher

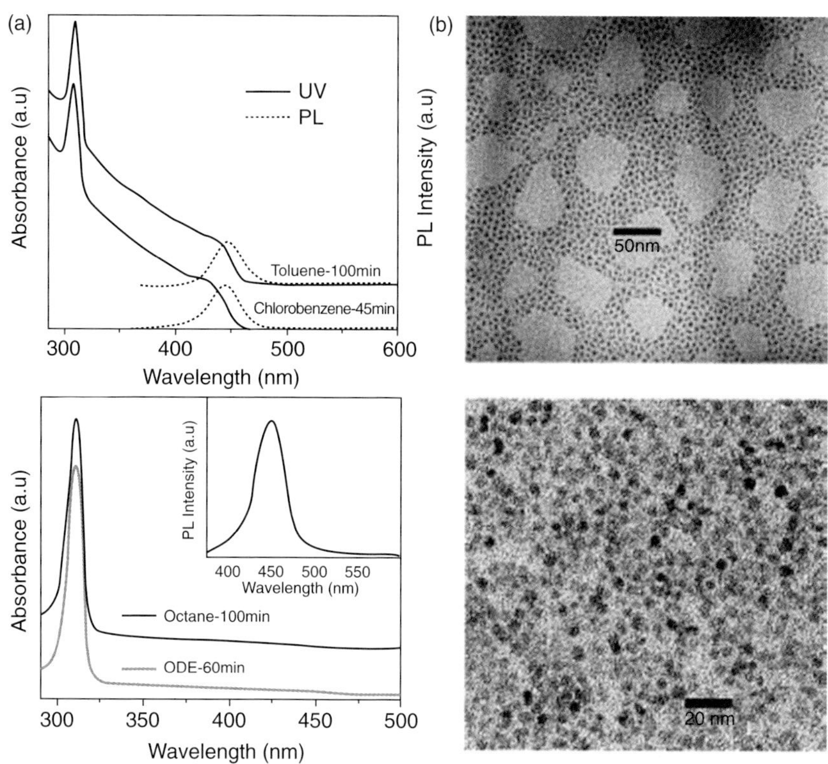

Figure 5.5 (a) UV/visible absorption and PL spectra of the nanocrystals with similar sizes synthesized in toluene and chlorobenzene (upper) and octane and 1-octadecene (lower). The inset shows the PL spectrum of the nanocrystals synthesized in 1-octadecene; (b) TEM images of the nanocrystals synthesized in toluene (upper) and 1-octadecene (lower). Reproduced with permission from Ref. [53]; © 2008, American Chemical Society.

solvent polarity favored formation of those nanocrystals with narrower size distributions and higher quantum yields.

5.3
Synthesis and Characterization of the Core–Shell QDs

Core–shell QDs have been prepared via the growth of a high band gap shell semiconductor material on the core material surface with enhanced luminescence and electrical properties [8, 54–56]. Panda *et al.* reported the synthesis of CdS/ZnS core–shell nanorods via the alternate electrostatic adsorption of Zn^{2+} and S^{2-} onto the CdS nanorod surface functionalized with citric acid (CA) [54]. The small and functionalized multivalent anions of CA encourage the formation of a uniform and finely grown core–shell material. Compared to the absorption band of the uncoated CdS nanorods, the absorption band edge of the core–shell material is slightly red-shifted due to the absorbance resulting from the leakage of excitons from the CdS nanorods into the ZnS shell [54, 57]. It should be noted that no emission from the ZnS shell material was observed. The increase in emission intensity of the core–shell structure suggests that tunneling of the charge carriers from the core to the shell was suppressed by the high band gap ZnS shell. As a result, a greater number of photogenerated electrons and holes will be confined inside the core, leading to a passivation of the nonradiative recombination sites on the core surface [8, 54, 56, 58]. As an alternative to organic fluorophores, core–shell QDs have been utilized as bioactive fluorescent probes in both cell imaging and evaluations of cytotoxicity [34, 55, 59, 60]. CdSe/CdS/ZnS double-shell nanorods with a high photochemical stability and high PL efficiency have been synthesized [55]. In this case, the CdSe nanocrystals first served as seeds to synthesize the "dot-in-a-rod" CdSe(dot)/CdS(rod) core–shell nanorods, after which one to four monolayers of wurtzite ZnS shell were grown epitaxially over the CdSe/CdS nanocrystals (Figure 5.6, left panels). The resultant core–shell and core–shell–shell structures displayed narrow length and diameter distributions. The PL quantum yields of the starting CdSe nanocrystals, the CdSe/CdS nanorods, and the CdSe/CdS/ZnS double-shell structures were estimated to be about 15%, 50%, and 75%, respectively. The more efficient PL of the double-shell nanorods resulted from a passivation of the surface states of CdS by the wide band-gap ZnS shell. The CdSe/CdS/ZnS double-shell nanorods exhibited a lower intracellular toxicity and a higher fluorescence signal in HeLa cell imaging (Figure 5.6, right panels), further demonstrating the potential of these semiconductor nanocrystals in biological applications.

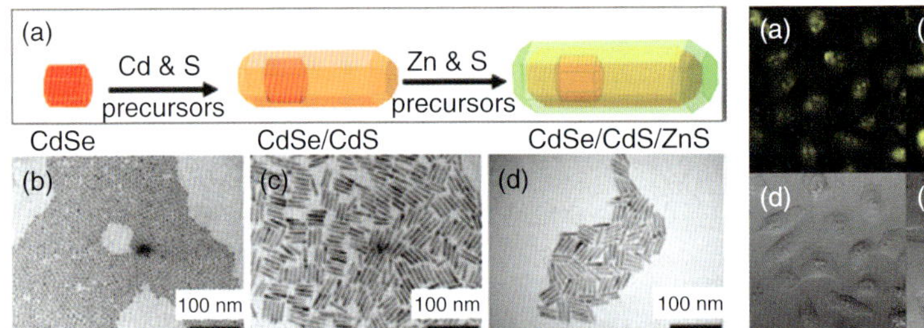

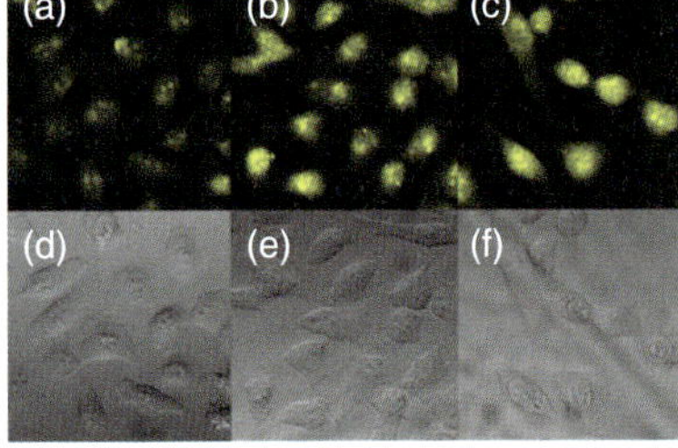

Figure 5.6 Left panels: (a) Schematic diagram showing the growth of CdS nanorods over the CdSe cores, and the growth of a ZnS shell over the resultant CdSe/CdS nanorods to form the double-shell CdSe/CdS/ZnS nanorods. TEM images of (b) starting CdSe cores, (c) CdSe/CdS nanorods, and (d) CdSe/CdS/ZnS nanorods. Right panels: Confocal images of HeLa cells incubated with the CdSe/ZnS QD sample (a and d), with the CdSe/CdS core–shell nanorod sample (b and e), and with the CdSe/CdS/ZnS double-shell nanorod sample (c and f) for 24 h at the same Cd concentration ($[Cd]_{sol} = 50\,\mu M$). Images a–c were acquired using a 488 nm excitation laser and a 605 ± 15 nm filter; images d–f are bright-field images. Reproduced with permission from Ref. [55]; © 2009, American Chemical Society.

5.4 Quantum Dot-Based Metal Ion Fluorescent Sensors

Due to their unique optical and electrical properties, QDs exhibit remarkably important advantages that include a high fluorescence quantum yield, and a high stability towards both photodegradation and chemical degradation [9, 61, 62]. Chen and coworkers have reported the synthesis of water-soluble and highly uniform L-cysteine-capped CdSe nanoparticles, and their applications for the detection of Hg^{2+} with a detection limit of approximately 4.5 nM [61]. The significant quenching of fluorescence of functionalized CdSe nanoparticles in the presence of Hg^{2+} originated from the strong affinity of Hg^{2+} as a "soft metal ion" to cysteine. Certain metal ions, such as Cd^{2+}, Zn^{2+}, Cu^{2+} and Fe^{3+}, have a profound influence on the fluorescence intensity of CdSe nanoparticles at a relatively high concentration. Such interference was caused via an interaction of these metal ions with the surface functional groups of nanoparticles, or by an inner filter effect. Recently, a novel fluorescent CdS-encapsulated DNA nanocomposite was synthesized which could be used for highly selective and sensitive Hg^{2+} analysis [63]. In this case, the nanocomposite was synthesized via the alternate adsorption of Cd^{2+} and S^{2-} onto the DNA template, confined in an agarose gel. The use of a gel matrix resulted in the formation of a rod-shaped nanocomposite, due to the migration of double-stranded (ds) DNA as long rods within the matrix. The CdS/DNA nanocomposite yielded a distinct emission at 330 nm, which differed from that of conventional CdS QDs (Figure 5.7a) [64–68]. According to the theory of the quantum confinement effect [69], the CdS coating contains small CdS particles with diameters smaller than 5.0 nm, and the nanoparticles can be resolved by using atomic force microscopy (AFM). The fluorescence of the CdS/DNA nanocomposite can be significantly quenched by trace amounts of Hg^{2+}, with quenching resulting from

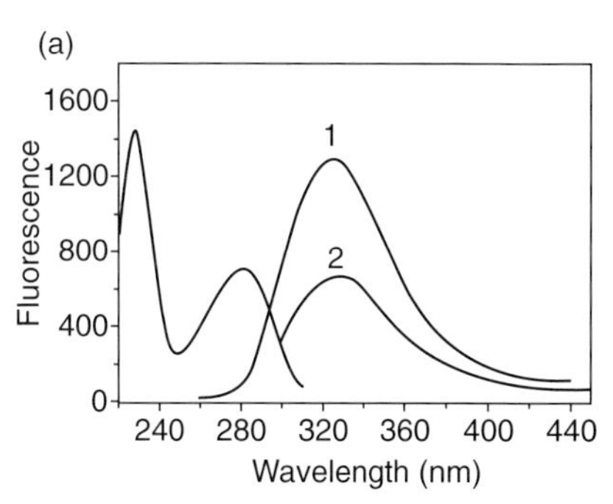

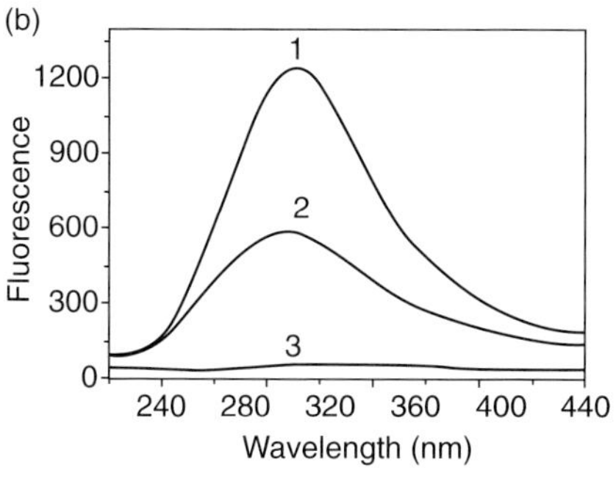

Figure 5.7 (a) Excitation spectrum (left curve) of 8.7 $\mu g\,ml^{-1}$ CdS/DNA nanocomposite solution measured with 330 nm as the emission wavelength and the emission spectra (right curves) collected with the excitation wavelengths at 228 nm (curve 1) and 280 nm (curve 2), respectively; (b) Fluorescence spectra of the CdS/DNA nanocomposite in the absence (curve 1) and presence of 20 μM (curve 2) and 200 μM Hg^{2+} (curve 3). The excitation wavelength was set at 228 nm. Reproduced with permission from Ref. [63]; © 2009, American Chemical Society.

a combined effect of formation of the much more insoluble HgS, the bridging S–Hg–S bonds at the surface, and the heavy-atom effect of Hg^{2+} (Figure 5.7b). The influence on the fluorescence quenching of the CdS/DNA nanocomposite of the other metal ions and of the common anions, was negligible. The nanocomposite was applied to the analysis of Hg^{2+} in a wastewater sample, with high sensitivity and selectivity. This synthetic approach may be extended to other types of DNA–semiconductor conjugates, with tailored properties for diverse applications.

5.5
Quantum Dot-Based Fluorescence Resonance Energy Transfer

Fluorescence resonance energy transfer (FRET) is a distance-dependent process in which the electronic excitation energy of a donor molecule is nonradiatively transferred to a nearby acceptor molecule [70–74]. The appreciable overlap between the emission spectrum of the donor and the absorption spectrum of the acceptor, and the limited distance (1–10 nm) between the donor and acceptor, are prerequisites for FRET to occur [70]. As a powerful technique, FRET has been used widely to study donor–acceptor distance distribution, receptor–ligand binding, intermolecular association, and macromolecule conformation [75–78]. Unfortunately, the use of conventional organic dyes as donor–acceptor complexes in FRET-based studies may often be limited due to crosstalk brought about by spectral overlap of the donor and acceptor emissions, and the narrower absorption spectra of the organic dyes [70]. Semiconductor QDs, with their aforementioned broad excitation spectra, narrow size-tunable emission spectra, high quantum yield, excellent photostability and chemical stability, have proven to be the ideal candidates for biological sensing and imaging [8–10, 24, 28, 75, 79–82].

He *et al.* reported the use of CdSe/ZnS QDs–single-stranded (ss) DNA–fluorescent dye conjugates as bioprobes to detect micrococcal nuclease (MNase) activity, with high specificity and sensitivity [83]. In this case, the bioprobes were constructed via complexation of the streptavidin-conjugated QDs with the biotinylated ssDNA, the 5′-end of which had been modified with 6-carboxy-X-rhodamine (ROX). FRET between the QDs and ROX caused the color of the conjugate to change from green to orange-red, whilst the cleavage of ssDNA by MNase into small fragments caused the color to change back to green. Thus, a simple and rapid method to monitor MNase activity in the culture medium of *Staphylococcus aureus*, and also to evaluate the pathogenicity of this microorganism, was developed. In addition to FRET changes between the QD and an appended fluorophore, on cleavage by nuclease or protease, FRET changes that occurred with variations in pH and chain growth, induced by a polymerase reaction, have also been demonstrated [11].

Adenosine triphosphate (ATP) is a multifunctional nucleotide, the role of which is to transport chemical energy within cells for metabolism [84]. Previously, FRET-based ATP biosensors have been developed by Deng and coworkers [85], whereby a capture DNA probe comprising an ATP aptamer hybridized with both 3′-Cy5-

labeled DNA and 3′-biotin-modified DNA was created. The biotin–streptavidin binding induced a close proximity between the streptavidin-conjugated QDs and the acceptor Cy5, which resulted in a FRET-sensitized Cy5 emission. The conformational change of the aptamer upon interaction with ATP led to a dissociation of 3′-Cy5-labeled DNA from the resultant complex, which caused the FRET-mediated signal to be greatly reduced. Both, trypsin and trypsin inhibitor (TI) have been determined in biological samples by using protein-conjugated QDs [86]. In this case, FRET occurred when the CdTe QDs became bound to rhodamine isothiocyanate that had been conjugated to bovine serum albumin (BSA). Whilst the digestion of BSA by trypsin led to a decrease in FRET efficiency, the inhibition of trypsin by TI led to a significant improvement in FRET. Moreover, the results obtained suggested that, by using this method, trypsin and TI concentrations as low as 10 pM and 250 pM, respectively, could be detected. The transfer of energy from Eu^{3+}–trisbipyridine (Eu–TBP) donors to QD acceptors has also been investigated in the presence and absence of serum [87]. Although FRET was generated via the formation of a Eu–TBP–QD complex, the unspecific binding induced by serum led to a fall in the overall FRET efficiency. Consequently, it appeared that QDs could only be used as acceptors when slow decay donors, such as europium, were used.

The interaction between QDs and fluorescent proteins was investigated by Hering *et al.* [88]. Previously, both green fluorescent protein 5 (GFP5) and a mutant (wild-type) GFP had been proposed to play important roles in improving bacterial expression [89]. The proteins were adsorbed onto the QD surface with a stoichiometry of 10 : 1, via electrostatic interaction. Subsequently, as the QD concentration in the GFP5–QD complex was increased (the concentration of GFP5 was fixed), a quenching of the PL of GFP5 occurred that was indicative of an energy-transfer process. It was unclear, however, whether the FRET had led to a quenching of the GFP5 fluorescence. Possible explanations for this situation included a charge transfer to the QDs, and a domination of the PL by the uncomplexed fraction of GFP5 [90, 91]. In contrast to the GFP5–QD system, a subtle FRET-sensitized emission from the GFP homologous red-emitting protein, HcRed1, was observed when QDs were complexed to HcRed1, and excited. This ability to monitor the formation of donor–acceptor pairs, using fluorescent proteins and QDs, might find potential application in the exploration of cellular events.

By taking advantage of the specific recognition between an antigen and an antibody, FRET has found extensive applications in immunoassays [92–96]. Indeed, it appears that FRET-based immunobiosensors have overcome the limitations inherent in conventional (labor-intensive) immunoassays, including the use of solid-phase plates for antigen/antibody immobilization, and the many binding and washing steps [95]. Recently, Grant and coworkers reported the details of an optical biosensor which uses gold nanoparticles (AuNPs) and QDs to detect the porcine reproductive and respiratory syndrome virus (PRRSV) [92]. For this, the biosensor assembly was fabricated by binding protein A-labeled AuNPs or QDs to a capture probe, SDOW17, which had been modified with a fluorescent dye that required specific recognition of the analyte. Subsequent binding between SDOW17 and PRRSV led to a conformational change of SDOW17 that increased the distance between the fluorescent dye and the nanoparticles, and in turn caused a decrease in energy transfer from the donor to the acceptor. When compared to the AuNP-based sensor, the QD biosensor proved to be robust and feasible for sensitive PRRSV detection. Similar studies have also been instigated to develop a method capable of detecting human cardiac troponin I, via a QD-based biosensor [93].

A one-step, homogeneous, open-sandwich fluoroimmunoassay based on FRET has been developed to monitor levels of estrogen receptor β (ER-β) [95]. The incubation of ER-β with a QD-labeled anti-ER-β monoclonal antibody and an Alexa Fluor-labeled anti-ER-β polyclonal antibody resulted in the formation of a sandwich complex which brought the QDs in close proximity to the Alexa Fluor. The FRET signal, measured using confocal microscopy, demonstrated a linear relationship with the ER-β concentration, and the detection limit was estimated at 0.05 nM.

Both, green-emitting QDs (gQDs) and red-emitting QDs (rQDs) have been utilized as donors and acceptors to study immune-reactions on live cell membranes, respectively [94]. In this case, HeLa cells were labeled with a mouse anti-human CD71 monoclonal antibody (anti-CD71) conjugated to rQDs. Through interactions between anti-CD71 and goat anti-mouse immunoglobulin G (IgG), a complex between the rQD-conjugated anti-CD71 and the gQD-labeled IgG was formed on the cell surface, such that FRET between the different-colored QDs occurred. This approach obviated the need for organic dyes, and provided a new method for the investigation of biomolecular interactions on live cells.

The sensing of DNA hybridization can be carried out based on the FRET between blue-luminescent CdTe QDs and Cy3-labeled ssDNA [97]. The water-soluble and negatively charged CdTe QDs, capped with thioglycolic acid ligands, were derivatized with a cationic polymer so as to acquire a positive surface charge ($CdTe^+$). An efficient energy transfer, from the QD donor to the dye acceptor, was then achieved via electrostatic and hydrophobic interactions of $CdTe^+$ with ssDNA. Upon incubation of the hybrids of $CdTe^+$ and Cy3-ssDNA with complementary ssDNA, a lower FRET efficiency was obtained that was induced by the more rigid DNA duplex structure. Thus, the hybridization event could be recognized by changes in FRET efficiency in the absence and presence of a sample DNA. Similar studies on the detection of DNA, based on the simple electrostatic interaction between cationic, PEGylated amine-functionalized QDs and negatively charged DNA, have been carried out by the group of Song [98]. In this case, the complexation with carboxytetramethylrhodamine (TAMRA)-modified ssDNA led to a quenching of the QDs' fluorescence. However, a recovery of such fluorescence, with a concomitant decrease in TAMRA fluorescence, was achieved upon hybridization with the target DNA.

Both, labeled and label-free detection of DNA hybridization was carried out on the basis of a QD-sensitized FRET signal [99, 100]. The QDs, modified with EG_3OH/EG_3COOH, were constructed by a ligand exchange of trioctylphosphine-oxide-capped QDs with a mixture of EG_3OH/EG_3COOH. The as-prepared QDs possessed a high stability, solubility, and resistance to the nonspecific adsorption of DNA [101, 102]. The compact, functional QD–DNA conjugate was formed by coupling the aminated target DNA to the activated EG_3OH/EG_3COOH-capped QDs. Hybridization of the Alexa 594-labeled DNA with the target DNA resulted in an energy transfer from the QDs to the fluorophore, after which the dye fluorescence signal was used to detect the labeled DNA probe. Alternatively, upon hybridization of the QD–DNA conjugate with the unlabeled DNA, an intercalation of ethidium bromide into the dsDNA occurred which facilitated an energy transfer from the QDs to ethidium bromide upon excitation, enabling quantification of the unlabeled DNA. The details of an alternative DNA-intercalating dye, BOBO-3, as an acceptor in QD-mediated FRET have also been reported [103]. Notably, the use of intercalating dyes as acceptors has proven to be simple and cost-effective, and does not require any of the analytes to be labeled.

Two-color nucleic acid detection based on the use of FRET has been reported by Krull and coworkers [104]. For this, two different types of QD – namely, smaller gQDs and larger rQDs that have been conjugated to probe oligonucleotides – have been used as energy donors. The probe oligonucleotides conjugated to the gQDs were complementary to the target sequences that were related to spinal muscular atrophy, whereas those conjugated to the rQDs recognized the DNA sequences used to diagnose the presence of a pathogen (*E. coli*). Hybridization of the conjugated probe oligonucleotides with Cy3- and Alexa647-labeled targets resulted in a FRET-sensitized emission from the Cy3 and Alexa647. The target concentrations, in the range of 0.1 to 1 μM and 10 to 100 nM, respectively, were determined when 1.0 and 0.06 μM of green and red QD–DNA conjugates were used, respectively. Thus, multicolor nucleic acid analyses could be carried out on a single sensor, with a single excitation source. Three-color hybrid nanoprobes of molecular beacons, conjugated to green, yellow and orange QDs for single nucleotide discrimination, have also been demonstrated with long-term optical stability and a small background signal [105].

The sensitive detection of nucleic acids in bulk solutions using FRET-based nanosensors, in which QDs were used as donors and organic dyes as acceptors, has certain limitations that include a low sensitivity and a low FRET efficiency, due mainly to the large size of the QDs and the great length of the nucleic acids [106]. In order to circumvent these problems, Johnson *et al.* reported the single-molecule detection of QD-based FRET in capillary flows [106]. By using again the highly specific streptavidin–biotin binding, the QD/DNA/Cy5 complex was formed between streptavidin-coated QDs and Cy5-labeled ssDNA or dsDNA. In the single-molecule detection in capillary flow, a simultaneous detection of photons emitted from QDs and Cy5 by individual donor and acceptor avalanche photodiodes was accomplished. As a result, more Cy5 and QDs bursts were observed from the QD/dsDNA/Cy5 complex than from the QD/ssDNA/Cy5 complex, thus enabling the rapid and sensitive detection of DNA or RNA. The enhanced FRET efficiency in the single-molecule detection resulted from the deformation of DNA in the capillary stream.

The development of a solid-phase biosensor with immobilized QDs would prove advantageous on the basis that it would be reusable, easier to fabricate, and more

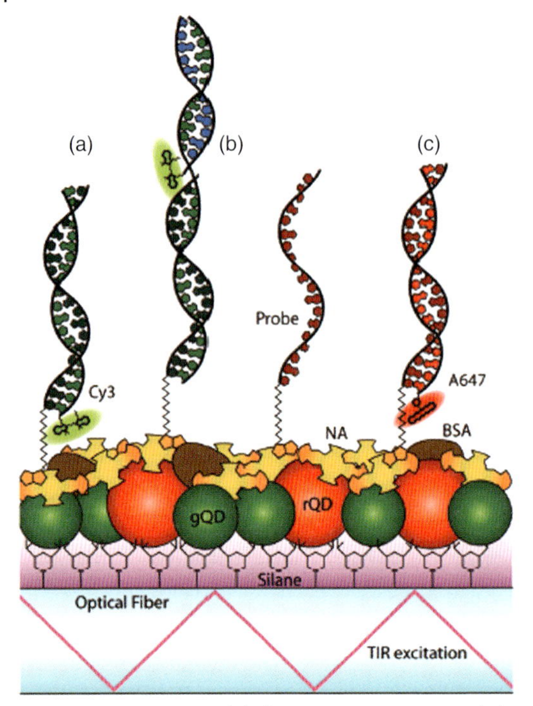

Figure 5.8 QD–FRET solid-phase hybridization assay design: a mixture of gQDs and rQDs were immobilized on an optical fiber. Probe oligonucleotides were immobilized on the QDs using a neutravidin bridge. BSA was used to passivate any remaining adsorption sites. (a) A direct assay was possible using the gQD–Cy3 FRET pair with a labeled target; (b) Target labeling was avoided with a sandwich assay; (c) A direct assay was also possible using the rQD–A647 FRET pair. By using mixed films of gQDs, rQDs and two probe oligonucleotides, it was possible to combine the pairs in parts (a) and (c) in a multiplexed assay. Reproduced with permission from Ref. [107]; © 2009, American Chemical Society.

environment friendly [107]. Krull *et al.* demonstrated a solid-phase DNA hybridization assay on an optical fiber via a QD–FRET-based approach (Figure 5.8) [107, 108]. In this case, gQDs and rQDs were immobilized onto optical fibers via a silane-based, multidentate surface ligand exchange [109], and this was followed by derivatization with a neutravidin layer and attachment of the biotinylated probe oligonucleotides. In order to minimize nonspecific adsorption, BSA was used to block any remaining adsorption sites. Hybridization with Cy3 (the acceptor for gQDs) and AlexaFluor647 (the acceptor for rQDs) -labeled target oligonucleotides resulted in FRET signals which could be used for multiplexed DNA assays of different target sequences. These assays remain functional in complex matrixes, and can be used to discriminate single nucleotide polymorphisms, with high sensitivity and selectivity.

The molecular dissociation of the complex between DNA and chitosan of different molecular weights has been monitored using FRET [110]. Here, QDs functionalized with amine groups were attached to the plasmid DNA in which the guanine residues were activated with *N*-bromosuccinimide to form QD-conjugated DNA (QD-DNA). Chitosan samples of different molecular weights were labeled with Texas Red-*N*-hydrosulfosuccinimide (TR-chitosan). The complex formation between plasmid DNA and chitosan resulted in a FRET signal, in which TR-chitosan was excited by the emission energy of the QD-DNA. However, as chitosan molecular weight was increased, the dissociation of the chitosan–DNA complex became more significant in an acid medium, such that no FRET occurred between QD-DNA and TR-chitosan. The high-molecular-weight chitosan–DNA complex showed a superior transfection efficiency in comparison with the low-molecular-weight complexes.

The serum assay of the tumor marker Mucin 1 (MUC1) is of potential significance for tumor diagnosis, due to its overexpression in almost all human epithelial cell adenocarcinomas [111–113]. By relying on the variation in secondary structure of the aptamer upon binding to MUC1, Yu *et al.* have developed a novel QD–FRET-based method for MUC1 detection (Figure 5.9) [114]. In the absence of MUC1, the MUC1 aptamer folds into a hairpin structure with a four base-pair stem and a three-thymine loop, which hinders hybridization of the QD-modified ssDNA with the aptameric region, the other end of which will pre-hybridize to the ssDNA which has been modified with a quencher. In the presence of MUC1, however, the binding of MUC1 to the aptameric region causes the hairpin structure formation to be demolished, and allows for a hybridization of the QD-modified ssDNA with the aptamer strand. Thus, the close proximity between QDs and the quencher

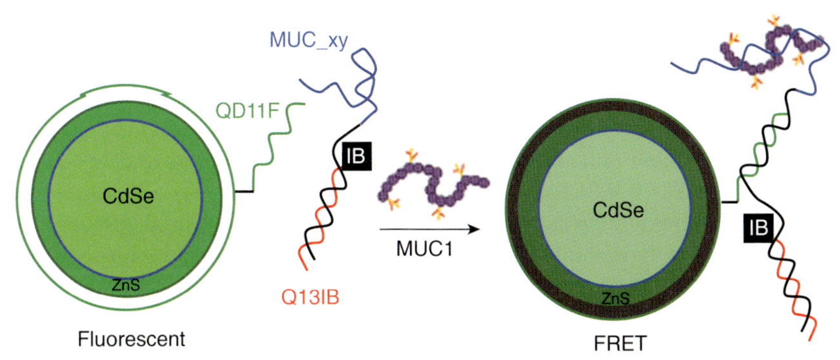

Figure 5.9 Schematic representation of the overall aptamer-based detection scheme of MUC1 peptides. In the absence of MUC1 peptides, the aptameric region (blue) will fold into its inherent secondary structure, which prevents the QD-tethered DNA strand (QD11F, shown in green) from hybridizing to MUC1_xy (black/blue). It should be noted that multiple strands may be conjugated to the QD surface, but only one is shown here for simplicity. The hybridization of Q13IB (red), the strand with a quencher, to the aptamer strand (MUC1_xy) always occurs. In the presence of MUC1 peptides, the aptameric region binds the peptide, hence unfolding the aptameric region to a full single strand. This allows both QD11F and Q13IB to hybridize such that FRET occurs, leading to a decrease in the fluorescence signal. Reproduced with permission from Ref. [114]; © 2009, American Chemical Society.

will result in FRET. The decrease in the fluorescence intensity of QDs was shown to correlate with the concentration of the MUC1 peptide, thus allowing for the detection of MUC1 in epithelial cancer cells. The aptamer/analyte binding mode for MUC1 has provided further insight into the use of this system in the possible early diagnosis of human tumors.

Typically, FRET occurs over distances in the range of 1–10 nm between the donor and acceptor [115–117]. However, long-range FRET processes for donor–acceptor distances up to 13 nm have been demonstrated with self-assembled donors comprising QDs and fluorescent proteins [115]. Assembly of the nanoconjugate involves the electrostatic interaction of QDs to an enhanced yellow fluorescent protein (EYFP)–ssDNA complex, followed by hybridization of the EYFP–ssDNA complex with complementary oligonucleotides modified with the Atto647-dye. Due to the large and spectrally broad absorption cross-section of QDs, as well as the high quantum yield of EYFP, the QD/EYFP subsystem can serve as a powerful donor with the FRET pathway of QD $\rightarrow$ EYFP $\rightarrow$ Atto647, thereby enabling long-distance FRET processes over 9–13 nm. The supramolecular donor system created from QDs and fluorescent proteins might find potential application in areas as diverse as long-range, FRET-based, high-throughput screening assays, multi-chromophore system fabrication, and protein–DNA assembly.

5.5.1 Chemiluminescence Resonance Energy Transfer

In contrast to FRET, chemiluminescence resonance energy transfer (CRET) occurs between a chemiluminescent donor and an acceptor molecule, via the oxidation of a luminescent substrate [82, 118, 119]. Ren and coworkers applied CRET to immunoreactions in which luminol was used as a donor and QDs as the acceptor [82]. In the case of the HRP-catalyzed luminol/hydrogen peroxide chemiluminescence reaction, CRET occurs by mixing luminol with BSA which has been conjugated to CdTe QDs and HRP-labeled anti-BSA in the presence of H_2O_2. In comparison to FRET, CRET is simple and does not involve the use of an excitation light source, although the use of H_2O_2 will lead to quenching of the QDs' photoluminescence. In future, it is possible that CRET might find applications in quantitative bioanalysis, and in cell and tissue imaging.

5.5.2 Bioluminescence Resonance Energy Transfer

Bioluminescence resonance energy transfer (BRET) occurs when bioluminescent proteins are present as donors and QDs as acceptors, and where the energy produced during a biochemical reaction can resonantly excite the QDs [120–125]. Rao *et al.* conducted a series of studies with QD–BRET-based sensors [120–123], in which a mutant of *Renilla* luciferase (Luc8) was used as the donor to produce light during the luciferase-catalyzed oxidation of the Luc8 substrate, coelenterazine [122]. Previously, these QD–BRET-based sensors have been used to monitor the

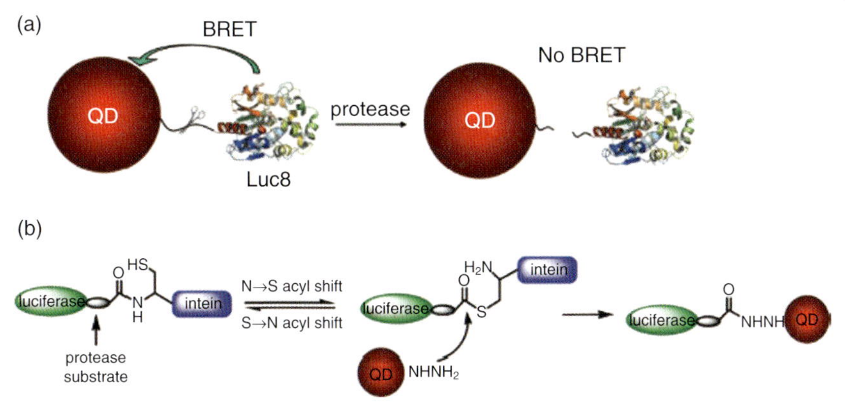

Figure 5.10 Design and preparation of QD-BRET-based nanosensors for MMP sensing. (a) Schematic of the nanosensor comprising a QD and luciferase protein linked to the QD through an MMP peptide substrate; (b) Intein-mediated site-specific conjugation of Luc8 fusion proteins to QDs. Reproduced with permission from Ref. [122]; © 2008, American Chemical Society.

activities of the matrix metalloproteinases (MMPs), a class of zinc-dependent endopeptidases the overexpression of which in human cancer correlates with advanced tumor stage, increased invasion and metastasis, and shortened survival [122, 126–130]. The fabrication of QD–BRET-based sensors involves the ligation of an MMP substrate to Luc8 and QDs, achieved via a protein-splicing molecule, intein (Figure 5.10). Coupling of the recombinant protein, through a genetic fusion of intein, the MMP substrate and Luc8 to the QDs which have been modified with hydrazide, brings the QDs and Luc8 into close proximity, leading in turn to a BRET signal. Cleavage of the MMP substrate by MMP, however, causes a decrease in energy transfer from the donor to the acceptor. Moreover, the feasibility of this nanosensor in multiplex protease assays has recently been demonstrated.

Other strategies to establish the QD–BRET system have been reported by the same group [120, 121, 131–133] and have included (but have not been limited to) the coupling of carboxylic acid groups on the QDs to amine groups on Luc8 [131, 132], a metal-mediated complexation between QDs and Luc8 [120], and the specific binding of the Halo Tag protein, fused via a donor protein, to the Halo Tag ligand tethered on the QDs [133]. The QD–BRET systems, by combining the advantages of QDs with the extreme sensitivity of bioluminescence imaging, hold great potential for multiplexed imaging *in vivo* [121, 123, 125, 131].

Rao *et al.* have also reported the imaging (both *in vitro* and *in vivo*, in mice) of C6 glioma cells labeled with QD655–Luc8–R9 (arginine 9-mer, a polycationic peptide) [131]. The aim of incorporating R9 into the nanoconjugate was to improve the efficiency of cell uptake [134]. The elimination of physical light excitation makes the bioluminescent QD probes ideal candidates for molecular imaging in complex sample media. A rapid, single-step nucleic acid assay based on BRET has been demonstrated in which two complementary oligonucleotide probes were labeled with *Renilla* luciferase and QDs, respectively [124]. In the presence of coelenterazine, the hybridization event resulted in a BRET signal whereby QD was excited via the emitted light of luciferase. However, competitive binding between the target DNA and the probe modified with QDs caused a disruption of the BRET signal, which allowed detection of the target at concentrations in the sample down to 20 nM.

5.6 Quantum Dot-Based Nanohybrids

In recent years, the construction of nanostructured hybrid materials has attracted increasing attention based on their widespread application in optical/electrical sensors, as light-emitting diodes and optoelectronic devices, and in biological devices [135–138]. Recently, ZnO QDs that were mediated and directed by surface-photografted poly(acrylic acid) (PAA) brushes have been fabricated [139], that have a diameter of 4–5 nm and exhibit clear quantum effects. Based on a biomimetic chemical approach, the straightforward preparation of patternable PAA/ZnO QD nanohybrids with large surface areas and arrays on flexible plastics has been demonstrated. These hybrid materials might find application in diverse areas such as polymer-supported advanced technologies, biomineralization, and macroelectronics. Likewise, QDs that were encapsulated by a polymer shell which could be conveniently tailored were synthesized via a reversible addition–fragmentation chain-transfer polymerization in miniemulsions [138]. The details of water-dispersible nanohybrids containing CdS nanocrystals embedded into the polymer domains, prepared via an *in situ* polymerization in the micellar system, have also been reported [140].

Previously, it has been shown that carbon nanotube (CNT)–QD nanohybrids exhibit novel optical properties, by combining the selective wavelength absorption properties of QDs with the efficient charge-transfer and electron-transport properties of CNTs [141]. The strategy of the *in situ* growth of QDs onto nonfunctionalized CNTs minimizes any structural disruption at the junctions, which in turn enables the synthesis of hybrid materials with a high quantum yield, a long lifetime, and a high dispersion in aqueous media. The ability of the nanohybrids to enhance the intrinsic fluorescence of polycyclic aromatic hydrocarbons (PAHs) has been demonstrated. The interaction between CdSe adsorbed onto CNTs and PAHs leads to an energy-transfer process from CdSe to the energy levels of PAHs. As a result, the decrease in the typical fluorescence of CdSe was accompanied by a concomitant increase in the intrinsic fluorescence of the PAHs. When the feasibility of these nanohybrids to detect trace levels of PAHs in river water samples was explored, the incorporation of single QDs into silver nanoparticles, so as to combine the surface plasmon resonance (SPR) effect of the metal nanoparticles with the quantum size effect of the semiconductor QDs, was found to constitute a new approach to the nanohybrid material systems [142]. Such hybrid materials could be constructed by covalently linking streptavidin-functionalized QDs to biotin-coated silver nanoparticles, thereby displaying a significant enhancement in fluorescence intensity and causing dramatic changes in the dynamic behavior. Sadly, however, these hybrid materials have lifetimes that are considerably shorter than their QD counterparts.

Antibody-coated QD bioconjugates incorporated within biodegradable polymeric nanospheres for live cell labeling and imaging have been demonstrated [143]. The encapsulation of protein-conjugated QDs within a poly(D, L-lactide-*co*-glycolide) nanosphere to form hybrid QD-nanocomposites (QDNCs) has been achieved via a double microemulsion procedure. Upon cell penetration, the polymer nanospheres are hydrolyzed, enabling a controllable release of the functionalized QD probes within the cytoplasmic space (Figure 5.11). This results in a minimal stress on the cell plasma membrane and minimal toxicity to the cell, and provides a universal intracellular nanoscale delivery platform to study complex biological processes in live cells. More recently, nanohybrids integrating magnetic nanoparticles and QDs have attracted much attention in areas such as drug deliv-

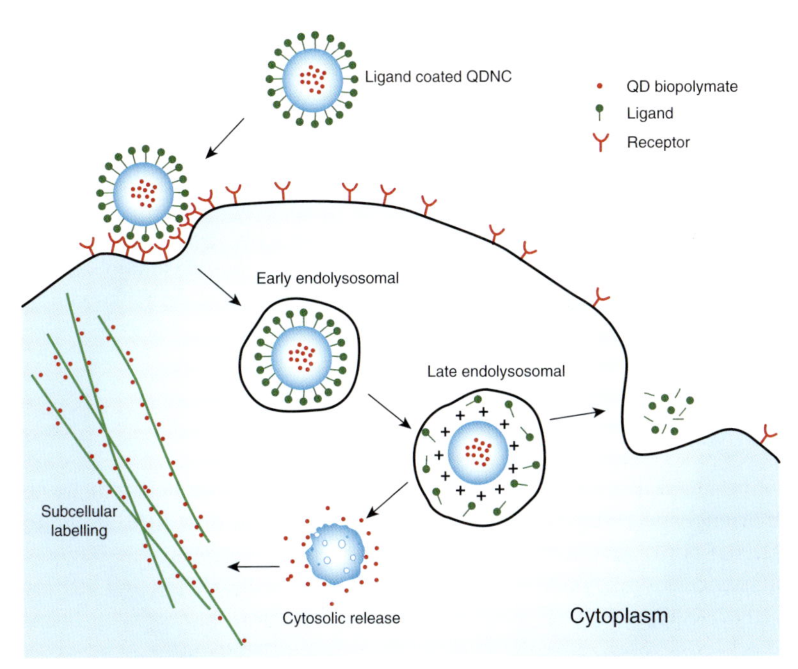

Figure 5.11 Mechanism of cytosolic delivery and subcellular targeting of QDNCs. Schematic representation of QDNC escape from the endolysosomal compartment upon cellular internalization with cytosolic release of the encapsulated cargo. Antibody-conjugated QDs can be delivered in this manner to allow the labeling of subcellular organelles or other molecular targets. Reproduced with permission from Ref. [143]; © 2008, American Chemical Society.

ery, bioimaging, and bioseparation [144]. For example, Fe_3O_4/CdTe nanohybrids conjugated with an anti-CEACAM8 antibody, which recognizes the CEACAM8 receptor on the HeLa cell membrane, have been used successfully for the immunolabeling and fluorescent imaging of HeLa cells [144]. Novel nanohybrids based on the combination of bacterial magnetic nanoparticles with QDs have also been developed for the targeting and identification of cancer cells [145].

5.7 Conclusions and Outlook

The unique electronic, optical, magnetic, and mechanical properties of QDs have resulted in their becoming ideal bioprobes for sensing and imaging applications. With sizes that are comparable to biomacromolecules, and an ability to resist photobleaching, denaturation of the biomolecules and changes in both pH and temperature, QD nanocomposites have been used successfully as fluorescent sensors for highly sensitive and selective assays. Moreover, by circumventing the limitations inherent in conventional organic dyes, such as the crosstalk brought about by spectral overlap between the donor and acceptor emissions, and the narrower absorption spectra, QD-based FRET biosensors are expected to hold great promise for many important applications, including donor–acceptor distance distribution, macromolecule conformation, receptor–ligand binding, DNA hybridization, antigen–antibody recognition, DNA–protein interaction, and the monitoring of environmental pollutants. To further improve sensitivity and FRET efficiency in bulk solutions, the development of solid-phase sensors with immobilized QDs is desirable, in that they are reusable, easier to fabricate, and environment-friendly. It is without doubt that, by addressing the issues regarding the reusability, toxicity and long-term stability of the QDs, a multitude of applications based on QD–biomolecule architectures will be developed and demonstrated in the near future.

Acknowledgments

The authors thank the National Science Foundation Center for Research Excellence in Science and Technology (NSF HRD-0932421) and the NIH-RIMI Program (P20-MD001824-01) at California State University, Los Angeles, the National Natural Science Foundation of China (Nos. 20775093 and 20975114), and SRF for ROCS, SEM. (No. [2009] 1001) for supporting these studies.

Abbreviations

anti-CD71	mouse anti-human CD71 monoclonal antibody
ATP	adenosine triphosphate
BRET	bioluminescence resonance energy transfer

BSA	bovine serum albumin
CA	citric acid
CFR	core formation ratio
CNT	carbon nanotube
CRET	chemiluminescence resonance energy transfer
ds	double-stranded
ER-β	estrogen receptor β
Eu-TBP	Eu^{3+}-trisbipyridine
EYFP	enhanced yellow fluorescent proteins
FRET	fluorescence resonance energy transfer
GFP5	green fluorescent protein 5
gQD	green-emitting QD
HRP	horseradish peroxidase
IgG	goat anti-mouse immunoglobulin G
Luc8	*Renilla* luciferase
MMP	matrix metalloproteinase
MNase	micrococcal nuclease
MUC1	tumor marker Mucin 1
OA	oleic acid
PAA	poly(acrylic acid)
PAH	polycyclic aromatic hydrocarbon
PC	phytochelatin
PL	photoluminescence
PRRSV	porcine reproductive and respiratory syndrome virus
QD-DNA	QD-conjugated DNA
QD–NC	QD–nanocomposite
QD	quantum dot
ROX	6-carboxy-X-rhodamine
rQD	red-emitting QD
SDOW17	anti-PRRSV monoclonal antibody
TAMRA	carboxytetramethylrhodamine
TEM	transmission electron microscopy
Texas Red-chitosan	Texas Red-*N*-hydrosulfosuccinimide
TI	trypsin inhibitor
Try	trypsin
UV-visible	Ultraviolet-visible

References

1 Alivisatos, A.P. (1996) Perspectives on the physical chemistry of semiconductor nanocrystals. *Journal of Physical Chemistry*, **100**, 13226–39.

2 Brus, L. (1991) Quantum crystallites and nonlinear optics. *Applied Physics A*, **A53**, 465–74.

3 Donega, C., Hickey, S.G., Wuister, S.F., Vanmaekelbergh, D. and Meijerink, A. (2003) Single-step synthesis to control the photoluminescence quantum yield and size dispersion of CdSe nanocrystals. *Journal of Physical Chemistry B*, **107**, 489–96.

4 Henglein, A. (1989) Small-particle research: physicochemical properties of extremely small colloidal metal and semiconductor particles. *Chemical Reviews*, **89**, 1861–73.

5 Wang, Y. and Herron, N. (1991) Nanometer-sized semiconductor clusters: materials synthesis, quantum size effects, and photophysical properties. *Journal of Physical Chemistry*, **95**, 525–32.

6 Weller, H. (1993) Colloidal semiconductor Q-particles: chemistry in the transition region between solid and molecular states. *Angewandte Chemie, International Edition in English*, **105**, 43–55.

7 Biju, V., Itoh, T., Anas, A., Sujith, A. and Ishikawa, M. (2008) Semiconductor quantum dots and metal nanoparticles: syntheses, optical properties, and biological applications. *Analytical and Bioanalytical Chemistry*, **391**, 2469–95.

8 Dabbousi, B.O., Rodriguez-Viejo, J., Mikulec, F.V., Heine, J.R., Mattoussi, H., Ober, R., Jensen, K.F. and Bawendi, M.G. (1997) (CdSe)ZnS core-shell quantum dots: synthesis and optical and structural characterization of a size series of highly luminescent materials. *Journal of Physical Chemistry B*, **101**, 9463–75.

9 Bruchez, M.J., Moronne, M., Gin, P., Weiss, S. and Alivisatos, A.P. (1998) Semiconductor nanocrystals as fluorescent biological labels. *Science*, **281**, 2013–16.

10 Medintz, I.L., Uyeda, H.T., Goldman, E.R. and Mattoussi, H. (2005) Quantum dot bioconjugates for imaging, labelling and sensing. *Nature Materials*, **4**, 435–46.

11 Suzuki, M., Husimi, Y., Komatsu, H., Suzuki, K. and Douglas, K.T. (2008) Quantum dot FRET biosensors that respond to pH, to proteolytic or nucleolytic cleavage, to DNA synthesis, or to a multiplexing combination. *Journal of the American Chemical Society*, **130**, 5720–5.

12 Medintz, I.L., Konnert, J.H., Clapp, A.R., Stanish, I., Twigg, M.E., Mattoussi, H., Mauro, J.M. and Deschamps, J.R. (2004) A fluorescence resonance energy transfer-derived structure of a quantum dot-protein bioconjugate nanoassembly. *Proceedings of the National Academy of Sciences of the United States of America*, **101**, 9612–17.

13 Medintz, I.L., Clapp, A.R., Mattoussi, H., Goldman, E.R., Fisher, B. and Mauro, J.M. (2003) Self-assembled nanoscale biosensors based on quantum dot FRET donors. *Nature Materials*, **2**, 630–8.

14 Rogach, A.L., Franzl, T., Klar, T.A., Feldmann, J., Gaponik, N., Lesnyak, V., Shavel, A., Eychmueller, A., Rakovich, Y.P. and Donegan, J.F. (2007) Aqueous synthesis of thiol-capped CdTe nanocrystals: state-of-the-art. *Journal of Physical Chemistry C*, **111**, 14628–37.

15 Andrey, L.R. (2008) *Semiconductor Nanocrystal Quantum Dots: Synthesis, Assembly, Spectroscopy and Applications*, Springer Wien, New York.

16 Pan, D., Jiang, S., An, L. and Jiang, B. (2004) Controllable synthesis of highly luminescent and monodisperse CdS nanocrystals by a two-phase approach under mild conditions. *Advanced Materials*, **16**, 982–5.

17 Pan, D., Wang, Q., Jiang, S., Ji, X. and An, L. (2005) Synthesis of extremely small CdSe and highly luminescent CdSe/CdS core-shell nanocrystals via a novel two-phase thermal approach. *Advanced Materials*, **17**, 176–80.

18 Pan, D., Zhao, N., Wang, Q., Jiang, S., Ji, X. and An, L. (2005) Facile synthesis and characterization of luminescent TiO_2 nanocrystals. *Advanced Materials*, **17**, 1991–5.

19 Zhao, N., Pan, D., Nie, W. and Ji, X. (2006) Two-phase synthesis of shape-controlled colloidal zirconia nanocrystals and their characterization. *Journal of the American Chemical Society*, **128**, 10118–24.

20 Zhu, J.J., Palchik, O., Chen, S.G. and Gedanken, A. (2000) Microwave assisted preparation of CdSe, PbSe, and Cu_{2-x} Se nanoparticles. *Journal of Physical Chemistry B*, **104**, 7344–7.

21 Rogach, A.L., Nagesha, D., Ostrander, J.W., Giersig, M. and Kotov, N.A. (2000) "Raisin bun"-type composite spheres of silica and semiconductor nanocrystals. *Chemistry of Materials*, **12**, 2676–85.

22 Trindade, T., Obrien, P. and Zhang, X.M. (1997) Synthesis of CdS and CdSe nanocrystallites using a novel single-molecule precursors approach. *Chemistry of Materials*, **9**, 523–30.

23 Rogach, A.L., Kornowski, A., Gao, M.Y., Eychmuller, A. and Weller, H. (1999) Synthesis and characterization of a size series of extremely small thiol-stabilized CdSe nanocrystals. *Journal of Physical Chemistry B*, **103**, 3065–9.

24 Murray, C.B., Norris, D.J. and Bawendi, M.G. (1993) Synthesis and characterization of nearly monodisperse CdE (E = sulfur, selenium, tellurium) semiconductor nanocrystallites. *Journal of the American Chemical Society*, **115**, 8706–15.

25 Peng, Z.A. and Peng, X. (2001) Mechanisms of the shape evolution of CdSe nanocrystals. *Journal of the American Chemical Society*, **123**, 1389–95.

26 Manna, L., Scher, E.C. and Alivisatos, A.P. (2000) Synthesis of soluble and processable rod-, arrow-, teardrop-, and tetrapod-shaped CdSe nanocrystals. *Journal of the American Chemical Society*, **122**, 12700–6.

27 Yu, W.W., Wang, Y.A. and Peng, X. (2003) Formation and stability of size-, shape-, and structure-controlled CdTe nanocrystals: ligand effects on monomers and nanocrystals. *Chemistry of Materials*, **15**, 4300–8.

28 Hines, M.A. and Guyot-Sionnest, P. (1996) Synthesis and characterization of strongly luminescing ZnS-capped CdSe nanocrystals. *Journal of Physical Chemistry*, **100**, 468–71.

29 Peng, Z.A. and Peng, X. (2001) Formation of high-quality CdTe, CdSe, and CdS nanocrystals using CdO as precursor. *Journal of the American Chemical Society*, **123**, 183–4.

30 Pan, B., He, R., Gao, F., Cui, D. and Zhang, Y. (2006) Study on growth kinetics of CdSe nanocrystals in oleic acid/dodecylamine. *Journal of Crystal Growth*, **286**, 318–23.

31 Qu, L., Peng, Z.A. and Peng, X. (2001) Alternative routes toward high quality CdSe nanocrystals. *Nano Letters*, **1**, 333–7.

32 Sapra, S., Rogach, A.L. and Feldmann, J. (2006) Phosphine-free synthesis of monodisperse CdSe nanocrystals in olive oil. *Journal of Material Chemistry*, **16**, 3391–5.

33 Fan, H., Leve, E.W., Scullin, C., Gabaldon, J., Tallant, D., Bunge, S., Boyle, T., Wilson, M.C. and Brinker, C.J. (2005) Surfactant-assisted synthesis of water-soluble and biocompatible semiconductor quantum dot micelles. *Nano Letters*, **5**, 645–8.

34 Mattoussi, H., Mauro, J.M., Goldman, E.R., Anderson, G.P., Sundar, V.C., Mikulec, F.V. and Bawendi, M.G. (2000) Self-assembly of CdSe-ZnS quantum dot bioconjugates using an engineered recombinant protein. *Journal of the American Chemical Society*, **122**, 12142–50.

35 Bae, W.K., Char, K., Hur, H. and Lee, S. (2008) Single-step synthesis of quantum dots with chemical composition gradients. *Chemistry of Materials*, **20**, 531–9.

36 Yu, W.W., Chang, E., Drezek, R. and Colvin, V.L. (2006) Water-soluble quantum dots for biomedical applications. *Biochemical and Biophysical Research Communications*, **348**, 781–6.

37 Kumar, A. and Jakhmola, A. (2009) RNA-templated fluorescent Zn/PbS (PbS + Zn^{2+}) supernanostructures. *Journal of Physical Chemistry C*, **113**, 9553–9.

38 Patel, A.A., Wu, F., Zhang, J., Torres-Martinez, C.L., Mehra, R.K., Yang, Y. and Risbud, S.H. (2000) Synthesis, optical spectroscopy and ultrafast electron dynamics of PbS nanoparticles with different surface capping. *Journal of Physical Chemistry B*, **104**, 11598–605.

39 Carney, C.K., Harry, S.R., Sewell, S.L. and Wright, D.W. (2007) Detoxification biominerals. Biomineralization I: crystallization and self-organization process. *Topics in Current Chemistry*, **270**, 155–85.

40 Dickerson, M.B., Sandhage, K.H. and Naik, R.R. (2008) Protein- and peptide-directed syntheses of inorganic materials. *Chemical Reviews*, **108**, 4935–78.

41 Bae, W.K. and Mehra, R.K. (1998) Properties of glutathione- and phytochelatin-capped CdS bionanocrystallites. *Journal of Inorganic Biochemistry*, **69**, 33–43.
42 Singhal, R.K., Anderson, M.E. and Meister, A. (1987) Glutathione, a first line of defense against cadmium toxicity. *The FASEB Journal*, **1**, 220–3.
43 Spoerke, E.D. and Voigt, J.A. (2007) Influence of engineered peptides on the formation and properties of cadmium sulfide nanocrystals. *Advanced Functional Materials*, **17**, 2031–7.
44 Wong, K.K.W. and Mann, S. (1996) Biomimetic synthesis of cadmium sulfide-ferritin nanocomposites. *Advanced Materials*, **8**, 928–32.
45 Iwahori, K., Yoshizawa, K., Muraoka, M. and Yamashita, I. (2005) Fabrication of ZnSe nanoparticles in the apoferritin cavity by designing a slow chemical reaction system. *Inorganic Chemistry*, **44**, 6393–400.
46 Sugimoto, T. (1987) Preparation of monodispersed colloidal particles. *Advances in Colloid and Interface Science*, **28**, 65–108.
47 Park, J., Privman, V. and Matijevic, E. (2001) Model of formation of monodispersed colloids. *Journal of Physical Chemistry B*, **105**, 11630–5.
48 Peng, X., Wickham, J. and Alivisatos, A.P. (1998) Kinetics of II-VI and III-V colloidal semiconductor nanocrystal growth: “focusing” of size distributions. *Journal of the American Chemical Society*, **120**, 5343–4.
49 Talapin, D.V., Rogach, A.L., Haase, M. and Weller, H. (2001) Evolution of an ensemble of nanoparticles in a colloidal solution: theoretical study. *Journal of Physical Chemistry B*, **105**, 12278–85.
50 Rogach, A.L., Talapin, D.V., Shevchenko, E.V., Kornowski, A., Haase, M. and Weller, H. (2002) Organization of matter on different size scales: monodisperse nanocrystals and their superstructures. *Advanced Functional Materials*, **12**, 653–64.
51 Qu, L., Yu, W.W. and Peng, X. (2004) In situ observation of the nucleation and growth of CdSe nanocrystals. *Nano Letters*, **4**, 465–9.
52 Bullen, C.R. and Mulvaney, P. (2004) Nucleation and growth kinetics of CdSe nanocrystals in octadecene. *Nano Letters*, **4**, 2303–7.
53 Pan, D., Ji, X., An, L. and Lu, Y. (2008) Observation of nucleation and growth of CdS nanocrystals in a two-phase system. *Chemistry of Materials*, **20**, 3560–6.
54 Datta, A., Panda, S.K. and Chaudhuri, S. (2007) Synthesis and optical and electrical properties of CdS/ZnS core/shell nanorods. *Journal of Physical Chemistry C*, **111**, 17260–4.
55 Deka, S., Quarta, A., Lupo, M.G., Falqui, A., Boninelli, S., Giannini, C., Morello, G., De Giorgi, M., Lanzani, G., Spinella, C., Cingolani, R., Pellegrino, T. and Manna, L. (2009) CdSe/CdS/ZnS double shell nanorods with high photoluminescence efficiency and their exploitation as biolabeling probes. *Journal of the American Chemical Society*, **131**, 2948–58.
56 Xie, R., Kolb, U., Li, J., Basche, T. and Mews, A. (2005) Synthesis and characterization of highly luminescent CdSe-core CdS/$Zn_{0.5}Cd_{0.5}S$/ZnS multishell nanocrystals. *Journal of the American Chemical Society*, **127**, 7480–8.
57 Wang, Y., Tang, Z., Correa-Duarte, M.A., Pastoriza-Santos, I., Giersig, M., Kotov, N.A. and Liz-Marzan, L.M. (2004) Mechanism of strong luminescence photoactivation of citrate-stabilized water-soluble nanoparticles with CdSe cores. *Journal of Physical Chemistry B*, **108**, 15461–9.
58 Kalele, S., Gosavi, S.W., Urban, J. and Kulkarni, S.K. (2006) Nanoshell particles: synthesis, properties and applications. *Current Science*, **91**, 1038–52.
59 Kirchner, C., Liedl, T., Kudera, S., Pellegrino, T., Javier, A.M., Gaub, H.E., Stoelzle, S., Fertig, N. and Parak, W.J. (2005) Cytotoxicity of colloidal CdSe and CdSe/ZnS nanoparticles. *Nano Letters*, **5**, 331–8.
60 Lee, J., Kim, J., Park, E., Jo, S. and Song, R. (2008) PEG-ylated cationic CdSe/ZnS QDs as an efficient intracellular labeling agent. *Physical Chemistry Chemical Physics*, **10**, 1739–42.
61 Chen, J., Gao, Y., Guo, C., Wu, G., Chen, Y. and Lin, B. (2008) Facile

synthesis of water-soluble and size-homogeneous cadmium selenide nanoparticles and their application as a long-wavelength fluorescent probe for detection of Hg(II) in aqueous solution. *Spectrochimica Acta Part A*, **69A**, 572–9.

62 Chan, W.C.W., Maxwell, D.J., Gao, X., Bailey, R.E., Han, M. and Nie, S. (2002) Luminescent quantum dots for multiplexed biological detection and imaging. *Current Opinion in Biotechnology*, **13**, 40–6.

63 Long, Y., Jiang, D., Zhu, X., Wang, J. and Zhou, F. (2009) Trace Hg analysis via quenching of the fluorescence of a CdS-encapsulated DNA nanocomposite. *Analytical Chemistry*, **81**, 2652–7.

64 Chen, Y. and Rosenzweig, Z. (2002) Luminescent CdS quantum dots as selective ion probes. *Analytical Chemistry*, **74**, 5132–8.

65 Hou, Y., Ye, J., Gui, Z. and Zhang, G. (2008) Temperature-modulated photoluminescence of quantum dots. *Langmuir*, **24**, 9682–5.

66 Lemon, B. and Crooks, R.M. (2000) Preparation and characterization of dendrimer-encapsulated CdS semiconductor quantum dots. *Journal of the American Chemical Society*, **122**, 12886–7.

67 Schild, H.G. (1992) Poly(N-isopropylacrylamide): experiment, theory and application. *Progress in Polymer Science*, **17**, 163–249.

68 Tan, Z., Zhang, F., Zhu, T. and Xu, J. (2007) Bright and color-saturated emission from blue light-emitting diodes based on solution-processed colloidal nanocrystal quantum dots. *Nano Letters*, **7**, 3803–7.

69 Alivisatos, A.P. (1996) Semiconductor clusters, nanocrystals, and quantum dots. *Science*, **271**, 933–7.

70 Willard, D.M., Carillo, L.L., Jung, J. and van Orden, A. (2001) CdSe-ZnS quantum dots as resonance energy transfer donors in a model protein-protein binding assay. *Nano Letters*, **1**, 469–74.

71 Stryer, L. (1978) Fluorescence energy transfer as a spectroscopic ruler. *Annual Review of Biochemistry*, **47**, 819–46.

72 Fairclough, R.H. and Cantor, C.R. (1978) The use of singlet-singlet energy transfer to study macromolecular assemblies. *Methods in Enzymology*, **48**, 347–79.

73 Wu, P. and Brand, L. (1994) Resonance energy transfer: methods and applications. *Analytical Biochemistry*, **218**, 1–13.

74 Selvin, P.R. (1995) Fluorescence resonance energy transfer. *Methods in Enzymology*, **246**, 300–34.

75 Pons, T., Medintz, I.L., Wang, X., English, D.S. and Mattoussi, H. (2006) Solution-phase single quantum dot fluorescence resonance energy transfer. *Journal of the American Chemical Society*, **128**, 15324–31.

76 Deniz, A.A., Dahan, M., Grunwell, J.R., Ha, T., Faulhaber, A.E., Chemla, D.S., Weiss, S. and Schultz, P.G. (1999) Single-pair fluorescence resonance energy transfer on freely diffusing molecules: observation of Forster distance dependence and subpopulations. *Proceedings of the National Academy of Sciences of the United States of America*, **96**, 3670–5.

77 Ha, T., Ting, A.Y., Liang, J., Caldwell, W.B., Deniz, A.A., Chemla, D.S., Schultz, P.G. and Weiss, S. (1999) Single-molecule fluorescence spectroscopy of enzyme conformational dynamics and cleavage mechanism. *Proceedings of the National Academy of Sciences of the United States of America*, **96**, 893–8.

78 Schobel, U., Egelhaaf, H.-J., Brecht, A., Oelkrug, D. and Gauglitz, G. (1999) New donor-acceptor pair for fluorescent immunoassays by energy transfer. *Bioconjugate Chemistry*, **10**, 1107–14.

79 Chan, W.C. and Nie, S. (1998) Quantum dot bioconjugates for ultrasensitive nonisotopic detection. *Science*, **281**, 2016–18.

80 Michalet, X., Pinaud, F.F., Bentolila, L.A., Tsay, J.M., Doose, S., Li, J.J., Sundaresan, G., Wu, A.M., Gambhir, S.S. and Weiss, S. (2005) Quantum dots for live cells, *in vivo* imaging, and diagnostics. *Science*, **307**, 538–44.

81 Alivisatos, P. (2004) The use of nanocrystals in biological detection. *Nature Biotechnology*, **22**, 47–52.

82 Huang, X., Li, L., Qian, H., Dong, C. and Ren, J. (2006) A resonance energy transfer between chemiluminescent donors and luminescent quantum dots as acceptors (CRET). *Angewandte Chemie, International Edition*, **45**, 5140–3.

83 Huang, S., Xiao, Q., He, Z., Liu, Y., Tinnefeld, P., Su, X. and Peng, X. (2008) A high sensitive and specific QDs FRET bioprobe for MNase. *Chemical Communications*, 5990–2.

84 Jose, D.A., Mishra, S., Ghosh, A., Shrivastav, A., Mishra, S.K. and Das, A. (2007) Colorimetric sensor for ATP in aqueous solution. *Organic Letters*, **9**, 1979–82.

85 Chen, Z., Li, G., Zhang, L., Jiang, J., Li, Z., Peng, Z. and Deng, L. (2008) A new method for the detection of ATP using a quantum-dot-tagged aptamer. *Analytical and Bioanalytical Chemistry*, **392**, 1185–8.

86 Liu, C.-W. and Chang, H.-T. (2007) Protein-conjugated quantum dots for detecting trypsin and trypsin inhibitor through fluorescence resonance energy transfer. *The Open Analytical Chemistry Journal*, **1**, 1–6.

87 Beck, M., Hildebrandt, N. and Loehmannsroeben, H.-G. (2006) Quantum dots as acceptors in FRET-assays containing serum. *Proceedings–SPIE the International Society for Optical Engineering*, **6191**, 61910X/1–X/8.

88 Hering, V.R., Gibson, G., Schumacher, R.I., Faljoni-Alario, A. and Politi, M.J. (2007) Energy transfer between CdSe/ZnS core/shell quantum dots and fluorescent proteins. *Bioconjugate Chemistry*, **18**, 1705–8.

89 Siemering, K.R., Golbik, R., Sever, R. and Haseloff, J. (1996) Mutations that suppress the thermosensitivity of green fluorescent protein. *Current Biology*, **6**, 1653–63.

90 Song, L., Jares-Erijman, E.A. and Jovin, T.M. (2002) A photochromic acceptor as a reversible light-driven switch in fluorescence resonance energy transfer (FRET). *Journal of Photochemistry and Photobiology A*, **150**, 177–85.

91 Nikoobakht, B., Burda, C., Braun, M., Hun, M. and El-Sayed, M.A. (2002) The quenching of CdSe quantum dots photoluminescence by gold nanoparticles in solution. *Photochemistry and Photobiology*, **75**, 591–7.

92 Stringer, R.C., Schommer, S., Hoehn, D. and Grant, S.A. (2008) Development of an optical biosensor using gold nanoparticles and quantum dots for the detection of Porcine Reproductive and Respiratory Syndrome Virus. *Sensors and Actuators B*, **134**, 427–31.

93 Stringer, R.C., Hoehn, D. and Grant, S.A. (2008) Quantum dot-based biosensor for detection of human cardiac troponin I using a liquid-core waveguide. *IEEE Sensors Journal*, **8**, 295–300.

94 Liu, T.-C., Zhang, H.-L., Wang, J.-H., Wang, H.-Q., Zhang, Z.-H., Hua, X.-F., Cao, Y.-C., Luo, Q.-M. and Zhao, Y.-D. (2008) Study on molecular interactions between proteins on live cell membranes using quantum dot-based fluorescence resonance energy transfer. *Analytical and Bioanalytical Chemistry*, **391**, 2819–24.

95 Wei, Q., Lee, M., Yu, X., Lee, E.K., Seong, G.H., Choo, J. and Cho, Y.W. (2006) Development of an open sandwich fluoroimmunoassay based on fluorescence resonance energy transfer. *Analytical Biochemistry*, **358**, 31–7.

96 Hildebrandt, N., Charbonniere, L.J. and Loehmannsroeben, H.-G. (2007) Time-resolved analysis of a highly sensitive Förster resonance energy transfer immunoassay using terbium complexes as donors and quantum dots as acceptors. *Journal of Biomedicine and Biotechnology*, **2007**, 1–6.

97 Peng, H., Zhang, L., Kjällman, T.H.M., Soeller, C. and Travas-Sejdic, J. (2007) DNA hybridization detection with blue luminescent quantum dots and dye-labeled single-stranded DNA. *Journal of the American Chemical Society*, **129**, 3048–9.

98 Lee, J., Choi, Y., Kim, J., Park, E. and Song, R. (2009) Positively charged compact quantum dot-DNA complexes for detection of nucleic acids. *ChemPhysChem*, **10**, 806–11.

99 Zhou, D., Ying, L., Hong, X., Hall, E.A., Abell, C. and Klenerman, D. (2008) A compact functional quantum dot-DNA conjugates: preparation, hybridization, and specific label-free DNA hybridization. *Langmuir*, **24**, 1659–64.

100 Wu, C.-S., Cupps, J.M. and Fan, X. (2009) Compact quantum dot probes for rapid and sensitive DNA detection using highly efficient fluorescence resonant energy transfer. *Nanotechnology*, **20**, 305502/1–7.

101 Roberts, C., Chen, C.S., Mrksich, M., Martichonok, V., Ingber, D.E. and Whitesides, G.M. (1998) Using mixed self-assembled monolayers presenting RGD and $(EG)_3OH$ groups to characterize long-term attachment of bovine capillary endothelial cells to surfaces. *Journal of the American Chemical Society*, **120**, 6548–55.

102 Susumu, K., Uyeda, H.T., Medintz, I.L., Pons, T., Delehanty, J.B. and Mattoussi, H. (2007) Enhancing the stability and biological functionalities of quantum dots via compact multifunctional ligands. *Journal of the American Chemical Society*, **129**, 13987–96.

103 Lim, T.C., Bailey, V.J., Ho, Y.-P. and Wang, T.-H. (2008) Intercalating dye as an acceptor in quantum-dot-mediated FRET. *Nanotechnology*, **19**, 075701/1–7.

104 Algar, W.R. and Krull, U.J. (2007) Towards multi-colour strategies for the detection of oligonucleotide hybridization using quantum dots as energy donors in fluorescence resonance energy transfer (FRET). *Analytica Chimica Acta*, **581**, 193–201.

105 Kim, J.H., Chaudhary, S. and Ozkan, M. (2007) Multicolor hybrid nanoprobes of molecular beacon conjugated quantum dots: FRET and gel electrophoresis assisted target DNA detection. *Nanotechnology*, **18**, 195105/1–7.

106 Zhang, C. and Johnson, L.W. (2006) Quantum dots-based fluorescence resonance energy transfer with improved FRET efficiency in capillary flows. *Analytical Chemistry*, **78**, 5532–7.

107 Algar, W.R. and Krull, U.J. (2009) Toward a multiplexed solid-phase nucleic acid hybridization assay using quantum dots as donors in fluorescence resonance energy transfer. *Analytical Chemistry*, **81**, 4113–20.

108 Algar, W.R. and Krull, U.J. (2009) Interfacial transduction of nucleic acid hybridization using immobilized quantum dots as donors in fluorescence resonance energy transfer. *Langmuir*, **25**, 633–8.

109 Algar, W.R. and Krull, U.J. (2008) Multidentate surface ligand exchange for the immobilization of CdSe/ZnS quantum dots and surface quantum dot-oligonucleotide conjugates. *Langmuir*, **24**, 5514–20.

110 Lee, J.I., Ha, K.-S. and Yoo, H.S. (2008) Quantum dots-assisted fluorescence resonance energy transfer approach for intracellular trafficking of chitosan/DNA complex. *Acta Biomaterialia*, **4**, 791–8.

111 Perey, L., Hayes, D.F., Maimonis, P., Abe, M., O'Hara, C. and Kufe, D.W. (1992) Tumor selective reactivity of a monoclonal antibody prepared against a recombinant peptide derived from the DF3 human breast carcinoma-associated antigen. *Cancer Research*, **52**, 2563–8.

112 Hiraga, Y., Tanaka, S., Haruma, K., Yoshihara, M., Sumii, K., Kajiyama, G., Shimamoto, F. and Kohno, N. (1998) Immunoreactive MUC1 expression at the deepest invasive portion correlated with prognosis of colorectal cancer. *Oncology*, **55**, 307–19.

113 Willsher, P.C., Xing, P.-X., Clarke, C.P., Ho, D.W. and McKenzie, I.F. (1993) Mucin 1 antigens in the serum and bronchial lavage fluid of patients with lung cancer. *Cancer*, **72**, 2936–42.

114 Cheng, A.K.H., Su, H., Wang, Y.A. and Yu, H.-Z. (2009) Aptamer-based detection of epithelial tumor marker mucin 1 with quantum dot-based fluorescence readout. *Analytical Chemistry*, **81**, 6130–9.

115 Lu, H., Schoeps, O., Woggon, U. and Niemeyer, C.M. (2008) Self-assembled donor comprising quantum dots and fluorescent proteins for long-range fluorescence resonance energy transfer. *Journal of the American Chemical Society*, **130**, 4815–27.

116 Sapsford, K.E., Berti, L. and Medintz, I.L. (2006) Materials for fluorescence resonance energy transfer analysis: beyond traditional donor-acceptor combinations. *Angewandte Chemie, International Edition*, **45**, 4562–88.

117 Lakowicz, J.R. (1999) *Principles of Fluorescence Spectroscopy*, 2nd edn, Kluwer/Plenum, New York.

118 Wang, H.-Q., Li, Y.-Q., Wang, J.-H., Xu, Q., Li, X.-Q. and Zhao, Y.-D. (2008) Influence of quantum dot's quantum yield to chemiluminescent resonance energy transfer. *Analytica Chimica Acta*, **610**, 68–73.

119 Waud, J.P., Fajardo, A.B., Sudhaharan, T., Trimby, A.R., Jeffery, J., Jones, A. and Campbell, A.K. (2001) Measurement of proteases using chemiluminescence-resonance-energy-transfer chimaeras between green fluorescent protein and aequorin. *The Biochemical Journal*, **357**, 687–97.

120 Yao, H., Zhang, Y., Xiao, F., Xia, Z. and Rao, J. (2007) Quantum dot/bioluminescence resonance energy transfer based highly sensitive detection of proteases. *Angewandte Chemie, International Edition*, **46**, 4346–9.

121 Xia, Z. and Rao, J. (2009) Biosensing and imaging based on bioluminescence resonance energy transfer. *Current Opinion in Biotechnology*, **20**, 37–44.

122 Xia, Z., Xing, Y., So, M.-K., Koh, A.L., Sinclair, R. and Rao, J. (2008) Multiplex detection of protease activity with quantum dot nanosensors prepared by intein-mediated specific bioconjugation. *Analytical Chemistry*, **80**, 8649–55.

123 Xing, Y., So, M.-K., Koh, A.L., Sinclair, R. and Rao, J. (2008) Improved QD-BRET conjugates for detection and imaging. *Biochemical and Biophysical Research Communications*, **372**, 388–94.

124 Cissell, K.A., Campbell, S. and Deo, S.K. (2008) Rapid, single-step nucleic acid detection. *Analytical and Bioanalytical Chemistry*, **391**, 2577–81.

125 Evanko, D. (2006) Bioluminescent quantum dots. *Nature Methods*, **3**, 240–1.

126 Fingleton, B. (2006) Matrix metalloproteinases: roles in cancer and metastasis. *Frontiers in Bioscience*, **11**, 479–91.

127 Egeblad, M. and Werb, Z. (2002) New functions for the matrix metalloproteinases in cancer progression. *Nature Reviews Cancer*, **2**, 161–74.

128 Brinckerhoff, C.E. and Matrisian, L.M. (2002) Matrix metalloproteinases: a tail of a frog that became a prince. *Nature Reviews Molecular Cell Biology*, **3**, 207–14.

129 Bachmeier, B.E., Iancu, C.M., Jochum, M. and Nerlich, A.G. (2005) Matrix metalloproteinases in cancer: comparison of known and novel aspects of their inhibition as a therapeutic approach. *Expert Reviews in Anticancer Therapy*, **5**, 149–63.

130 Overall, C.M. and Kleifeld, O. (2006) Tumour microenvironment-opinion: validating matrix metalloproteinases as drug targets and anti-targets for cancer therapy. *Nature Reviews Cancer*, **6**, 227–39.

131 So, M.-K., Xu, C., Loening, A.M., Gambhir, S.S. and Rao, J. (2006) Self-illuminating quantum dot conjugates for *in vivo* imaging. *Nature Biotechnology*, **24**, 339–43.

132 So, M.-K., Loening, A.M., Gambhir, S.S. and Rao, J. (2006) Creating self-illuminating quantum dot conjugates. *Nature Protocols*, **1**, 1160–4.

133 Zhang, Y., So, M.-K., Loening, A.M., Yao, H., Gambhir, S.S. and Rao, J. (2006) HaloTag protein-mediated site-specific conjugation of bioluminescent proteins to quantum dots. *Angewandte Chemie, International Edition*, **45**, 4936–40.

134 Gao, X., Cui, Y., Levenson, R.M., Chung, L.W.K. and Nie, S. (2004) *In vivo* cancer targeting and imaging with semiconductor quantum dots. *Nature Biotechnology*, **22**, 969–76.

135 Lee, J., Sundar, V.C., Heine, J.R., Bawendi, M.G. and Jensen, K.F. (2000) Full color emission from II-VI semiconductor quantum dot-polymer composites. *Advanced Materials*, **12**, 1102–5.

136 Larson, D.R., Zipfel, W.R., Williams, R.M., Clark, S.W., Bruchez, M.P., Wise, F.W. and Webb, W.W. (2003) Water-soluble quantum dots for multiphoton fluorescence imaging *in vivo*. *Science*, **300**, 1434–7.

137 Smith, A.M. and Nie, S. (2004) Chemical analysis and cellular imaging with quantum dots. *Analyst*, **129**, 672–7.

138 Esteves, A.C.C., Hodge, P., Trindade, T. and Barros-Timmons, A.M.M.V. (2007)

Preparation of nanocomposites by reversible addition-fragmentation chain transfer polymerization from the surface of quantum dots in miniemulsion. *Journal of Polymer Science Part A-1, Polymer Chemistry*, **47**, 5367–77.

139 Zou, S., Bai, H., Yang, P. and Yang, W. (2009) A biomimetic chemical approach to facile preparation of large-area, patterned, ZnO quantum dot/polymer nanocomposites on flexible plastics. *Macromolecular Chemistry and Physics*, **210**, 1519–27.

140 Emin, S., Sogoshi, N., Nakabayashi, S., Villeneuve, M. and Dushkin, C. (2009) Growth kinetics of CdS quantum dots and synthesis of their polymer nano-composites in CTAB reverse micelles. *Journal of Photochemistry and Photobiology A: Chemistry*, **207**, 173–80.

141 Carrillo-Carrión, C., Simonet, B.M. and Valcárcel, M. (2009) Carbon nanotube-quantum dot nanocomposites as new fluorescence nanoparticles for the determination of trace levels of PAHs in water. *Analytica Chimica Acta*, **652**, 278–84.

142 Fu, Y., Zhang, J. and Lakowicz, J.R. (2009) Silver-enhanced fluorescence emission of single quantum dot nanocomposites. *Chemical Communications*, 313–15.

143 Kim, B.Y.S., Jiang, W., Oreopoulos, J., Yip, C.M., Rutka, J.T. and Chan, W.C.W. (2008) Biodegradable quantum dot nanocomposites enable live cell labeling and imaging of cytoplasmic targets. *Nano Letters*, **8**, 3887–92.

144 Sun, P., Zhang, H., Liu, C., Fang, J., Wang, M., Chen, J., Zhang, J., Mao, C. and Xu, S. (2010) Preparation and characterization of Fe_3O_4/CdTe magnetic/fluorescent nanocomposites and their applications in immuno-labeling and fluorescent imaging of cancer cells. *Langmuir*, **26**, 1278–84.

145 Maeda, Y., Yoshino, T. and Matsunaga, T. (2009) Novel nanocomposites consisting of *in vivo*-biotinylated bacterial magnetic particles and quantum dots for magnetic separation and fluorescent labeling of cancer cells. *Journal of Materials Chemistry*, **19**, 6361–6.

6
Gold–Polymer Nanocomposites for Bioimaging and Biosensing

Nobuo Uehara and Tsutomu Nagaoka

6.1
Introduction

Gold–polymer nanocomposites (hereafter referred to as gold nanocomposites) are nanosized composites that consist of a gold core coated with a polymer shell. Indeed, the first colloidal gold solution produced by Faraday in 1857 was a gold nanocomposite solution stabilized with a polymer [1]. This solution, which is stored at the Royal Institute of Great Britain, has kept its red color until the present day. Faraday recognized that gold existed as very small discrete particles in the solution, and investigated the relationship between the solution's color and the particle size of the gold. The origin of the relationship between gold compounds and the life sciences began during the era of the European alchemists. Although metallic gold and gold compounds had been believed to possess medical benefits until the early nineteenth century, today only one gold salt is used medically, to treat rheumatism. During the early twentieth century, Lange conducted the first verifiable investigations into gold sols in the life sciences, in which a variation of the Zsigmondy flocculation test was used to identify degenerated proteins [2].

Significant progress in the use of metal nanoparticles in the life sciences has been inspired by recent developments in nanotechnology, that have revealed the importance of metal nanoparticles for nanodevices and as building blocks in the creation of nanostructures. The rapid development of nanotechnologies such as lithography, which has been further accelerated in the semiconductor industries, has allowed research groups to manipulate top-down patterning at nanoscale levels. Meanwhile, supermolecular chemistry has enabled the fabrication of biomimetic architectures in the mesoscopic region. As with metal nanoparticles, gold nanoparticles and nanocomposites have been used primarily as building blocks, not only in nanoengineering but also in the life sciences, due to their versatile properties such as chemical stability (i.e., low toxicity), stable development of color, and ease of preparation and modification. Whilst many excellent reports have been made recently on the application of gold nanoparticles to the life sciences [3–10], very few reports exist on the functionality of gold nanocomposites in this area,

Nanomaterials for the Life Sciences Vol.8: Nanocomposites. Edited by Challa S. S. R. Kumar

ISBN: 978-3-527-32168-1

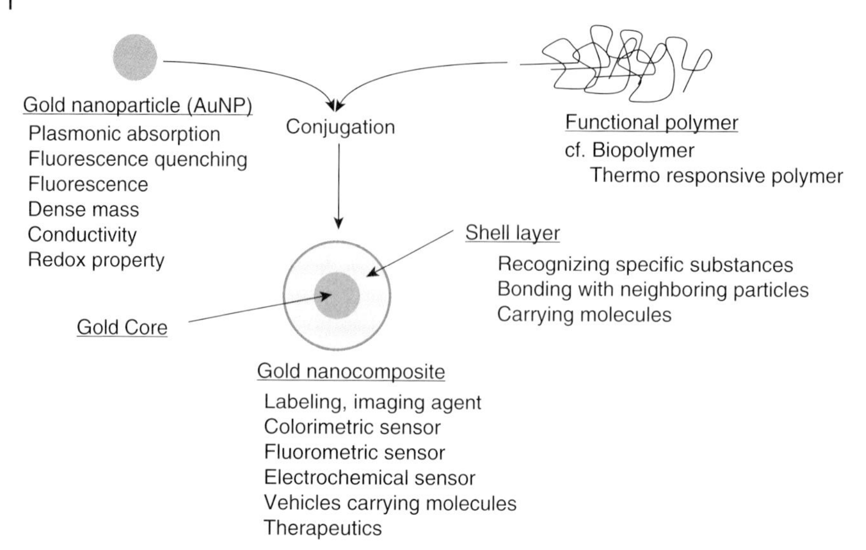

Figure 6.1 The preparation of gold nanocomposites, and their applications to the life sciences.

despite their versatility. Thus, the aim of this chapter is to review recent developments in gold nanocomposites for biosensing and bioimaging.

The inherent properties of a nanosized gold core and the functions of a conjugated polymer in a gold nanocomposite are shown schematically in Figure 6.1. The properties of nanosized gold cores, such as the absorption of visible light, fluorescence quenching, electrochemical activity, mass enhancement, and light–heat conversion are desirable features for applications in the life sciences. As conjugated polymers serve not only to stabilize gold nanocomposites in solution, but also offer several other attractive functions such as phase transition and recognition, the sophisticated combination of gold nanoparticles with functional polymers to form gold nanocomposites has allowed the development of a variety of versatile platforms. For example, the properties of light–heat conversion and the transportation of molecules have been used in therapeutics and for the development of drug-delivery systems, respectively. Although the medical application of gold nanocomposites has been actively investigated [11–15], such applications will not be discussed in this chapter as they are described in detail elsewhere in this book. Instead, attention will be focused on recent developments in the application of gold nanocomposites to biosensing and bioimaging for practical diagnostics. To the authors' knowledge, this is the first review to have focused on the analytical aspects of the development of gold nanocomposites for applications in the life sciences.

Gold nanocomposites have been studied intensively as diagnostic tools because gold cores develop a stable red color due to their surface plasmon resonance (SPR). Although, in the past, diagnostics with gold nanocomposites were based mainly on monotone assay systems, a novel breakthrough for a colorimetric assay using gold nanocomposites was inspired by Mirkin and coworkers, who developed DNA assay systems based on sandwich-like aggregates in which the oligonucleotides could bridge gold nanocomposites [16]. Subsequently, two types of gold nanocomposite were prepared by conjugation with different single-strand oligonucleotides, in which the gold nanocomposites were stabilized by the attached oligonucleotides that, in turn, acted as negatively charged soluble polymers. When a target DNA was hybridized that possessed sequences complementary to the attached oligonucleotides at both ends, the gold nanocomposites became bridged through the hybridized oligonucleotides; this resulted in an aggregation of the gold nanocomposites and a change in the color of the solution, from red to blue. Based on this colorimetric change, the target DNA could be assayed with the naked eye. These investigations not only inspired the fabrication of DNA assays with gold nanocomposites, but also stimulated a series of studies of colorimetric readout based on the bridging-type aggregation of gold nanocomposites [17–20].

In this chapter, following a discussion of the preparation of gold nanocomposites, details are provided of the recent developments in imaging and assay systems, where gold nanocomposites function as the key materials.

6.2 Fabrication of Gold Nanocomposites

6.2.1 General Aspects

The conjugation of a polymer shell with the surface of a gold core is a crucial step in preparing gold nanocomposites. The different methods used to prepare gold nanocomposites are shown in Figure 6.2, and include:

- *Grafting-from*, in which a polymer chain is extended from the surface of the gold core.
- *Grafting-to*, where a gold core is generated in polymer aggregates.
- *Post-modification*, which involves the conjugation of as-prepared gold nanoparticles (AuNPs) with the polymer.

As-prepared AuNPs are required for the "grafting from" and "post-modification" methods. When considering the following modification steps, the most suitable method of synthesizing AuNPs for life-science applications is in a homogeneous aqueous phase. The generation of AuNPs in aqueous solution is easily achieved by using the correct reductant that will reduce gold salts to a zero-valent state. In this respect, citrates have become one of the most commonly used reductants, as they serve not only as a reductant but also as a stabilizer for the generated AuNPs. Both, the concentration and type of stabilizer will influence the particle size and the morphology of the AuNPs. Moreover, stabilizers adsorbed onto AuNPs can serve as scaffolds at which the polymer chains are extended, via the "grafting-from" method, and play an important role in the "post-modification" method through replacement with polymer ligands.

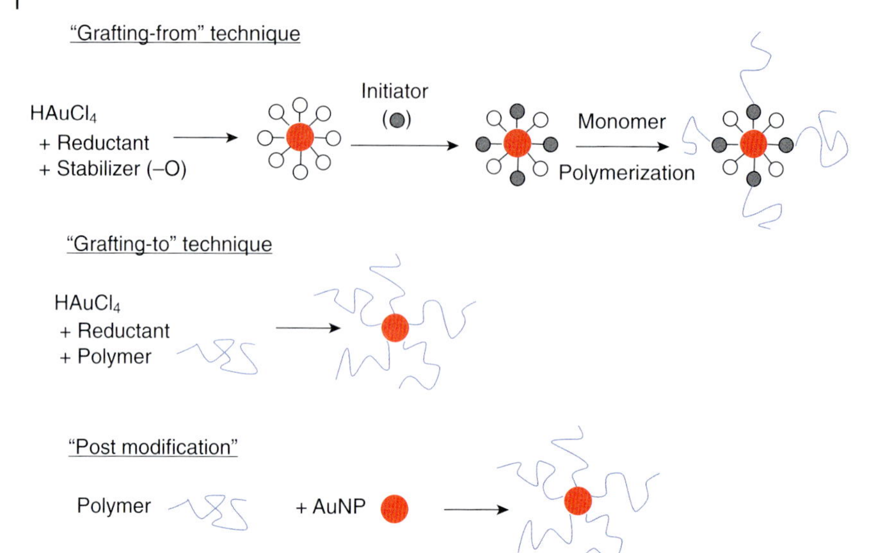

Figure 6.2 Schematic depiction of the preparation of gold nanocomposites through "Grafting-from," "Grafting-to," and "Post modification".

A two-phase synthesis has also been recognized as an important method for preparing AuNPs, although this requires the use of water-immiscible organic solvents that will be unfavorable for bioassays. In a two-phase synthesis, a water-immiscible organic phase containing an alkanethiol is contacted with an aqueous phase containing gold salts and reductants. The AuNPs generated in the aqueous phase are then extracted into the organic phase and stabilized by the alkanethiol. Although, highly monodispersed AuNPs can be prepared using this technique, in order to be practically applicable the method requires a re-dispersion of the AuNPs into the aqueous phase in order to remove the organic solvent.

Nonchemical methods, such as physical synthesis via laser irradiation, arc-discharge and flame ionization, in addition to biological syntheses using accumulated AuNPs in microorganisms, have also been developed. Details of the preparation and application of gold nanocomposites are available in several excellent reviews [21–24].

Polymers which form the shells of gold nanocomposites play crucial roles in the recognition of analytes, the formation of bridging structures, the catch-and-release of substances, and the stabilization of gold nanocomposites. Biopolymers such as peptides and oligonucleotides, as well as artificial polymers such as poly(*N*-isopropylacrylamide), have been used as the shell constituents in gold nanocomposites. Poly(*N*-isopropylacrylamide) and its derivatives, which are referred to as thermoresponsive polymers, can alter their morphologies reversibly according to the phase transition induced by thermal stimuli. Thermoresponsive polymers themselves have been used in the medical sciences as drug-delivery carriers and tissue-engineering materials, on the basis of their versatility [25–27].

The stability of gold nanocomposites is a crucial factor for their application as building blocks for nanoarchitecture, or as nanoprobes in bioassays. Polymers with functional groups containing sulfur atoms (e.g., a thiol group) will bond chemically to the surface of the gold core, thus stabilizing the resultant gold nanocomposites for long periods of time. In contrast, polymers without sulfur atoms interact with the surface of the gold core to form physically adsorbed gold nanocomposites; these tend to be displaced by sulfur-containing substances, resulting in their liberation from the gold nanocomposites. Although the instability of gold nanocomposites has been recognized as a shortcoming, a novel platform has been fabricated using such instability as an active function. The group of the present author (N.U.) prepared gold nanocomposites composed of gold cores and a thermoresponsive polymer, poly(*N*-isopropylacrylamide-*co*-methacrylate) as a model substance for the platform. Cysteine (a sulfhydryl amino acid), when added to the reaction mixture, replaced the physically adsorbed polymers on the surface of AuNPs such that a thermal stimulus caused the solution to be heated. This induced a phase transition in the polymer to facilitate polymer liberation, which led in turn to an aggregation and precipitation of the AuNPs. The extinction at 520 nm, which was ascribed to a plasmon band of discrete AuNPs, was decreased with increasing cysteine concentration [28].

6.2.2
"Grafting-From" Modifications

A modification technique that elongates polymer chains on the surface of planar gold substrates has been applied to the "grafting-from" modification of AuNP surfaces [29]. The main advantages of this technique are that: (i) the thickness of the polymer shell can be controlled easily with living polymerization; (ii) various polymer architectures such as mesh, comb, and block can be designed; and (iii) a high efficiency of polymer introduction can be achieved [23, 24].

A typical protocol for the "grafting-from" technique is shown in Figure 6.3. Prior to polymerization, initiators with sulfur atoms are introduced onto the surface of

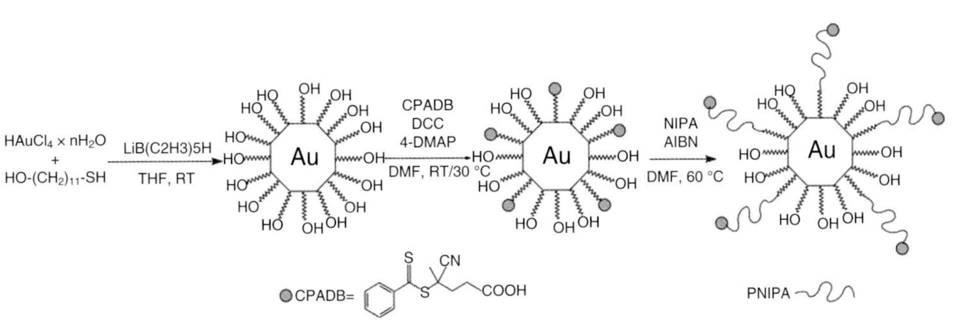

Figure 6.3 Preparation of gold nanocomposite conjugated with poly(*N*-isopropylacrylamide) (PNIPA), using the "grafting-from" technique. The macro-RAFT (reversible addition–fragmentation chain-transfer agent) (4-cyanopentanoic acid dithiobenzoate; CPADB) was introduced to stabilizer, followed by RAFT polymerization. Reproduced with permission from Ref. [30]; © American Chemical Society.

the gold cores, after which the polymer chains are extended via the reversible addition–fragmentation chain-transfer (RAFT) reaction so as to provide gold nanocomposites of the core–shell type [30]. Initiators with a disulfide group have frequently been used in graft polymerization, as successive protocols can be conducted in the same aqueous phase. Crosslinking with polymer networks not only strengthens the grafted shell layers but also increases the resistance to detachment from the gold surface [31, 32].

Biopolymers such as oligonucleotides and peptides can also be introduced with "grafting-from" methods. In order to introduce single-strand oligonucleotide chains, primers with a thiol group were attached to the surface of gold cores, followed by an extension of the single-stranded (ss) DNA chains via rolling circle polymerization, using DNA polymerase [33]. Due to the small the size of the primer, an effective introduction of ssDNA chains could be achieved, as compared to the "post-modification" method (see below). In order to graft peptides onto gold cores, sulfhydryl amines are introduced onto the surface of gold cores, followed by an elongation of peptide chains with a ring-opening polymerization (ROP) reaction. Higuchi and coworkers [34] demonstrated the introduction of poly(gamma-methyl L-glutamate-*co*-L-glutamic acid) grafted onto gold cores to form bioconjugating gold nanocomposites. The grafted peptide had a molecular weight of approximately 73 kDa and an α-helical structure in aqueous solution.

6.2.3 "Grafting-To" Modifications

The "grafting-to" modification, which refers to the evolution of gold cores in polymer aggregates, is another practical method for preparing gold nanocomposites. The advantages of this method are the ease of synthesis and the extended availability of polymers. As a single-pot reaction may easily be conducted for successive procedures of the "grafting-to" modification, laborious separation steps can be eliminated. In addition, the availability of various polymers allows the textural designs of gold nanocomposites to be expanded. For example, gold nanocomposites prepared with a highly monodispersed polymer will possess uniform shell layers [35].

Polymers terminated by a sulfur-containing group (e.g., dithioester, trithioester, thiol, thioether, disulfide), which can be prepared by radical polymerization using chain-transfer reagents containing sulfur atoms [36], provide chemically bonded shell layers with gold nanocomposites [37, 38]. On the other hand, when polymers without sulfur atoms were used, the generated gold cores were conjugated with the polymers through multipoint physical adsorption [39]. Various polymers, including poly(*N*-vinylpyrrolidine) (PVP) [40–42], poly(vinylpyridine) [43], poly(ethyleneglycol) (PEG) [44], poly(vinyl alcohol) (PVA) [45], poly(vinyl methylether) [46], chitosan [47], poly(ethyleneimine) (PEI) [48, 49], and poly(dialyl dimethylammonium) [50], have been examined for conjugation, based on their easy availability. The coated polymer layers in the composites are prone to detachment due to physical adsorption. As with the "grafting-from" modification, crosslinking with polymer networks is also effective in preventing the detachment of physically adsorbed polymers. Another promising strategy to overcome detachment is to use either unimolecular micelles [51, 52] or stabilized micelles with crosslinking networks [53], both of which serve as stable molecular templates. Although conventional polymer micelles can be used as molecular templates for preparing gold nanocomposites [54], the stability of the gold nanocomposites is affected by that of the micelles which, in turn, is influenced by the polymer concentration, the solution temperature, and the ionic strength. When polymer micelles that provide gold nanocomposites with shell layers become unstable, the gold nanocomposites disrupt the aggregation of the gold cores.

6.2.4 Post-Modification

The conjugation of as-prepared AuNPs with as-prepared polymers is the most common and easiest way to prepare gold nanocomposites, mainly because the method is simple and effective at avoiding uncertain factors, such as the size and molecular weight distributions of AuNPs. However, "post-modification" has certain drawbacks, such as a low efficiency of polymer introduction due to steric hindrance, and unintended adsorption through polymer functional groups. In addition to the other modifications described here, SH-terminated polymers form covalently bonded layers that surround the gold cores, whilst polymers without thiol groups form physically adsorbed layers.

6.3 Imaging and Labeling with Gold Nanocomposites

6.3.1 Optical Imaging

The conjugation of biofunctional materials with AuNPs to form gold nanocomposites has shown much promise for the preparation of bioimaging agents. Gold nanocores in gold–polymer nanocomposites demonstrate the inherent optical properties of AuNPs, but are influenced weakly by the attached polymer layers. Spherical AuNPs evolve absorption bands in the visible region of the spectrum due to their surface plasmon band, while discrete AuNPs with particle sizes of approximately 10 nm display a reddish color, the maximum extinction wavelength of which is at about 520 nm. An increase in the particle size of AuNPs shifts the spectral bands bathochromically [55, 56]. On the other hand, the extinction spectra for rod-shaped AuNPs can be altered up to the near-infrared (NIR) region (700–900 nm) by tuning the aspect ratio. Theoretical discussions on the relationship between the absorption maxima and the particle size or shape of AuNPs have been considered, using Mie theory. (For a detailed consideration of the relationship, which is beyond the scope of this chapter, several excellent reports and reviews

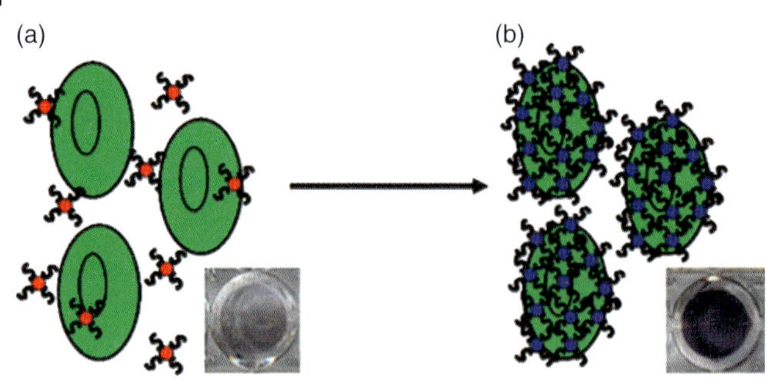

Figure 6.4 Schematic representation of colorimetric assay with aptamer-conjugated gold nanocomposites. (a) Before accumulation; (b) After accumulation on the cell surfaces. Adapted from Ref. [66] with permission from the American Chemical Society.

are recommended [57–60].) The stable color development that arises from the SPR has advantages over antiphotobleaching. In addition, the chemical inertness of gold cores contributes to the low cytotoxicity of the nanocomposites, and to their stable existence in body fluids when used as bioimaging agents.

Gold nanocomposites conjugated with lectin [61], biotin [62] and antibodies [63] have been used as bioimaging agents that possess specific recognition functions, based on immune interactions. Meanwhile, based on the observation that aptamers (which include engineered peptides) recognize specific sites on the cell surface, the visualization of HeLa cells was demonstrated using gold nanocomposites attached to aptamers having short transmembrane sequences [64, 65]. Discrete gold nanocomposites, conjugated with aptamers that induce acute leukemia cell accumulation onto target cells, resulted in the staining of cells with a blue-purple coloration (see Figure 6.4) [66]. There was also an increase in the extinction of the gold nanocomposites (up to 40 000 cells), which could be observed with the naked eye as an increase in the density of the blue coloration.

The NIR region is important for optical imaging due to biotransparency, and also to achieve noninvasive detection (although it is not suitable for assaying with the naked eye). The coupling of photoacoustic detection with the NIR ray represents a promising technique for noninvasive detection [67]. The detection of acoustic sounds generated from the absorption of a chopped laser light with rod-shaped gold nanocomposites can provide structural information on early-stage prostate cancer. For example, Pan and coworkers [62] demonstrated the imaging of mouse blood vessels with photoacoustic tomography using nanocomposites (referred to as "nanobeacons") that were composed of thiol-functionalized AuNPs, phospholipids, and biotin.

Fluorescence imaging has long been used for practical diagnosis, due to its sensitivity and the possibility to combine it with other techniques, such as microscopy. The greater feasibility of fluorescence imaging, compared to absorption-based imaging, is due to the excellent signal-to-noise ratio (SNR) and clear contrast.

As AuNPs with particle diameters below about 10 nm or above about 30 nm are fluorescent, they have been used as inorganic fluorescent dyes. Unlike the case with organic fluorescent dyes, fluorescent AuNPs resist photobleaching, which allows the use of a laser light source [63]. Fluorescent gold nanocomposites conjugated with the anti-epidermal growth factor receptor (anti-EGFR) were applied as an imaging agent to visualize skin cancer cells, by using a fluorescence correction spectrometer coupled to a microscope [68]. With photobleaching of any background cellular auto-fluorescence prior to fluorostaining, the images became very clear and displayed extreme improvements in contrast. Biotransparent NIR rays can be used for the two-photon excitement of gold nanorods. The staining of epidermal cancer cells with rod-like gold nanocores conjugated with anti-EGFR provided a fluorescence intensity that was three orders of magnitude larger than that obtained without gold nanorod staining.

6.3.2
Magnetic Resonance Imaging, X-Ray Imaging, and Surface-Enhanced Raman Scattering Imaging

Although gold nanocomposites have been used as vehicles to transport multiple gadolinium chelates into selective cellular targets, the use of gold cores did not result in any enhancement of the magnetic resonance imaging (MRI). Based on the fact that complexes formed by gadolinium and diethylenetriamine penta-acetic acid (DTPA, a polyaminocarboxylic acid) have been used as typical enhancers for MRI, gold nanocomposites conjugated with dithiolated DTPA oligomerized in the nanocomposite shell were designed as vehicles for gadolinium complexes [69].

Gold nanocomposites have also shown much promise as contrast agents for X-ray imaging, as the X-ray absorption coefficient of AuNPs is larger than that of iodine (a common X-ray absorber). The shortcomings of Ultravist (iopromide, a common X-ray imaging agent) include the induction of renal toxicity, a rapid and extensive catabolism and high vascular permeability, all of which could be overcome by using gold nanocomposites [70]. Gold nanocomposites consisting of a 30 nm gold core and a PEG shell exhibited antibiofouling properties, as well as an X-ray absorption coefficient which was 5.7-fold higher than those for Ultravist. As gold nanocomposites have been shown to accumulate in rat hepatomas, by staining hepatomas with gold nanocomposites it was possible to clearly contrast hepatomas from their neighboring, normal, cells.

Raman scattering may be extremely enhanced at or near silver and AuNPs surfaces (this situation is referred to as surface-enhanced Raman scattering; SERS), although this is usually weak for reasons of inelasticity. Inspired by a report that DNA could be assayed with SERS signals from silver-stained gold nanocomposites bound to substrates through a target DNA [71], SERS have been used in combination with gold nanocomposites in life sciences applications [72, 73]. Notably, gold nanocomposites conjugated with PEGs have been used as SERS labeling agents for the detection of tumors *in vivo* [74].

6.3.3
Gold Nanocomposites as Indicators

Gold nanocomposites are useful not only as imaging agents, but also as indicators, owing to their chemical inertness and the stable color development of the gold cores. Assay kits for simple diagnoses consist of gold nanocomposites (used as an indicator), a biofunctionalized white substrate (such as nitrocellulose), and a porous silica gel. The analytes act as bridging agents to link the gold nanocomposites with the functional groups on the substrates, and this results in a staining of the substrates. As the amount of analytes is increased, the amount of adsorbed gold nanocomposites is also increased, and this change can be read either with the naked eye or by using a spectrophotometer. Antigen–antibody interactions [75], aptamer–protein interactions [76] and DNA hybridization [77, 78] have each been investigated for the formation of bridging structures, with an aim to fabricate lateral diagnostic kits. The formation of bridging structures involving with analytes is discussed in the following section.

6.4
Colorimetric Sensors

6.4.1
General Aspects

As noted in Section 6.3, solutions of discrete AuNPs with particle sizes between approximately 10 nm and 50 nm display a reddish color, the maximum extinction wavelength of which lies at about 520 nm. The extinction bands evolved by the surface plasmon bands of AuNPs are influenced not only by their particle size, but also by their morphology. When AuNPs aggregate, their interparticle distances become shorter than their particle sizes; this causes interparticle plasmon coupling, leading to the evolution of a new extinction band at a longer wavelength and a change in the color of the solution to blue-purple. A relationship between the maximum wavelength and particle size or morphology of gold cores in gold nanocomposites can be explained with Mie theory. The change in color is the fundamental basis for colorimetric sensors using gold nanocomposites, while the rapid change in solution color caused by the morphological change of gold cores has allowed the fabrication of various colorimetric sensor platforms for assaying with the naked eye [79–85]. These are classified according to the directions of changes in the solution color – that is, red-to-blue and blue-to-red – as a result of the morphological change in the nanocomposites gold cores. At attention in this chapter is focused on biosensing and bioimaging with gold–polymer nanocomposites, colorimetric sensing using AuNPs functionalized with small molecules will not be described. However, details on AuNPs functionalized with small molecules are available in some excellent reviews [3–10].

6.4.2
Red-to-Blue Sensors

Red-to-blue colorimetric sensors are based on a change in morphology of gold cores from the discrete state to the aggregation state. Two representative mechanisms concerning the aggregation of gold nanocomposites have been investigated; the first mechanism is based on the formation of bridging structures between gold nanocomposites, and the second is based on spontaneous aggregation induced by salting-out effects.

6.4.2.1 Crosslinking Aggregation

Five schemes of gold nanocomposites aggregation through bridging structures composed of analytes and functional groups of the gold nanocomposites have been recognized (see Figure 6.5): System A, the hybridization of oligonucleotides; (ii) System B, the formation of chelating bonds via metal ions; System C, the formation of supermolecular structures; and (iv) System D, a specific recognition based on biological interactions. This classification is not rigorous, however, and intermediate interactions among them often serve as the bridging structures of gold

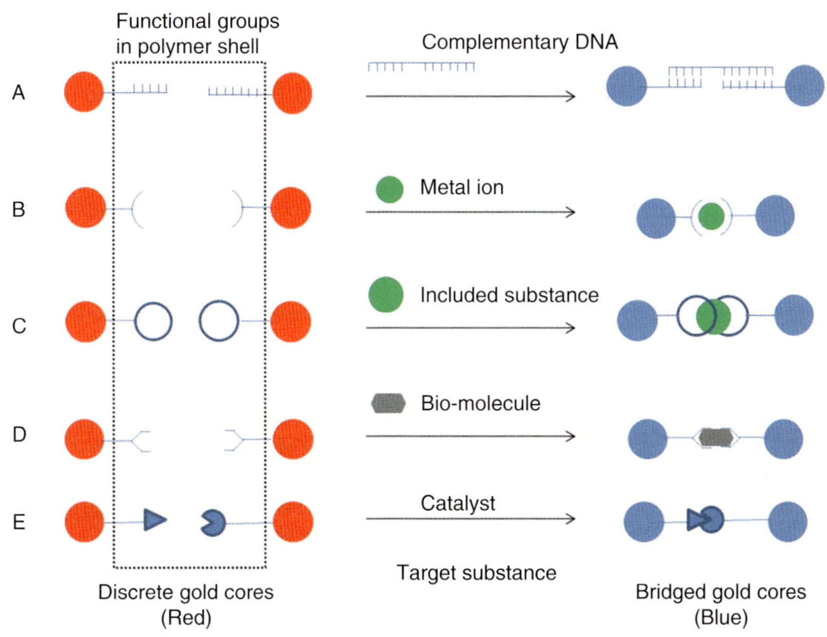

Figure 6.5 Bridging structures composed of hybridized DNA (system A), chelating bond (system B), supermolecule (system C), biological interaction (system D), and covalent bond between gold cores (system E) in gold nanocomposites for red-to-blue sensors.

nanocomposites. The formation and disruption of bridging structures have provided fundamental platforms with gold nanocomposites, not only for colorimetric sensors but also for other useful readout tools such as fluorometry, quartz crystal microbalances, and cantilevers.

6.4.2.1.1 **System A: Hybridization of Oligonucleotides** Bridging structures through double-strand oligonucleotides were first applied to a DNA assay with gold nanocomposites by Mirkin and coworkers [16]. The hybridization of a target DNA with a complementary ssDNA forms a sandwich-type aggregation of gold nanocomposites through a double-strand bridge. For colorimetric assays of DNA, two types of gold nanocomposite probe were prepared individually by conjugating AuNPs with two corresponding ssDNAs, the sequences of which were complementary to those at both ends of the target DNA. Hybridization of the target DNA with the two ssDNAs attached with gold nanocomposites resulted in an aggregation that changed the solution color from red to blue-purple. Heating a solution of the aggregates induced dehybridization of the double-strand oligonucleotides bridging gold nanocomposites into single strands, causing disruption of the aggregation and a colorimetric change from blue-purple to red. The colorimetric change caused by the phase transition of the oligonucleotides was reversible. Mirkin and coworkers claimed [86] that the phase transition of the oligonucleotides conjugated with gold nanocomposites occurred more sharply than unmodified oligonucleotides, due to a dense loading of the oligonucleotides on the AuNPs.

The platform based on hybridization has been applied not only to DNA assays but also to assays of substances that affect the stability of the bridging double-stranded DNAs (dsDNAs). Thymidine–thymidine mismatches in dsDNAs made them unstable. As the Hg^{2+} ion formed a complex with thymidines to form 1:2 chelates, the coordination of Hg^{2+} with thymidine–thymidine mismatches stabilized the double strands, leading to an increase in the melting temperature (T_m) of the double strands. An increase in concentration of $1\,\mu mol\,l^{-1}$ of Hg^{2+} corresponded to an increase in T_m by about 5°, thus providing an easy method of determining the Hg^{2+} concentration [79, 80]. Calcium ion-stabilized nanoarchitectures were composed of guanine (G)-rich oligonucleotides [81]. When calcium ions were bound to cavities of guanine quartets formed from four G-rich oligonucleotides, the quartets were stabilized through the included calcium ions to form a nanoarchitecture, leading to the aggregation of gold nanocomposites conjugated with the G-rich oligonucleotides. Copper ions were assayed indirectly through a catalytic reaction that produced bridging DNAs from DNAzymes [82]. DNAzymes were cleaved by copper ions catalytically to generate fragmented oligonucleotides that bridged gold nanocomposite probes through hybridization. Foreign metal ions such as Zn^{2+}, Mn^{2+}, Co^{2+}, Ni^{2+}, Ca^{2+}, and Pb^{2+} at $0.1\,mmol\,l^{-1}$ did not facilitate the reaction.

6.4.2.1.2 **System B: Bridging with Coordination Bonds** Heavy-metal ions also act as linkers of chitosan layers that cover gold cores in the gold nanocomposites [83]. Chitosan-coated gold nanocomposites were stabilized through the hydration of hydroxyl groups in chitosan chains; the subsequent complex formation of nickel and copper ions with the amino groups of chitosan promoted an interparticle linkage of the gold nanocomposites, resulting in their aggregation.

6.4.2.1.3 **System C: Bridging with Supermolecular Bonds** Calsequestrin, which is the most abundant calcium-binding protein, formed nanoarchitectures with calcium ions. As conformation of the nanoarchitectures was controlled by the concentration of calcium ions, gold nanocomposites conjugated with calsequestrin could be used as the building blocks of architectures controlled by calcium ions [84]. Polymerized calsequestrin with a calcium ion concentration more than $1\,\text{mmol}\,\text{l}^{-1}$ caused the formation of gold nanocomposite aggregates. Although EDTA may be used to mask the calcium ions so as to facilitate a disruption of the aggregate, a reversible change between aggregation and disruption can be achieved by the repeated addition of EDTA and calcium ions.

6.4.2.1.4 **System D: Biological Recognition** In addition to the hybridization of oligonucleotides, other biological recognition functions with high specificity – such as antigen–antibody interactions, avidin–biotin interactions, lectin–sugar interactions and interactions involving aptamers – have provided corresponding platforms of colorimetric assays with gold nanocomposites (see Figure 6.5, system D).

Antigen–antibody interactions (referred to as "immune-interactions") have been investigated previously [85, 87–89]. Initially, in 1999, Shenton and coworkers [90] showed that gold nanocomposites conjugated with immunoglobulin (Ig) E or IgG antibodies that exhibited specificities towards dinitrophenyl (DNP) and biotin, were able to recognize bisdinitrophenyl (DNP–DNP) and biotinylated dinitrophenyl (biotin–DNP); this resulted in aggregation and a consequent colorimetric change. Zeng and coworkers synthesized single-chain fragment-variable (scFv) recombinant antibodies that contained a cysteine moiety in the scFv linker region in order to fabricate gold nanocomposites with AuNPs, and applied them to a colorimetric immunoassay [88]. Addition of the antigen rabbit IgG induced an aggregation of the gold nanocomposites, resulting in a colorimetric change. The ratio of the extinction at 520 nm to that at 620 nm exhibited a linear relationship at IgG concentrations below $100\,\text{nmol}\,\text{l}^{-1}$, with a detection of limit $1.7\,\text{nmol}\,\text{l}^{-1}$.

Avidin–biotin interactions have been among the most useful tools in bioengineering and the life-sciences [91]. Fitzmaurice first demonstrated the value of avidin–biotin interaction for the bridging structure between gold nanocomposites [92]. The aggregation of gold nanocomposites induced by avidin–biotin interactions was reversible, and could be controlled by adjusting the avidin : biotin molar ratio [93]. In one study, kinase activity was monitored indirectly with the gold nanocomposites, based on the avidin–biotin interaction [94]. In this case, the gold nanocomposites conjugated with peptides were biotinylated via a kinase-catalyzed reaction with biotin–ATP. The biotinylated nanocomposites were then mixed with avidin-modified gold nanocomposites so as to induce an aggregate formation, via avidin–biotin interaction, that led to a colorimetric change (from red to blue). As the amounts of biotinylated gold nanocomposites generated by the catalytic reaction were controlled by the kinase activity, the colorimetric change corresponded to the activity of the kinase inhibitor. A morphological control of rod-shaped gold nanocomposites was achieved through the avidin–biotin interaction [95, 96]. Whilst the introduction of biotin disulfide to the edges of gold nanorods induced end-to-end linkages to form nanorod chains, the introduction of biotin to the whole surface of the nanorods induced side-to-side assemblies of the nanorods.

Lectin–sugar interactions are also important in biological recognition. Gold nanocomposites conjugated with carbohydrates are stabilized via the hydration of hydroxyl groups in the carbohydrates [97–103]. When the carbohydrate-stabilized gold nanocomposites bind to lectin, RCA 120 or concanavalin A (Con A), the resultant aggregation of gold nanocomposites can be used as a platform for colorimetric sensors to assay sugars or lectins. *Cholera toxin* was assayed based on a platform which used gold nanocomposites conjugated with a thiolated lactose derivative [104]. As the lactose derivative mimicked the GM_1 ganglioside (the receptor to which cholera toxin binds in the small intestine), the cholera toxin would accumulate the thiolated lactose derivative introduced on the gold nanocomposites through the lectin–sugar interaction, which would cause the gold nanocomposites to aggregate. This allowed the cholera toxin to be assayed within 10 min, with a detection limit of $54\,\mathrm{nmol\,l^{-1}}$ ($3\,\mathrm{mg\,l^{-1}}$). *Antimicrobial susceptibility* was also estimated with a platform based on sugar–lectin interactions, where gold nanocomposites conjugated with dextran were aggregated with carbohydrate and Con A [105]. The growth of *Escherichia coli* reduced the carbohydrate content of the growth medium, and prevented aggregation through the sugar–lectin interaction. In the absence of *E. coli*, however, the gold nanocomposites were able to aggregate (due to a lack of carbohydrate consumption), leading to a bathochromic spectral shift of the gold nanocomposites. When the antibiotic, ampicillin, was added to the growth medium, *E. coli* growth was inhibited and gold nanocomposite aggregation facilitated. Thus, the antimicrobial susceptibility of *E. coli* to ampicillin could be assessed by monitoring the spectral shift of the gold nanocomposites.

Aptamers are synthetic, single-stranded nucleic acids or peptides that can bind to target molecules with high affinity and specificity [106]. Several types of aptamer have been designed to bind proteins, cell surfaces, small molecules and metal ions, or to be cleaved by metal ions or other substances. Based on the binding properties of aptamers, red-to-blue colorimetric probes have been developed in which gold nanoparticles are conjugated with aptamers to assay metal ions [107], enzymes [76, 108], peptides [109], and cells [66]. Based on their cleavage properties, blue-to-red colorimetric probes and fluorometric sensors have also been designed with gold nanocomposites conjugated with aptamers (see below).

6.4.2.2 **Non-Crosslinking Aggregation**

Non-crosslinking aggregation has become an alternative approach to exploit red-to-blue colorimetric sensors of gold nanocomposites. The enthalpic destabilization of gold nanocomposites induces their spontaneous aggregation. Citric acid reduction, which has been most commonly used in preparing AuNPs, introduces nega-

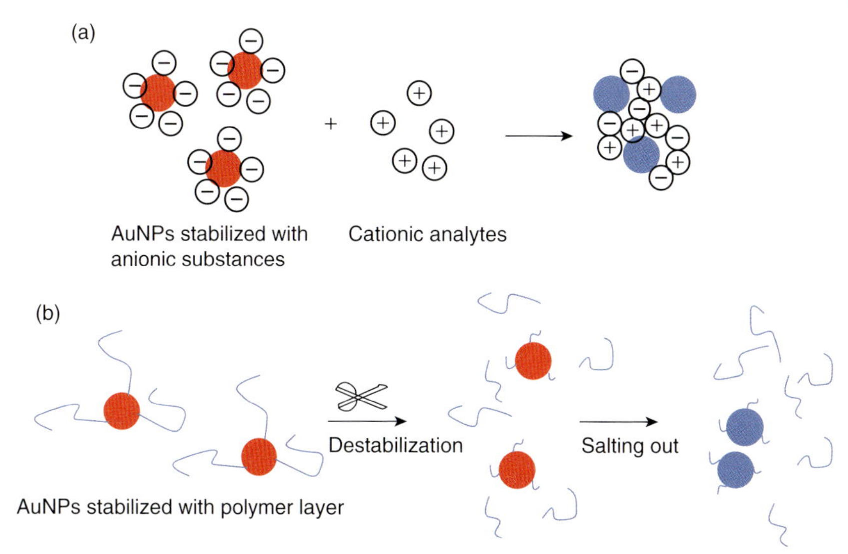

Figure 6.6 Spontaneous aggregation of gold nanocomposites based on (a) cancellation of the surface charges, and (b) salting out assisted with entropic destabilization.

tive charges onto the AuNP surfaces through the adsorption of anionic citrate to stabilize the formed AuNPs. The enthalpic destabilization of gold nanocomposites was achieved through the cancellation of negative charges on the gold nanocomposites (Figure 6.6a), or the shrinkage of electric double-layers surrounding the gold nanoparticles caused by salting out (Figure 6.6b).

Cancellation of the negative charges on the gold nanocomposites was simply attained by adding cationic substances. For example, human serum albumin (HSA), an anionic protein, was examined as the shell layer of gold nanocomposites for a lysozyme assay [110]. The addition of high-pI proteins such as lysozyme, α-chymotrypsinogen A, and conalbumin canceled the negative charges of the gold nanocomposites, resulting in aggregation and colorimetric change. A the pI value of proteins and the solution pH largely governed the selectivity of the cancellation of surface charge, the achievement of selectivity for specific proteins was difficult in this assay.

In contrast to canceling the surface charges of the gold nanoparticles, shrinking the electric double layer represents a promising approach to serve platforms of colorimetric sensors with selectivity. Increasing the ionic strength of a solution by adding salts (e.g., sodium chloride) compresses the electric double layer to destabilize the gold nanocomposites, resulting in their spontaneous aggregation [111]. Introducing controllable mechanisms that can stabilize the gold nanocomposites can facilitate the designing of selective colorimetric assays. The conjugation of single-strand oligonucleotides (ssDNA) that act as anionic polymers with AuNPs has been studied for this purpose [112]. Negative charges, as well as the flexibility of ssDNA, serve to provide stability to the gold nanocomposites. Hybridization with the complementary ssDNA forms rigid double-strand oligonucleotides (dsDNA) to reduce their flexibility, causing entropic instability of the gold nanocomposites. The instability facilitates assembly of the gold nanocomposites with the aid of salting out. As single nucleotide polymorphisms (SNPs) form no stable dsDNA, the addition of SNPs kept the gold dispersed. Based on the different responses between the complementary DNA and the SNPs, discrimination of SNPs was achieved with the naked eye.

The control of aptamer morphology was utilized to exploit a platform based on spontaneous aggregation. The inclusion of adenosine or lead ions with aptamers linked to AuNPs caused their structures to fold, and led to a stabilization of the AuNPs to resist salting out [113]. On the other hand, as unfolded aptamers do not stabilize the gold nanocomposites, the addition of salt induced a spontaneous aggregation. Dynamic light scattering (DLS) measurements indicated that the unfolded aptamers shrank to decrease the thickness of their layer surrounding AuNPs, whereas the folded aptamers retained their layer, without shrinkage. The results were inconsistent with the stabilization of ssDNA described above [112]. The roles of oligonucleotides conjugated with gold nanocomposites are complicated and unclear at this time.

The extension of hydrophilic polymer chains from the AuNP surfaces represents another means of introducing stability to the AuNPs [30]. He and coworkers [114] investigated a DNA assay based on polymerization, in which a target DNA bridged an initiator-bonded ssDNA with a probe DNA linked to gold nanocomposites through hybridization, forming sandwich-type conjugates. This was followed by atom transfer radical polymerization (ATRP) to graft a poly(oligo(ethyleneglycol) methacrylate) chain, which extended the hydrophilic ethyleneglycol units from the gold surface and stabilized the resulting nanocomposites due to an entropic repulsion of the oligo(ethyleneglycol) chain. Without polymerization, a spontaneous aggregation of the gold nanocomposites was induced by salting out. The copper ions essential in facilitating the ATRP complexed with DNA on the AuNP surfaces to facilitate a spontaneous aggregation. Consequently, complexation of the copper ions assisted the spontaneous aggregation induced by salting out.

6.4.3 Blue-to-Red Sensors

The principle underlying blue-to-red sensors is based on changes in the plasmon bands of gold nanocomposites, from an assembled state to a dispersed state. The advantages of blue-to-red sensors are high reproducibility and quantitative properties; both benefits are due to the stability of the resulting gold nanocomposites with high dispersiveness. In contrast, red-to-blue sensors require a strict control of the experimental conditions, because the aggregated gold nanocomposites are prone to precipitation; that is, it is difficult to maintain the assembled state of gold nanocomposites in solution. Therefore, spectral bands of the aggregated gold nanocomposites gradually decrease with increased standing time, causing problems in obtaining constant spectra for the aggregated gold nanocomposites.

In addition to naked dsDNAs, those dsDNAs bound with gold cores at their ends exhibit a reversible phase transition between hybridization and dehybridization. This causes reversible changes in the morphology of the gold nanocomposites between assembly and disassembly, which can be read out through colorimetric changes between blue and red, respectively. As noted above, the phase transition of the dsDNA with a gold core in the nanocomposites is more distinct than that of the unmodified dsDNA [79]. The intercalation of DNA binding molecules increases the phase transition temperature, as the phase transition is influenced by the stability of double-strands. These properties facilitate the use of an assay to estimate the affinity of DNA binding molecules with the gold nanoparticles bridged with dsDNA [115]. The assay was conducted via a competitive intercalation of assayed molecules between hair-pin-type DNA and gold core-linked dsDNA. The affinity of the intercalation was estimated with the phase transition temperature of the gold nanocomposites.

Based on the fact that aptamers are cleaved by specific substances [107], blue-to-red colorimetric probes with gold nanocomposites have also been designed with aptamers. The designed probes are gold nanocomposites in which individual AuNPs are linked with one another via aptamers cleaved with adenosine [116]. The cleavage of aptamers with adenosine caused disruption of the interparticle bridging structures and disassembly of the conjugates, leading to a blue-to-red colorimetric change (see Figure 6.7).

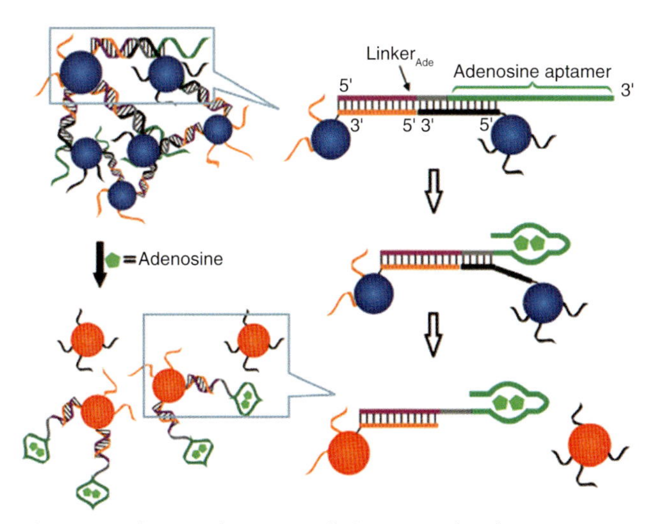

Figure 6.7 Schematic description of adenosine-induced disassembly of AuNPs linked by a DNA containing an adenosine aptamer. Each aptamer binds two adenosine molecules to disassemble the AuNPs. Reproduced with permission from Ref. [116]; © American Chemical Society.

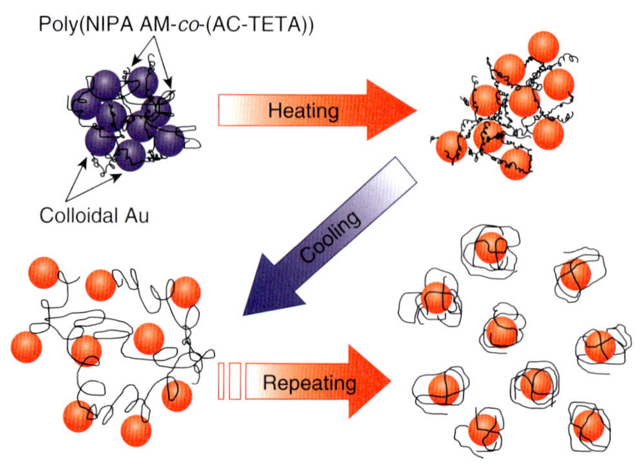

Figure 6.8 Thermal stimuli-induced disassembly of aggregated gold nanocomposites conjugated with a thermoresponsive polymer. Reproduced with permission from Ref. [117]; © American Chemical Society.

The group of the present author (N.U.) has developed another type of red-to-blue sensors without disrupting bridging structures [117, 118]. For this, gold nanocomposites were prepared by the conjugation of AuNPs with a thermoresponsive copolymer possessing poly(ethyleneamine). Although the resultant gold nanocomposites aggregated through the poly(ethyleneamine) groups, no sedimentation was observed due to a stabilizing effect of the conjugated polymers. The change in aggregated gold nanocomposites conjugated with the thermoresponsive copolymer caused by thermal stimuli is shown schematically in Figure 6.8. Heating the solution induced the phase transition, causing shrinkage of the polymer chains adsorbed onto the AuNPs surface due to dehydration. The shrunken polymers on the AuNPs expanded the interparticle distance such that the aggregated gold nanocomposites disassembled, leading to change in the solution color to red. Cysteine, a sulfhydryl amino acid, inhibited the disassembly by replacing the polymer adsorbed on the gold nanocomposites. As the concentration of cysteine added to the solution was increased, the resulting solution exhibited a gradation from red to blue purple. The gradation could be quantified with the $L^*a^*b^*$ color coordinates to determine the cysteine concentration.

6.5 Fluorometric Systems with Gold Nanocomposites

Fluorometric detection has been used as a powerful tool in bioanalysis (e.g., in immunoassays and labeling) owing to its sensitivity, selectivity, and versatility. Gold nanocomposites have also played two important roles in fluorometric bioanalysis – one role as a fluoropher, and another as a fluorescence quencher. A role for gold nanocomposites that act as fluorophers was discussed in Section 6.4, labeling and imaging). In this section, attention will be focused on the role of gold nanocomposites as fluorescent quenchers in bioassays. In addition to distance-dependent fluorescence quenching techniques, other advanced topics such as time-resolved fluorometry and fluorescent polarization are also discussed.

6.5.1 Fluorescence Quenching

Those AuNPs with a size greater than 10 nm serve as fluorescence quenchers through both energy transfer and electron transfer. The fluorometric quenching constant of AuNPs is several orders of magnitude greater than that of typical small-molecule quenchers [119–122]. Fluorescence quenching caused by energy transfer and electron transfer occurs when the distance between a fluoropher and quencher is smaller than 2 nm [123, 124]. Thus, a precise control of the proximity of a fluoropher to a quencher will allows the use of fluorescence quenching systems that employ nanoconjugates composed of gold nanocomposites and fluorophers. The recovery of the inherent fluorescence of fluorophers resulting from the disruption of bridging structures between the fluorophers and gold cores in the gold nanocomposites is shown in Figure 6.9.

6.5.1.1 System A: Disruption of dsDNA Linker

As stated above, double-strand oligonucleotides have been investigated as promising linkers of gold cores in nanocomposites, owing to certain feasible characteristics such as the precise design of a sequence and reversible control between hybridization and dehybridization. The inhibition of fluorescence quenching in gold nanocomposites can be achieved by replacement with a complementary oligonucleotide or a target DNA [125, 126]. Probe nanoconjugates were prepared by hybridizing gold nanocomposites with single-strand oligonucleotides with a fluoropher having the complement. The gold nanocomposites linked with the

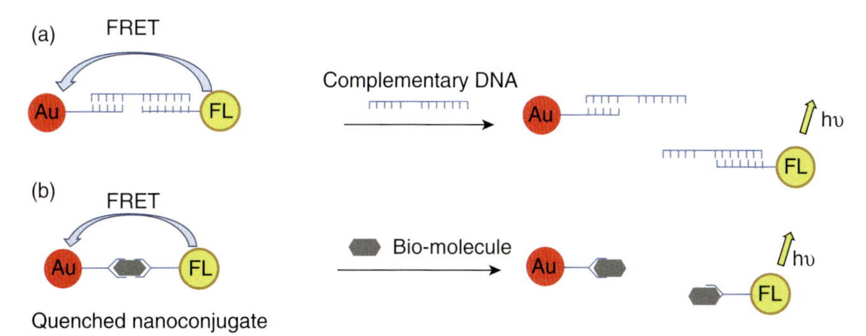

Figure 6.9 Recovery of inherent fluorescence due to breakdown of fluorescence resonance energy transfer (FRET) by disruption of bridging structures between gold nanocomposites and fluoropher. Disruptions result from interactions with (a) a complementary DNA, and (b) a biomolecule.

fluoropher in the nanoconjugates quenched fluorescence through energy transfer. A target DNA that possessed the complementary sequence replaced the fluoropher so as to liberate it, and this resulted in the recovery of fluorescence.

6.5.1.2 System B: Disruption of Biological Interaction

Various interactions involving aptamers as well as antigen–antibody [127] and avidin–biotin [128] interactions have also provided bridging structures to connect gold nanocomposites and fluorophers. For example, Chang and coworkers [129] prepared gold nanocomposites conjugated with platelet-derived growth factor (PDGF)-binding aptamers, followed by the immobilization of cationic fluorophers on the aptamer through affinity. The immobilized fluorophers were quenched by the fluorescence resonance energy transfer (FRET) from the nanocomposites. The PDGF competed with the immobilized fluorophers to replace and liberate them, leading to a recovery of the fluorescence. Based on the competing interaction, another PDGF assay probe with a fluorescent gold nanocluster attached to PDGF and gold nanocomposites conjugated to the PDGF receptor was further exploited [130]. Interaction between PDGF and the PDGF receptor facilitated an aggregation of the fluorescent gold nanocluster and the gold nanocomposites to form quenched conjugates. The addition of PDGF or the PDGF receptor competed with the PDGF–PDGF receptor interaction to disrupt the conjugates and release the fluorescent gold nanocluster, which could be monitored fluorometrically. The limits of detection for PDGF and the PDGF receptor were $80\,\text{pmol}\,\text{l}^{-1}$ and $0.25\,\text{nmol}\,\text{l}^{-1}$, respectively.

6.5.1.3 System C: Cleavage of Linking Aptamer and Peptide

One facile characteristic of aptamers – that specific substances cleave the corresponding aptamers – has also been applied to fluorescence quenching systems as well as blue-to-red colorimetric systems. In order to assay adenosine and cocaine, Lu and coworkers [131] prepared a sensor solution containing two types of gold nanocomposite in which two forms of quantum dot (QD) were linked with AuNPs through an adenosine aptamer and a cocaine aptamer. The maximum emission wavelengths of the QDs linked with the adenosine aptamer and the cocaine aptamer were 525 nm and 585 nm, respectively. Moreover, as the emission spectra did not overlap, a simultaneous assay of adenosine and cocaine could be achieved with a one-pot reaction.

6.5.2 Time-Resolved Fluorometry and Fluorescence Polarization

Time-resolved fluorometry has been applied to a bioassay using gold nanocomposites, based on the fact that the Plasmon–fluoropher interaction influences not only the life-time but also the energy transfer of an excited fluoropher. The probe conjugates for time-resolved fluorometry were prepared by combining avidin-bonded gold nanocomposites with biotinylated fluorophers, followed by introducing a biotinylated antibody to the conjugates [132]. The additional conjugation of protein with the probe conjugates through another antigen–antibody interaction reduced the life-time of the excited state of the fluoropher. The linear relationship between the time-constant of fluorescence decay and the concentration of bovine serum albumin (BSA) – the examined protein – was obtained at picomolar levels of BSA. The time-resolved fluorometry with the probe conjugates was expected to apply to fluorescence imaging.

Fluorescence polarization has also been applied to the assay of biomolecules, based on the formation of bridging structures [133]. The conjugation of AuNPs with fluorophers increased the volume and mass of the resulting conjugates, leading to a suppression of molecular rotation. As the rotation of fluorescent molecules influences fluorescence polarization, the formation of bridging structures via analytes can be read out with the fluorescence polarization. An assay of Hg^{2+} ions was demonstrated with changes in fluorescence polarization, based on the fact that the chelation of Hg^{2+} ions stabilized the thymine–thymine (T-T) mismatch. Gold nanocomposite probes conjugated with ssDNA were hybridized with a fluoropher attached to ssDNA in which six T-T mismatches were strategically placed. The Formation of T–Hg^{2+}–T complexes stabilized the hybridized DNA, resulting in a substantial increase in fluorescence polarization. The detection limit of Hg^{2+} was 0.2 ppb, with a good selectivity that tolerated 1000-fold concentrations of other metal ions.

6.6
Electrochemical Applications of Gold Nanoparticles and Their Composites

To date, many reviews have been prepared on the subject of nanoparticle-based electroanalysis (e.g., [134, 135]) consequently, those published between 2007 and 2009 are mainly discussed at this point, detailing the most recent developments in this field.

Historically, AuNPs have been exploited as tags or reporters in a variety of science and engineering fields, but more recently have found applications as the scaffolds for providing nanofunctionalities. As with electrochemistry, metal nanoparticles may also be expected to provide high catalytic activities when assembled on electrode surfaces, and in this respect many reports have described the catalytic behavior of electrodes in conjugation with other nanomaterials, such as carbon nanotubes (CNTs). Today, AuNPs have become a suitable material of the first concern for those scientists and engineers who wish to build nanoarchitectures in their research products.

6.6.1
AuNPs Assembled on Electrode Surfaces for Enhanced Electrode Responses

6.6.1.1 Conducting Polymer/AuNP Composites on Electrodes

In order to obtain the functionality of interest on electrode surfaces, nanoparticles must be correctly immobilized on surfaces, in stable fashion. Clearly, the immobilization techniques used will be important, and many different versions have been reported. Conducting polymers (CPs), which may be suitable materials as matrices to immobilize metal nanoparticles (NPs), have certain important advantages over other polymeric materials for the immobilization of metal NPs:

- The conductivities of CPs are compatible with the electrochemical events, which occur on electrode assemblies, by mediating charge transfer between metal NPs and electrodes.
- Metal NPs can be easily incorporated into the polymers, either by doping during polymerization or by anion exchange after polymerization.
- CPs have been extensively studied over the past three decades, such that research groups currently may have easy access to a wealth of relevant information.

CP/NP composite films can be prepared in many different ways, with the metal NPs either being immobilized on the CP surfaces (Figure 6.10a–c) or embedded in the polymer textures (Figure 6.10d–f) [136–139]. The doping of NPs occurs by the anionic charge accumulating on their surfaces (Figure 6.10e). Conducting polymers become positively charged upon polymerization and, hence, insert anionic compounds to retain their electroneutrality. AuNPs are often negatively charged by the adsorption of anionic protecting molecules, and this simple charge

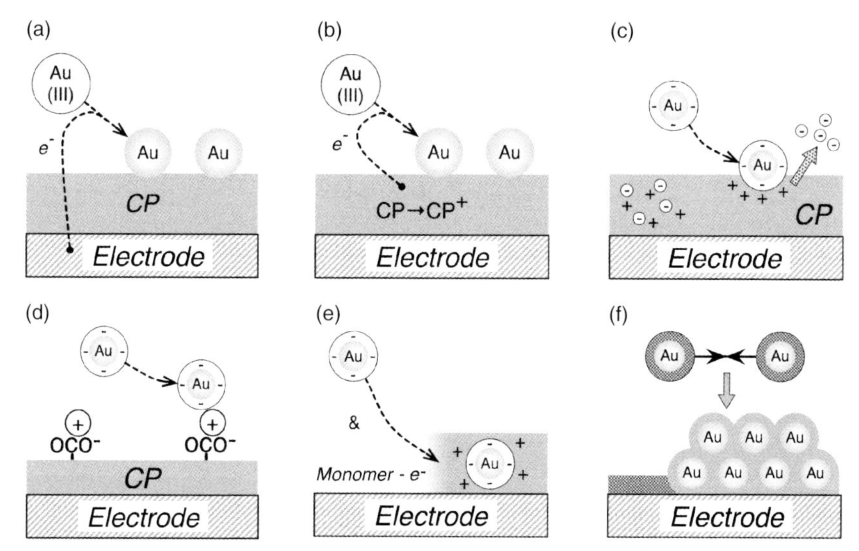

Figure 6.10 Preparation techniques of conducting polymer/AuNP composites. Preparation from Au(III) by: (a) electron transfer from the electrode; (b) reduction with a conducting polymer (CP); (c) preparation by anion exchange on a CP; (d) assembling with electrostatic interactions on a derivatized CP; (e) preparation by direct doping of the AuNP; and (f) electrochemical coupling of monomer-protected AuNP composites on a monomer-modified electrode.

relationship leads to the uptake of NPs in the CP textures during polymerization. With this approach, metal NPs may be separately prepared by using well-established techniques that allow a fine control of particle size with a narrow size distribution. Polypyrrole (PPy) films embedding AuNPs were electrosynthesized on glassy carbon electrodes from solutions containing the pyrrole monomer and the particle (ca. 5 nm) [140]. Subsequent scanning electron microscopy (SEM) observations suggested that the nanoparticles might have been the nuclei for the pyrrole polymerization. The deposition mechanisms of PPy and PPy/AuNP were also studied by *in situ* atomic force microscopy (AFM) measurements [141]. In this way, composite materials of poly(3,4-ethylenedioxithiophene) inserting AuNPs (ca. 5 nm) protected with bulky anionic molecules, such as tannic acid, have been electrochemically prepared on Pt disk electrodes [142]. A mixture of a monomer and AuNP was electro-oxidized, in constant current mode, to dope the particle in the polymer texture. Transmission electron microscopy (TEM) observations, coupled with energy-dispersive X-ray spectroscopy (EDS), revealed the presence of a well-dispersed AuNP structure throughout the polymer film. Interestingly, differential pulse voltammetry studies showed that the oxidation of ascorbate was almost totally suppressed on the composite film, due to an electrostatic shield effect that had been induced by a negatively charged cloud surrounding the AuNP. Suppression of the current on this electrode would provide an important advantage for electroanalytical applications in biomedical fluids.

It has been recognized that NPs can be adsorbed onto CPs by a mechanism similar to anion exchange (Figure 6.10c). As an example, a polyaniline (PANI) film (25 nm thickness), which had been electrodeposited galvanostatically onto an indium-tin oxide (ITO) electrode, was dip-coated with AuNP in a citrate-protected AuNP dispersion for 2 h [143]. The optical absorption spectrum of the film showed bands for both the AuNP plasmon and oxidized PANI. When PPy films were electrodeposited onto ITO electrodes, and AuNPs dissolved in CH_2Cl_2 then self-assembled on the PPy/ITO electrode [144], the color of the electrode became blue after AuNP treatment, showing a plasmon band at ~610 nm.

The chemical deposition of metal NPs occurs on the surfaces of CP films by the transfer of electrons, either directly from the CPs or from the electrode [145] (Figure 6.10a,b). Consequently, the metal ions are reduced to metal NPs and deposited onto the CP surfaces. Metal ion reduction has been reported to change the oxidation state of PANI, from an emeraldine to a more oxidized pernigraniline base form [146]. When Wang *et al.* studied metal NP deposition (Ag, Au, Pt) on acid-doped and undoped emeraldine (salt and base), a variety of morphologies was found for the three metals, depending on the doping level and the acids used. As with Au, doped PANI produced evenly dispersed submicrometer-sized, cauliflower-like Au particles which were in fact conglomerates of smaller NPs with a rice-grain morphology [147]. PPy nanotubes decorated with AuNPs by this *in situ* reduction technique were reported, and a blue-shift of the absorption spectra of PPy-nanotubes after loading of the AuNP suggested that PPy was partially over-oxidized with $HAuCl_4$ [148].

6.6.1.2 Applications of Conducting Polymer/AuNP Composites on Electrodes

The CP/NP composites exhibit excellent catalytic activities for electroanalysis. For example, PPy nanowires were prepared electrochemically on a glassy carbon electrode (GCE), followed by the deposition of AuNPs on the nanowire by reducing $HAuCl_4$ (Au(III)) during repeated voltammetric cycles. This prepared composite electrode, AuNP/PPy/GCE, showed excellent responses for hydrazine and hydroxylamine oxidations in comparison to those on GCE, PPy/GCE, and AuNp/GCE [149]. An electrode modified with PANI nanotubes was also reported. The AuNPs were deposited electrochemically onto the assembly from dissolved Au(III), after which an amino-terminated probe DNA was anchored onto the AuNP for sequence-specific DNA detections [150]. A more elaborate system using PANI composites has been reported for glucose biosensing. Citrate-protected AuNPs which were approximately 5 nm in size were added to AgCl/PANI nanocomposites. The latter took up about 30 nm AgCl particles, with the composites obtained showing an excellent electrochemical behavior, even in the neutral pH region [151].

The derivatization of CPs to produce a scaffold for further functionalization has been investigated for the detection of DNA and proteins (Figure 6.10d) [152, 153]. For example, Shiddiky *et al.* used a GCE modified with a carboxylated CP film (poly-5,2′:5′,2″-terthiophene-3′-carboxylic acid) to link, covalently, a dendrimer terminated with 32 amino groups [152]. In this configuration, the AuNPs could be fixed on the periphery of the dendrimer to fix a thiol-terminated DNA probe or an antibody (anti-human IgG), which was eventually linked to DNA- or avidin-linked hydrazine labels after target binding, respectively. The hydrazine labels allowed detection of the target DNA and protein to be amplified down to 450 aM and 4.0 fg ml^{-1}, respectively. A less-complex approach has been reported for human interleukin (IL)-5 immunosensing; in this case, poly(pyrrolepropylic acid) (PPa) was used for the covalent immobilization of antibodies [154], and PPy–PPa–AuNP composite films were synthesized on GCEs. The film was grown electrochemically on the electrode, from a mixture of pyrrole, pyrrolepropylic acid, and AuNP, followed by 1-ethyl-3-(3-dimethylaminopropyl)-carbodiimide (EDC) activation to bind the probe antibodies onto the film [153]. The cyclic voltammogram of the PPy–PPa–Au composite showed a better reversibility than that of the PPy–PPa composite, while a detection limit of 10 fg ml^{-1} for human IL-5 was reported using electrochemical impedance spectroscopy.

CPs can be used not only for electrode materials, but also for magnetic beads; magnetic beads coated with PPy were further decorated with AuNP to provide electrocatalytic activities for biorelated molecules such as ascorbic acid [155]. PANI–AuNP–AgCl composite particles immobilizing an antibody have been reported for the immunosensing of a low-density lipoprotein [156].

6.6.1.3 Other Composite Materials for Electrochemical Measurements

Recently reported biosensors often utilize more than two types of nanomaterial. In many cases, various types of synergistic effect are to be expected from the combined use, although the role of each material is often not very clearly defined.

In order to avoid such ambiguity, the discussion here is confined to nanocomposites consisting of fewer than three nanoelements.

Chitosan is a biocompatible poly-aminosaccharide that possesses a cationic charge in a pH range less than its pK_a (ca. 6.5), and can be used as an excellent inclusion matrix. A number of electroanalytical applications have been reported in combination with CNTs, metal NPs, enzymes, and antibodies. In this case, metal NPs can be deposited on the electrode surfaces via an *in situ* electrochemical reduction, or they may be included directly in the film as the NP (as discussed above) [157–159]. An interesting approach has been proposed to prepare nanosized Hg electrodes using a chitosan–carbon NP composite material [160]. Hanging mercury drop electrodes have long been used in the stripping analyses of trace metals, based on a reproducible surface renewal and a high hydrogen overpotential, although the toxic nature of mercury has clearly hampered many applications. Consequently, much effort has been made to reduce the amount of mercury in working electrodes. Rassaei *et al.* prepared electrodes by exposing mercury nanodroplets (ca. 5.4 nm) that had been trapped in the pores of a composite film which consisted of negatively charged carbon NPs (ca. 8 nm) and chitosan [160]. As chitosan has good binding (coordination) properties for many metal ions (through its amine group), it is able to hold mercury ions in the composite pores. This allows the nanodroplet electrode to be renewed more than ten times, from the nanodroplet to the mercury ion (and *vice versa*), simply by switching the potential. Moreover, there is no significant loss of mercury from the composite film following the stripping experiments.

In recent years, *sol–gel processes* have offered a means to achieve the low-temperature encapsulation of biomolecules in three-dimensional (3-D) networks. In particular, thiolated silicon alkoxides have been used to self-assemble metal NPs in the silica network [161–163]. The AuNPs in the sol–gel network can provide both a conducting nanosite for charge transfer and a scaffold for biomolecules. Molecularly imprinted polymers (MIPs) are artificial enzymes that can be used as sensors, even in harsh chemical and physical environments [164]. Recently, the details of an MIP which embedded AuNPs grown on an ITO electrode were reported. In this case, the AuNPs were anchored onto an ITO electrode that was subsequently coated with a sol which had been imprinted with imipramine, an antidepressant drug. A twofold increase in differential pulse voltammetric response was observed when the AuNPs were introduced into the MIP film. It has been suggested that electron transfer through the electrically insulated MIP films is facilitated on loading of the AgNPs [165].

Recently, CNTs have been used extensively in electroanalysis, with several metal and semiconductor NPs having been introduced successfully onto CNTs. Many biosensors have been reported using these CNTs [166, 167], and a more sophisticated device consisting of Pt and Au-coated Pt nanotubes decorated on CNT networks has been reported for highly sensitive biochemical detections [168]. Willner *et al.* have reported oligoaniline-crosslinked composites of AuNPs for glucose sensing (Figure 6.10f) [169]. For this, the AuNPs were protected with an aniline monomer and sulfonate for self-doping; similarly, glucose oxidase and an Au electrode were decorated with an aniline monomer. The AuNP, enzyme and electrode, all of which had been modified with the monomer, were then reacted to form a composite film on the electrode, which could then be used for glucose sensing.

6.6.2 Composite Nanoparticles and Their Applications as Labels

6.6.2.1 Quantification Strategies

For many years, AuNPs have been used as markers for a number of bioanalytical and medical purposes. Some reported strategies where AuNPs have been used as labels, or to provide scaffolds for labels, are shown in Figure 6.11. In these examples, it is clear that both the immuno- and DNA-binding events behave in similar fashion during the electroanalysis. In many cases, sandwich immunoassays have been preferred to detect antigens, while in the case of DNA analysis the target DNA was hybridized with capture and reporter DNA so as to immobilize the AuNP. It should be noted here that the analyte responses arise only when the target analytes exist in the samples, to make quantification possible. In these examples, redox-active molecules are produced either directly or indirectly by the AuNPs; they are mediated either by the AuNP surface (Figure 6.11a) [170, 171] or by an enzyme, such as horseradish peroxidase (Figure 6.11b) [172, 173].

In the past, much effort has been expended to develop NP-based composite materials. In the example of Figure 6.11a, the metal NPs can be modified with another functional material (e.g., Prussian blue), the aim being to obtain an electrocatalytic activity for hydrogen peroxide reduction [171]. Redox-active molecules, such as a metal complex and metal semiconductor NP, can be used as a reporter molecule, which binds to DNA either electrostatically on its phosphate sites, or covalently with the thiolated end of a DNA strand [174, 175]. An adenosine triphosphate sensor was reported using AuNP-cored dendrimers, which interact with phosphates at their peripheries [176].

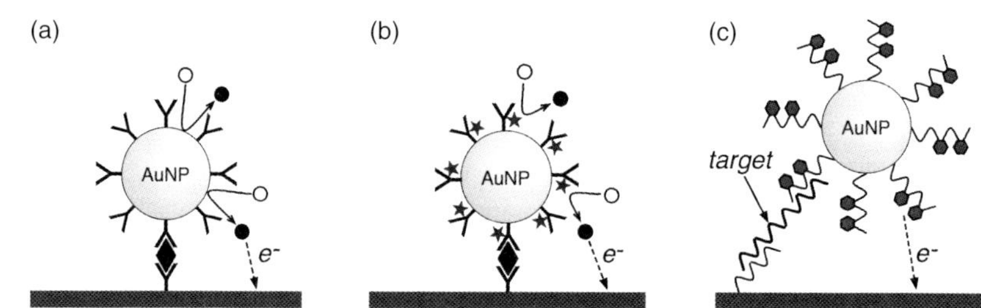

Figure 6.11 Quantification strategies using AuNPs in bioanalytical applications through mediation of the third molecule (a) with AuNP, (b) with labeled enzyme on AuNP, and (c) through redox reaction with molecules immobilized on AuNP.

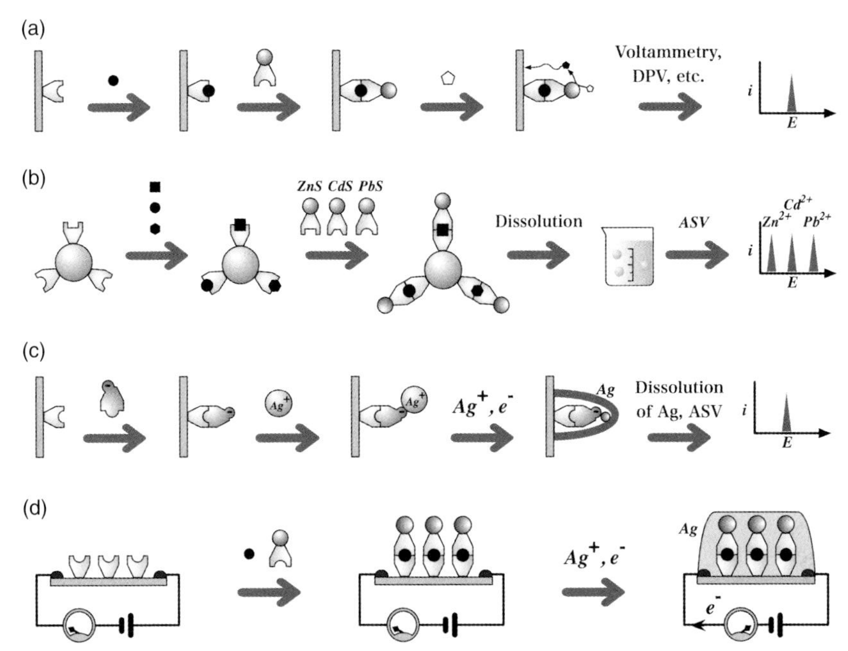

Figure 6.12 Different assay techniques using metal NPs. (a) Electroanalysis on a modified electrode; (b) Electroanalysis by coded semiconductor NPs using magnetic beads; (c) Sensitization with Ag staining; (d) Conductivity analysis by metallization of the electrode gap.

The details of voltammetric quantification procedures using these particles are shown in Figure 6.12a. The first ligand is immobilized on the electrode surface, followed by the addition of an analyte (antigen or targeted DNA) to the ligand. A second ligand and a redox-active reporter molecule are then added prior to any voltammetric measurements. When a reporter molecule has already been assembled on the particle, as in the case of Figure 6.11c, no extra reporter molecule is required.

For electroanalysis, AuNP-labeled antibodies have been used for the sensitive detection of analytes in conjunction with anodic stripping voltammetry (ASV) [177, 178]. Some illustrative examples of this, using metal NPs as labels or reporters during the electroanalysis, are shown in Figure 6.12b. Magnetic beads covered with plastic materials have frequently been used to separate the analyte from solution, after which the bead may be further decorated with one or more functional materials to provide activity for a specific molecule. Here, the analyte is first collected on magnetic beads decorated with a proper ligand (antibody, etc.), and then subjected to binding to another ligand linking the metal NP. Finally, the magnetic beads collecting the analyte are separated from the sample solution; this is followed by dissolution of the metal NP in an acidic medium for subsequent ASV measurements, whereby a high sensitivity is attained via the built-in concentration procedure for metal ions on the electrode [177–179]. Although, for these applications the AuNPs can be used as alternating materials, they should be dissolved in a corrosive medium (e.g., acidic HBr/Br_2 solution) prior to ASV measurements. In fact, rather than using AuNPs with such a corrosive medium, semiconductor NPs have become a better choice for many electrochemical bioassays [180]. Another benefit of using semiconductor particles is that a multicomponent analysis can be made by employing different metal semiconductor particles [180, 181]. As shown in Figure 6.12b, three different analytes were reacted with probes carrying ZnS, CdS and PbS nanocrystals to produce three voltammetric peaks assigned to each analyte. An immunoassay and DNA analysis have also been made with this coding technique. It has been reported that, in addition to the above three metals, CuS can be used in this protocol [181].

The metal NPs may further enhance the sensitivity of the ASV technique (Figure 6.12c). In this case, anionic groups carried with an analyte (such as DNA) can adsorb Ag^+ by cation exchange, and the analyte will subsequently be metalized with Ag by electroless plating [182, 183]. The Ag^+, which is most likely reduced to Ag during the very early stages of reduction, will function as a catalyst (nucleus) for the plating, to trigger the transferring electrons to be accepted from a reducing agent (such as hydroquinone) to Ag^+ ions in the solution. The metal layer thus formed will enhance the sensitivity of the assay when the electrode is subjected to ASV. It should be noted that the metallization is used differently in a conductivity assay (Figure 6.12d) [184, 185], where a ligand is anchored on an insulating substrate, the structures of which consist of a pair of metal electrodes separated by micrometers. After binding an analyte and a secondary ligand carrying AuNP on the first ligand, the metal NP will function as a catalyst (as in Figure 6.12c), and the analyte binding to the ligand can be detected by a conductivity measurement. This technique is considered useful for a high-density array platform, as it enables the quantification of several analytes to be made simultaneously. Indeed, by using this technique it was possible to conduct analyses for both DNA and protein. In connection with this technique, one of the present authors (T.N.) developed a conductivity-based, highly sensitive system fabricated with nanogapped AuNP arrays, such that DNA detection could be achieved without using the Ag staining procedure [186–189] (cf. see Chapter 5 in Volume 5 of this Monograph Series for further details [190]).

Technically, electrochemical impedance spectroscopy [191] provides an effective means for electrochemical immunoassay and aptasensing [156, 174, 192, 193]. This technique allows the use of a redox-inactive ligand, without any modification using label molecules (Figure 6.13). The electron transfer resistance, R_{et}, which may be related to the standard electron-transfer rate constant, can be easily evaluated from the semicircular impedance spectra. Cyclic voltammetry can also be used to detect such a change by the peak separation, but access to the kinetic parameter is less straightforward [194]. Binding of the analyte and secondary ligand with the first ligand immobilized on the electrode surface often introduces a bulky structure on the electrode which may retard the discharge of well-defined redox molecules, such as $[Fe(CN)_6]^{3/4-}$. These less-favorable conditions in electron

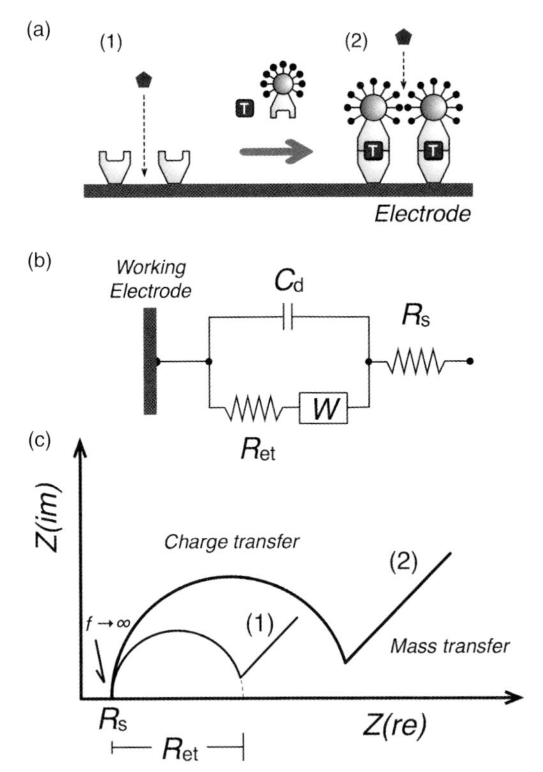

Figure 6.13 Outline of the metalloimmunoassay. (a) Quantification procedures; (⬟) marker molecule, (T) targeted analyté; (b) Schematic diagram of equivalent circuit for the working electrode surface; (c) Nyquist plots of the electrodes that are (1) unbound and (2) bound with the target molecule and secondary ligand. C_d, double-layer capacitance; R_s, solution resistance; R_{et}, electron-transfer resistance; W, Warburg impedance. Z(im) and Z(re) are the impedances of the imaginary and real parts.

transfer will result in an augmentation of R_{et}, thus allowing an evaluation of the analyte concentration. The increase in R_{et} may be related to a longer distance in electron transfer between the point of discharge and the electrode surface. Indeed, it is known that the electron-transfer rate decays exponentially with the distance, provided that the process can be attributed to the tunneling of electrons across an energy barrier [195]. The femtomolar-level quantification of thrombin using an aptamer-functionalized AuNP has been reported using this technique [192].

6.6.2.2 **Sensitivity Enhancements**

Sensitivity enhancement by metal NPs has been reported in many different ways, of which some illustrative examples are shown in Figure 6.14. In the first approach, a magnetic bead modified with a primary ligand is reacted with a targeted analyte to immobilize the analyte on the bead (Figure 6.14a). The bead is then re-reacted with a secondary ligand, in which the AuNPs are modified with a large number

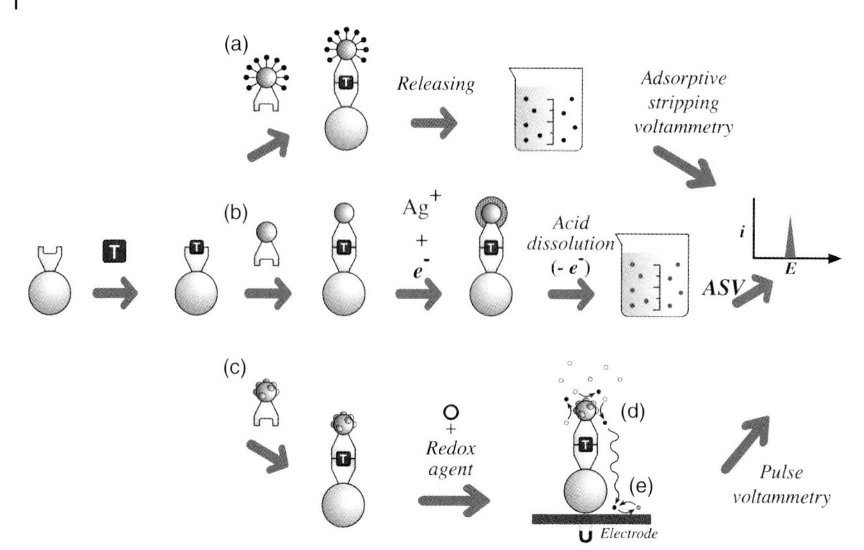

Figure 6.14 Sensitivity augmentation with AuNP composites. (a) Coding with DNA bases; (b) Ag staining; (c) Mediations coupled on the electrode surface.

of redox-active reporter molecules [196, 197]. Next, the bead is removed from the solution and the reporter released into a solution after decomposing the complex. Finally, the solution involving the reporter is subjected to a voltammetric analysis. If the reporter adsorbs onto the electrode, then stripping voltammetry can be used to obtain the maximum sensitivity performance. Technically, the reporter must be quantitatively adsorbed onto the electrode to attain sufficient accuracy and sensitivity in this framework; if this condition is not met, then the electrolyte and medium employed during the voltammetric measurements should be carefully chosen to ensure that the adsorption as thorough as possible. Despite such technical difficulty, this technique is potentially useful for barcode analysis. As with DNA-decorated probes, guanine and adenine bases are redox-active and can be used to code for different analytes.

As an alternative to the scheme shown in Figure 6.14a, Ag sensitization is frequently used, coupled with the ASV technique, and several such reports of this technique have been made to date [179, 197, 198]. As noted above, the method relies on electroless plating on the metal NP, which primarily mediates electrons from a reducer (hydroquinone) to a silver ion, after which the silver layer formed on the NP is dissolved in an acidic medium for subsequent ASV measurement. The first report of Wang *et al.* (2001) noted that the signal amplitude for this Ag staining protocol was 125-fold greater than that for the AuNP dissolution protocol [179]. Unfortunately, the Ag sensitization technique is prone to result in nonselective plating when used for devices fabricated on electrodes. This occurs because the electrodes are good sources of electrons, and act as catalysts in the Ag-staining procedure. Consequently, any nonselective staining that occurs on the electrode will significantly augment the background levels during ASV measurements, leading to a poor SNR on quantification. In order to overcome such a problem, various techniques have been described in which the electrode surfaces are deactivated with appropriate coatings. Alternatively, an ITO electrode can be used, which is more tolerant to the Ag deposition [199, 200].

More elaborate systems (Figure 6.14c) were reported to have adopted both the catalytic reaction on AuNP (Figure 6.14d) and the electrochemical amplification on an electrode surface (Figure 6.14e) [170, 201]. In this case, *p*-nitrophenol was first added as a parent reporter molecule, and then converted on AuNP to *p*-aminophenol, with $NaBH_4$ as a reducing agent (Figure 6.14d). The *p*-aminophenol is catalytically oxidized to *p*-quinone imine with a ferrocenyl-tethered dendrimer immobilized on an ITO electrode as the mediator (Figure 6.14b). A target mouse IgG of 0.7 aM was detected by these protocols, using differential pulse voltammetry.

6.7 Perspective

Among metal nanoparticles, AuNPs have undergone intensive investigation, due not only to their easy and reproducible preparation but also to the excellent controllability of their particle size and shape. In addition, as AuNPs possess characteristics such as chemical inertness, plasmonic absorption, fluorescence, electrochemical activity, and dense mass, they have played important roles as "smart" materials in the life sciences. Gold nanocomposites composed of gold nanocores and polymer shells are highly functional nanomaterials which exhibit versatile properties arising from the characteristics of both nanoparticles and functional polymers. Functional polymers such as biopolymers have provided gold nanocomposites with versatile biological functions such as recognition and specific binding–functions that are crucial to both labeling and imaging. In addition to biological functions that arise from biopolymers, thermoresponsive polymers have introduced reversible, stimulus-controlled morphological changes in gold nanocomposites (assembly versus disassembly). These changes can be output through changes in the physical properties of the gold cores. In this chapter, a summary has been provided of the evolving research on gold nanocomposites in the life sciences, in addition to a review of the role of these materials in bioassays. The versatility of gold nanocomposites has led to the creation of *in vitro* assays with various platforms. Whilst there is little room to develop new functions for the physical properties of AuNPs, the development of functional polymers conjugated with AuNPs continues to show promise for the exploitation of novel functional materials. Compared to the number of *in vitro* studies conducted, there have been few *in vivo* applications of gold nanocomposites, and this points to novel research fields in the life sciences that are yet to be explored. The strategic development of new functional polymers and their corresponding gold nanocomposites should facilitate new discoveries and breakthroughs in the life sciences.

References

1 Faraday, M. (1857) The Bakerian lecture–experimental relations of gold (and other metals) to light. *Philosophical Transactions*, **147**, 145–81.

2 Lang, C.F.A. (1912) Under die Ausflockung von Goldsol durch Liquor cerebrospinalis. *Berliner Klinische Wochenschrift*, **49**, 897–901.

3 Moeller, R. and Fritzsche, W. (2006) Biosensors based on gold nanoparticle labeling. *Annual Review of Nano Research*, **1**, 429–66.

4 Castaneda, M.T., Alegret, S. and Merkoci, A. (2007) Gold nanoparticles in DNA and protein analysis, in *Comprehensive Analytical Chemistry*, vol. **49** (eds S. Alegret and A. Merkoci), Elsevier B.V., Ch. 18, pp. 941–58.

5 White, K.A. and Rosi, N.L. (2008) Gold nanoparticle-based assays for the detection of biologically relevant molecules. *Nanomedicine*, **3**, 543–53.

6 Robert, W. (2008) The use of gold nanoparticles in diagnostics and detection. *Chemistry Society Reviews*, **37**, 2028–45.

7 Zhao, W., Brook, M.A. and Li, Y. (2008) Design of gold nanoparticle-based colorimetric biosensing assays. *ChemBioChem*, **9**, 2363–71.

8 Chen, P.C., Mwakwari, S.C. and Oyelere, A.K. (2008) Gold nanoparticles: from nanomedicine to nanosensing. *Nanotechnology, Science and Applications*, **1**, 45–66.

9 You, C.-C., Agasti, S.S. and Rotello, V.M. (2008) Chemical and biological sensing using gold nanoparticles, in *Nano and Microsensors for Chemical and Biological Terrorism Surveillance* (ed. J.B.-H. Tok), Royal Society of Chemistry, pp. 29–59.

10 Russ, A.W., Massey, M. and Krull, U.J. (2009) The application of quantum dots, gold nanoparticles and molecular switches to optical nucleic-acid diagnostics. *Trends in Analytical Chemistry*, **28**, 292–306.

11 Ghosh, P., Han, G., De, M., Kim, C.K. and Rotello, V.M. (2008) Gold nanoparticles in delivery applications. *Advances in Drug Delivery Reviews*, **60**, 1307–15.

12 Sperling, R.A., Rivera, G.P., Zhang, F., Zanella, M. and Parak, W.J. (2008) Biological applications of gold nanoparticles. *Chemistry Society Reviews*, **37**, 1896–908.

13 Pissuwan, D., Valenzuela, S.M. and Cortie, M.B. (2008) Prospects for gold nanorod particles in diagnostic and therapeutic applications. *Biotechnology and Genetics Engineering Reviews*, **25**, 93–112.

14 Cai, W., Gao, T., Hong, H. and Sun, J. (2008) Applications of gold nanoparticles in cancer nanotechnology. *Nanotechnology, Science and Applications*, **1**, 17–32.

15 Muthu, M.S. and Singh, S. (2009) Targeted nanomedicines: effective treatment modalities for cancer, AIDS and brain disorders. *Nanomedicine*, **4**, 105–18.

16 Elghanian, R., Storhoff, J.J., Mucic, R.C., Letsinger, R.L. and Mirkin, C.A. (1997) Selective colorimetric detection of polynucleotides based on the distance-dependent optical properties of gold nanoparticles. *Science*, **277**, 1078–81.

17 Storhoff, J. and Mirkin, C.A. (1999) Programmed materials synthesis with DNA. *Chemical Reviews*, **99**, 1849–62.

18 Thaxton, C.S. and Mirkin, C.A. (2004) DNA-gold-nanoparticle conjugates, in *Nanobiotechnology* (eds C.M. Niemeyer and C.A. Mirkin), Wiley-VCH Verlag GmbH, Weinheim, pp. 288–307.

19 Rosi, N.L. and Mirkin, C.A. (2005) Nanostructures in biodiagnosis. *Chemistry Reviews*, **105**, 1547–62.

20 Thaxton, C.S., Georganopoulou, D.G. and Mirkin, C.A. (2006) Gold nanoparticle probes for the detection of nucleic acid targets. *Clinica Chimica Acta*, **363**, 120–6.

21 Li, D., He, Q. and Li, J. (2009) Smart core/shell nanocomposites: intelligent polymers modified gold nanoparticles. *Advances in Colloid and Interface Science*, **149**, 28–38.

22 Huo, Q. and Worden, J.G. (2007) Monofunctional gold nanoparticles: synthesis and applications. *Journal of Nanoparticle Research*, **9**, 1013–25.

23 Shan, J. and Tenhu, H. (2007) Recent advances in polymer protected gold nanoparticles: synthesis, properties and applications. *Chemical Communications*, 4580–98.

24 Ofir, Y., Samanta, B. and Rotello, V.M. (2008) Polymer and biopolymer mediated self-assembly of gold nanoparticles. *Chemical Society Reviews*, **37**, 1814–25.

25 Liu, R., Fraylich, M. and Saunders, B.R. (2009) Thermoresponsive copolymers: from fundamental studies to applications. *Colloid and Polymer Science*, **287**, 627–43.

26 Schmaljohann, D. (2006) Thermo- and pH-responsive polymers in drug delivery. *Advances in Drug Delivery Reviews*, **58**, 1655–70.

27 Nakayama, M. and Okano, T. (2006) Intelligent thermoresponsive polymeric micelles for targeted drug delivery. *Journal of Drug Delivery Science Technology*, **16**, 35–44.

28 Okubo, K., Shimada, T., Shimizu, T. and Uehara, N. (2007) Simple and selective sensing of cysteine using gold nanoparticles conjugated with a thermoresponsive copolymer having carboxyl groups. *Analytical Sciences*, **23**, 85–90.

29 Kaholek, M., Lee, W.-K., Ahn, S.-J., Ma, H., Caster, K.C., LaMattina, B. and Zauscher, S. (2004) Stimulus-responsive poly(*N*-isopropylacrylamide) brushes and nanopatterns prepared by surface-initiated polymerization. *Chemistry of Materials*, **16**, 3688–96.

30 Raula, J., Shan, J., Nuopponen, M., Niskanen, A., Jiang, H., Kauppinen, E.I. and Tenhu, H. (2003) Synthesis of gold nanoparticles grafted with a thermoresponsive polymer by surface-induced reversible-addition-fragmentation chain-transfer polymerization. *Langmuir*, **19**, 3499–504.

31 Kim, D.J., Kang, S.M., Kong, B., Kim, W.-J., Paik, H.-J., Choi, H. and Choi, I.S. (2005) Formation of thermoresponsive gold nanoparticle/PNIPAAm hybrids by surface-initiated, atom transfer radical polymerization in aqueous media. *Macromolecular Chemistry and Physics*, **206**, 1941–6.

32 Li, D., He, Q., Cui, Y., Wang, K., Zhang, X. and Li, J. (2007) Thermosensitive copolymer networks modify gold nanoparticles for nanocomposite entrapment. *Chemistry–A European Journal*, **13**, 2224–9.

33 Zhao, W., Gao, Y., Kandadai, S.A., Brook, M.A. and Li, Y. (2006) DNA polymerization on gold nanoparticles through rolling circle amplification: towards novel scaffolds for three-dimensional periodic nanoassemblies. *Angewandte Chemie, International Edition*, **45**, 2409–13.

34 Higuchi, M., Ushiba, K. and Kawaguchi, M. (2007) Structural control of peptide-coated gold nanoparticle assemblies by the conformational transition of surface peptides. *Journal of Colloid and Interface Science*, **308**, 356–63.

35 Shan, J., Nuopponen, M., Jiang, H., Kauppinen, E. and Tenhu, H. (2003) Preparation of poly(*N*-isopropylacrylamide)-monolayer-protected gold clusters: synthesis methods, core size, and thickness of monolayer. *Macromolecules*, **36**, 4526–33.

36 Wang, Z., Tan, B., Hussain, I., Schaeffer, N., Wyatt, M.F., Brust, M. and Cooper, A.I. (2007) Design of polymeric stabilizers for size-controlled synthesis of monodisperse gold nanoparticles in water. *Langmuir*, **23**, 885–95.

37 Liu, Y., Shipton, M.K., Ryan, J., Kaufman, E.D., Franzen, S. and Feldheim, D.L. (2007) Synthesis, stability, and cellular internalization of gold nanoparticles containing mixed peptide-poly(ethylene glycol) monolayers. *Analytical Chemistry*, **79**, 2221–9.

38 Aqil, A., Qiu, H., Greisch, J.-F., Jerome, R., Pauw, E.D. and Jerome, C. (2008) Coating of gold nanoparticles by thermosensitive poly(*N*-

isopropylacrylamide) end-capped by biotin. *Polymer*, **49**, 1145–53.

39 Sakai, T. and Alexandridis, P. (2004) Single-step synthesis and stabilization of metal nanoparticles in aqueous pluronic block copolymer solutions at ambient temperature. *Langmuir*, **20**, 8426–30.

40 Yang, S., Zhang, T., Zhang, L., Wang, S., Yang, Z. and Ding, B. (2007) Continuous synthesis of gold nanoparticles and nanoplates with controlled size and shape under UV irradiation. *Colloids and Surfaces A*, **296**, 37–44.

41 Grace, A.N. and Pandian, K. (2006) One pot synthesis of polymer protected gold nanoparticles and nanoprisms in glycerol. *Colloids and Surfaces A*, **290**, 138–42.

42 Liu, Q., Liu, H., Zhou, Q., Liang, Y., Yin, G. and Xu, Z. (2006) Synthesis of nearly monodispersive gold nanoparticles by a sodium diphenylamine sulfonate reduction process. *Journal of Materials Science*, **41**, 3657–62.

43 Dong, H., Fey, E., Gandelman, A. and Jones, W.E. Jr (2006) Synthesis and assembly of metal nanoparticles on electrospun poly(4-vinylpyridine) fibers and poly(4-vinylpyridine) composite fibers. *Chemistry of Materials*, **18**, 2008–11.

44 Zhang, P., Jiang, X., Zhang, X., Zhang, W. and Shi, L. (2006) Formation of gold@polymer core-shell particles and gold particle clusters on a template of thermoresponsive and pH-responsive coordination triblock copolymer. *Langmuir*, **22**, 9393–6.

45 Sakamoto, M., Tachikawa, T., Fujitsuka, M. and Majima, T. (2006) Acceleration of laser-induced formation of gold nanoparticles in a poly(vinyl alcohol) film. *Langmuir*, **22**, 6361–6.

46 Bhattacharjee, R.R., Chakraborty, M. and Mandal, T.K. (2006) Reversible association of thermoresponsive gold nanoparticles: polyelectrolyte effect on the lower critical solution temperature of poly(vinyl methyl ether). *Journal of Physical Chemistry B*, **110**, 6768–75.

47 Wang, B., Chen, K., Jiang, S., Reincke, F., Tong, W., Wang, D. and Gao, C. (2006) Chitosan-mediated synthesis of gold nanoparticles on patterned poly(dimethylsiloxane) surfaces. *Biomacromolecules*, **7**, 1203–9.

48 Note, C., Kosmella, S. and Koetz, J. (2006) Poly(ethyleneimine) as reducing and stabilizing agent for the formation of gold nanoparticles in w/o microemulsions. *Colloids and Surface A*, **290**, 150–6.

49 Sun, X., Dong, S. and Wang, E. (2006) One-step polyelectrolyte-based route to well-dispersed gold nanoparticles: synthesis and insight. *Materials Chemistry and Physics*, **96**, 29–33.

50 Chen, H., Wang, Y., Wang, Y., Dong, S. and Wang, E. (2006) One-step preparation and characterization of PDDA-protected gold nanoparticles. *Polymer*, **47**, 763–6.

51 Filali, M., Meier, M.A.R., Schubert, U.S. and Gohy, J.-F. (2005) Star-block copolymers as templates for the preparation of stable gold nanoparticles. *Langmuir*, **21**, 7995–8000.

52 Li, J., Shi, L., An, Y., Li, Y., Chen, X. and Dong, H. (2006) Reverse micelles of star-block copolymer as nanoreactors for preparation of gold nanoparticles. *Polymer*, **47**, 8480–7.

53 Liu, S., Weaver, J.V.M., Save, M. and Armes, S.P. (2002) Synthesis of pH-responsive shell cross-linked micelles and their use as nanoreactors for the preparation of gold nanoparticles. *Langmuir*, **18**, 8350–7.

54 Jewrajka, S.K. and Chatterjee, U. (2006) Block copolymer mediated synthesis of amphiphilic gold nanoparticles in water and an aqueous tetrahydrofuran medium: an approach for the preparation of polymer-gold nanocomposites. *Journal of Polymer Science Part A: Polymer Chemistry*, **44**, 1841–54.

55 Kim, D., Park, S., Lee, J.H., Jeong, Y.Y. and Jon, S. (2007) Antibiofouling polymer-coated gold nanoparticles as a contrast agent for *in vivo* X-ray computed tomography imaging. *Journal of the American Chemical Society*, **129**, 7661–5.

56 Njoki, P.N., Lim, I.-I.S., Mott, D., Park, H.-Y., Khan, B., Mishra, S., Sujakumar, R., Luo, J. and Zhong, C.-J. (2007) Size correlation of optical and spectroscopic

properties for gold nanoparticles. *Journal of Physical Chemistry C*, **111**, 14664–9.

57 Zhong, Z., Patskovskyy, S., Bouvrette, P., Luong, J.H.T. and Agedanken, A. (2004) The surface chemistry of au colloids and their interactions with functional amino acids. *Journal of Physical Chemistry B*, **108**, 4046–52.

58 Ghosh, S.K. and Pal, T. (2007) Interparticle coupling effect on the surface plasmon resonance of gold nanoparticles: from theory to applications. *Chemical Reviews*, **107**, 4797–862.

59 Jain, P.K., Huang, X., El-Sayed, I.H. and El-Sayed, M.A. (2007) Review of some interesting surface plasmon resonance-enhanced properties of noble metal nanoparticles and their applications to biosystems. *Plasmonics*, **2**, 107–18.

60 Khlebtsov, N.G. (2008) Determination of size and concentration of gold nanoparticles from extinction spectra. *Analytical Chemistry*, **80**, 6620–5.

61 Wang, J., Duan, T., Sun, L., Liu, D. and Wang, Z. (2009) Functional gold nanoparticles for studying the interaction of lectin with glycosyl complex on living cellular surfaces. *Analytical Biochemistry*, **392**, 77–82.

62 Pan, D., Pramanik, M., Senpan, A., Yang, X., Song, K.H., Scott, M.J., Zhang, H., Gaffney, P.J., Wickline, S.A., Wang, L.V. and Lanza, G.M. (2009) Molecular photoacoustic tomography with colloidal nanobeacons. *Angewandte Chemie, International Edition*, **48**, 4170–3.

63 He, H., Xie, C. and Ren, J. (2008) Nonbleaching fluorescence of gold nanoparticles and its applications in cancer cell imaging. *Analytical Chemistry*, **80**, 5951–7.

64 Sun, L., Liu, D. and Wang, Z. (2008) Functional gold nanoparticle-peptide complexes as cell-targeting agents. *Langmuir*, **24**, 10293–7.

65 Javier, D.J., Nitin, N., Levy, M., Ellington, A. and Richards-Kortum, R. (2008) Aptamer-targeted gold nanoparticles as molecular-specific contrast agents for reflectance imaging. *Bioconjugate Chemistry*, **19**, 1309–12.

66 Medley, C.D., Smith, J.E., Tang, Z., Wu, Y., Bamrungsap, S. and Tan, W. (2008) Gold nanoparticle-based colorimetric assay for the direct detection of cancerous cells. *Analytical Chemistry*, **80**, 1067–72.

67 Agarwal, A., Huang, S.W., O'Donnell, M., Day, K.C., Day, M., Kotov, N. and Ashkenazi, S. (2007) Targeted gold nanorod contrast agent for prostate cancer detection by photoacoustic imaging. *Journal of Applied Physics*, **102**, 064701.

68 Durr, N.J., Larson, T., Smith, D.K., Korgel, B.A., Sokolov, K. and Ben-Yakar, A. (2007) Two-photon luminescence imaging of cancer cells using molecularly targeted gold nanorods. *Nano Letters*, **7**, 941–5.

69 Debouttière, P.-J., Roux, S., Vocanson, F., Billotey, C., Beuf, O., Favre-Réguillon, A., Lin, Y., Pellet-Rostaing, S., Lamartine, R., Perriat, P. and Tillement, O. (2006) Design of gold nanoparticles for magnetic resonance imaging. *Advanced Functional Materials*, **16**, 2330–9.

70 Kim, D., Park, S., Lee, J.H., Jeong, Y.Y. and Jon, S. (2007) Antibiofouling polymer-coated gold nanoparticles as a contrast agent for *in vivo* X-ray computed tomography imaging. *Journal of the American Chemical Society*, **129**, 7661–5.

71 Cao, Y.W., Jin, R. and Mirkin, C.A. (2002) Nanoparticles with Raman spectroscopic fingerprints for DNA and RNA detection. *Science*, **297**, 1536–40.

72 El-Kouedi, M. and Keating, C.D. (2004) *Biofunctionalized Nanoparticles for Surface-Enhanced Raman Scattering and Surface Plasmon Resonance. Nanobiotechnology* (eds C.M. Niemeyer and C.A. Mirkin), Wiley-VCH Verlag GmbH, Weinheim, pp. 429–43.

73 Tang, H.-W, X., Yang, B., Kirkham, J. and Smith, D.A. (2007) Probing intrinsic and extrinsic components in single osteosarcoma cells by near-infrared surface-enhanced Raman scattering. *Analytical Chemistry*, **79**, 3646–53.

74 Qian, X., Peng, X.-H., Ansari, D.O., Yin-Goen, Q., Chen, G.Z., Shin, D.M., Yang, L., Young, A.N., Wang, M.D. and Nie, S. (2008) *In vivo* tumor targeting and spectroscopic detection with surface-enhanced Raman nanoparticle tags. *Nature Biotechnology*, **26**, 83–90.

75 Ma, Z. and Sui, S.-F. (2002) Naked-eye sensitive detection of immunoglubulin G by enlargement of au nanoparticles *in vitro*. *Angewandte Chemie, International Edition*, **41**, 2176–9.

76 Xu, H., Mao, X., Zeng, Q., Wang, S., Kawde, A.-N. and Liu, G. (2009) Aptamer-functionalized gold nanoparticles as probes in a dry-reagent strip biosensor for protein analysis. *Analytical Chemistry*, **81**, 669–75.

77 Glynou, K., Ioannou, P.C., Christopoulos, T.K. and Syriopoulou, V. (2003) Oligonucleotide-functionalized gold nanoparticles as probes in a dry-reagent strip biosensor for DNA analysis by hybridization. *Analytical Chemistry*, **75**, 4155–60.

78 Storhoff, J.J., Marla, S.S., Bao, P., Hagenow, S., Mehta, H., Lucas, A., Garimella, V., Patno, T., Buckingham, W., Cork, W. and Müller, U.R. (2004) Gold nanoparticle-based detection of genomic DNA targets on microarrays using a novel optical detection system. *Biosensors and Bioelectronics*, **19**, 875–83.

79 Lee, J.-S., Han, M.S. and Mirkin, C.A. (2007) Colorimetric detection of mercuric ion (Hg^{2+}) in aqueous media using DNA-functionalized gold nanoparticles. *Angewandte Chemie, International Edition*, **46**, 4093–6.

80 Li, L., Li, B., Qi, Y. and Jin, Y. (2009) Label-free aptamer-based colorimetric detection of mercury ions in aqueous media using unmodified gold nanoparticles as colorimetric probe. *Analytical and Bioanalytical Chemistry*, **393**, 2051–7.

81 Wu, Z.-S., Guo, M.-M., Shen, G.-L. and Yu, R.-Q. (2007) G-rich oligonucleotide-functionalized gold nanoparticle aggregation. *Analytical and Bioanalytical Chemistry*, **387**, 2623–6.

82 Liu, J. and Lu, Y. (2007) Colorimetric Cu^{2+} detection with a ligation DNAzyme and nanoparticles. *Chemical Communications*, 4872–4.

83 Sugunan, A., Thanachayanont, C., Dutta, J. and Hilborn, J.G. (2005) Heavy-metal ion sensors using chitosan-capped gold nanoparticles. *Science and Technology of Advances Materials*, **6**, 335–40.

84 Kim, S., Park, J.W., Kim, D., Kim, D., Lee, I.-H. and Jon, S. (2009) Bioinspired colorimetric detection of calcium(II) ions in serum using calsequestrin-functionalized gold nanoparticles. *Angewandte Chemie, International Edition*, **48**, 4138–41.

85 Thanh, N.T.K. and Rosenzweig, Z. (2002) Development of an aggregation-based immunoassay for anti-protein a using gold nanoparticles. *Analytical Chemistry*, **74**, 1624–8.

86 Jin, R., Wu, G., Li, Z., Mirkin, C.A. and Schatz, G.C. (2003) What controls the melting properties of DNA-linked gold nanoparticle assemblies? *Journal of the American Chemical Society*, **125**, 1643–54.

87 Hirsch, L.R., Jackson, J.B., Lee, A., Halas, N.J. and West, J.L. (2003) A whole blood immunoassay using gold nanoshells. *Analytical Chemistry*, **75**, 2377–81.

88 Liu, Y., Liu, Y., Mernaugh, R.L. and Zeng, X. (2009) Single chain fragment variable recombinant antibody functionalized gold nanoparticles for a highly sensitive colorimetric immunoassay. *Biosensors and Bioelectronics*, **24**, 2853–7.

89 Wang, C., Chen, Y., Wang, T., Ma, Z. and Su, Z. (2007) Biorecognition-driven self-assembly of gold nanorods: a rapid and sensitive approach toward antibody sensing. *Chemistry of Materials*, **19**, 5809–11.

90 Shenton, W., Davis, S.A. and Mann, S. (1999) Directed self-assembly of nanoparticles into macroscopic materials using antibody-antigen recognition. *Advanced Materials*, **11**, 449–52.

91 Grunwald, C. (2008) A brief introduction to the streptavidin-biotin system and its usage in modern surface based assay. *Zeitschrift für Physikalische Chemie*, **222**, 789–821.

92 Connolly, S. and Fitzmaurice, D. (1999) Programmed assembly of gold nanocrystals in aqueous solution. *Advanced Materials*, **11**, 1202–5.

93 Aslan, K., Luhrs, C.C. and Prez-Luna, V.H. (2004) Controlled and reversible aggregation of biotinylated gold nanoparticles with streptavidin. *Journal of Physical Chemistry B*, **108**, 15631–9.

94 Wang, Z., Levy, R., Fernig, D.G. and Brust, M. (2006) Kinase-catalyzed modification of gold nanoparticles: a new approach to colorimetric kinase activity screening. *Journal of the American Chemical Society*, **128**, 2214–15.

95 Caswell, K.K., Wilson, J.N., Bunz, U.H.F. and Murphy, C.J. (2003) Preferential end-to-end assembly of gold nanorods by biotin-streptavidin connectors. *Journal of the American Chemical Society*, **125**, 13914–15.

96 Gole, A. and Murphy, C.J. (2005) Biotin-streptavidin-induced aggregation of gold nanorods: tuning rod-rod orientation. *Langmuir*, **21**, 10756–62.

97 Toyoshima, M. and Miura, Y. (2009) Preparation of glycopolymer-substituted gold nanoparticles and their molecular recognition. *Journal of Polymer Science Part A: Polymer Chemistry*, **47**, 1412–21.

98 Schofield, C.L., Mukhopadhyay, B., Hardy, S.M., McDonnell, M.B., Field, R.A. and Russell, D.A. (2008) Colorimetric detection of *Ricinus communis* Agglutinin 120 using optimally presented carbohydrate-stabilized gold nanoparticles. *Analyst*, **133**, 626–34.

99 Sato, Y., Murakami, T., Yoshioka, K. and Niwa, O. (2008) 12-Mercaptododecyl β-maltoside-modified gold nanoparticles: specific ligands for concanavalin A having long flexible hydrocarbon chains. *Analytical and Bioanalytical Chemistry*, **391**, 2527–32.

100 Schofield, C.L., Haines, A.H., Field, R.A. and Russell, D.A. (2006) Silver and gold glyconanoparticles for colorimetric bioassays. *Langmuir*, **22**, 6707–11.

101 Takae, S., Akiyama, Y., Otsuka, H., Nakamura, T., Nagasaki, Y. and Kataoka, K. (2005) Ligand density effect on biorecognition by pegylated gold nanoparticles: regulated interaction of RCA120 lectin with lactose installed to the distal end of tethered peg strands on gold surface. *Biomacromolecules*, **6**, 818–24.

102 Ma, Y., Li, N., Yang, C. and Yang, X. (2005) One-step synthesis of amino-dextran-protected gold and silver nanoparticles and its application in biosensors. *Analytical and Bioanalytical Chemistry*, **382**, 1044–8.

103 Lee, S. and Perez-Luna, V.H. (2005) Dextran-gold nanoparticle hybrid material for biomolecule immobilization and detection. *Analytical Chemistry*, **77**, 7204–11.

104 Schofield, C.L., Field, R.A. and Russell, D.A. (2007) Glyconanoparticles for the colorimetric detection of cholera toxin. *Analytical Chemistry*, **79**, 1356–61.

105 Nath, S., Kaittanis, C., Tinkham, A. and Perez, J.M. (2008) Dextran-coated gold nanoparticles for the assessment of antimicrobial susceptibility. *Analytical Chemistry*, **80**, 1033–8.

106 Lu, Y., Liu, J. and Mazumdar, D. (2009) Nanoparticles/dip stick, in *Nucleic Acid and Peptide Aptamers*, vol. **535** (ed. G. Mayer), Humana Press, Totowa, NJ, USA, pp. 223–39.

107 Slocik, J.M., Zabinski, J.S. Jr, Phillips, D.M. and Naik, R.R. (2008) Colorimetric response of peptide-functionalized gold nanoparticles to metal ions. *Small*, **4**, 548–51.

108 Pavlov, V., Xiao, Y., Shlyahovsky, B. and Willner, I. (2004) Aptamer-functionalized Au nanoparticles for the amplified optical detection of thrombin. *Journal of the American Chemical Society*, **126**, 11768–9.

109 Huang, C.C., Huang, Y.F., Cao, Z., Tan, W. and Chang, H.T. (2005) Aptamer-modified gold nanoparticles for colorimetric determination of platelet-derived growth factors and their receptors. *Analytical Chemistry*, **77**, 5735–41.

110 Chen, Y.-M., Yu, C.-J., Cheng, T.-L. and Tseng, W.-L. (2008) Colorimetric detection of lysozyme based on electrostatic interaction with human serum albumin-modified gold nanoparticles. *Langmuir*, **24**, 3654–60.

111 Lim, I.-I.S., Ip, W., Crew, E., Njoki, P.N., Mott, D., Zhong, C.-J., Pan, Y. and Zhou, S. (2007) Homocysteine-mediated reactivity and assembly of gold nanoparticles. *Langmuir*, **23**, 826–33.

112 Sato, K., Onoguchi, M., Sato, Y., Hosokawa, K. and Maeda, M. (2006)

Non-cross-linking gold nanoparticle aggregation for sensitive detection of single-nucleotide polymorphisms: optimization of the particle diameter. *Analytical Biochemistry*, **350**, 162–4.

113 Zhao, W., Chiuman, W., Lam, J.C.F., McManus, S.A., Chen, W., Cui, Y., Pelton, R., Brook, M.A. and Li, Y. (2008) DNA aptamer folding on gold nanoparticles: from colloid chemistry to biosensors. *Journal of the American Chemical Society*, **130**, 3610–18.

114 Zheng, W. and He, L. (2009) Particle stability in polymer-assisted reverse colorimetric DNA assays. *Analytical and Bioanalytical Chemistry*, **393**, 1305–13.

115 Hurst, S.J., Han, M.S., Lytton-Jean, A.K.R. and Mirkin, C.A. (2007) Screening the sequence selectivity of DNA-binding molecules using a gold nanoparticle-based colorimetric approach. *Analytical Chemistry*, **79**, 7201–5.

116 Liu, J. and Lu, Y. (2007) Non-base pairing DNA provides a new dimension for controlling aptamer-linked nanoparticles and sensors. *Journal of the American Chemical Society*, **129**, 8634–43.

117 Shimada, T., Ookubo, K., Komuro, N., Shimizu, T. and Uehara, N. (2007) Blue-to-red chromatic sensor composed of gold nanoparticles conjugated with thermoresponsive copolymer for thiol sensing. *Langmuir*, **23**, 11225–32.

118 Uehara, N., Fujita, M. and Shimizu, T. (2009) Colorimetric assay of aminopeptidase N activity based on inhibition of the disassembly of gold nano-composites conjugated with a thermo-responsive copolymer. *Analytical Sciences*, **25**, 267–73.

119 Nerambourg, N., Werts, M.H.V., Charlot, M. and Blanchard-Desce, M. (2007) Quenching of molecular fluorescence on the surface of monolayer-protected gold nanoparticles investigated using place exchange equilibria. *Langmuir*, **23**, 5563–70.

120 Dulkeith, E., Morteani, A.C., Niedereichholz, T., Klar, T.A., Feldmann, J., Levi, S.A., van Veggel, F.C.J.M., Reinhoudt, D.N., Möller, M. and Gittins, D.I. (2002) Fluorescence quenching of dye molecules near gold nanoparticles: radiative and nonradiative effects. *Physical Review Letters*, **89**, 203002.

121 Huang, T. and Murray, R.W. (2002) Quenching of $[Ru(bpy)_3]^{2+}$ fluorescence by binding to Au nanoparticles. *Langmuir*, **18**, 7077–81.

122 Fan, C., Wang, S., Hong, J.W., Bazan, G.C., Plaxco, K.W. and Heeger, A.J. (2003) Beyond superquenching: hyper-efficient energy transfer from conjugated polymers to gold nanoparticles. *Proceedings of the National Academy of Sciences of the United States of America*, **100**, 6297–301.

123 Ghosh, S.K. and Pal, T. (2009) Photophysical aspects of molecular probes near nanostructured gold surfaces. *Physical Chemistry Chemical Physics*, **11**, 3831–44.

124 Algar, R.W., Massey, M. and Krull, U.J. (2009) The application of quantum dots, gold nanoparticles and molecular switches to optical nucleic-acid diagnostics. *Trends in Analytical Chemistry*, **28**, 292–306.

125 Seferos, D.S., Giljohann, D.A., Hill, H.D., Prigodich, A.E., Mirkin, C.A. and Nano-Flares (2007) Probes for transfection and mRNA detection in living cells. *Journal of the American Chemical Society*, **129**, 15477–9.

126 Li, Y.-T., Liu, H.-S., Lin, H.-P. and Chen, S.-H. (2005) Gold nanoparticles for microfluidics-based biosensing of PCR products by hybridization-induced fluorescence quenching. *Electrophoresis*, **26**, 4743–50.

127 Aslan, K. and Perez-Luna, V.H. (2006) Nonradiative interactions between biotin-functionalized gold nanoparticles and fluorophore-labeled antibiotin. *Plasmonics*, **1**, 111–19.

128 Gu, J.-Q., Shen, J., Sun, L.-D. and Yan, C.-H. (2008) Resonance energy transfer in steady-state and time-decay fluoro-immunoassays for lanthanide nanoparticles based on biotin and avidin affinity. *Journal of Physical Chemistry C*, **112**, 6589–93.

129 Huang, C.-C., Chiu, S.-H., Huang, Y.-F. and Chang, H.-T. (2007) Aptamer-functionalized gold nanoparticles for turn-on light switch detection of platelet-derived growth factor. *Analytical Chemistry*, **79**, 4798–804.

130 Huang, C.-C., Chiang, C.-K., Lin, Z.-H., Lee, K.-H. and Chang, H.-T. (2008) Bioconjugated gold nanodots and nanoparticles for protein assays based on photoluminescence quenching. *Analytical Chemistry*, **80**, 1497–504.

131 Liu, J., Lee, J.H. and Lu, Y. (2007) Quantum dot encoding of aptamer-linked nanostructures for one-pot simultaneous detection of multiple analytes. *Analytical Chemistry*, **79**, 4120–5.

132 Freddi, S., D'Alfonso, L., Collini, M., Caccia, M., Sironi, L., Tallarida, G., Caprioli, S. and Chirico, G. (2009) Excited-state lifetime assay for protein detection on gold colloids-fluorophore complexes. *Journal of Physical Chemistry C*, **113**, 2722–30.

133 Ye, B.-C. and Yin, B.-C. (2008) Highly sensitive detection of mercury(II) ions by fluorescence polarization enhanced by gold nanoparticles. *Angewandte Chemie, International Edition*, **47**, 8386–9.

134 Pingarrón, J.M., Yáñez-Sedeño, P. and González-Cortés, A. (2008) Gold nanoparticle-based electrochemical biosensors. *Electrochimica Acta*, **53**, 5848–66.

135 Wang, J. (2007) Nanoparticle-based electrochemical bioassays of proteins. *Electroanalysis*, **19**, 769–76.

136 Sih, B.C. and Wolf, M.O. (2005) Metal nanoparticle-conjugated polymer nanocomposites. *Chemical Communications*, 3375–84.

137 Niu, L., Li, Q.H., Wei, F.H., Chen, X. and Wang, H. (2003) Formation optimization of platinum-modified polyaniline films for the electrocatalytic oxidation of methanol. *Synthetic Metals*, **139**, 271–6.

138 Sarma, T.K., Chowdhury, D., Paul, A. and Chattopadhyay, A. (2002) Synthesis of Au nanoparticle-conductive polyaniline composite using H_2O_2 as oxidising as well as reducing agent. *Chemical Communications*, 1048–9.

139 Umeda, R., Awaji, H., Nakahodo, T. and Fujihara, H. (2008) Nanotube composites consisting of metal nanoparticles and polythiophene from electropolymerization of terthiophene-functionalized metal (Au, Pd) nanoparticles. *Journal of the American Chemical Society*, **130**, 3240–1.

140 Chen, W., Li, C.M., Chen, P. and Sun, C.Q. (2006) Electrosynthesis and characterization of polypyrrole/Au nanocomposite. *Electrochimica Acta*, **52**, 1082–6.

141 Chen, W., Li, C.M., Yu, L., Lu, Z. and Zhou, Q. (2008) In situ AFM study of electrochemical synthesis of polypyrrole/Au nanocomposite. *Electrochemistry Communications*, **10**, 1340–3.

142 Terzi, F., Zanardi, C., Martina, V., Pigani, L. and Seeber, R. (2008) Electrochemical, spectroscopic and microscopic characterisation of novel poly(3,4-ethylenedioxythiophene)/gold nanoparticles composite materials. *Journal of Electroanalytical Chemistry*, **619–620**, 75–82.

143 Leroux, Y., Eang, E., Fave, C., Trippe, G. and Lacroix, J.C. (2007) Conducting polymer/gold nanoparticle hybrid materials: a step toward electroactive plasmonic devices. *Electrochemistry Communications*, **9**, 1258–62.

144 Zotti, G., Vercelli, B. and Berlin, A. (2008) Gold nanoparticle linking to polypyrrole and polythiophene: monolayers and multilayers. *Chemistry of Materials*, **20**, 6509–16.

145 Xu, P., Han, X., Zhang, B., Mack, N.H., Jeon, S.H. and Wang, H.L. (2009) Synthesis and characterization of nanostructured polypyrroles: morphology-dependent electrochemical responses and chemical deposition of Au nanoparticles. *Polymer*, **50**, 2624–9.

146 Li, W., Jia, Q.X. and Wang, H.L. (2006) Facile synthesis of metal nanoparticles using conducting polymer colloids. *Polymer*, **47**, 23–6.

147 Wang, H.L., Li, W., Jia, Q.X. and Akhadov, E. (2007) Tailoring conducting polymer chemistry for the chemical deposition of metal particles and clusters. *Chemistry of Materials*, **19**, 520–5.

148 Xu, J.J., Hu, J.C., Quan, B.G. and Wei, Z.X. (2009) Decorating polypyrrole nanotubes with Au nanoparticles by an in situ reduction process. *Macromolecular Rapid Communications*, **30**, 936–40.

149 Li, J. and Lin, X. (2007) Electrocatalytic oxidation of hydrazine and hydroxylamine at gold nanoparticle-polypyrrole nanowire modified glassy carbon electrode. *Sensors and Actuators, B: Chemistry*, **126**, 527–35.

150 Feng, Y., Yang, T., Zhang, W., Jiang, C. and Jiao, K. (2008) Enhanced sensitivity for deoxyribonucleic acid electrochemical impedance sensor: gold nanoparticle/polyaniline nanotube membranes. *Analytica Chimica Acta*, **616**, 144–51.

151 Yan, W., Feng, X., Chen, X., Hou, W. and Zhu, J.J. (2008) A super highly sensitive glucose biosensor based on Au nanoparticles-AgCl@polyaniline hybrid material. *Biosensors and Bioelectronics*, **23**, 925–31.

152 Shiddiky, M.J.A., Rahman, M.A. and Shim, Y.B. (2007) Hydrazine-catalyzed ultrasensitive detection of DNA and proteins. *Analytical Chemistry*, **79**, 6886–90.

153 Chen, W., Lu, Z. and Li, C.M. (2008) Sensitive human interleukin 5 impedimetric sensor based on polypyrrole-pyrrolepropylic acid-gold nanocomposite. *Analytical Chemistry*, **80**, 8485–92.

154 Tolani, S.B., Craig, M., DeLong, R.K., Ghosh, K. and Wanekaya, A.K. (2009) Towards biosensors based on conducting polymer nanowires. *Analytical and Bioanalytical Chemistry*, **393**, 1225–31.

155 Zhang, H., Zhong, X., Xu, J.J. and Chen, H.Y. (2008) Fe_3O_4/Polypyrrole/Au nanocomposites with core/shell/shell structure: synthesis, characterization, and their electrochemical properties. *Langmuir*, **24**, 13748–52.

156 Yan, W., Chen, X., Li, X., Feng, X. and Zhu, J.J. (2008) Fabrication of a label-free electrochemical immunosensor of low-density lipoprotein. *Journal of Physical Chemistry B*, **112**, 1275–81.

157 Yang, G., Chang, Y., Yang, H., Tan, L., Wu, Z., Lu, X. and Yang, Y. (2009) The preparation of reagentless electrochemical immunosensor based on a nano-gold and chitosan hybrid film for human chorionic gonadotrophin. *Analytica Chimica Acta*, **644**, 72–7.

158 Wu, B.Y., Hou, S.H., Yin, F., Li, J., Zhao, Z.X., Huang, J.D. and Chen, Q. (2007) Amperometric glucose biosensor based on layer-by-layer assembly of multilayer films composed of chitosan, gold nanoparticles and glucose oxidase modified Pt electrode. *Biosensors and Bioelectronics*, **22**, 838–44.

159 Kang, X., Mai, Z., Zou, X., Cai, P. and Mo, J. (2007) A novel glucose biosensor based on immobilization of glucose oxidase in chitosan on a glassy carbon electrode modified with gold-platinum alloy nanoparticles/multiwall carbon nanotubes. *Analytical Biochemistry*, **369**, 71–9.

160 Rassaei, L., Sillanpää, M., Edler, K.J. and Marken, F. (2009) Electrochemically active mercury nanodroplets trapped in a carbon nanoparticle–chitosan matrix. *Electroanalysis*, **21**, 261–6.

161 Wang, J., Pamidi, P.V.A. and Zanette, D.R. (1998) Self-assembled silica gel networks. *Journal of the American Chemical Society*, **120**, 5852–3.

162 Jia, J., Wang, B., Wu, A., Cheng, G., Li, Z. and Dong, S. (2002) A method to construct a third-generation horseradish peroxidase biosensor: self-assembling gold nanoparticles to three-dimensional sol-gel network. *Analytical Chemistry*, **74**, 2217–23.

163 Taheri, A., Noroozifar, M. and Khorasani-Motlagh, M. (2009) Investigation of a new electrochemical cyanide sensor based on Ag nanoparticles embedded in a three-dimensional sol-gel. *Journal of Electroanalytical Chemistry*, **628**, 48–54.

164 Tokonami, S., Shiigi, H. and Nagaoka, T. (2009) Review: micro- and nanosized molecularly imprinted polymers for high-throughput analytical applications. *Analytica Chimica Acta*, **641**, 7–13.

165 Du, D., Chen, S., Cai, J., Tao, Y., Tu, H. and Zhang, A. (2008) Recognition of dimethoate carried by bi-layer electrodeposition of silver nanoparticles and imprinted poly-o-phenylenediamine. *Electrochimica Acta*, **53**, 6589–95.

166 Chen, S., Yuan, R., Chai, Y., Zhang, L., Wang, N. and Li, X. (2007)

Amperometric third-generation hydrogen peroxide biosensor based on the immobilization of hemoglobin on multiwall carbon nanotubes and gold colloidal nanoparticles. *Biosensors and Bioelectronics*, **22**, 1268–74.

167 Jiang, H.J., Zhao, Y., Yang, H. and Akins, D.L. (2009) Synthesis and electrochemical properties of single-walled carbon nanotube-gold nanoparticle composites. *Materials Chemistry and Physics*, **114**, 879–83.

168 Claussen, J.C., Franklin, A.D., Haque, A.U., Marshall Porterfield, D. and Fisher, T.S. (2009) Electrochemical biosensor of nanocube-augmented carbon nanotube networks. *ACS Nano*, **3**, 37–44.

169 Yehezkeli, O., Yan, Y.M., Baravik, I., Tel-Vered, R. and Willner, I. (2009) Integrated oligoaniline-cross-linked composites of Au nanoparticles/glucose oxidase electrodes: a generic paradigm for electrically contacted enzyme systems. *Chemistry–A European Journal*, **15**, 2674–9.

170 Das, J., Aziz, M.A. and Yang, H. (2006) A nanocatalyst-based assay for proteins: DNA-free ultrasensitive electrochemical detection using catalytic reduction of p-nitrophenol by gold-nanoparticle labels. *Journal of the American Chemical Society*, **128**, 16022–3.

171 Qiu, J.D., Peng, H.Z., Liang, R.P., Li, J. and Xia, X.H. (2007) Synthesis, characterization, and immobilization of Prussian blue-modified au nanoparticles: application to electrocatalytic reduction of H_2O_2. *Langmuir*, **23**, 2133–7.

172 Ambrosi, A., Castañeda, M.T., Killard, A.J., Smyth, M.R., Alegret, S. and Merkoçi, A. (2007) Double-codified gold nanolabels for enhanced immunoanalysis. *Analytical Chemistry*, **79**, 5232–40.

173 Mani, V., Chikkaveeraiah, B.V., Patel, V., Gutkind, J.S. and Rusling, J.F. (2009) Ultrasensitive immunosensor for cancer biomarker proteins using gold nanoparticle film electrodes and multienzyme-particle amplification. *ACS Nano*, **3**, 585–94.

174 Li, W., Nie, Z., Xu, X., Shen, Q., Deng, C., Chen, J. and Yao, S. (2009) A sensitive, label free electrochemical aptasensor for ATP detection. *Talanta*, **78**, 954–8.

175 Hu, K., Liu, P., Ye, S. and Zhang, S. (2009) Ultrasensitive electrochemical detection of DNA based on PbS nanoparticle tags and nanoporous gold electrode. *Biosensors and Bioelectronics*, **24**, 3113–19.

176 Daniel, M.C., Aranzaes, J.R., Nlate, S. and Astruc, D. (2005) Gold-nanoparticle-cored polyferrocenyl dendrimers: modes of synthesis and functions as exoreceptors of biologically important anions and re-usable redox sensors. *Journal of Inorganic and Organometallic Polymers*, **15**, 107–19.

177 Dequaire, M., Degrand, C. and Limoges, B. (2000) An electrochemical metalloimmunoassay based on a colloidal gold label. *Analytical Chemistry*, **72**, 5521–8.

178 Wang, J., Xu, D.K., Kawde, A.N. and Polsky, R. (2001) Metal nanoparticle-based electrochemical stripping potentiometric detection of DNA hybridization. *Analytical Chemistry*, **73**, 5576–81.

179 Wang, J., Polsky, R. and Xu, D.K. (2001) Silver-enhanced colloidal gold electrochemical stripping detection of DNA hybridization. *Langmuir*, **17**, 5739–41.

180 Wang, J., Liu, G. and Merkoçi, A. (2003) Electrochemical coding technology for simultaneous detection of multiple DNA targets. *Journal of the American Chemical Society*, **125**, 3214–15.

181 Liu, G., Wang, J., Kim, J. and Jan, M.R. (2004) Electrochemical coding for multiplexed immunoassays of proteins. *Analytical Chemistry*, **76**, 7126–30.

182 Braun, E., Eichen, Y., Sivan, U. and Ben-Yoseph, G. (1998) DNA-templated assembly and electrode attachment of a conducting silver wire. *Nature*, **391**, 775–8.

183 Danscher, G. and Rytter Norgaard, J.O. (1983) Light microscopic visualization of colloidal gold on resin-embedded tissue. *Journal of Histochemistry and Cytochemistry*, **31**, 1394–8.

184 Velev, O.D. and Kaler, E.W. (1999) In situ assembly of colloidal particles into miniaturized biosensors. *Langmuir*, **15**, 3693–8.

185 Park, S.J., Taton, T.A. and Mirkin, C.A. (2002) Array-based electrical detection of DNA with nanoparticle probes. *Science*, **295**, 1503–6.

186 Tokonami, S., Shiigi, H. and Nagaoka, T. (2008) Characterization and DNA sensing properties of nanogapped array electrodes. *Journal of the Electrochemical Society*, **155**, J105–9.

187 Tokonami, S., Shiigi, H. and Nagaoka, T. (2008) Open bridge-structured gold nanoparticle array for label-free DNA detection. *Analytical Chemistry*, **80**, 8071–5.

188 Nagaoka, T., Shiigi, H. and Tokonami, S. (2007) Highly sensitive and selective chemical sensing techniques using gold nanoparticle assemblies and superstructures. *Bunseki Kagaku*, **56**, 201–11.

189 Shiigi, H., Tokonami, S., Yakabe, H. and Nagaoka, T. (2005) Label-free electronic detection of DNA-hybridization on nanogapped gold particle film. *Journal of the American Chemical Society*, **127**, 3280–1.

190 Tokonami, S., Shiigi, H. and Nagaoka, T. (2010) Gold nano films – synthesis, characterization and potential biomedical applications, in *Nanostructured Thin Films and Surfaces* (ed. C.S.S.R. Kumar), Wiley-VCH Verlag GmbH, Weinheim, pp. 175–202.

191 Bard, A.J. and Faulkner, L.R. (2001) *Electrochemical Methods: Fundamentals and Applications*, John Wiley & Sons, Inc., pp. 368–88.

192 Deng, C., Chen, J., Nie, Z., Wang, M., Chu, X., Chen, X., Xiao, X., Lei, C. and Yao, S. (2009) Impedimetric aptasensor with femtomolar sensitivity based on the enlargement of surface-charged gold nanoparticles. *Analytical Chemistry*, **81**, 739–45.

193 Yang, G.J., Huang, J.L., Meng, W.J., Shen, M. and Jiao, X.A. (2009) A reusable capacitive immunosensor for detection of *Salmonella* spp. based on grafted ethylene diamine and self-assembled gold nanoparticle monolayers. *Analytica Chimica Acta*, **647**, 159–66.

194 Nagaoka, T. and Okazaki, S. (1985) Electron-transfer kinetics of quinones at solid electrodes in aprotic solvents at low temperatures. *Journal of Physical Chemistry*, **89**, 2340–4.

195 Bard, A.J. and Faulkner, L.R. (2001) *Electrochemical Methods: Fundamentals and Applications*, John Wiley & Sons, Inc., pp. 624–7.

196 Wang, J., Liu, G.D., Munge, B., Lin, L. and Zhu, Q.Y. (2004) DNA-based amplified bioelectronic detection and coding of proteins. *Angewandte Chemie, International Edition*, **43**, 2158–61.

197 Chu, X., Fu, X., Chen, K., Shen, G.L. and Yu, R.Q. (2005) An electrochemical stripping metalloimmunoassay based on silver-enhanced gold nanoparticle label. *Biosensors and Bioelectronics*, **20**, 1805–12.

198 de la Escosura-Muñiz, A., Maltez-da Costa, M. and Merkoçi, A. (2009) Controlling the electrochemical deposition of silver onto gold nanoparticles: reducing interferences and increasing the sensitivity of magnetoimmuno assays. *Biosensors and Bioelectronics*, **24**, 2475–82.

199 Cai, H., Wang, Y.Q., He, P.G. and Fang, Y.H. (2002) Electrochemical detection of DNA hybridization based on silver-enhanced gold nanoparticle label. *Analytica Chimica Acta*, **469**, 165–72.

200 Lee, T.M.H., Li, L.L. and Hsing, I.M. (2003) Enhanced electrochemical detection of DNA hybridization based on electrode-surface modification. *Langmuir*, **19**, 4338–43.

201 Selvaraju, T., Das, J., Han, S.W. and Yang, H. (2008) Ultrasensitive electrochemical immunosensing using magnetic beads and gold nanocatalysts. *Biosensors and Bioelectronics*, **23**, 932–8.

7
Design and Applications of Genetically Engineered Nanocomposites

Gopal Abbineni and Chuanbin Mao

7.1
Introduction

Genetic engineering has proven to be an invaluable tool in molecular biology, allowing the direct manipulation of genes in organisms and resulting in the display of desired guest peptides or proteins on the surface of biological structures such as bacteria, viruses, and yeast cells. Additionally, synthetic nanomaterials such as nanoparticles are very attractive materials because of their quantum size, large surface-to-volume ratio, and novel physical or chemical properties that differ from those of their bulk counterparts. With recent advances in nanoscience coupled to an increased understanding of molecular biology, it is now possible to combine the power of genetic tools, which can be applied to the biological systems, with synthetic nanomaterials to form genetically engineered nanocomposites. Such a feat has granted numerous applications, ranging from electronics and materials sciences to biomedicine [1]. Several genetically modifiable biological structures have been used in developing nanocomposites, including M13 bacteriophage, *Salmonella typhimurium* flagella, yeasts [2], insect cells [3], and the tobacco mosaic virus (TMV) [4]. The discussion in this chapter emphasizes the uses of M13 bacteriophage and *Salmonella* bacterium flagella to form genetically engineered nanocomposites.

Genetically modified bacterial flagella and bacteriophage are attractive in developing nanotechnology, because they have several advantages over other biological structures. For example, they can be genetically engineered to bear different surface chemistries by displaying peptides on the exterior surface, and mass-produced by amplifying bacteria using microbiological methods. They can, potentially, serve as templates for nanomaterials synthesis and assembly, as building blocks for self-assembly into higher-order structures, as carriers for delivering drugs or genes, and as platforms for selecting target-recognizing peptides. An increasing demand for progressively advanced electrical and biomedical devices has prompted the development of new nanoscale constructs that utilize

Nanomaterials for the Life Sciences Vol.8: Nanocomposites. Edited by Challa S. S. R. Kumar

ISBN: 978-3-527-32168-1

engineered filamentous bacteriophage and flagella. Compared to other biological structures, bacteriophage and flagella can be employed to build a random peptide library from which target-specific peptides can be selected. Therefore, in this chapter attention will be focused on the use of bacteriophage and flagella in developing novel nanocomposites.

Following an overview of the structure, biology and genetic engineering of two typical genetically modifiable biomacromolecules (filamentous bacteriophage and flagella), the applications of genetically engineered biomacromolecules will be introduced for the formation of one-dimensional (1-D), two-dimensional (2-D) and three-dimensional (3-D) nanocomposites. Some examples of the application of these materials in device fabrications, medicine and sensing will be provided. Finally, future developments in genetically engineered nanocomposites will be discussed. Although the use of biomacromolecules in nanotechnology has been reviewed previously [5–7], the aim of this chapter is to briefly overview the structures of bacteriophage and flagella, their genetic modification, and their assembly into nanocomposites. The applications of genetically engineered nanocomposites will also be highlighted, and their future prospects discussed.

7.2 Genetically Modifiable Biomacromolecules for the Design of Genetically Engineered Nanocomposites

Genetic engineering enables the introduction of site-specific structural and/or chemical functionality into a biomolecule, which in turn allows numerous applications and potential uses in the nanosciences. With a multitude of applications and concomitant error-free production, filamentous bacteriophage and bacterial flagella represent excellent examples of the elegance of genetic engineering. As discussed below, M13 bacteriophage and bacterial flagella possess unique features that can be exploited when developing nanotechnology.

7.2.1 Bacteriophage as a Bacteria-Specific Virus

The M13 bacteriophage is a filamentous virus that infects a wide variety of Gram-negative bacteria, and is composed of circular single-stranded DNA (ssDNA) that is approximately 950 nm in length and 6.5 nm in diameter. Filamentous bacteriophages belong to the Ff class, which includes the M13, f1, fd, and ft strains. These particular viruses consist of long cylindrical protein capsids enclosing the ssDNA genome of ~6500 nucleotides [8]. The capsid is composed of ~2700 copies of major coat proteins (pVIII), with approximately five copies of pIII and pVI proteins at one end of the bacteriophage and approximately five copies of pVII and pIX proteins at the other end. The surface-exposed N terminus of any protein on the phage can be genetically modified with a desired foreign peptide. These proteins can be used as excellent scaffolds for genetic modification, as their structures permit facile genetic manipulation. The structure of a typical M13 bacteriophage is shown in Figure 7.1.

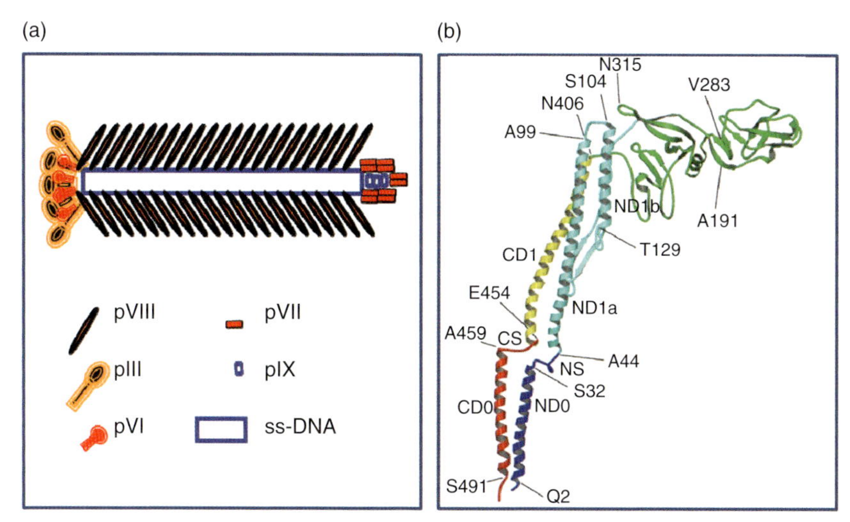

Figure 7.1 (a) Structure of native filamentous bacteriophage; (b) Ribbon structure of flagellin. Reproduced with permission from Ref. [9].

The construction of a genetically engineered bacteriophage generally involves two approaches. The first (more traditional) approach uses a phagemid/phage as a vector. An oligonucleotide sequence encoding a desired foreign peptide is spliced into one of the phage coat protein genes (as shown in Figure 7.2a), which allows the foreign peptide to be displayed on the exposed surface of the bacteriophage. The second (more current) approach utilizes combinatorial biology to form type 3 and type 8 phage display libraries [10, 11]. A phage library consists of multibillion bacteriophages generated by fusing random peptides with the N terminus of every copy of the major coat proteins, pVIII, or the minor coat proteins, pIII. If the random peptides are present on pIII proteins, it is termed a "type 3 library"; likewise, if the random peptides are present on pVIII proteins, it is termed a "type 8 library." These phage libraries are used to select specific amino acid residues against a variety of substrates by using affinity selection. For example, Figure 7.3 shows the use of type 3 and 8 libraries to screen for a peptide displayed on a bacteriophage which is specific for a particular nanomaterial. Phage display libraries are used to explore interactions between either two proteins, nanomaterials (inorganic, organic and biological) and proteins, or peptides and short ligands. The design of a genetically engineered bacteriophage is shown in Figure 7.2a.

7.2.2
Bacterial Flagella as Genetically Modifiable Protein Nanotubes

Salmonella typhimurium is a rod-shaped, Gram-negative bacterium with a length of approximately 4 μm and a diameter of approximately 1.3 μm. Each bacterium bears many *flagella*; these are supramolecular structures that extend from the

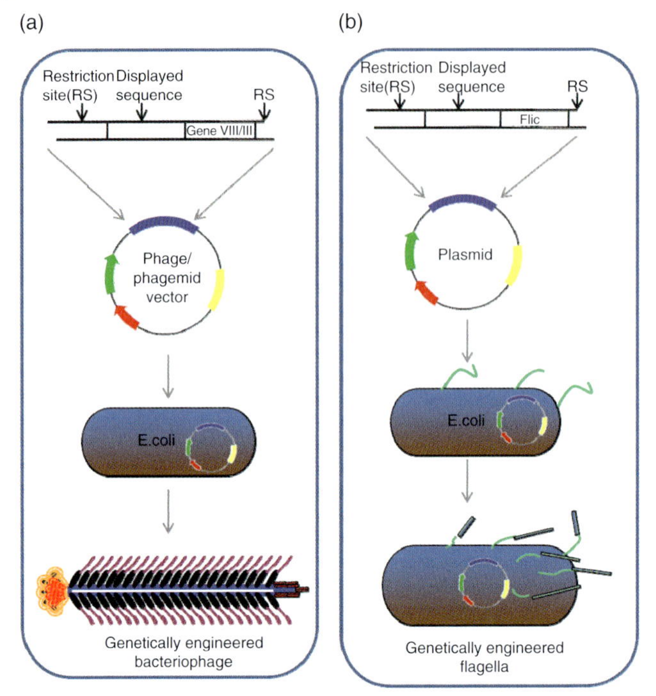

Figure 7.2 General schemes for designing (a) genetically engineered bacteriophage, and (b) genetically engineered flagella.

bacterial cytoplasm to the exterior surface and help the bacterium to "swim." The three main structural elements of the flagellum include a basal body, a hook, and a filament. The flagellar filament itself can be pictured as a protein nanotube with a uniform outer diameter of approximately 12–25 nm and an inner diameter of about 2 nm [9]. The structure of a typical bacterial flagellum is shown in Figure 7.1b [12]. Each flagellum is constructed by the self-assembly of about 30 different proteins that are helically assembled from about 20 000 copies of identical protein subunits known as *flagellin*, encoded by the gene FliC. Flagellin are globular proteins composed of four different domains, namely D0, D1, D2, and D3 [9, 13]. The surface-exposed domains of the flagella (namely D2 and D3) are highly variable, and can be genetically modified with desired foreign peptides. The results of previous studies have shown that up to approximately 302 amino acid residues can be displayed in the functional form on the variable region of FliC [14, 15].

As a complement to phage display, bacterial surface display (and, more specifically, flagellum display) has been very successful in recent years [16]. Previously, several groups have used the flagella surface as a scaffold to display several different peptide loops with various functionalities [17–20]. The construction of genetically engineered flagella is based on the insertion of foreign DNA into the variable region of the gene, *FliC*. In the past, genetically engineered flagella have served

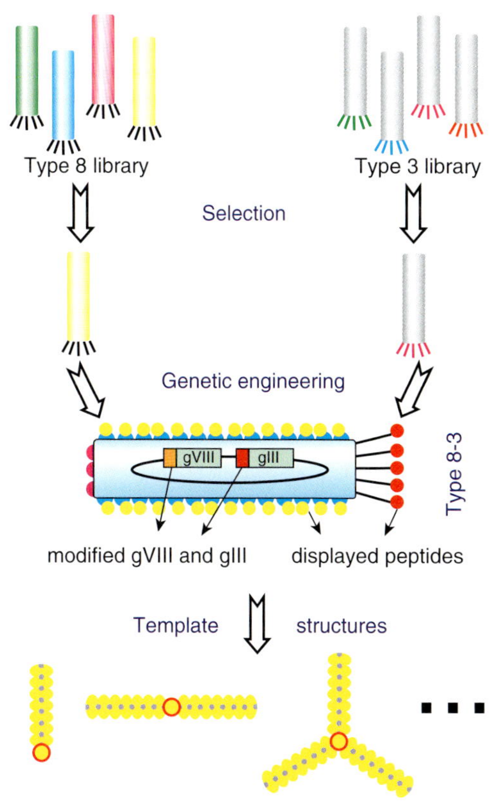

Figure 7.3 General scheme for engineering and characterization of type 8 and type 3 phage on nanomaterials. Reproduced with permission from Ref. [40].

as ideal candidates for the development of recombinant vaccines [21], in the characterization of protein–protein interactions [22], and as excellent templates for ordered arrays of nanomaterials [13]. The design of genetically engineered flagella is shown schematically in Figure 7.2b.

7.3 Nanocomposites Built from Genetically Engineered Biomacromolecules

7.3.1 One-Dimensional Nanocomposites Templated by Filamentous Biotemplates

Due to their low dimensions and high aspect ratios, 1-D nanomaterials possess excellent physical properties [23]. Indeed, it is due to their unique physical and chemical characteristics that 1-D nanomaterials such as nanotubes [24], nanorods and nanowires [25] have gained tremendous importance in the revolutionary era of nanobiotechnology. During this time, particular emphasis has been placed on the fabrication of novel nanodevices with current limitations in bulk industrial production [26]. Generally, the growth of nanomaterials in a 1-D orientation can be accomplished by two approaches: (i) an approach that includes *in situ* growth, where nanomaterials are grown in parallel to a substrate; and (ii) an approach that includes "post growth," where the nanomaterials assemble together and form a functional structure [27]. One-dimensional nanomaterials can be produced by a variety of methods, including template-based synthesis [28], hydrothermal synthesis [29], H_2-assisted thermal evaporation [27], crystallization [30], and surfactant-mediated synthesis [31]. The following discussion will focus on the formation of nanomaterials, using genetically engineered filamentous bacteriophage and flagella as biotemplates.

7.3.1.1 Inorganic Nanomaterials on Genetically Engineered Biotemplates

The monodisperse appearance and genetically modifiable surface chemistry of both bacterial flagella and bacteriophage have been successfully exploited in the assembly of inorganic nanomaterials on engineered biotemplates. The insertion of highly cationic/anionic amino acid residues greatly increases the density of positive/negative charge on the flagella or phage surface, leading to an electrostatic repulsion between neighboring residues and a consequent straightening of the biotemplate. This property is instrumental for the 1-D growth of inorganic nanomaterials [32]. Such genetically engineered flagella templates have been used successfully to form both metal nanotubes [33] and metal nanowires [33]. Likewise, studies on genetically engineered bacteriophage templates have allowed the production of semiconductor nanowires [34].

The result of previous studies have shown that certain amino acids exhibit a strong binding affinity to the surface of inorganic materials [35, 36]. Consequently, inorganic material specificity was controlled by displaying specific amino acid residues on the surface of filamentous bacteriophage, by using genetic engineering [37]. With this concept in mind, additional studies have shown that several other metals and inorganic nanomaterials can be assembled, in similar fashion, on engineered filamentous bacteriophage and flagellar templates. A detailed summary of peptide specificity to various inorganic nanoparticles, using genetically engineered flagella and phage biotemplates, is provided in Table 7.1. It has been shown recently that phage displaying ($[MHGKTQATSGTIQS]_3$) [38] and, previously VSGSSPDS, can bind specifically to the surface of gold due to a preferential binding of the rich hydroxyl groups to gold lattices [39]. Huang *et al.* have demonstrated gold- and streptavidin-binding amino acid residues on the pVIII and pIII proteins of filamentous bacteriophage, and successfully assembled Au and CdSe nanocrystals into ordered 1-D arrays that served as a template to nucleate highly conductive nanowires [40]. A series of transmission electron microscopy (TEM) images of the nucleation of gold on the surface of engineered bacteriophage to form a highly conductive gold nanowire, is shown in Figure 7.4a–c. This engineered nanoconstruct may serve as a template for electronic components in nanoelectronics.

Table 7.1 Amino acids displayed on engineered flagella and bacteriophage for binding or nucleating inorganic nanomaterials.

Template	Inorganic nanoparticle(s)	Resulting genetically engineered nanocomposites	Reference
Flagella displayed with histidine loop	Au, Cu, Cd, Pd	Flagella–Au, Cu, Cd nanotube bundles	[33]
		Flagella–Pd nanoparticles	
Flagella displayed with aspartic and glutamic acid loop	Co, Ag	Ordered array of Co nanoparticles on flagella nanotubes	[33]
		Flagella–Ag nanowires	
Flagella displayed with arginine-lysine loop	$CaCO_3$	Biomineralization of $CaCO_3$ on genetically engineered flagella	[43]
Flagella displayed with lysine-cysteine loop	SiO_2	Silica flagella nanotube with uniform pore size of ~14 nm	[32]
Bacteriophage displayed with EPGHDAVP on major coat protein, pVIII	Co	Co^{2+}–phage fibrous material further mineralization of Co–Pt nanoparticles	[41]
Phage displayed with HNKHLPSTQPLA on minor coat protein, pIII	Fe–Pt	Crystallization of Fe–Pt nanoparticles on phage template	[46]
Phage displayed with CNAGDHANC on minor coat protein, pIII	Co–Pt	Nucleation and growth of Co–Pt nanoparticles on phage template	[46]
Phage displayed with CNNPMHQNC on minor coat protein, pIII	ZnS	Nucleation and orientation of ZnS on phage template to form nanowires	[34]
Phage displayed with SLTPLTTSHLRS on minor coat protein, pIII	CdS	Nucleation and orientation of CdS on phage template to form nanowires	[47]
Phage displayed with VPSSGPQDTRTT on minor coat protein, pIII	Al and mild steel	Metal-binding phage	[48]
Phage displayed with ERNDMTHV on major coat protein, pVIII	Hydroxyapatite	Nucleation of hydroxyapatite on phage template	[49]

In another study, Lee *et al.* [41] used engineered bacteriophage as a mechanical framework to assemble inorganic (Co–Pt) hybrid materials. The initial step was to apply a type 8 phage library to the identification of a peptide that could selectively bind to cobalt ions. Later, the filamentous bacteriophage was directionally

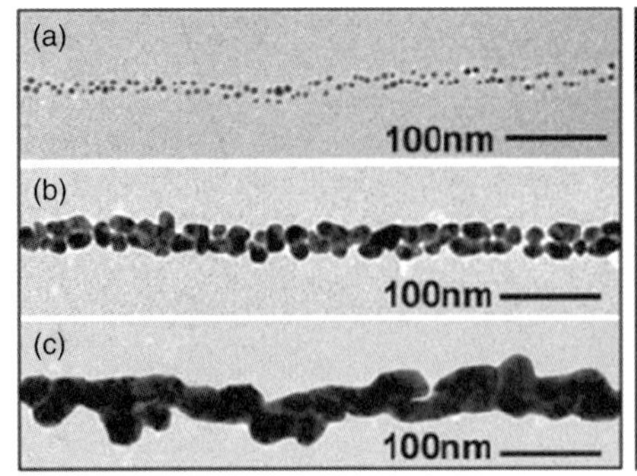

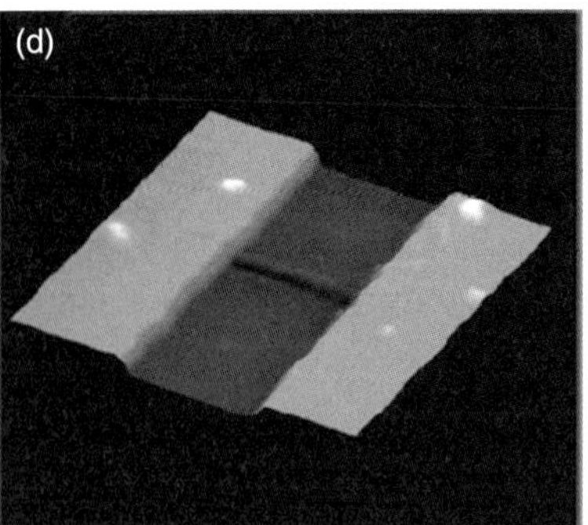

Figure 7.4 Engineered nanocomposites of virus-gold nanowires. (a–c) Transmission electron microscopy images of the progressive growth of continuous gold nanowires template by gold nanoparticle arrays on pVIII proteins of filamentous bacteriophage: (a) before electroless deposition, (b) at 3 min deposition, and (c) at 5 min deposition; (d) Three-dimensional plot of atomic force microscopy image of a two-terminal device based on a virus–gold nanowire construct. Images reproduced with permission from Ref. [40].

organized into fibrous structures when interacted with cobalt ions, after which the Co–Pt alloys were successfully synthesized on the fibrous scaffold. A schematic representation of the engineered virus–Co–Pt assembly, and TEM images of nucleated Co–Pt nanoparticles on filamentous phage are shown in Figure 7.5a–c. These engineered nanocomposites exhibited extraordinary magnetic properties, due to the presence of nanophase Co–Pt alloys.

Likewise, Mao *et al.* have exploited the self-assembling structure of engineered bacteriophage to nucleate CdS and ZnS nanocrystals so as to form nanowires. In these studies, a type 3 phage library was applied to obtain peptides that would bind selectively to CdS and ZnS nanocrystals. The screened and isolated bacteriophage was then used successfully as a template for the nucleation and orientation of ZnS or CdS nanowires [34]. When the bacteriophage was engineered with dual peptides, however, a heterogeneous nucleation was observed. Subsequently, when the engineered bacteriophage were incubated with Zn^{2+}, Cd^{2+} and S^{-2} ions at a molar ratio of 1 : 1 : 2, the phage were nucleated with ZnS and CdS nanocrystals and followed a peptide-templated growth mechanism similar to the biomineralization that is observed in Nature [34]. The TEM images of phage–CdS nanowires and phage–CdS–ZnS hybrid nanowires are shown in Figure 7.6a–c. It would appear that genetically engineered nanocomposites which use bacteriophage as a template for the orientation of single and multiple material nanocrystals have great potential in the fabrication of nanodevices [42].

More recently, the group of Muralidharan has been actively involved in genetically modifying the surface of flagella with varying functional groups such as anionic, cationic, thiol and imidazole groups, the aim being to synthesize novel hybrid nanomaterials and nanotubes [43]. In a previous studies, the same group genetically engineered the surface of flagella with an aspartate–glutamate peptide

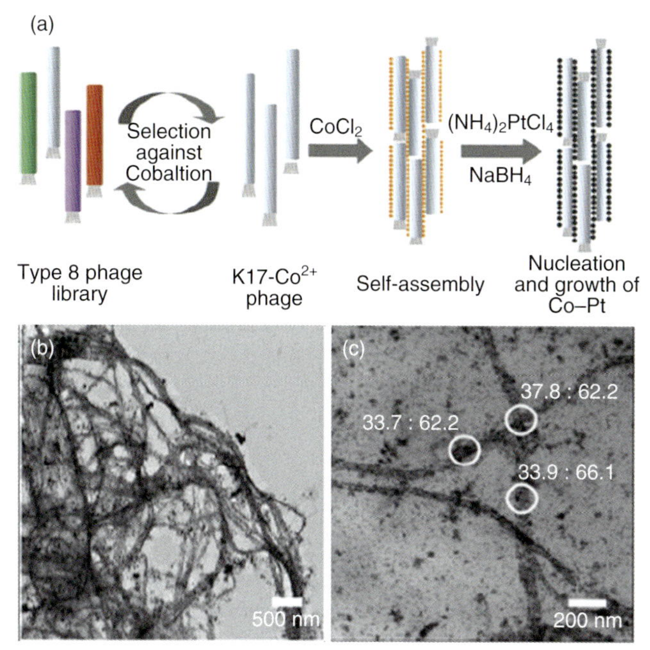

Figure 7.5 Engineered nanoconstruct of virus–Co–Pt nanowires. (a) Cartoon showing the selection and assembly of cobalt ions on the surface of an engineered bacteriophage; (b) TEM image of bacteriophage–cobalt ions complex; (c) TEM image of bundled phage with Co–Pt nanoparticles. Images reproduced with permission from Ref. [41].

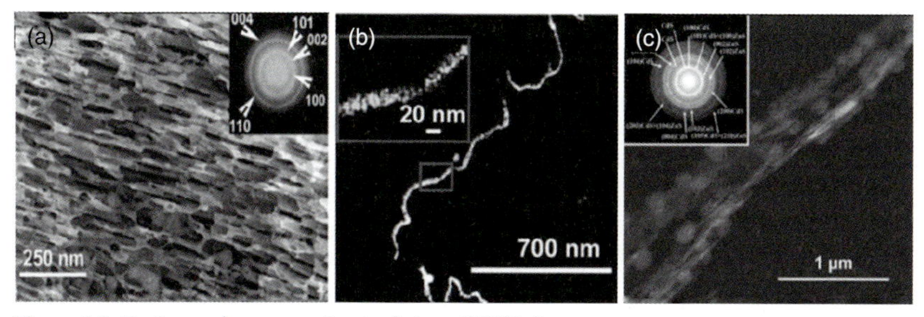

Figure 7.6 Engineered nanoconstruct of virus–CdS/ZnS nanowires. (a) TEM image of layered structure of virus directing ZnS nanocrystal synthesis; (b) Annular dark-field scanning TEM (ADF-STEM) image of CdS-virus nanowires; (c) High-angle annular dark-field scanning TEM (HAADF-STEM) image of viral ZnS–CdS hybrid layered structure. Images reproduced with permission from Ref. [34].

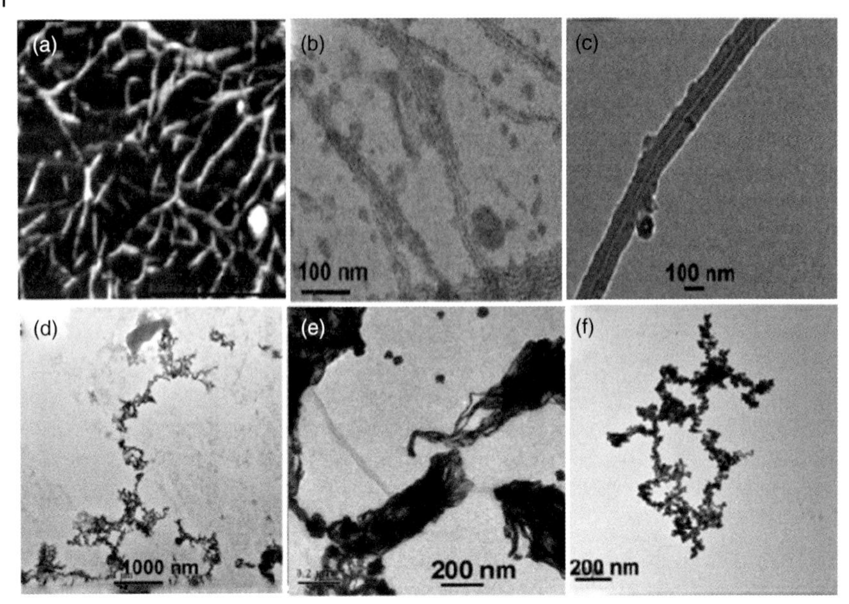

Figure 7.7 Engineered nanoconstruct of flagella-inorganic nanomaterials (a) AFM image of mineralized calcium carbonate on engineered flagella; (b) TEM image of gold nanoparticles synthesized on engineered flagella; (c) TEM image of copper nanotubes synthesized on engineered flagella; (d–f) TEM images of cobalt, silver, and palladium nanowires synthesized on engineered flagella. Images reproduced with permission from Refs [33, 43].

loop. When the engineered flagella were later deposited on a silanized quartz surface, this resulted in the self-assembly of flagella on protonated amine groups through electrostatic interactions [43]. The flagella layer was then covered with a polymer, poly(ethyleneimine) (PEI), and again deposited with flagella through electrostatic interactions. This approach was later used to demonstrate the biomineralization of calcium carbonate. The resultant hybrid engineered nanoconstruct, composed of inorganic–biomineralized materials, demonstrated tremendous potential for applications as durable materials such as cartilage and bones. An atomic force microscopy (AFM) image showing the mineralization of calcium carbonate on engineered flagella is shown in Figure 7.7a. Likewise, several other groups have reduced gold and copper on the surface of a histidine loop displayed on flagella to form nanotube structures; the TEM images of the formed nanotubes are shown in Figure 7.7b,c Other inorganic nanoparticles such as copper, platinum and palladium, have also been synthesized on the surface of engineered flagella so as to form nanowires with suitable catalytic applications in electronics. Typical TEM images of these formed nanowires are shown in Figure 7.7d,e.

Recently, engineered flagellar templates were used to synthesize highly ordered engineered nanotubes, namely that of silica/flagella. Notably, it was possible to genetically engineer the surface of flagella with different sequences, such as KKCC

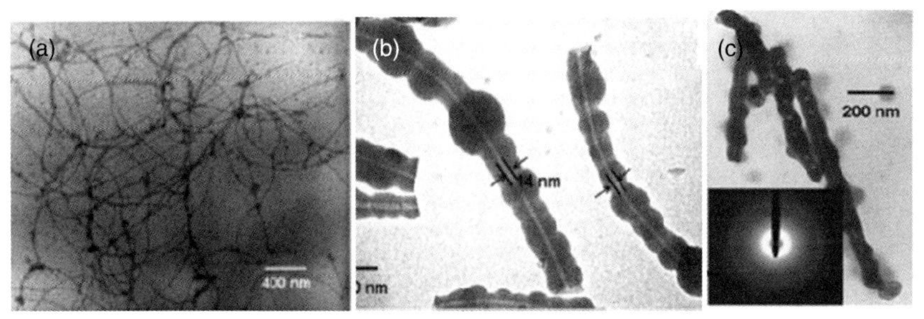

Figure 7.8 Engineered nanoconstruct of flagella-silica nanotubes. (a) TEM image of wild-type flagella negatively stained with 1% uranyl acetate; (b) TEM image of flagella–silica nanotube; (c) Silica nanotubes calcinated at 800 °C for 1 h, showing the formation of nanohole formation. Images reproduced with permission from Ref. [32].

(K-lysine, C-cysteine) and K_{10} C_{10} (ten lysine and ten cysteine residues). The genetically engineered flagella were detached from the bacterial surface by simple vortexing and then used for the assembly of silica nanotubes, by adding aminopropyltriethoxysilane (APTES) and tetra-ethyl orthosilane (TEOS) as precursors to the aqueous flagellar solution. The electrostatic absorption of APTES on the surface of amino acid residues on the flagella resulted in the hydrolysis of APTES to form silicic acid. The silicic acid formed could then act as nuclei for the subsequent silica growth, allowing the formation of silica fibers. The prepared hybrid flagella/silica fibers exhibited uniform and precise pore sizes under TEM. Once the silica sheaths has been formed, the biotemplate was calcinated at high temperatures, and this resulted in the formation of silica nanotubes. The engineered nanocomposites (silica/flagella nanotubes) exhibited distinct inner and outer functionalities. The TEM images of flagella and the flagella/silica nanocomposite structures prepared in this way are shown in Figure 7.8a–c. The formed silica nanotubes have possible medical applications, in particular for drug and gene delivery [44, 45].

The results of above-described studies confirmed that flagella and filamentous bacteriophage could both serve as excellent templates for nucleating inorganic nanoparticles. They also confirmed the ease with which peptides could be displayed on the surface of bacteriophage and flagella. This would, in turn, permit the site-specific nucleation of inorganic materials on the flagella/bacteriophage surfaces, and lead to the creation of an inorganic–organic hybrid nanomaterial. It would appear, therefore, that the filamentous genetically modifiable biomacromolecule represents an ideal template for the assembly of materials into a 1-D array.

7.3.1.2 **Organic Nanomaterials on Genetically Engineered Biotemplates**

In general, the tissues of human beings are assembled from biomacromolecules, and mainly composed of proteins, carbohydrates and lipids, under genetic control in physiological conditions [50]; this results in the synthesis of highly ordered

Table 7.2 Amino acids displayed on genetically engineered bacteriophage and flagella for interacting with organic nanomaterials.

Template	Organic material	Resulting genetically engineered nanocomposites	Reference
Bacteriophage displayed with lysine residue	Polyacrylamide	Opaque gel-like polymers	[51]
Phage displayed with THRTSTLDYFVI on minor coat protein, pIII	Chlorine-doped polypyrrole (PpyCl)	Phage specific to PpyCl	[53]
Phage displayed with EEEEEEEE on major coat protein, pVIII	Zinc phthalocyanine (ZnPc)-loaded liposomes	Self-assembly of ZnPc-loaded liposomes on M13 bacteriophage	[64]
Flagella displayed with EEEEEEEE sequence	Zinc phthalocyanine (ZnPc)-loaded liposomes	Self-assembly of ZnPc-loaded liposomes on flagella nanotubes	[65]
Flagella displayed with aspartic acid and glutamic acid	Aniline	Formation of polyaniline nanotubes	[67]

nanoscale structures from available molecules. For many years, chemists have been attempting to mimic Nature's assembly processes in order to benefit from the outstanding properties of these highly organized nanostructures. Although direct assembly has not yet been possible, many of the limitations encountered have been overcome by using templates to pre-organize arrays of nanomaterials that can, in turn, react to form biological molecules such as proteins and nucleic acids. Amino acids are also displayed on the surface of bacteriophage and flagella for interacting with organic nanomaterials (Table 7.2), which can be exploited for building organic nanomaterials.

The first step in building an organic nanomaterial on a biological template is to identify the short peptide sequences (rather than large proteins) that are specific for an organic material, through a process of combinatorial mutagenesis. This step is often followed by the self-assembly of an organic material on the protein template [39]. Here, the main objective is to retain the bulk properties of the biological template, while modifying the surface properties so as to obtain the desired specificity and recognition. It is first necessary, however, to understand the interaction between short protein residues and organic molecules, followed by the use of a developed method to produce functional hybrid systems. In the pharmaceutical industry, this understanding between the polymer and protein conjugates will surely result in the rapid development of new drugs.

In comparison to the assembly of inorganic nanoparticles on engineered flagella and phage templates, the assembly of organic nanomaterials on biotemplates is

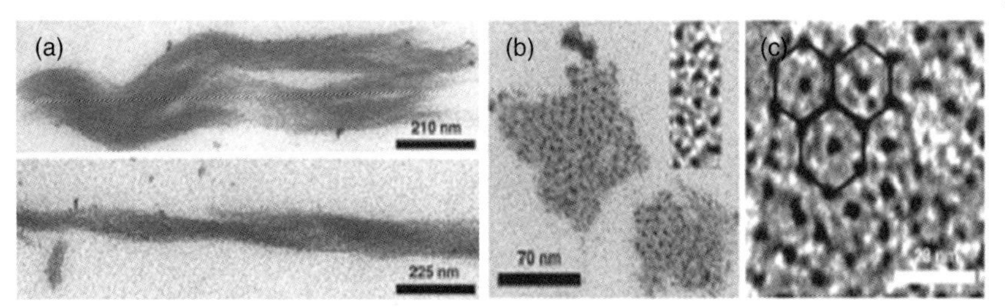

Figure 7.9 Engineered nanoconstruct of phage–polymer bundles. (a) TEM image of phage fibers; (b) TEM image of flagella–silica nanotube; (c) End-on view of a phage bundle, showing hexagonal close packing of the phage–polymer conjugates. Images reproduced with permission from Ref. [51].

currently very limited. Recently, Willis *et al.* [51] have envisioned the use of chemically modified bacteriophage as a template to conjugate macromolecules before polymerization. In this study, it was proved that organic polymers (namely, polyacrylamide) could self-assemble on the surface of a bacteriophage so as to form ordered helical bundles with an hexagonal close packing. Following chemical modification of the surface-exposed single lysine residue on the filamentous bacteriophage with Traut's reagent and addition of the methylacrylamide monomer, polymerization was complete in 15 min at 4 °C, and resulted in the creation of a gel-like polymer [52]. Although the resultant phage–polymer gel was very rigid, it had excellent mechanical properties and showed ordered arrays of nanopores. TEM images of the phage–polymer bundles, and of the nanopores between the bundles, are shown in Figure 7.9a–c. This engineered nanoconstruct, utilizing a phage–polymer, exhibited remarkable physical properties that could be used for construction applications.

Previous studies conducted by Belcher and colleagues have shown that genetically engineered bacteriophage provide an excellent scaffold for the systematic incorporation and oriented assembly of organic polymers. Sanghvi *et al.* [53] have developed a unique strategy for functionalizing chlorine-doped polypyrrole (PpyCl), an electrically conductive polymer that demonstrates potential applications in medical sensing, including drug delivery [54], nerve regeneration [55, 56], and biosensors [57, 58]. In order to identify the genetically engineered phage specific to the PpyCl polymer, a PpyCl film was first formed by using electrochemical methods [59, 60]. This was followed by incubation with an M13 phage library (having 10^9different random peptides on the pIII protein) [61–63] , followed by screening for peptide inserts that exhibited a high specificity and stability with PpyCl. Based on the study results, a THRTSTLDYFVI peptide sequence on the minor coat protein (pIII) of the M13 bacteriophage was found to be specific to PpyCl. It was also shown that the N-terminal end of the selected peptide possessed hydrophilic residues that rendered it polar-soluble. In contrast, the C terminus contained an aspartic acid residue that provided an anionic nature to the peptide, with resultant binding to the cationic PpyCl. The hydrophobic residues surrounding the aspartic acid were shown to provide support for folding and binding of the peptide towards PpyCl. This strategy represents a versatile approach for the development of bioactive hybrid materials (phage–organic materials) for potential future medical applications. There is also a great potential in the coming years for more intriguing research on polymer-based engineered templates.

Recently, a new strategy was developed to create genetically engineered nanocomposites made from bacteriophage and flagellar protein nanotubes, for use as potential drug carriers [64, 65]. In this case, the surface of the bacteriophage and flagella were made anionic (via genetic engineering) by their displaying eight glutamic acid residues. Liposomes were then loaded with a drug, in this case zinc phthalocyanine (ZnPc), a well-known photosensitizer that is used for the photodynamic therapy (PDT) of malignant cells [66]. The cationic liposomes loaded with ZnPc were later self-assembled onto engineered biotemplates (phage/flagella) by using electrostatic interactions. When the biotemplate was additionally co-displayed with cancer cell-targeting peptides and liposome-interacting peptides, the engineered constructs were also capable of targeting specific cancer cells. TEM images of the self-assembly of liposomes on genetically engineered bacteriophage and flagella are shown in Figure 7.10a–d. These genetically engineered biotemplates not only prevented liposomes from aggregating, but also increased the uptake of drugs at the target site.

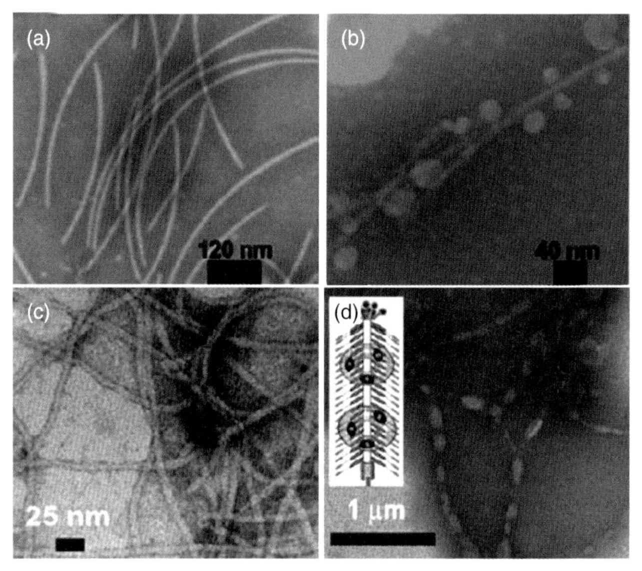

Figure 7.10 Engineered nanoconstruct of self-assembled liposomes on flagella/phage-bundles. (a) TEM image of flagella fibers; (b) TEM image of self-assembled liposomes on flagella template; (c) TEM image of filamentous phage fibers; (d) TEM image of self-assembled liposomes on phage template. Images reproduced with permission from Refs [64, 65].

Recently, the group of Muralidharan have used engineered flagella as a nucleation site for the assembly of organic nanomaterials to form functional nanowires. In a previous study, the same group modified the surface of flagella with an aspartate–glutamate loop so as to initiate the formation of aniline nanowires. In this case, the purified flagella exhibited a high negative charge which later initiated the polymerization of aniline to form polyaniline nanowires [67]. The flagella–polyaniline nanowires were formed due to ionic interactions and π-π stacking between the charged surface of the flagella and polyaniline. Polyaniline is an extensively investigated polymer nanomaterial that can be used for the development of electrically conductive nanowires, as well as sensing applications. The resulting polymerized flagella can act (potentially) as conductive nanowires.

Although, in recent years, the number of studies in which organic nanomaterials have been assembled along biotemplates has steadily increased, their practical applications remain limited. In fact, very few investigations have been conducted on the growth of polymers along flagellar templates, despite this area having great potential for providing future applications in the field of materials science. Wang *et al.* have demonstrated the formation of organic materials on other filamentous viruses [68, 69]. Clearly, further investigations are required to form organic functional materials on biomacromolecules, with the subsequent formation of novel nanostructures.

7.3.2
Three-Dimensional Assembly of Genetically Engineered Biomacromolecules and Nanomaterials

As the alignment of nanomaterials into one dimension is insufficient for the creation of functional devices, it is highly desirable to assemble and communicate nanomaterials into 3-D architectures for this purpose [70]. Similarly, it is essential that, for medical applications, molecular scaffolds are developed with 3-D matrices. The physical and chemical properties of any nanomaterial depend on its size and shape, and its organization into 3-D architectures [71, 72]. The assembly of nanomaterials into 3-D structures may result in materials with fascinating electrical, chemical and optical properties that are vastly different from those of the bulk materials, and are necessary for the development of functional devices [73]. Recent advancements in science have greatly simplified the imaging of 3-D structures, using the sophisticated TEM and reconstruction methods currently available. Likewise, several physical and chemical assembly techniques have been adopted to direct nanomaterials into multicomponent 3-D structures towards a variety of surfaces. One of the most versatile methods for assembling 3-D structures is the layer-by-layer (LBL) approach, that involves the alternate deposition of complementarily charged species to construct multilayered nanomaterials [74, 75].

At this point, only the formation of functional 3-D structures utilizing the functionalities and capabilities of bacteriophage will be discussed, because the assembly of 3-D nanomaterials based on flagella has been rarely studied (at least, to the present authors' knowledge). Recently, the 3-D assembly of nanomaterials on bacteriophage surfaces has opened new doors for the design of high-performance batteries. To date, only limited studies have been conducted regarding the formation of 3-D structures using flagella as a template. At present, three methods have been used successfully to create 3-D structures by using genetically engineered bacteriophage, namely LBL self-assembly, a solvent-assisted capillary method, and lyotropic liquid crystalline self-assembly.

7.3.2.1 Layer-by-Layer Assembly into 3-D Structures

The technique of LBL assembly involves the alternate deposition of positively and negatively charged species, so as to create functional and tunable materials through controlled electrostatic interactions [76]. Molecular mobility is a key component involved in the ordering of nanoscopic structures into supramolecular macroscopic systems, and attractive and repulsive electrostatic interactions are often used in materials science to induce molecular mobility. One major advantages of electrostatic assembly when creating engineered nanocomposites is that the inherent biological function of the biomolecule is retained.

Yoo *et al.* used the inherent charge of the bacteriophage surface and oppositely charged polymers to demonstrate the formation of a 2-D monolayer of phage on a polyelectrolyte multilayer, via an interdiffusion-induced electrostatic LBL assembly [77]. For this, genetically engineered M13 viruses were self-assembled on a polyelectrolyte multilayer, using the polymeric interdiffusion principle. The electrostatic interactions between the oppositely charged electrolytes caused the virus to separate and to self-assemble into a 2-D monolayer on the multilayered surface. The inherent steric constraints in the charged amino acid residues of the M13 virus, along with the two oppositely charged electrolytes, caused the virus to "float" and to become ordered on the multilayered surface. In this case, the authors chose two oppositely charged polymers, the polycationic linear-polyethyleneimine (LPEI) and the polyanionic polyacrylic acid (PAA), based on the inherent biocompatibility, high conductivity and electrochemical properties of these materials to form a multilayered film. A detailed illustration of the self-assembly of viruses on the polyelectrolyte multilayer, using the interdiffusion phenomenon, is shown in Figure 7.11a–c. The repulsive interactions between the negatively charged viruses induced a spontaneous viral monolayer on the surface of the polyelectrolyte multilayer. The electrostatic interactions between the viruses and the polyelectrolyte multilayer were entropically favored to form an ordered layer.

In order to demonstrate the interdiffusion and ordering process, a polyelectrolyte of about 5 nm thickness was deposited onto a silicon substrate, followed by the absorption of viruses onto the layers. When a layer of LPEI had been deposited on top of the viruses, they were seen to diffuse throughout the LPEI layer. In contrast, when a counterion PAA was deposited on the virus, no diffusion was observed, which confirmed the fact that LPIE plays a crucial role in driving interdiffusion. The further growth or nucleation of inorganic nanocrystals on the formed 2-D viral scaffolds resulted in highly dense and ordered structures which were shown to be prerequisite materials for the assembly of 3-D nanodevices.

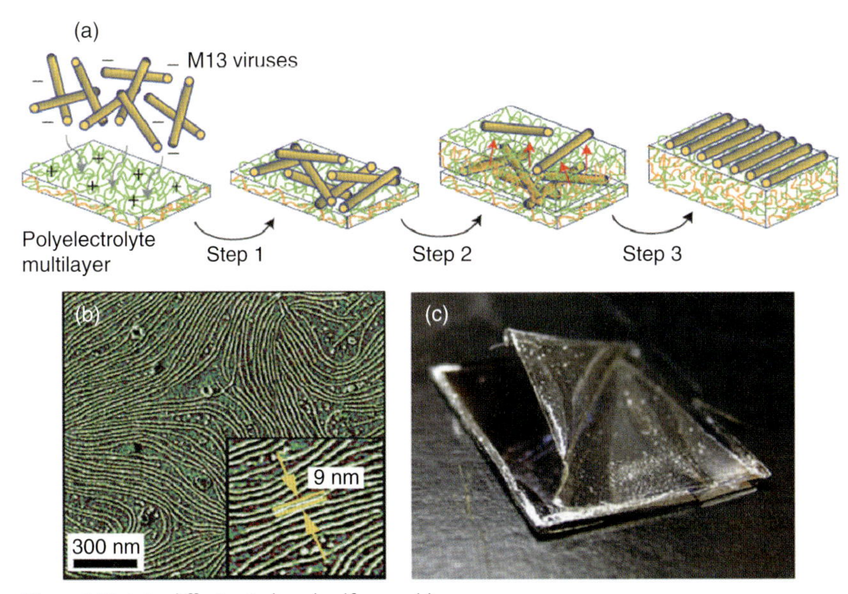

Figure 7.11 Interdiffusion-induced self-assembly. (a) Schematic procedure of M13 viruses selectively bound and ordered on a polyelectrolyte multilayer (PEM); (b) Closely packed monolayer of engineered virus on PEM; (c) An assembled virus monolayer on $(LPEI/PAA)_{100.5}$, forming a free-standing film. Images reproduced with permission from Ref. [77].

Possible practical applications for these structures can be found in electronics, magnetic, and optics; notably, the scaffolds might be used to nucleate and grow nanoparticles into assembled nanowires, with several possible applications in electronics [62].

7.3.2.2 **Solvent-Assisted Capillary Molding into 3-D Assemblies**

Previously, it has been shown that capillary forces play a major role in the self-organization of nanomaterials into ordered 3-D structures. For example, these capillary forces assist in the aggregation of colloidal nanoparticles when they are dropped onto a rough surface such as silica [78, 79]. Capillary action involves three major steps: (i) the adherence of a nanoparticle suspension onto a substrate due to charge interactions; (ii) displacement of the nanoparticles from suspension and adherence to the substrate as a result of drying; (iii) the reorganization of nanoparticles at the interface of the substrate when drying is complete. The columbic repulsions between similarly charged species prevent the dense packing of nanoparticles at the interface [79]. This method is especially useful when using a substrate which is thicker than 100 nm. The anisotropism, flexibility and heterofunctional nature of filamentous bacteriophage make these viruses suitable candidates for the construction of 3-D nanodevices.

Recently, Yoo *et al.* electrostatically assembled engineered viruses onto a polyelectrolyte multilayer, using the solvent-assisted capillary method [76]. For this, a patterned polyelectrolytic template was first prepared with a suitable mobile cation within the polymer. A negatively charged engineered viral solution was then captured onto the polymer, which resulted in an ordered monolayer patterning of viruses on the polymer surface. The closed monolayered packing of viruses was due mainly to the enhanced surface mobility of cationic moiety in the polyelectrolytic multilayers, as a result of capillary action. This solvent-based capillary approach can serve as an excellent alternative tool for patterning LBL assemblies, avoiding the use of chemically patterned surface chemistry [80]. A schematic representation and AFM image of the solvent-assisted capillary molding technique, and spontaneous ordering of viruses on the polyelectrolytic multilayers, is shown in Figure 7.12a–d, where the multilayered polymers are composed of alternate depositions of LPEI and PAA. Further progress of this multilayered film assembly, when conjugated with suitable ligands, might lead to the production of functional biological detection devices.

7.3.3 Genetically Engineered Nanocomposite Films

Genetically engineered nanocomposite films are composed mainly of two active components: genetically engineered biomolecules; and synthetic nanomaterials. Both, filamentous bacteriophage and flagella can form lyotropic liquid crystalline structures. Yet, the inherent ability of genetically engineered filamentous biomacromolecules to organize into liquid crystals, and their intrinsic tropism towards bacteria to produce in bulk, have led to these biomacromolecules becoming an attractive proposition for nanocomposite film preparation. If the bacteriophage or flagella carrying nanocomponents are able to self-assemble into the lyotropic liquid crystal, then a film with an ordered structure will result. For example, when the ZnS-binding peptides are displayed on the bacteriophage surface, the concentration of the ZnS-bound bacteriophage can be adjusted so as to induce the formation of a liquid crystalline structure. Finally, a liquid crystalline suspension can be evaporated to form a free-standing film, where the ZnS nanocrystals are ordered by the filamentous bacteriophage [62]. Because target-specific peptides can be displayed on the surface of bacteriophage or flagella, such a liquid crystalline 3-D construct could demonstrate target specificity, such that its application in biosensing might be feasible.

Recently, the fiber-like morphology and ability to display cell-recognizing peptides on the surface of genetically engineered bacteriophage have been exploited to generate film-like scaffolds for tissue regeneration [69, 81]. Merzlyak *et al.* reported the genetic modification of the major coat protein of filamentous bacteriophage with RGD (arginine-glycine-aspartic acid; a tripeptide that binds to the integrin receptors present on all mammalian cells) and IKVAV peptides, to produce a fibrous scaffold for cell growth [81]. The successful growth of neural progenitor cells on these scaffolds, and the fruitful implantation of such scaffolds

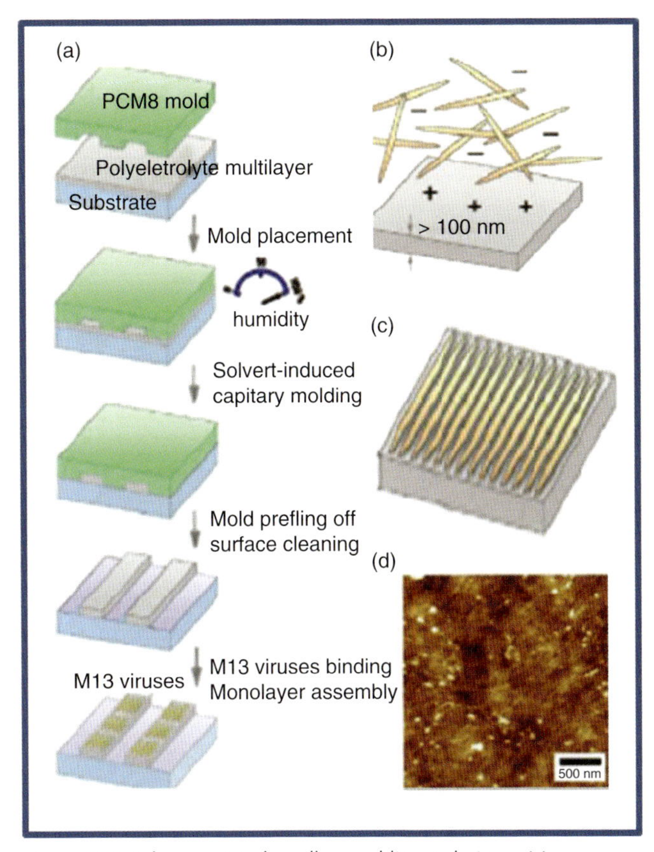

Figure 7.12 Solvent-assisted capillary molding technique. (a) Schematic procedure of M13 viruses selectively bound and ordered on a polyelectrolyte multilayer (PEM); (b) Electrostatic interaction of bacteriophage with cationic moiety in polymer; (c) Ordered viral assembly on mobility-rich PEM surface; (d) AFM image showing adsorption of virus onto the PEM surface. Images reproduced with permission from Ref. [76].

in an *in vivo* system, might lead to the creation of a new approach towards treating spinal cord injuries in animals. Likewise, Rong *et al.* have recently prepared similar phage films with RGD sequences on the filamentous bacteriophage surface to culture NIH-3T3 fibroblasts [69]. It would appear, therefore, that genetically engineered biomolecules might shortly be used for tissue-regeneration applications.

Most recently, the present authors' group employed an LBL assembly method to construct multilayered genetically engineered nanocomposite films [82], whereby films were prepared from alternate layers of genetically engineered M13 bacteriophage (displayed with four arginine residues) and negatively charged 8 nm gold nanoparticles. This genetically engineered nanocomposite was also demonstrated successfully for use as a humidity sensor [82]. It would appear that these genetically engineered biomacromolecules may be used to form films with nanofibrillar structures. Hence, genetically engineered nanocomposite films show great potential, on the basis that they incorporate the unique physical properties of the nanocrystalline phase with the bioselective efficiency of the engineered biomolecule.

7.4 Applications of Genetically Engineered Nanocomposites

7.4.1 Engineered Nanocomposites for Device Applications

Electronics is one branch of science where new innovations are constantly sought, with the society of today demanding products with increased memory, operating speed and efficiency, yet in smaller and more compact forms. Examples range from smaller cell phones with greater functionality, to lighter and more powerful lap tops, to highly sophisticated medical equipment. Recent progress involving the construction of engineered nanocomposites has attracted much attention to the field of nanobiotechnology, notably with regards to their potential use in batteries and electronics.

Whilst previously, electrical biochip technology was used to detect and quantify viruses in human infections [83], today viruses are used as scaffolds or templates for the assembly of nanomaterials. Moreover, this technology is finding numerous potential applications, including that of nanoelectronics. Although it has been shown previously that biomolecules such as oligonucleotides (DNA) [84] and peptide sequences [85] can be functionalized and incorporated into transistors, memory devices and electrodes for batteries [86], only recently has the importance of bacteriophage in electronics been realized.

Metallic *lithium* is vital for the manufacture of the small but highly efficient batteries that, today, are referred to a "lithium ion" batteries. It is strongly believed that nanomaterials increase the electrochemical properties of Li ion batteries in comparison to bulk materials [87]. The use of filamentous M13 bacteriophage in the fabrication of battery devices was first demonstrated by the group of Belcher, when phage were configured to prepare Li ion battery electrodes with improved storage and rate capabilities [1, 88]. Subsequently, the inherent quality of viruses was used to assemble them into the multiple length scales required for use in battery production.

For these studies, a dual display system was used to produce engineered M13 bacteriophage, the major coat proteins (pVIII) of which were displayed with tetra-glutamate that would interact with silver nanoparticles and serve as a template on which anhydrous iron phosphate (a-$FePO_4$) could form nanowires. The minor coat proteins (pIII) of the bacteriophage were displayed with single wall carbon nano-tube (SWCNT) -binding peptides through π-stacking interactions. The resultant

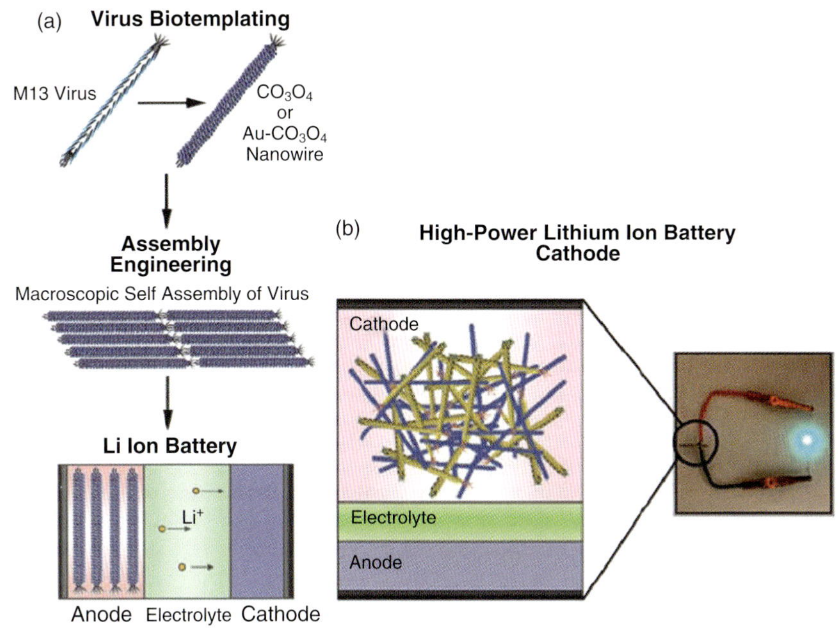

Figure 7.13 Genetically engineered nanocomposites for electronics. (a) Schematic illustration of genetically engineered bacteriophage-enabled synthesis and assembly of nanowires as electrodes for lithium ion batteries; (b) Photograph of a lithium ion battery constructed from a virus, used to power a green LED. Images reproduced with permission from Refs [88] and [1].

genetically engineered virus was used as a scaffold for the synthesis and assembly of materials for high-power batteries [1]. The main aim of these studies was to increase the cathode electronic conductivity, by increasing the electrical contact between neighboring active materials. A photograph of a virus-enabled Li ion battery powering a green light-emitting diode (LED), is shown in Figure 7.13b.

In a previous investigation, the same group used engineered bacteriophage as a scaffold when designing cobalt oxide (Co_3O_4) nanowire electrodes [88]. To accomplish this, bi-functional viruses were used which had been genetically modified with gold-binding peptides on the minor coat proteins (pIII), and tetra-glutamate residues on the major coat proteins, which served as excellent templates for nanowire formation. The resultant virus, incorporating the hybrid Au–Co_3O_4, exhibited excellent electrochemical properties. These cobalt oxide-based electrodes also exhibited a threefold better performance than currently available carbon-based anodes. The capacity of the Au–CO_3O_4 viral hybrid was 30% enhanced when compared to a CO_3O_4 nanowire alone, formed on the M13 bacteriophage. A schematic representation of the genetically engineered bacteriophage-enabled synthesis, and the assembly of nanowires used to produce electrodes in lithium ion batteries, are shown in Figure 7.13a. These developments suggest that very soon, within the electronics industry, engineered biotemplates might well serve as excellent scaffolds in the design and assembly of nanoscale components with improved performances.

7.4.2
Engineered Nanocomposites for Medical Applications

Since the first identification of monoclonal antibodies, tremendous progress has been made in the development of antibody therapies against cancer, the main limitations having been their large size and possible adverse immune response [5]. More recently, however, studies using phage nanobiotechnology have shown that phage might serve as ideal substitutes for antibodies in the near future. For example, despite previous studies suggesting that it would be impossible to detect the live formation of amyloid plaques in affected human brains, significant contributions from Solomon *et al.* [89] have indicated that filamentous bacteriophage could serve as an antibody probe to monitor amyloid plaque formation in Alzheimer's disease. Most importantly, these studies represent a clear benchmark, in that engineered bacteriophage can serve as an efficient and nontoxic delivery vector to the brain. Evolving from these findings, several studies have demonstrated the importance of bacteriophage in medical applications [90, 91], including the development of phage-based therapeutics to attenuate the effects of cocaine in the central nervous system [92]. Indeed, the results of the latter study suggest that phage-based treatments might, in time, help to eliminate many syndromes associated with drug abuse.

The inherent quality of phage is enhanced when conjugated with nanoparticles to form engineered nanocomposites. Using this concept, Hagit Bar *et al.* unveiled some broader medical applications using engineered nanocomposites, a primary benefit being the demonstration of a novel anti-cancer therapy that incorporated a genetically engineered phage–nanoparticle complex. In this case, the major coat proteins of filamentous bacteriophage were chemically conjugated with a drug via cleavable bonds, while the minor coat proteins (pIII) were modified to exhibit target cell specificity by conferring an antibody molecule [93] that would ensure that the drug was delivered to the target site.

Ultimately, doxorubicin was coupled to the major coat proteins, and a targeting moiety to the minor coat proteins, that could bind selectively to either the epidermal growth factor receptor (EGFR) or ErbB2 receptors. This phage-conjugated nanoconstruct was successful in delivering doxorubicin to the tumor site. A cartoon showing how the drug was coupled to the major coat proteins, and the antibody to the minor coat, is shown in Figure 7.14. Recently, it has also been shown that drug-loaded liposomes on a genetically engineered phage and flagellar surface can serve as novel nanocarriers for targeted drug delivery [64, 65]. Taken together, the results of these studies suggest that engineered biotemplates may serve as excellent scaffolds in the design and assembly of nanoscale components used for improved drug delivery vehicles in medical applications.

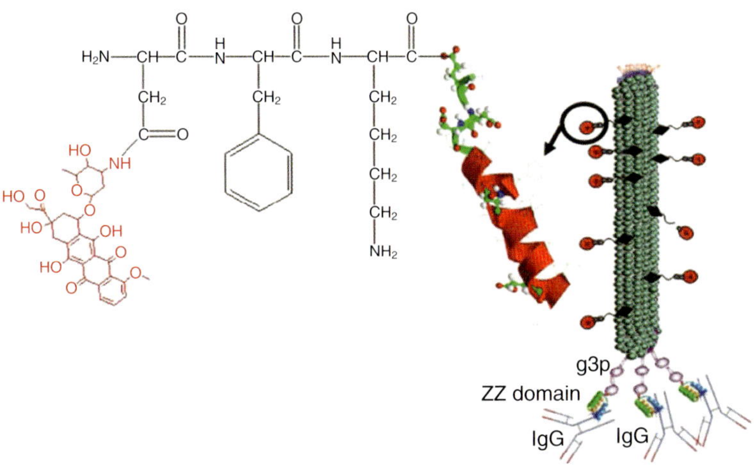

Figure 7.14 Genetically engineered nanocomposites for drug delivery. For medical applications, the drug is conjugated to major coat proteins, while the antibody is coupled to minor coat proteins. Reproduced with permission from Ref. [93].

7.4.3
Engineered Nanoconstructs for Sensing Applications

With advancements in nanobiotechnology, the demand for viral assemblies to be used as platforms for sensors has increased. Sensors are analytical devices that provide quantitative and/or semi-quantitative information regarding the analyte [94]. In general, sensors are composed of three main components – the receptor, the transducer, and the display (measurable output). Recent developments in science have greatly facilitated the availability of biosensors containing a biological component as the receptor, closely associated with the transducer that converts biological interactions into measurable outputs. Several biomolecules, including oligonucleotides (DNA) [95], proteins [96], bacteria [97] and, more recently, viruses [82, 98], have been used successfully as excellent receptors for developing biosensors. In addition to their high chemical and thermal stability, high selectivity and sensitivity, bacteriophage – when enabled by a phage display technique – represent ideal candidates for sensing applications.

Earlier, Yoo *et al.* [76] showed that self-assembled patterns of viruses on polyelectrolyte multilayers could serve as a platform for the development of biological sensors used in protein detection. To verify this concept, biotin-labeled antibodies were added to a self-assembled virus film, with the binding of biotin to the virus surface being confirmed by AFM. Subsequently, fluorescent dye-labeled streptavidin confirmed specific binding to the biotin-coated layer. This virus-based detection system is illustrated schematically in Figure 7.15a.

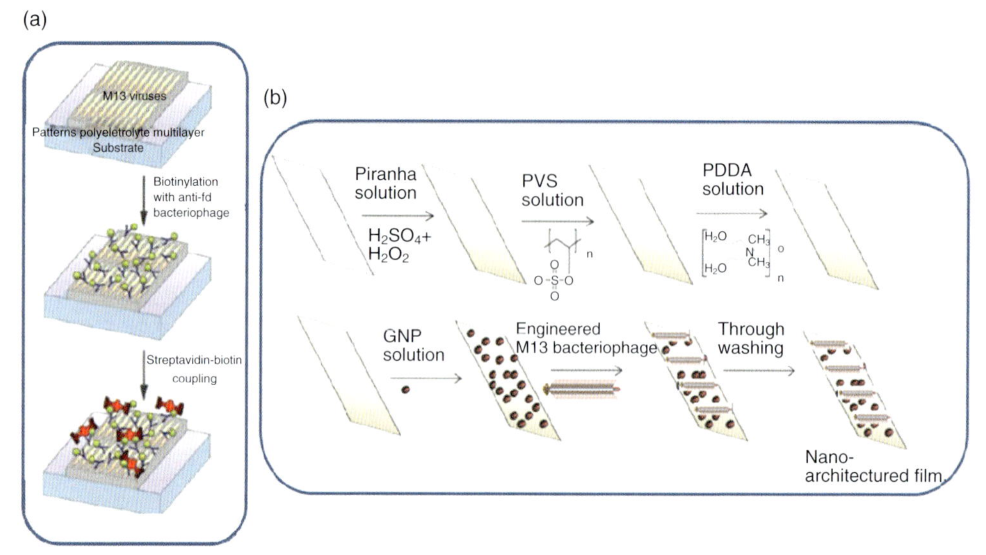

Figure 7.15 Schematics showing the development of (a) a virus-based biosensor and (b) a virus-based humidity sensor. Image (a) reproduced with permission from Ref. [76].

Recently, the present authors' group reported the LBL assembly of genetically engineered bacteriophage and AuNPs to serve as a humidity sensor [82]. For this, a method was developed to fabricate a nanocomposite film. The filamentous bacteriophage was genetically engineered to display tetra-arginine residues $(Arg)_4$ on the side wall, so as to create a positively charged outer surface that could interact directly with negatively charged AuNPs. Because of the localized surface plasmon resonance (SPR) exhibited by the AuNPs, it was possible to obtain an SPR spectrum that was tunable to changes in the environmental humidity. The design of engineered nanocomposites used for humidity sensing is shown in Figure 7.15b.

These multilayered nanoarchitectured films were prepared on a quartz slide that had been precoated with bilayers of anionic polyvinyl sulfonate potassium salt (PVS) and cationic poly(diallyldimethylammonium chloride) (PDDA), collectively referred to as PVS/PDDA. The anionic AuNPs and cationic M13 viruses were adsorbed onto the films by immersing the slide in an anionic AuNP solution and the tetraArg–M13 solution alternately, as shown in Figure 7.15b. The process was repeated for 30 min, with 10 min rinses in distilled water between immersions. The performance of the sensor was observed within a range of 19% to 86% relative humidity (RH) at 25 °C. These results showed clearly that, as the RH decreased from 86% to 19%, the λ_{max} of the SPR spectra was shifted towards the right, and the response was linear throughout the observed range. Based on these findings, it was suggested that genetically engineered biomacromolecules could also serve as a template for designing detection devices and humidity sensors with improved sensing abilities compared to traditional sensors.

7.5 Summary and Outlook

In this chapter, an overview has been provided of the processes used to construct genetically engineered nanocomposites, using bacteriophage and flagella as examples. The incorporation of genetically modifiable biomacromolecules into nanotechnology will enable the production of novel nanocomposites. In addition, peptides with target-specific properties can be displayed on both bacteriophage and flagella. Hence, any construct bearing the bacteriophage or flagella may possess the target-specificity that is highly desirable in targeted disease diagnosis and treatment, as well as biosensing. Both, bacteriophage and flagella have been used successfully to template the formation and assembly of nanomaterials, with the resultant nanocomposites finding potential applications in many fields, including electronics, medicine, and sensing.

Although a host of novel nanoassembled structures are currently undergoing investigation, where the bacteriophage or flagella serve as active components based on their genetically modifiable surface chemistry, self-assembly, nanoscale size and error-free production, it is also clear that other biological macromolecules and structures – including microtubules, other viruses, DNA, and bacteria – may be employed in the creation of nanocomposites with roles in medicine, electronics and sensing [94, 99–102]. Bacteriophage and flagella are unique, however, compared to other biological structures, in that target-specific peptides can be selected from a multibillion random peptide library by using phage display and flagellar display techniques. Clearly, genetically engineered nanocomposites will continue to show major advances in many fields of research, indicating a very high potential for future applications.

Today, nanocomposites using genetically modifiable substrates such as bacteriophage and flagella form the basis of a revolutionary and exciting field, with possible applications in nanobiotechnology-related industries. Yet, further progress will be essential to develop and apply genetically engineered nanocomposites in the future. First, functional organic polymers and genetically modifiable biomacromolecules must be integrated to form novel functional constructs. Second, additional studies must be conducted to develop genetically engineered nanocomposites into nanomedicines, bioimaging contrast enhancement agents, and biosensors capable of recognizing specific targets such as cells, tissues, and toxins. Finally – and perhaps the most daunting challenge – genetically engineered biomacromolecules can be employed in the creation of practical electronic devices. But, in order to meet these demands, research groups from diverse areas such as the materials sciences, cancer biology, biochemistry and pure biotechnology must combine their knowledge and experience if humankind is to benefit from this vast and promising field.

Acknowledgments

The authors are grateful for financial support from The National Science Foundation, The National Institutes of Health, The Oklahoma Center for the Advancement of Science & Technology, and The Department of Defense Breast Cancer Research Program.

References

1 Lee, Y.J., Yi, H., Kim, W.-J., Kang, K., Yun, D.S., Strano, M.S., Ceder, G. and Belcher, A.M. (2009) Fabricating genetically engineered high-power lithium-ion batteries using multiple virus genes. *Science*, **324**, 1051–5.

2 Feldhaus, M.J., Siegel, R.W., Opresko, L.K., Coleman, J.R., Feldhaus, J.M., Yeung, Y.A., Cochran, J.R., Heinzelman, P., Colby, D., Swers, J., Graff, C., Wiley, H.S. and Wittrup, K.D. (2003) Flow-cytometric isolation human antibodies from a nonimmune *Saccharomyces cerevisiae* surface display library. *Nature Biotechnology*, **21**, 163–70.

3 Wang, Y., Rubtsov, A., Heiser, R., White, J., Crawford, F., Marrack, P. and Kappler, A.J.W. (2005) Using a baculovirus display library to identify MHC class I mimotopes. *Proceedings of the National Academy of Sciences of the United States of America*, **102**, 2476–81.

4 McCormick, A.A. and Palmer, K.E. (2008) Genetically engineered tobacco virus as nanoparticle vaccines. *Expert Review of Vaccines*, **7**, 33–41.

5 Petrenko, V.A. (2008) Evolution of phage display: from bioactive peptides to bioselective nanomaterials. *Expert Opinion on Drug Delivery*, **5**, 825–36.

6 Lee, T.J., Schwartz, C. and Guo, P.X. (2009) Construction of bacteriophage Phi29 DNA packaging motor and its applications in nanotechnology and therapy. *Annals of Biomedical Engineering*, **37**, 2064–81.

7 Khataee, H.R. and Khataee, A.R. (2009) Advances in F0F1-ATP synthase biological protein nanomotor: from mechanism and strategies to potential applications. *Nano*, **4**, 55–67.

8 Arap, M.A. (2005) Phage display technology–applications and innovations. *Genetics and Molecular Biology*, **28**, 1–9.

9 Yonekura, K., Maki-Yonekura, S. and Namba, K. (2003) Complete atomic model of the bacterial flagellar filament by electron cryomicroscopy. *Nature*, **424**, 643–50.

10 Petrenko, V.A., Smith, G.P., Gong, X. and Quinn, T. (1996) A library of organic landscapes on filamentous phage. *Protein Engineering*, **9**, 797–801.

11 Armstrong, N., Adey, N.B., McConnell, S.J. and Kay, B.K. (1996) *Phage Display of Peptides and Proteins: A Laboratory Manual* (eds B.K. Kay, J. Winter and J. McCafferty), Academic Press, San Diego [Online].

12 Yonekura, K., Maki-Yonekura, S. and Namba, K. (2002) Growth mechanism of the bacterial flagellar filament. *Research in Microbiology*, **153**, 191–7.

13 Kumara, M.T., Srividya, N., Muralidharan, S. and Tripp, B.C. (2006) Bioengineered flagella protein nanotubes with cysteine loops: self-assembly and manipulation in an optical trap. *Nano Letters*, **6**, 2121–9.

14 Beatson, S.A., Minamino, T. and Pallen, M.J. (2006) Variation in bacterial flagellins: from sequence to structure. *Trends in Microbiology*, **14**, 151–5.

15 Westerlund-Wikstrom, B., Tanskanen, J., Virkola, R., Hacker, J., Lindberg, M., Skurnik, M. and Korhonen, T.K. (1997) Functional expression of adhesive peptides as fusions to *Escherichia coli* flagellin. *Protein Engineering*, **10**, 1319–26.

16 Majander, K., Korhonen, T.K. and Westerlund-Wikstrom, B. (2005)

Simultaneous display of multiple foreign peptides in the FliD capping and FliC filament proteins of the *Escherichia coli* flagellum. *Applied and Environmental Microbiology*, **71**, 4263–8.

17 Lu, Z., Murray, K.S., Van Cleave, V., LaVallie, E.R., Stahl, M.L. and McCoy, J.M. (1995) Expression of thioredoxin random peptide libraries on the *Escherichia coli* cell surface as functional fusions to flagellin: a system designed for exploring protein–protein interactions. *Biotechnology*, **13**, 366–72.

18 Lu, Z., Tripp, B.C. and McCoy, J.M. (1998) Displaying libraries of conformationally constrained peptides on the surface of *Escherichia coli* as flagellin fusions. *Methods in Molecular Biology*, **87**, 265–80.

19 Tripp, B.C., Lu, Z., Bourque, K., Sookdeo, H. and McCoy, J.M. (2001) Investigation of the "switch-epitope" concept with random peptide libraries displayed as thioredoxin loop fusions. *Protein Engineering*, **14**, 367–77.

20 Lu, Z., LaVallie, E.R. and McCoy, J.M. (2003) Using bio-panning of FLITRX peptide libraries displayed on *E. coli* cell surface to study protein–protein interactions. *Methods in Molecular Biology*, **205**, 267–80.

21 Newton, S.M.C., Jacob, C.O. and Stocker, B.A.D. (1989) Immune response to cholera toxin epitope inserted in *Salmonella* flagellin. *Science*, **244**, 70–2.

22 Westerlund-Wikström, B. (2000) Peptide display on bacterial flagella: principles and applications. *International Journal of Medical Microbiology*, **290**, 223–30.

23 Liu, H., Li, Y., Xiao, S., Gan, H., Jiu, T., Li, H., Jiang, L., Zhu, D., Yu, D., Xiang, B. and Chen, Y. (2003) Synthesis of organic one-dimensional nanomaterials by solid-phase reaction. *Journal of the American Chemical Society*, **125**, 10794–5.

24 Iijima, S. (1991) Helical microtubules of graphitic carbon. *Nature*, **354**, 56–8.

25 Ma, Y., Qi, L., Shen, W. and Ma, J. (2005) Selective synthesis of single-crystalline selenium nanobelts and nanowires in micellar solutions of nonionic surfactants. *Langmuir*, **21**, 6161–4.

26 Chen, Y., Li, C.P., Chen, H. and Chen, Y. (2006) One-dimensional nanomaterials synthesized using high-energy ball milling and annealing process. *Science and Technology of Advanced Materials*, **7**, 839–46.

27 Hu, P., Liu, Y., Fu, L., Cao, L. and Zhu, D. (2004) Self-assembled growth of ZnS nanobelt networks. *Journal of Physical Chemistry B*, **108**, 936–8.

28 Huczko, A. (2000) Template-based synthesis of nanomaterials. *Applied Physics A*, **70**, 365–76.

29 Luo, Y. (2008) Versatile hydrothermal synthesis of one-dimensional composite structures. *Solid State Communications*, **148**, 516–20.

30 Yu, Q., Ou, H.D., Song, R.Q. and Xu, A.W. (2006) The effect of polyacrylamide on the crystallization of calcium carbonate: synthesis of aragonite single-crystal nanorods and hollow vatarite hexagons. *Journal of Crystal Growth*, **286**, 178–83.

31 Murphy, C.J., Gole, A.M., Hunyadi, S.E. and Orendorff, C.J. (2006) One-dimensional colloidal gold and silver nanostructures. *Inorganic Chemistry*, **45**, 7544–54.

32 Wang, F., Li, D. and Mao, C. (2008) Genetically modifiable flagella as templates for silica fibers: from hybrid nanotubes to 1D periodic nanohole arrays. *Advanced Functional Materials*, **18(C)**, 1–7.

33 Kumara, M.T., Tripp, B.C. and Muralidharan, S. (2007) Self-assembly of metal nanoparticles and nanotubes on bioengineered flagella scaffolds. *Chemistry of Materials*, **19**, 2056–64.

34 Mao, C., Flynn, C.E., Hayhurst, A., Sweeney, R., Qi, J., Georgiou, G., Iverson, B. and Belcher, A.M. (2003) Viral assembly of oriented quantum dot nanowires. *Proceedings of the National Academy of Sciences of the United States of America*, **100**, 6946–51.

35 Peelle, B.R., Krauland, E.M., Wittrup, K.D. and Belcher, A.M. (2005) Design criteria for engineering inorganic material-specific peptides. *Langmuir*, **21**, 6929–33.

36 Seker, U.O.S., Wilson, B., Sahin, D., Tamerler, C. and Sarikaya, M. (2009)

Quantitative affinity of genetically engineered repeating polypeptides to inorganic surfaces. *Biomacromolecules*, **10**, 250–7.

37 Armstrong, N., Adey, N.B., Connell, S.J.M. and Kay, B.K. (1996) *Phage Display of Peptides and Proteins: A Laboratory Manual*, Academic Press, San Diego.

38 So, C.R., Kulp, J.L., Oren, E.E., Zareie, H., Tamerler, C., Evans, J.S. and Sarikaya, M. (2009) Molecular recognition and supramolecular self-assembly of a genetically engineered gold binding peptide on Au{111}. *ACS Nano*, **3**, 1525–31.

39 Sarikaya, M., Tamerler, C., Jen, A.K.Y., Schulten, K. and Baneyx, F. (2003) Molecular biomimetics: nanotechnology through biology. *Nature Materials*, **2**, 577–85.

40 Huang, Y., Chiang, C.-Y., Lee, S.K., Gao, Y., Hu, E.L., Yoreo, A.D. and Belcher, A.M. (2005) Programmable assembly of nanoarchitectures using genetically engineered viruses. *Nano Letters*, **5**, 1429–34.

41 Lee, S.-K., Yun, D.S. and Belcher, A.M. (2006) Cobalt ion mediated self-assembly of genetically engineered bacteriophage for biomimetic Co–Pt hybrid material. *Biomacromolecules*, **7**, 14–17.

42 Wu, Y., Fan, R. and Yang, P. (2002) Block-by-Block growth of single-crystalline Si/SiGe superlattice nanowires. *Nano Letters*, **2**, 83–6.

43 Kumara, M.T., Brian, T.C. and Muralidharan, S. (2007) Layer by layer assembly of bioengineered flagella protein nanotubes. *Biomacromolecules*, **8**, 3718–22.

44 Ma, H., Tarr, J., DeCoster, M.A., McNamara, J., Caruntu, D., Chen, J.F., O'Connor, C.J. and Zhou, W.L. (2009) Synthesis of magnetic porous hollow silica nanotubes for drug delivery. *Journal of Applied Physics*, **105**, 1–3.

45 Son, S.J., Bai, X. and Lee, S.B. (2007) Inorganic hollow nanoparticles and nanotubes in nanomedicine: drug/gene delivery applications. *Drug Discovery Today*, **12**, 650–6.

46 Reiss, B.D., Mao, C., Solis, D.J., Ryan, K.S., Thomson, T. and Belcher, A.M. (2004) Biological routes to metal alloy ferromagnetic nanostructures. *Nano Letters*, **4**, 1127–32.

47 Flynn, C.E., Mao, C., Hayhurst, A., Williams, J.L., Georgiou, G., Iverson, B. and Belcher, A.M. (2003) Synthesis and organization of nanoscale II–VI semiconductor materials using evolved peptide specificity and viral capsid assembly. *Journal of Materials Chemistry*, **13**, 2414–21.

48 Zuo, R., Örnek, D. and Wood, T.K. (2005) Aluminum- and mild steel-binding peptides from phage display. *Applied Microbiology and Biotechnology*, **68**, 505–9.

49 Modali, S., Abbineni, G., Jayanna, P., Petrenko, V. and Mao, C. (2008) Evolutionary selection of bone mineral hydroxyapatite binding peptide using landscape phage library, in *Nanotechnology 2008: Life Sciences, Medicine & Bio Materials*, CRC press, Houston, pp. 465–7.

50 Tamerler, C. and Sarikaya, M. (2009) Genetically designed peptide-based molecular materials. *ACS Nano*, **3**, 1606–15.

51 Willis, B., Eubanks, L.M., Wood, M.R., Janda, K.D., Dickerson, T.J. and Lerner, A.R.A. (2008) Biologically templated organic polymers with nanoscale order. *Proceedings of the National Academy of Sciences of the United States of America*, **105**, 1416–19.

52 Marvin, D., Welsh, L., Symmons, M., Scott, W. and Straus, S. (2006) Molecular structure of fd (f1, M13) filamentous bacteriophage refined with respect to X-ray fibre diffraction and solid-state NMR data supports specific models of phage assembly at the bacterial membrane. *Journal of Molecular Biology*, **355**, 294–309.

53 Sanghvi, A.B., Miller, K.P.-H., Belcher, A.M. and Schmidt, C.E. (2005) Biomaterials functionalization using a novel peptide that selectively binds to a conducting polymer. *Nature Materials*, **4**, 496–502.

54 Konturri, K., Pentti, P. and Sundholm, G. (1998) Polypyrrole as a model membrane for drug delivery. *Journal of Electroanalytical Chemistry*, **453**, 231–8.

55 Schmidt, C.E., Shastri, V.R., Vacanti, J.P. and Langer, R. (1997) Stimulation of neurite outgrowth using an electrically conducting polymer. *Proceedings of the National Academy of Sciences of the United States of America*, **94**, 8948–53.

56 Valentini, R.F., Vargo, T.G., Gardella, J.A.J. and Aebischer, P. (1992) Electrically conductive polymeric substrates enhance nerve fibre outgrowth *in vitro*. *Biomaterials*, **13**, 183–90.

57 Vidal, J.C., Garcia, E. and Castillo, J.R. (1999) In situ preparation of a cholesterol biosensor: entrapment of cholesterol oxidase in an overoxidized polypyrrole film electrodeposited in a flow system: determination of total cholesterol in serum. *Analytica Chimica Acta*, **385**, 213–22.

58 Cui, X., Hetke, J.F., Wiler, J.A., Anderson, D.J. and Martin, D.C. (2001) Electrochemical deposition and characterization of conducting polymer polypyrrole/PSS on multichannel neural probes. *Sensors and Actuators A*, **93**, 8–18.

59 Diaz, A.F., Castillo, J.A., Logan, J.A. and Lee, W.Y. (1981) Electrochemistry of conductive polypyrrole films. *Journal of Electroanalytical Chemistry*, **129**, 115–32.

60 Tamm, J., Hallik, A., Alumaa, A. and Sammelselg, V. (1997) Electrochemical properties of polypyrrole/sulphate films. *Electrochimica Acta*, **42**, 2929–34.

61 Whaley, S.R., English, D.S., Hu, E.L., Barbara, P.F. and Belcher, A.M. (2000) Selection of peptides with semiconductor binding specificity for directed nanocrystal assembly. *Nature*, **405**, 665–8.

62 Lee, S.-W., Mao, C., Flynn, C.E. and Belcher, A.M. (2002) Ordering of quantum dots using genetically engineered viruses. *Science*, **296**, 892–5.

63 Kehoe, J.W. and Kay, B.K. (2005) Filamentous phage display in the new millennium. *Chemical Reviews*, **104**, 4056–72.

64 Ngweniform, P., Abbineni, G., Cao, B. and Mao, C. (2009) Self assembly of drug-loaded liposomes on genetically engineered target-recognizing M13 phage: a novel nanocarrier for targeted drug delivery. *Small*, **5**, 1963–9.

65 Ngweniform, P., Li, D. and Mao, C. (2009) Self-assembly of drug-loaded liposomes on genetically engineered protein nanotubes: a potential anti-cancer drug delivery vector. *Soft Materials*, **5**, 954–6.

66 Visona, A., Cuomo, V., Zanetti, G., Lisiani, L., Pagnan, A. and Jori, G. (1989) Accumulation of zinc-phthalocyanine (Zn-Pc) in the aorta of atherosclerotic rabbits. *Lasers in Medical Science*, **4**, 167–70.

67 Kumara, M.T., Muralidharan, S. and Tripp, B.C. (2007) Generation and characterization of inorganic and organic nanotubes on bioengineered flagella of mesophilic bacteria. *Journal of Nanoscience and Nanotechnology*, **7**, 2260–72.

68 Bruckman, M.A., Niu, Z., Li, S., Lee, L.A., Varazo, K., Nelson, T.L., Lavigne, J.J. and Wang, Q. (2007) Development of nanobiocomposite fibers by controlled assembly of rod-like tobacco mosaic virus. *Nanobiotechnology*, **3**, 31–9.

69 Rong, J., Lee, L.A., Li, K., Harp, B., Mello, C.M., Niu, Z. and Wang, Q. (2008) Oriented cell growth on self-assembled bacteriophage M13 thin films. *Chemical Communications*, 5185–7.

70 Lu, Y. and Liu, J. (2007) Smart nanomaterials inspired by biology: dynamic assembly of error-free nanomaterials in response to multiple chemical and biological stimuli. *Accounts of Chemical Research*, **40**, 315–23.

71 Petrova, H., Lin, C.-H., Hu, M., Chen, J., Siekkinen, A.R., Xia, Y., Sader, J.E. and Hartland, G.V. (2007) Vibrational response of Au–Ag nanoboxes and nanocages to ultrafast laser-Induced heating. *Nano Letters*, **7**, 1059–63.

72 Allred, D.B., Cheng, A., Sarikaya, M., Baneyx, F. and Schwartz, D.T. (2008) Three-dimensional architecture of inorganic nanoarrays electrodeposited through a surface-layer protein mask. *Nano Letters*, **8**, 1434–8.

73 Ling, X.Y., Phang, I.Y., Reinhoudt, D.N., Vancso, G.J. and Huskens, J. (2008) Supramolecular layer-by-layer assembly of 3D multicomponent nanostructures via multivalent molecular recognition. *International Journal of Molecular Sciences*, **9**, 486–97.

74 DeLongchamp, D.M., Kastantin, M. and Hammond, P.T. (2003) High-contrast electrochromism from layer-by-layer polymer films. *Chemistry of Materials*, **15**, 1575–86.

75 Lowman, G. and Hammond, P. (2005) Solid-state dye-sensitized solar cells combining a porous TiO_2 film and a layer-by-layer composite electrolyte. *Small*, **1**, 1070–3.

76 Yoo, P.J., Nam, K.T., Belcher, A.M. and Hammond, P.T. (2008) Solvent-assisted patterning of polyelectrolyte multilayers and selective deposition of virus assemblies. *Nano Letters*, **8**, 1081–9.

77 Yoo, P.J., Nam, K.T., Qi, J., Lee, S.-K., Park, J., Belcher, A.M. and Hammond, P.T. (2006) Spontaneous assembly of viruses on multilayered polymer surfaces. *Nature Materials*, **5**, 234–40.

78 Kralchevsky, P.A. and Denkov, N.D. (2001) Capillary forces and structuring in layers of colloid particles. *Current Opinion in Colloid and Interface Science*, **6**, 383–401.

79 Dutta, J. and Hofmann, H. (2003) Self organization of colloidal nanoparticles, in *Encyclopedia of Nanoscience and Nanotechnology*, vol. **X** (ed. H.S. Nalwa), American Scientific Publishers, Stevenson Ranch, CA, pp. 1–23.

80 Park, J. and Hammond, P.T. (2004) Multilayer transfer printing for polyelectrolyte multilayer patterning: direct transfer of layer-by-layer assembled micropatterned thin films. *Advanced Materials*, **16**, 520–5.

81 Merzlyak, A., Indrakanti, S. and Lee, S.-W. (2009) Genetically engineered nanofiber-like viruses for tissue regenerating materials. *Nano Letters*, **9**, 846–52.

82 Liu, A.H., Abbineni, G. and Mao, C.B. (2009) Nanocomposite films assembled from genetically engineered filamentous viruses and gold nanoparticles: nanoarchitecture- and humidity-tunable surface plasmon resonance spectra. *Advanced Materials*, **21** (9), 1001–5.

83 Łos, M., Łos, J.M., Blohm, L., Spillner, E., Grunwald, T., Albers, J., Hintsche, R. and Wegrzyn, G. (2005) Rapid detection of viruses using electrical biochips and anti-virion sera. *Letters in Applied Microbiology*, **40**, 479–85.

84 Seeman, N. and Lukeman, P. (2005) Nucleic acid nanostructures: bottom-up control of geometry on the nanoscale. *Reports on Progress in Physics*, **68**, 237–70.

85 Zhang, S.G. (2003) Fabrication of novel biomaterials through molecular self-assembly. *Nature Biotechnology*, **21**, 1171–8.

86 Nam, K.T., Wartena, R., Yoo, P.J., Liau, F.W., Lee, Y.J., Chiang, Y.-M., Hammond, P.T. and Belcher, A.M. (2008) Stamped microbattery electrodes based on self-assembled M13 viruses. *Proceedings of the National Academy of Sciences of the United States of America*, **105**, 17227–31.

87 Arico, A.S., Bruce, P., Scrosati, B., Tarascon, J.-M. and Schalkwijk, W.V. (2005) Nanostructured materials for advanced energy conversion and storage devices. *Nature Materials*, **4**, 366–77.

88 Nam, K.T., Kim, D.-W., Yoo, P.J., Chiang, C.-Y., Meethong, N., Hammond, P.T., Chiang, Y.-M. and Belcher, A.M. (2006) Virus-enabled synthesis and assembly of nanowires for lithium ion battery electrodes. *Science*, **312**, 885–8.

89 Frenkel, D. and Solomon, B. (2002) Filamentous phage as vector-mediated antibody delivery to the brain. *Proceedings of the National Academy of Sciences of the United States of America*, **99**, 5675–9.

90 Sweeney, R.Y., Park, E.Y., Iverson, B.L. and Georgiou, G. (2006) Assembly of multimeric phage nanostructures through leucine zipper interactions. *Biotechnology and Bioengineering*, **95**, 539–45.

91 Petty, N.K., Evans, T.J., Fineran, P.C. and Salmond, G.P. (2007) Biotechnological exploitation of bacteriophage research. *Trends in Biotechnology*, **25**, 7–15.

92 Carrera, M.R.A., Kaufmann, G.F., Mee, J.M., Meijler, M.M., Koob, G.F. and Janda, K.D. (2004) Treating cocaine addiction with viruses. *Proceedings of the National Academy of Sciences of the United States of America*, **101**, 10416–21.

93 Bar, H., Yacoby, I. and Benhar, I. (2008) Killing cancer cells by targeted drug-carrying phage nanomedicines. *BMC Biotechnology*, **8**, 1–14.

94 Mao, C., Liu, A. and Cao, B. (2009) Virus-based chemical and biological sensing. *Angewandte Chemie, International Edition*, **48**, 2–23.

95 Gagnon, Z., Senapati, S., Gordon, J. and Chang, H.C. (2008) Dielectrophoretic detection and quantification of hybridized DNA molecules on nano-genetic particles. *Electrophoresis*, **29** (24), 4808–12.

96 Diercksa, A.H., Ozinskya, A., Hansen, C.L., Spotts, J.M., Rodrigueza, D.J. and Aderema, A. (2009) A microfluidic device for multiplexed protein detection in nano-liter volumes. *Analytical Biochemistry*, **386**, 30–5.

97 Panda, B.R., Singh, A.K., Ramesh, A. and Chattopadhyay, A. (2008) Rapid estimation of bacteria by a fluorescent gold nanoparticle-polythiophene composite. *Langmuir*, **24**, 11995–2000.

98 Souza, G.R., Christianson, D.R., Staquicini, F.I., Ozawa, M.G., Snyder, E.Y., Sidman, R.L., Miller, J.H., Arap, W. and Pasqualini, R. (2006) Networks of gold nanoparticles and bacteriophage as biological sensors and cell-targeting agents. *Proceedings of the National Academy of Sciences of the United States of America*, **103**, 1215–20.

99 Fonoberov, V.A. and Balandin, A.A. (2005) Phonon confinement effects in hybrid virus-inorganic nanotubes for nanoelectronic applications. *Nano Letters*, **5**, 1920–3.

100 Corash, L. and Hanson, C. (1992) Photoinactivation of viruses and cells for medical applications. *Blood Cells*, **18**, 3–5.

101 Baker, N.A., Sept, D., Joseph, S., Holst, M.J. and McCammon, J.A. (2001) Electrostatics of nanosystems: application to microtubules and the ribosome. *Proceedings of the National Academy of Sciences of the United States of America*, **98**, 10037–41.

102 Pancrazio, J.J., Whelan, J.P., Borkholder, D.A., Ma, W. and Stenger, D.A. (1999) Development and application of cell-based biosensors. *Annals of Biomedical Engineering*, **27**, 697–711.

8
Multifunctional Nanocomposites for Biomedical Applications

Seungjoo Haam, Kwangyeol Lee, Jaemoon Yang, and Yong-Min Huh

8.1
Introduction

Successful medical practice comprises an early and accurate diagnosis of the disease, a patient condition-specific treatment, an accurate assessment of the treatment, and the application of preventive measures against potential complications after cure. Whilst, ideally, "personalized medicine" refers to medical care that is tailored to individual patients, in practice most patients will be categorized to various subgroups under the umbrella of a general disease. An example is that of breast cancer, where patients are cataloged into several subgroups depending on molecular markers and/or signatures reflecting the underlying tumor biology, as well as anatomical staging. Those patients within the same subgroup would be subjected to the same treatment, including conventional and/or molecular-targeted therapy. The combined efforts from a variety of research areas, in order to meet the ideal goal of personalized medicine, have led to fervent omics research being carried out so as to catalog the patient groups. Concomitant investigations have included: the development of *in vitro* diagnostic sensor technologies; the development of new *in vivo* imaging modalities and the combination of different existing imaging techniques; the creation of *in vivo* imaging contrast agents and new contrast mechanisms; the identification of drugs pertinent to the patients' conditions; and target-specific and controlled-release drug administration systems. Notably, many of these fields have benefited greatly from recent developments in nanotechnology [1–9]. Selected examples include the selective enrichment of proteome samples with magnetic nanoparticles, the application of non-photobleachable quantum dots (QDs) in molecular imaging [10], carbon nanotube (CNT)- or nanowire-based sensors for biomolecule detection [11], metal oxide (e.g., MnO and $MnFe_2O_4$) nanoparticles as magnetic resonance imaging (MRI) contrast agents [12, 13], antibody-attached nanocarriers for targeted drug delivery [14, 15], nanoparticle-based gene-transfection [16], and the photothermal treatment of cancer cells with Au nanoshells or nanorods [17–19]. The nanomaterials adopted in these approaches appear, very easily, to recognize and interact with a vast range of biological molecules of interest, mainly due to their comparable sizes.

Nanomaterials for the Life Sciences Vol.8: Nanocomposites. Edited by Challa S. S. R. Kumar

ISBN: 978-3-527-32168-1

Whilst cancer remains a major medical challenge, recent years have witnessed a fall in cancer-related mortality, due mainly to a better understanding of tumor biology and improved diagnostic devices and treatments [1, 20, 21]. Cataloged patients following both tumor biology-driven and classical anatomy-driven classification are subjected to personalized cancer therapy, namely the administration of patient condition-specific drugs (molecularly targeted drugs), in addition to other conventional therapies such as chemotherapy/radiotherapy and/or surgery. While this new paradigm shift in cancer treatment, and the recent successes leading to a fall in mortality, have been very welcome, there remain many hurdles to be overcome before cancer treatment can be considered "successful." Accordingly, recent nanomedical technological advances have been focused on topics such as simultaneous multimodal imaging to document molecular events for accurate patient cataloging at the earliest possible stage, a simultaneous diagnosis and treatment for early intervention, and drug targeting and release control so as to circumvent multiple drug resistance. Invariably, these advances have adopted nanocomposites in which multiple functions have been incorporated, since biologically relevant nanocomposites (which have a relatively large size of 3~200 nm under physiological conditions when compared to molecular drugs) can easily require a wide variety of payloads if chemicals and nano-objects are to provide desirable functions such as molecular imaging and the eradication of cancer.

In this chapter, the various functional components of bio/medicinal nanocomposites, and their required properties, are described. Details of notable recent advances in the formulation of multifunctional nanocomposites, which have led to major technological thrusts, are also provided. Finally, consideration is given to the direction in which this field is heading, and the technical hurdles that must be overcome if these biomedical applications are to be successful.

8.2 Anatomy of Multifunctional Nanocomposites

Previously, several excellent reviews have provided an anatomic description of the multifunctional nanocomposites used for biomedical applications [9, 22, 23] Following Ferrari's classification, the multifunctional nanoparticle can be separated into at least three parts – a biomedical payload, a carrier, and a biological surface modifier – depending on their respective roles (Figure 8.1) [1]:

- The *biomedical payload* refers to imaging agents [e.g., organic dyes, QDs, MRI contrast agents, computed tomography (CT) contrast agents, etc.] and/or therapeutic agents (anticancer drugs, DNA, siRNA, etc).
- The *carrier* provides protection to the loaded contents under physiological conditions during delivery to the desired site.
- *Biological surface modifiers* provide target specificity to the whole nanocomposites, as well as biocompatibility.

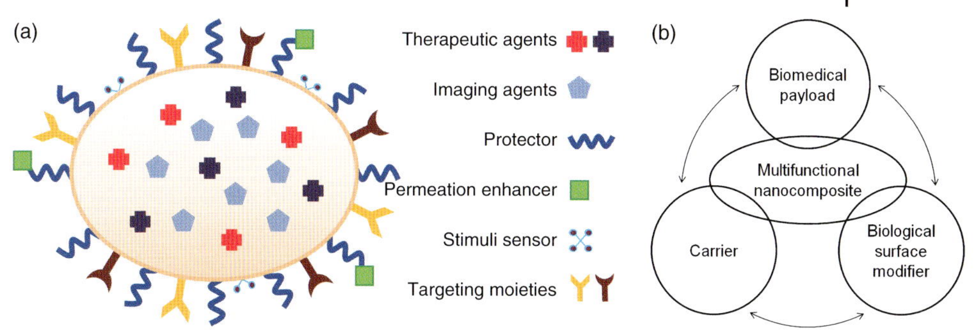

Figure 8.1 (a) Schematic representation of a multifunctional nanocomposite. The biomedical payloads (therapeutic agents and imaging agents) are loaded into a carrier, which is coated with protector such as polyethylene glycol (PEG) to avoid unwanted premature uptake by macrophages. A permeation enhancer, stimuli sensor, and targeting moiety can be conjugated with the nanocomposite; (b) The major three components of multifunctional nanocomposites.

Details of the components of multifunctional nanocomposites are listed in Table 8.1.

8.2.1 The Biomedical Payload

The targeted delivery of anticancer drugs to the tumor site, and their subsequent controlled release, has long been sought in cancer research. In particular, lipid-based or amphiphilic polymer-based micelles can be used to protect hydrophobic drugs (which are very poorly soluble in the blood) within the micelle's hollow structure, so as to deliver a large quantity of the drug to the tumor target. In this way, polymeric nanoparticles can greatly increase intracellular drug concentrations in cancer cells while avoiding toxicity in normal cells. Anticancer drugs can also be conjugated to the functional groups on the polymer backbone or surface functional groups of dendrimers. One notable example is the *N*-(2-hydroxypropyl)-methacrylamide copolymer–Gly-Phe-Leu-Gly-doxorubicin conjugates [24, 25]. In this case, the peptide linker (which is stable in the bloodstream) is cleaved by the lysosomal enzyme cathepsin B upon endocytic cellular uptake. While doxorubicin, camptothecin, paclitaxel and platinate have each been assessed as payloads on the nanocarrier, almost any type of drug – whether hydrophobic or hydrophilic – can be loaded into or onto the liposomal and micellar nanocarriers.

Previously, DNAs, short hairpin RNAs (shRNAs) and small interfering RNAs (siRNAs) have been loaded onto the nanocarrier. While viral vectors are frequently used for gene transfection [26], the need to augment gene-transfection efficiency by *in vivo* imaging and to reduce cell toxicity led to the use of other functional nanocarriers for gene delivery. For example, a poly(ethyleneimine) (PEI) layer on magnetic nanoparticles for *in vivo* MRI results in a positive charge on the outermost surface, which can in turn be used to strongly fix negatively charged DNA strands

Table 8.1 Materials used in multifunctional nanocomposites.

Category	Compositions	Role
Biomedical payload	Imaging agents (organic/inorganic fluorophores, magnetic materials, metal (gold/silver) nanoparticles) Therapeutic agents (anticancer drugs, DNA, RNA, photothermal/photodynamic materials)	Molecular imaging Separation Induction of apoptosis/necrosis Gene upregulation/downregulation
Carrier	Organic nanostructures (lipid, natural/synthetic polymers) Inorganic nanostructures (metal nanoparticles, ceramic nanoparticles, carbon nanostructures) Organic–inorganic nanocomposites	Protection of payloads Carrying to target site Release control of drug/gene Biocompatibility enhancing Solubilization Linker Stimuli-sensitive
Biological surface modifier	Antibody Peptide/Protein Aptamer Small molecules	Targeted delivery Uptake enhancer Signaling transduction Stimuli-sensitive

[27, 28]. The ligands affixed to the surface of inorganic nanoparticles (such as iron oxide nanoparticles) can also be used to hold siRNA molecules; these are short, double-stranded nucleic acid molecules that can act as mediators of RNA interference (RNAi) within the cytoplasm of cells [29–32]. In siRNA delivery, however, the rapid degradation of genes by exonucleases or endonucleases, combined with difficulties in crossing the cell membrane, pose significant technical challenges for the clinical application of siRNA-based therapeutics. One interesting approach employs biodegradable $CaPO_4$ nanoparticle to protect the siRNA molecules until they reach the cancer cells; upon arrival, and under low extracellular pH conditions, the $CaPO_4$ content is removed to reveal the therapeutic siRNA molecules [33].

Whilst a variety of organic dyes have been used to monitor the molecular events in biological systems, the presently available molecular imaging probes are photobleachable, toxic, and rapidly removed from the body. Organic dyes, when incorporated into SiO_2 spheres or polymer bead-based nanocarriers, are more stable under physiological conditions due to their being protected by the encapsulating layer, and this may provide reliable optical information for a prolonged period [34, 35]. For example, Cy5.5 dye molecules can be conjugated to the backbone of a methoxy-PEG-protected poly-L-lysine copolymer to provide information regarding tumor molecular biology [36].

In the past, QDs have been used extensively as biological imaging and labeling probes, and seem to possess several advantages over conventional organic dye molecules. These include a narrow photoluminescence signal, a high quantum yield, low photobleaching, and resistance to degradation under physiological condition [37–39]. Moreover, the optical emission signal from QDs can be greatly modulated by controlling their size and composition. Thus, QDs, when entrapped in a nanocarrier system, can provide useful optical information about biological events. While these features clearly offer excellent *in vitro* research opportunities, the *in vivo* medical application of these materials appears to be severely limited by the intrinsic toxicity of the metal content, such as Cd in QDs.

Magnetic metal oxide nanoparticles have recently attracted much attention due to their potential applications for the *in vivo* diagnosis of tumors, in stem cell trafficking and proteome enrichment [6, 13, 40, 41]. Recently, the thermal decomposition of metal precursors (such as metal acetylacetonate or metal oleate) has led to the formation of highly crystalline, highly monodisperse nanoparticles with much enhanced $emu\,g^{-1}$ values, which can be advantageously used as ultrasensitive T_2 and/or T_2^* contrast agents in high-resolution MRI [13, 42]. Similarly prepared MnO nanoparticles can result in T_1 contrast agents for high-resolution MRI [43], via yet another image-enhancing mechanism. In a recent study, a strict guideline was set for the quantitative analysis of nanoparticle single crystallinity, and some synthetic methods for nanoparticle synthesis have appeared to produce nanoparticles with excellent single crystallinity [44]. In order to combine biocompatibility with targeting ability, these nanoparticles should either be surface-modified by water-solubilizing ligands, or wrapped with natural or synthetic polymers such as proteins, poly(ethylene glycol) (PEG), or polysaccharides.

Other inorganic payloads commonly employed include gold nanoparticles (AuNPs) for CT imaging [45], gold nanorods for optical imaging and hyperthermal therapy [18], and rare earth metal-based nanoparticles for T_1 MRI contrast agents [46, 47] and up-conversion optical probes [48].

8.2.2 The Carrier

The choice of carrier material is critical for successful biomedical applications. In order to the ensure safe delivery of a biological payload to the target site, various organic carriers (liposomes/polymersomes, polymeric nanoparticles, and dendrimers) and inorganic carriers (metal/ceramic nanostructures and CNTs) have been prepared. These carriers can load functional payloads either by physical entrapment or chemical conjugation, and are usually further surface-modified to provide target-specificity and compatibility under physiological conditions. The structures of various nanocarriers are shown in Figure 8.2.

A hollow or porous structure is required for the entrapment; while the formation of soft lipid-derived liposomes [49, 50] and amphiphilic block-*co*-polymer-based hollow structures [51] concur with the entrapment of payloads, the formation of

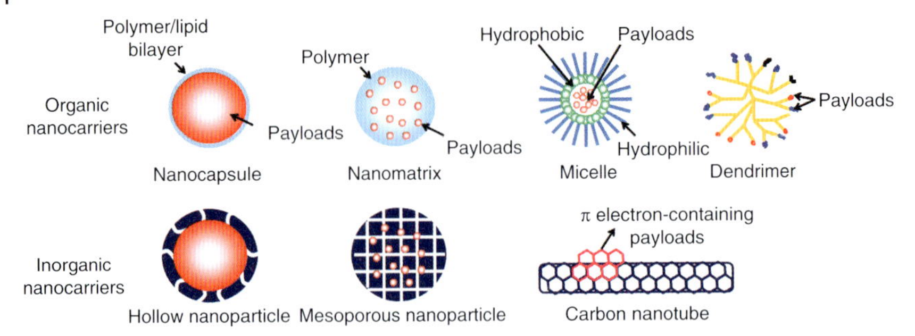

Figure 8.2 Various organic/inorganic carriers for biomedical applications.

rigid inorganic hollow containers does not have to be synchronized with the payload entrapment. The payload conjugation to a nanocarrier requires suitable intermolecular interactions, such as π–π interaction and covalent bond formation between the carrier and payload molecules. In particular, the efficacy of drugs that are either conjugated to or entrapped by the carrier, is significantly different from that of conventionally administered drugs, due mainly to significant differences in the drug release profiles.

Polymers are by far the most extensively studied materials used for carriers. A polymeric nanocarrier can be derived from synthetic polymers such as poly(lactic acid) (PLA) and poly(lactic-*co*-glycolic acid) (PLGA), or from natural polymers such as chitosan and collagen [22, 52–56]. The encapsulated drugs can be released in a controlled manner via degradation of the polymer under physiological conditions, by diffusion through the polymer matrix, by swelling followed by diffusion, or in response to changes in the local environment (e.g., pH) [57–59]. The polymers can either be conjugated to drugs or used without further modification so as to encapsulate payloads such as drugs for therapeutic purposes, and metal oxide nanoparticles and QDs for imaging purposes. Amphiphilic block-*co*-polymers can form micelles with a hydrophobic core and a hydrophilic surface, while the hydrophobicity of the core can present an ideal medium to hold highly hydrophobic drugs or imaging contrast agent nanoparticles [60–63] that can also be further surface-modified for targeted drug delivery. Inherently, polymer-based carriers suffer from the inhomogeneity of molecular weight, leading to grossly polydisperse nanocarrier sizes; this may greatly affect the drug release profile, leading to an unpredictable drug efficacy.

Lipid-based nanocarriers have shown particular promise, due to their biocompatibility and outstanding ability to contain both hydrophilic and hydrophobic drugs. Well-known lipid membrane modification methods [64, 65], by using readily available, naturally occurring molecules can lead to a variety of carrier systems with desired targeting abilities and surface charges for effective drug delivery. In spite of these obvious advantages, lipid-based nanocarriers may cause serious problems in practical applications, however. For example, upon intravenous injection, the particles are rapidly cleared from the bloodstream by the reticuloendothelial system (RES) defense mechanism, regardless of the particle composition [1, 5, 8, 22, 37, 53, 66, 67]. Furthermore, the instability of lipid-based carriers in the bloodstream might lead to a premature drug release before the target has been reached, and/or an undesired burst causing release of the drugs. An advantage here is that polymer and lipid molecules can be combined to form a carrier structure with a polymer core and a lipid shell, and this can be used to great effect by releasing two different drugs on a temporal basis [68]. For example, after having reached the targeted tumor site, a lipid shell could first release an anti-angiogenesis agent; the polymeric core could then release an anticancer drug in response to a local shortage of oxygen, which is the signature of tumor tissues.

Dendrimers, which have a tree-like hierarchical molecular structure, have two distinct structural features [69–73]. The central part usually contains voids into which payloads can be loaded, while the surface contains numerous identical functional groups for effective surface modification with targeting moieties, imaging agents, and drugs. Dendritic carriers are very beneficial due to their relatively small size (<5 nm) allowing a rapid clearance from the bloodstream via the kidneys and $-NH_2$ or –COOH-derived high solubility under physiological conditions. However, the high cost of production for dendrimers poses a significant challenge to their use as nanocarriers in the clinical situation.

In recent years, a number of inorganic nanocarriers with hollow or porous structures have been developed, including porous SiO_2 nanoparticles [34, 74–80], hollow gold nanocubes or spheres [17, 81, 82], calcium phosphate nanoparticles [33, 83–87], and magnetic oxide nanocontainers [88–90]. Likewise, CNTs can be surface-modified to possess functional groups such as –COOH to which drugs can be anchored; alternatively, they can be used without surface modification by employing the strong hydrophobic interaction between the CNT surface and the payload molecules/surfactants [11, 91–93]. Inorganic nanocarriers normally require a further surface modification to render them highly soluble in the body fluids. Yet, inorganic nanocarriers are not free from criticism, particularly from the aspect of biodegradability and biocompatibility. For successful biomedical applications, there should be pathways for biodegradation of inorganic nanocarriers, and a safe removal or an assimilation of any decomposition products from the bloodstream. Recently, the fecal removal of intravenously administered PEGylated CNTs has been demonstrated in animal studies [92]. While the main role of these inorganic nanocarriers has been the loading of functional contents, such as imaging agents and drugs, the carrier itself might be designed to possess useful functions. For example, hollow Fe_3O_4 nanoshells prepared via the calcination of $FeOOH@SiO_2$, followed by H_2-assisted reduction at elevated temperature and SiO_2 shell removal, can be used as both a drug-loading container and a T_2 MRI contrast agent (Figure 8.3a) [94]. Recently, hollow Mn_3O_4 nanoshells which had been prepared by the oxidation/etching of MnO nanoparticles in aqueous solution were used as both a drug-loading container and a T_1 MRI contrast agent (Figure 8.3b) [88]. In contrast, following exposure to near-infrared (NIR) radiation, the hollow structures of gold were shown to produce heat that could be used as photothermal therapy to kill cancer cells by local hyperthermia (Figure 8.3c) [17, 82].

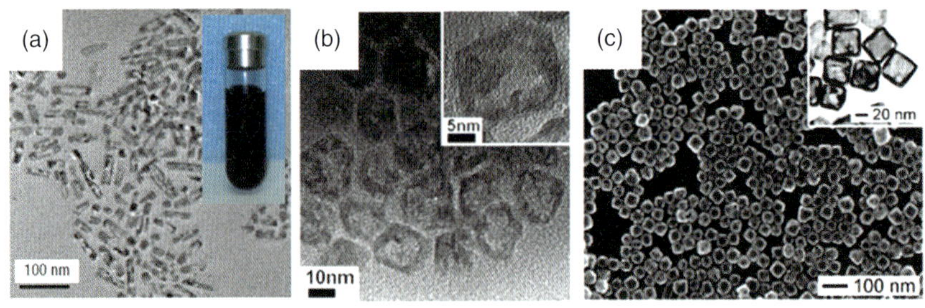

Figure 8.3 Inorganic hollow nanocontainers. (a) Transmission electron microscopy (TEM) image of iron oxide nanocapsules. Inset: Photograph of the iron oxide. Reproduced with permission from Ref. [94]; © Nature Publishing Group; (b) TEM image of hollow Mn_3O_4 prepared after immersion of MnO nanoparticle in water for five days. Reproduced from Ref. [88]; (c) Scanning electron microscopy image of Au nanocages prepared by refluxing an aqueous solution containing both silver nanocubes and $HAuCl_4$. The inset shows a TEM image of the Au nanocages. Reproduced with permission from Ref. [17]; © American Chemical Society.

8.2.3 The Biological Surface Modifier

As tumors grow very rapidly, and require a constant high intake of nutrients from the blood, the tumor tissues create their vasculature in a haphazard manner, and this results in the formation of "leaky" blood vessels [15, 20, 68]. Fortunately, the tumor tissues can easily be accessed by the nanocarriers through these leaky blood vessels, by employing an enhanced permeation and retention (EPR) effect. In fact, the innate structural property of tumor tissue seems to welcome the attention of the nanocarrier systems, even when the nanocarriers seem unconcerned in actively seeking the tumor target. Although the EPR effect enables a "passive" targeted delivery of drugs, the amount of drug that is undelivered (and thus nonspecifically biodistributed) may be significant, such that an "active" targeted delivery of drugs, with a consequent reduced cell toxicity, would be highly desirable. The surface of the nanocarriers can be modified with targeting agents capable of actively seeking the target tissues; those most frequently employed at present include antibodies and their fragments, aptamers, or other receptor ligands such as peptides, vitamins, and carbohydrates.

Antibody-based targeting has been clinically proven in many cases, and several FDA-approved antibodies are currently available. For example, trastuzumab (Herceptin), an anti-HER2 monoclonal antibody (mAb) that binds to ErbB2 receptors, can be used for the targeting and treatment of breast cancer [22, 95–99]. Bevacizumab (Avastin), an anti-VEGF mAb, that inhibits the factor responsible for the growth of new blood vessels, can be used to target and treat colorectal cancer [100]. Recently, it was demonstrated in a mouse model that anti-HER2 mAb, when attached to polymer-based micelles containing magnetic nanoparticles, could successfully guide the nanocarrier to the breast cancer tissue [12, 17, 26, 101]. Although both whole or fragments of antibodies can be used for targeting, it is generally accepted that the use of a whole antibody can lead to a higher target specificity, due to stronger binding. Furthermore, the whole antibody can be stored for prolonged periods without any loss in structural integrity, which makes it more attractive than the fragmented antibody [102, 103].

Aptamers, which are short single-stranded DNA or RNA oligonucleotides, can serve as useful targeting agents, because they are capable of binding to a wide variety of targets, including intracellular proteins, transmembrane proteins, soluble proteins, carbohydrates, and small-molecule drugs [104–108]. As aptamers are very stable molecules, their conjugation to the nanocarrier can be much simpler than using an antibody (which is a protein) for the targeting agent. Recently, aptamer-modified docetaxel-encapsulated nanoparticles demonstrated a good targeting ability for antigens on the surface of prostate cancer cells, leading to a desired efficacy *in vivo* [109].

The fact that cancer cells require a high input of nutrients leads to an overexpression of receptors involved with nutritive processes, including folate receptors, transferrin receptors, and epidermal growth factor receptors (EGFRs) [18, 26, 110–118]. Hence, molecules such as growth factors and vitamins, which can specifically target these receptors, may serve as targeting moieties for the nanocarriers. Unfortunately, however, the targeted receptors will also be present in healthy cells in large quantities, leading (potentially) to an unwanted and nonspecific binding of the nanocarriers to normal tissues. By adopting a similar reasoning, adhesion molecules such as integrins (and their ligand peptide, RGD) and extracellular matrix (ECM) components could be used as targeting moieties for nanocarriers to target fast-growing tumor tissues with overexpressed cell adhesion integrins and ECM receptors [119–124]. However, the nonspecific binding of growth factors or ECM components may, again, pose a significant challenge to the clinical application of these materials in targeted delivery.

In order to effectively target desired tissues, the nanocarriers should possess a biocompatibility (a long circulation time so as to evade the body's immune system, low toxicity, and high solubility) that will be greatly affected by the nature of the surface moieties. For example, liposomes coated with hydroxyapatite or PEG have shown prolonged circulation times and lowered immune responses, as well as evading detection and destruction by the immune system, thus greatly increasing their probable use for drug targeting [26, 125–127]. It was also shown that, whilst the PEG-coating of CNTs led to their increased stability in aqueous systems, branched-PEG coating was much superior to linear-PEG coating in terms of circulation time and cell toxicity [92]. Certain surface-modifying polymers may exhibit biodegradability so as to further reduce any adverse side effects caused by the administration of nanocarriers.

A nanocarriers, on arrival at the target tissue, must be transported into the cells in order to obtain the desired drug efficacy. Both, the size and surface charge of the nanocarrier play important roles in the internalization of drugs [128, 129]. In particular, the cell membrane exhibits an anionic charge due to its bilayer phospholipid structure, such that the cationic surface charge of the nanocarrier can enhance the efficiency of cellular uptake [130–132]. Physiological barriers, such as capillaries that include organ-specific patterns (e.g., the blood–brain barrier) may also inhibit the passage of large-sized nanocarriers [133–135]. Thus, the surface charge of a nanocarrier (which relates to the identity of the polymer and the nanocarrier size, which relates to the polymer's molecular weight) must be carefully considered when designing a nanocarrier to overcome physiological barriers such as the cellular membrane.

8.3 Types of Nanomaterial Multifunctionality

A careful selection of the three major components of nanocomposites – biomedical payload, carrier, and biological surface modifier – can lead to multifunctions that hitherto have been either unknown or impossible with molecular counterparts. The versatility of nanocomposites, with regards to multimodal imaging, theragnosis, multitherapy, and activatable smart nanocomposites, are described in the following sections.

8.3.1 Multimodal Imaging

8.3.1.1 Multicolor Optical Coding

Fluorescent-tagged biomolecules and cells can easily be sorted so as to reveal critical information concerning molecular events within a biological system. For this purpose, a variety of fluorescent entities, including organic dyes, QDs, single-walled CNTs and AuNPs, have been developed [10, 18, 92, 136]. For example, fluorescence resonance energy transfer (FRET), using two or more fluorescent materials can be used to obtain information concerning the physical distances between different fluorophores. In fact, FRET has been observed for all possible combinations of fluorophores, such as gold–organic dye, organic dye–organic-dye, and QD–organic dye [137–139].

A recent report involved the formation of polymer-based nanoparticles which contained two types of fluorescent dye (Figure 8.4a) [140]. One of the dyes, perylene diimide, emits a strong green fluorescence, while the other dye, spiro-pyran, fluoresces only weakly. The spiro-form dye can be converted, by irradiation with ultraviolet (UV) light, to the open-ring (mero) from, which fluoresces strongly to emit an intense red light. The mero form reverts slowly to the spiro form under visible light. The absorption band of the mero form coincides with the emission band of the perylene dye such that, under blue light excitation, the polymer nanoparticle produces a strong green light. However, upon UV irradiation the polymer nanoparticle emits a strong red fluorescence due to FRET between the dyes. This ability to produce dual colors by a simple photoswitching may be advantageous for applications in optical imaging, when a background color strongly interferes with one dye color, or when the authenticity of an observed color in a biosystem is questioned.

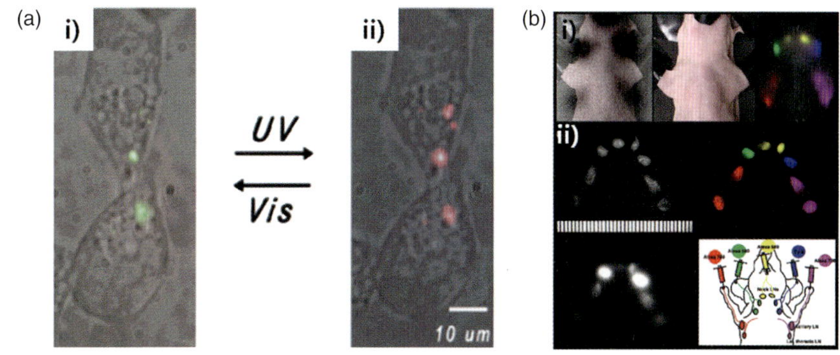

Figure 8.4 (a) Dual-color fluorescent nanoparticles were transported into HEK-293 cells using liposomes as delivery vehicles; these internalized polymer nanoparticles can be selectively highlighted with either green or red fluorescence. A short UV pulse switches the highlighted green spots (i) to vivid red fluorescence (ii), while visible light reverses the process. Reproduced with permission from Ref. [140]; © American Chemical Society; (b) *In vivo* dual-modal, five-color lymphatic drainage imaging with the ability to visualize five distinct lymphatic drainages. *In vivo* multiexcitation spectral fluorescence (i, right) and post-mortem *in situ* radionuclide (i, left) images of a mouse injected with five distinct nanoprobes. Five primary draining lymph nodes were simultaneously visualized with different colors through the skin in the *in vivo* spectral fluorescence image, and are more quantitatively seen in the radionuclide image. *Ex vivo* spectral fluorescence (ii, right) and radionuclide images of eight draining lymph nodes (ii, bottom) correlate well to the *in vivo* imaging. Reproduced with permission from Ref. [136]; © American Chemical Society.

Another interesting use was reported by Kim *et al.*, in which multiple drug substrates each with the Cy5.5 dye were attached to a central AuNP [137]. The Cy5.5-substrate/AuNP did not fluoresce due to a combination of static quenching and FRET. The substrate could be effectively cleaved in the presence of a suitable enzyme (in this case matrix metalloprotease, MMP), at which point the released dyes Cy5.5 and AuNP began to fluoresce, reflecting the action of the enzyme on the substrate. When the Cy5.5-substrate/AuNP was administered to mice bearing a squamous cell carcinoma (which overexpresses MMP-2), tumor targeting by the nanoparticles and subsequent optical imaging due to the release of fluorophores was not demonstrated.

In order to change colors, the QD sizes must be changed accordingly, although this might cause differences in nanoparticle movement in various body fluids; however, in this case, a dendrimer might provide a simple solution (Figure 8.4b) [136]. A sixth-generation polyamidoamine (PAMAM) with an ethylenediamine core has been attached to several different dyes (namely, Cy5, Alex660, Alex680, Alex700, and Alex750) to produce a series of different color codes but with minimal change in particle size. The dendrimers with different colors were used successfully in lymphatic imaging (Figure 8.4b). In addition to this optical information, the conjugation of ^{111}In to the dendrimer provided *in vivo* radionuclide images that were consistent with results obtained with optical imaging.

8.3.1.2 Optical/T_1 MR Bimodal Image Probes

Gadolinium(III) (Gd^{3+}), when protected by a ligand, provides by far the best T_1 MRI contrast ability among the various MR-active metal ions [141, 142]. In order to maximize T_1 contrast ability, several strategies have been developed to consolidate many gadolinium ion centers in a confined space, such as polymer nanoparticle backbone, dendrimer, and silica. These consolidating platforms invariably contain many surface-functional groups to link chelating ligands for the encapsulation of gadolinium ions. For example, the $-NH_2$ groups of polymer or dendrimer and the –OH groups of silica can be advantageously employed to form –NHCO- or $-(O)_3Si$- linkages, respectively [143–147]. In an extreme case for increasing the number of active gadolinium centers, the Gd_2O_3 nanoparticle is formed and the surface metal atoms provide the necessary T_1 contrast effect [47].

Yet, various other imaging agents can also be conjugated to the Gd^{3+} ion-consolidated platform. In one study, Cy5.5 dyes were conjugated to enable both MRI and fluorescence imaging for sentinel lymph node and lymphatic drainage [145]. The silica platform, when surface-modified by Gd complexes, can contain other nanoparticle(s), such as a cadmium-based QD [147] or luminescent ruthenium complexes [143], to form a bimodal image probe both for fluorescence imaging and MRI. Gd-containing dendrimers can be conjugated to a QD to form again fluorescence imaging and MRI agents [146]. The QD itself might be conjugated by Gd-containing paramagnetic lipid molecules to produce dual optical/MRI probes [148]. Such a probe, prepared from poly-L-lysine, conjugated with rhodamine and Gd–DOTA (gadolinium-tetra-azacyclododecanetetra-acetic acid) complexes, may also be loaded with the therapeutic enzyme, bacterial cytosine deaminase, to treat human breast cancer cells, albeit *in vitro* [144].

8.3.1.3 Optical/T_2 MR Bimodal Image Probes

Highly sensitive T_2 contrast effects in high-resolution MRI are obtained with iron oxide nanoparticles, manganese ferrite nanoparticles, and FePt nanoparticles [13, 40, 149]. The conjugation of these superparamagnetic nanoparticles with fluorophores also leads to the formation of dual optical/MRI probes. The conjugation strategies and fluorophore selection vary greatly, and include: formation of the heterodimer CdSe–Fe_3O_4 [150]; the formation of giant vesicles containing CdSe/ZnS QDs and γ-Fe_2O_3 nanoparticles [151]; the formation of heterodimer Au–Fe_3O_4 nanoparticles [152]; the encapsulation of dye-doped silica and formation and subsequent attachment of Fe_3O_4 nanoparticles [153]; magnetic nanoparticles coated with pyrene-conjugated amphiphilic copolymers [154]; iron oxide nanoparticles with crosslinked polymeric shells and subsequent conjugation with Cy5.5 [155]; and iron oxide nanoparticles terminated with single-walled CNTs [156] (see Table 8.2).

8.3.2 Theragnostics: Simultaneous Diagnosis and Therapy

8.3.2.1 Optical Imaging/Therapy

Amphiphilic polymer-based micelles can incorporate hydrophobic anticancer drugs such as Taxol, cisplatin, camptothecin, docetaxel, and doxorubicin within

Table 8.2 Various optical/T2 MRI bimodal probes. TEM images reproduced with permission from Refs [150–153, 155, 156]; © American Chemical Society), and Ref. [154]; © Elsevier.

Superparamagnetic nanoparticle	Optical imaging materials	Scheme	TEM image	Reference
Fe_3O_4	Quantum dot (CdSe)	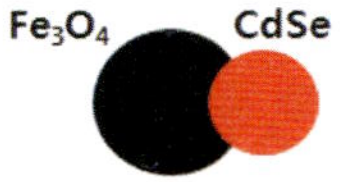 Fe_3O_4 CdSe	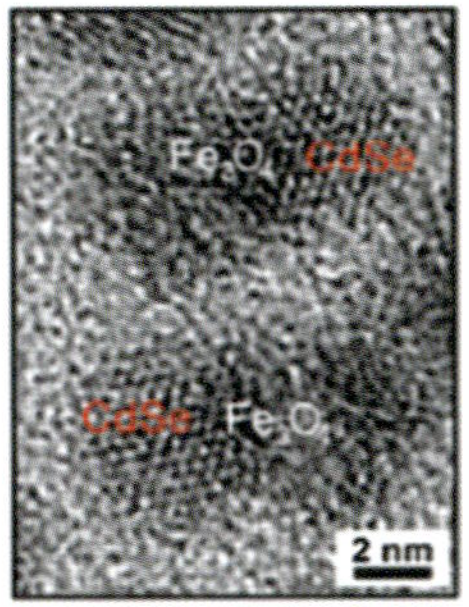 2 nm	[150]
γ-Fe_2O_3	Quantum dot (CdSe/ZnS)	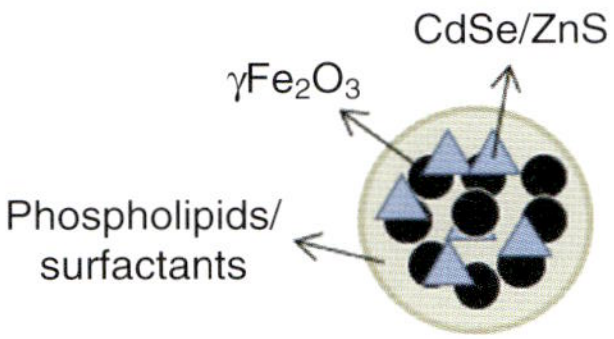 CdSe/ZnS γFe_2O_3 Phospholipids/ surfactants	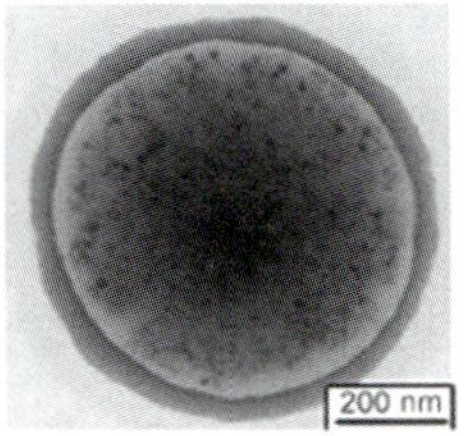 200 nm	[151]
Fe_3O_4	Au nanoparticle	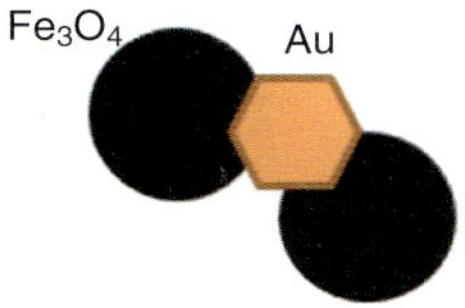 Fe_3O_4 Au	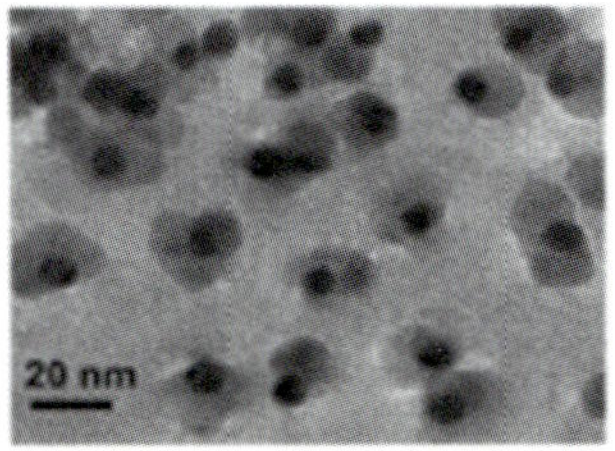 20 nm	[152]

Table 8.2 *Continued*

Superparamagnetic nanoparticle	Optical imaging materials	Scheme	TEM image	Reference
$MnFe_2O_4$	Organic dye (fluorescein isothiocyanate)	FITC, $MnFe_2O_4$, Silica	100 nm, 30 nm	[153]
$MnFe_2O_4$	Organic dye (Pyrene)	Pyrene, $MnFe_2O_4$, Polymer	40 nm	[154]

Fe_3O_4	Organic dye (Cyanin 5.5)	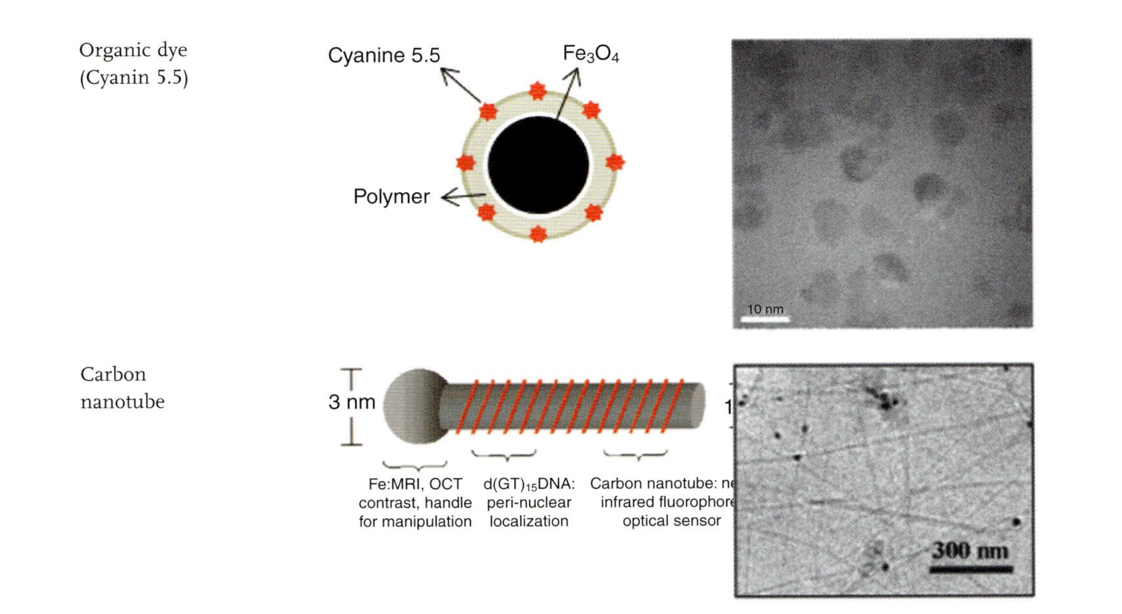	[155]
Fe_3O_4	Carbon nanotube		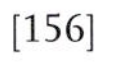 [156]

their hydrophobic cores. When, for example, camptothecin was incorporated into the hydrophobic core of the hydrophobically modified glycol chitosan, its stability under physiological conditions was greatly enhanced. Conjugation of the polymer backbone with Cy5.5 enabled an *in vivo* monitoring of the circulation time and biodistribution of polymer–drug conjugates, as well as drug accumulation in the tumor tissues, using NIR fluorescence spectroscopy [157]. In fact, NIR fluorophores such as Cy5.5, when excited at light wavelengths of 650 to 900 nm, presents one major advantage over conventional optical probes excited by visible lights, namely that the body is "transparent" in the NIR window due to a minimal absorption of light by the tissues [19].

With all controversy over toxicity issues set aside, QDs can provide useful optical imaging information. For example, a QD–aptamer conjugate developed by Bagalkot *et al.* was shown capable of incorporating molecules of the anticancer drug doxorubicin within the double-stranded RNA aptamers [158]. The fluorescence of the doxorubicin and QDs was quenched by the FRET mechanism. Upon release of the drug, however, the fluorescence signals of the QDs and doxorubicin were "turned on," which enabled the process of drug delivery to be monitored. It has been reported that QDs can serve as photosensitizers to form singlet oxygen in order to induce apoptosis in target tissues [159]. However, although QDs can clearly be used as optical probes in this way, the toxicity issue of heavy-metal ions such as Cd^{2+} must be addressed before considering their application in medical fields.

Gold nanorod exhibits two surface plasmon absorption peaks: a strong, long-wavelength peak due to the longitudinal oscillation of electrons; and a weak, short-wavelength peak due to transverse electronic oscillations [18, 160]. The long-wavelength peak can be shifted into the NIR region by extending the gold nanorod length. Subsequent NIR irradiation will lead to a surface plasmon resonance (SPR)-induced heating of the gold nanorods following their internalization into cells, causing local hyperthermia. A cell containing gold nanorods can also be examined using simple, dark-field microscopy. Recently, Huang *et al.* prepared anti-EGFR-antibody-conjugated gold nanorods and incubated them with cancer cells [18]. Subsequent NIR irradiation led to an effective killing of cancer cells containing gold nanorods, but had no effect on normal cells. Whilst the gold nanorods appeared to exhibit the required biocompatibility, the route of their elimination from the body must be established before any practical medical application.

8.3.2.2 **MR Imaging/Therapy**

Magnetic metal oxide nanoparticles, prepared by the thermal decomposition of metal-organic precursors, provide a greatly enhanced sensitivity in T_2-weighted MRI, when compared to previously used metal oxide nanoparticles formed by the coprecipitation method. An interesting correlation between the emu g^{-1} value and T_2 relaxivity was observed, with the best T_2 contrast effect achieved using a nanoparticle composition of $MnFe_2O_4$, with the highest emu g^{-1} [13]. Indeed, the correlation between emu g^{-1} and T_2 relaxivity has led to the T_2 MRI modality attracting much research interest. Recently, MnO nanoparticles, again prepared by thermal decomposition, were used successfully as a T_1 MRI contrast agent to fuel the renewed interest in high-resolution MRI [43]. In short, metal oxide nanoparticles, prepared via the thermal decomposition route, have opened the way to successful molecular imaging based on high-resolution MRI which, previously, was considered to be impossible.

The solubilizing protective coating on magnetic nanoparticles can be conjugated by anticancer drug molecules, with or without targeting moieties. An example was reported by Kohler *et al.*, who prepared $-NH_2$-terminated superparamagnetic nanoparticles that could bind to –COOH-terminated PEGylated methotrexate molecules [161]. In this case, the methotrexate molecules served as targeting moieties for the folate receptor that had been overexpressed by cancer cells. Although the chemical linkage between methotrexate and the magnetic nanoparticles was very stable under physiological conditions, upon internalization into cancer cells the drug was released by the effects of a low pH and intracellular enzymes, so as to induce apoptosis of the cancer cells. The drug–magnetic nanoparticle conjugate showed a markedly reduced toxicity towards normal cells, indicating a drastic difference in the cellular environments of normal and cancer cells.

Recently, a magnetic cluster formulation was reported which consisted of hydrophobic superparamagnetic nanoparticles trapped within an amphiphilic, polymer-based micelle [12, 162]. This allowed the T_2 MRI sensitivity to be greatly enhanced, simply by increasing the number of magnetic nanoparticles in the cluster. Moreover, the anticancer drug doxorubicin could be incorporated concomitantly into the hydrophobic core. Subsequently, when the antibody-assisted targeted delivery of drug-loaded magnetic particles to tumor tissue was monitored using high-resolution T_2-weighted MRI, there was a drastic reduction in tumor growth rate when the drug was released into the tumor tissue. Recently, hollow nanostructures of magnetic oxide Fe_3O_4 and Mn_3O_4 were prepared in which the hollow interior could be used for drug loading [88, 94]. In this case, both nanocontainers, when modified with a solubilizing protective polymer layer, could be used not only as MRI contrast agents (T_2 and T_1, respectively) but also as drug delivery vehicles. Although both agents were shaped differently from conventional superparamagnetic nanoparticles, their targeted delivery of drugs and real-time MR monitoring of drug efficacy proved to be identical.

MR-responsive magnetic nanoparticles can be conjugated with siRNAs, to be used for gene-silencing purposes [119]. In a recent study conducted by Cheon *et al.*, $MnFe_2O_4$ nanoparticles were coated with bovine serum albumin (BSA) and used as a core material. The primary amine groups of the BSA were converted into pyridyldisulfide groups by treatment with *N*-succinimidyl-3-(2-pyridyldithio) propionate (SPDP), whereby the –SH-terminated PEG–RGD and –SH-terminated siRNA-Cy5.5 were linked to the nanoparticle, forming a disulfide bond. In this case, the RGD moieties served the targeting role, while the siRNAs provides the required gene-silencing role to suppress expression of the green fluorescent protein (*GFP*) gene. The targeting efficiency towards integrin-overexpressed cells was monitored using MRI, while the expression level of *GFP* was monitored using optical microscopy. Jon *et al.* reported that superparamagnetic nanoparticles with thermally crosslinked polymer shells containing Si–OH could also be used as a siRNA delivery vehicle, which in turn allowed the tumor to be monitored *in vivo*, using MRI [163]. Likewise, when PEI surface-modified magnetic nanoparticles were loaded with therapeutic genes to treat human mesenchymal stromal cells, the location of the gene-transfected stromal cells was monitored *in vivo*, again using an MRI modality [28, 130].

8.3.2.3 Optical Coherence Tomography Imaging/Therapy

Optical coherence tomography (OCT) permits the cross-sectional, subsurface imaging of biological tissues, at micrometer-scale resolution. Typically, gold nanoshells were used as OCT imaging contrast agents, on the basis of their biocompatibility and easily detected backscattering of NIR light [19, 164]. When mice administered gold nanoshells were subjected to NIR laser irradiation, hyperthermia due to SPR of the gold nanoshell induced a major reduction in tumor progression that could be followed using OCT [165].

8.3.3 Combination Therapy

Whilst, today, many human diseases are regarded as systemic disorders, it is generally considered that cancer comprises a group of diseases, largely on the basis that the metabolism of cancer tissues is highly complex and leads to multiple symptoms [166]. It has long been known that combination therapies, with the simultaneous administration of two or more drugs, represents an effective approach to the treatment of cancer [167]. Typically, a combination of chemotherapeutic agents with different mechanisms of action and with additive or synergistic effects is administered, the aim being to overcome the drug resistance that causes major problems in cancer treatments, [168, 169]. It should be noted also that combinations of treatment modalities, such as surgery, radiation therapy, photodynamic therapy, gene therapy, and/or chemotherapy, are also widely considered in order to achieve a synergistic therapeutic efficacy.

Well-tailored nanocarriers enable the simultaneous delivery of the required multiple medications, and include: (i) therapeutic antibodies combined with chemotherapeutic drugs [12, 63, 97]; (ii) combinations of different chemotherapeutic drugs [170–172]; (iii) chemotherapeutic drugs with genes (DNAs, siRNAs or shRNAs) [31, 173–175]; and (iv) photothermal or photodynamic agents with chemotherapeutic drugs [81, 176]. Several representative combination therapies using nanoparticles are described in the following sections.

8.3.3.1 Therapeutic Antibodies/Chemotherapeutic Drugs

Combinations of therapeutic antibodies, such as trastuzumab and cetuximab, which target for HER-2/neu and EGFR, respectively, have been used successfully as "molecularly targeted drugs," particularly for the treatment of breast and ovarian cancers. The human EGFR (ErbB-1) and HER2/neu (ErbB-2) are members of a family of transmembrane receptor tyrosine kinases that regulate cell growth, survival and differentiation via growth signal transduction pathways. The above-mentioned antibodies exhibit significant anti-tumoral activity, even with single administration, but demonstrate an even stronger synergistic therapeutic capability with currently available chemotherapeutic drugs, at well-defined molecular situations. Hence, the use of a therapeutic antibody in a nanoparticle-based combination therapy will create dual benefits: (i) to provide the active targeting of a drug in a cell-, tissue-, or disease-specific manner; and (ii) to inhibit the growth of cancer cells [95]. Unfortunately, very few studies have been conducted to examine this synergistic therapeutic efficacy towards cancer when using therapeutic antibody-conjugated drug nanocarriers. Consequently, multifunctional magneto-polymeric nanohybrids (MMPNs) have recently been prepared for synergistic cancer treatment efficacy, and also for MRI purposes [12, 99]. The MMPNs were composed of magnetic nanocrystals, in the core of which was installed the anticancer drug doxorubicin, an amphiphilic block copolymer-based micelle carrier, and a conjugated Her-2/neu on the surface to facilitate both tumor-targeting and the treatment of metastatic breast cancer. A synergistic efficacy in cancer treatment was successfully demonstrated both *in vitro* and *in vivo* (Figure 8.5). In another related study, therapeutic antibody-conjugated polymeric gold nanoshells were prepared to provide multiple therapies: (i) a photothermal effect from gold-layered nanoshells; (ii) a polymeric core (PLGA) to release a chemotherapeutic drug; and (iii) cetuximab conjugated onto the outer ends of the PEG, as a therapeutic antibody [176].

8.3.3.2 **Combinations of Different Chemotherapeutic Drugs**

Several investigations of the nanoparticle-based delivery of multiple chemotherapeutic drugs have been reported to achieve synergistic actions of drugs after cellular uptake, through (ideally) the use of controlled drug release characteristics and the optimized pharmacokinetic profiles of the different drugs, in order to reduce injection times and adverse side effects [170, 171]. In this case, amphiphilic block *co*-polymer-based micelle-type nanoparticles were used as carriers of hydrophobic anticancer drugs. Although these nanocarriers successfully demonstrated an enhanced cytotoxic activity of two simultaneously delivered drugs by synergistic actions, the efficacy of these drugs in the systemic treatment of specific tumors *in vivo* has yet to be proven. Nonetheless, clinical trials are currently under way to investigate the application of nanoparticulate, albumin-bound paclitaxel, in combination with gemcitabine [170].

8.3.3.3 **Combinations of Chemotherapeutic Drugs and Genes**

The co-delivery of drugs and genetic components (DNAs, siRNAs or shRNAs) should have superior therapeutic effects compared to the use of the individual treatments alone [16, 31, 173–175]. Extensive studies have revealed that chemotherapeutic agents react directly with nuclear DNA for cellular apoptosis, while genes induce the additive cytotoxic effects or increase the chemosensitivity of target cancer cells. The siRNA-mediated silencing of Bcl-2, an anti-apoptotic gene,

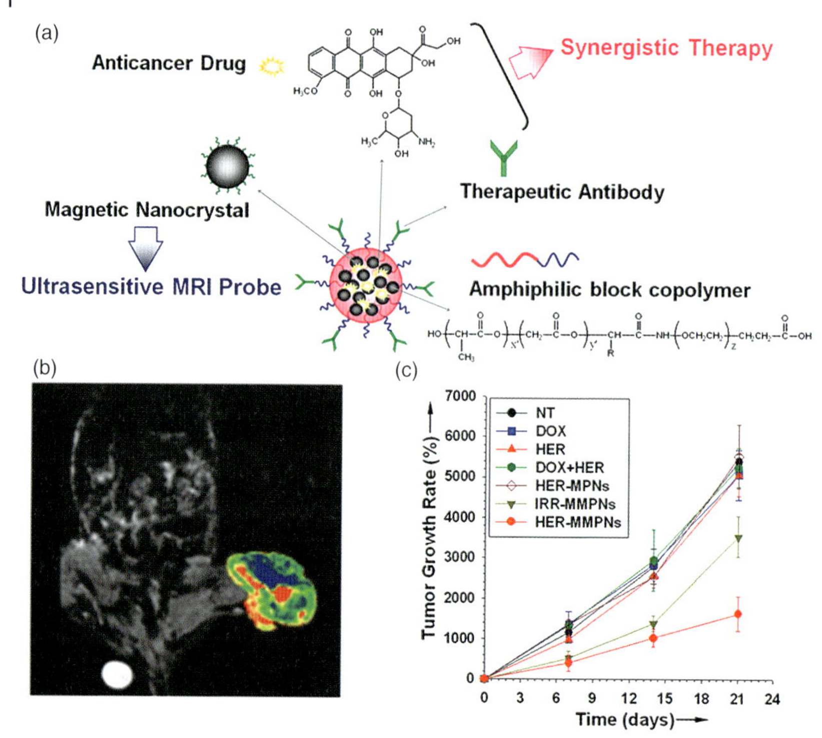

Figure 8.5 (a) Schematic illustration for the fabrication of multifunctional magneto-polymeric nanohybrid (MMPNs); (b) MR images and their tumor region of cancer-targeting events of HER-MMPNs in NIH3T6.7 cells implanted in mice after intravenous injection; (c) Comparative therapeutic efficacy study in an *in vivo* model. Reproduced with permission from Ref. [12].

when synergized with even small amounts of drugs (doxorubicin or paclitaxel) was successful in inducing cancer cell death [177]. The structural designs of the nanocarriers for the co-delivery of drug and gene have been categorized as:

- *Core–shell-type nanocarriers*, composed of hydrophobic and hydrophilic parts for the loading of both chemotherapeutic drugs and genes (Figure 8.6a,b).
- *Ionic polyplexes* which contain chemotherapeutic agents and genes, where an ionic polymer chain is conjugated with chemotherapeutic agents and the polyplex is formulated using counterionic polymer chains for gene loading (Figure 8.6c).

When the *core–shell-type nanocarriers* are delivered to target cells, the loaded chemotherapeutic agents can be continuously released in the intracellular region [16, 31]. Moreover, loaded genes of the cationic outer shell can transfect via a proton sponge effect from the cationic polymer chain. A notable example of this is the co-delivery

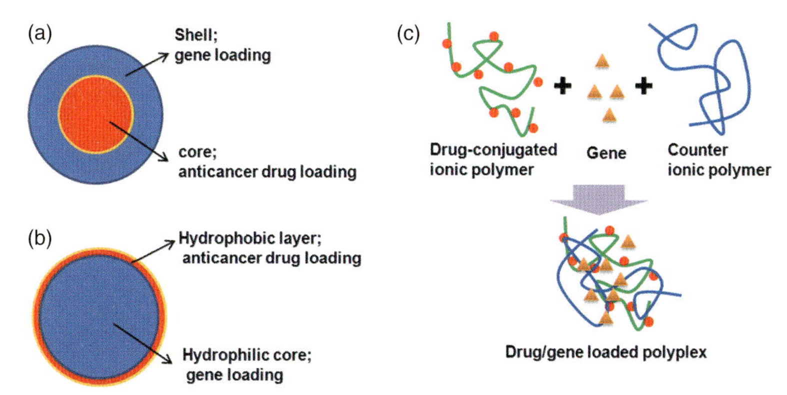

Figure 8.6 Schematic illustration of nanocarrier designs for the co-delivery of chemotherapeutic agents and genes.

of paclitaxel with an interleukin-12 (IL-12)-encoding plasmid by cationic core–shell nanoparticles, which synergistically suppresses cancer growth. In addition, the delivery of siRNAs against Bcl-2, using cationic micelle-type nanoparticles, caused the cancer cells to be sensitized towards paclitaxel, whereas siRNA alone showed no significant cytotoxicity (Figure 8.6). Thus, any synergistic effect associated with the co-delivery of paclitaxel and siRNA might be obtained because a suppression of the anti-apoptotic activity of Bcl-2 by siRNA would cause the cells to be more sensitive to paclitaxel. Other core–shell type nanocarriers, such as PEI–lipid nanocomplex, poly(ethylene oxide)-modified poly(beta-amino ester)/poly(ethylene oxide)-modified poly(epsilon-caprolactone) nanoparticles, and (MPEG-*b*-P[NDAPM-*co*-(HEMA-PCL)]) polymers, have also been developed for simultaneous drug and gene delivery.

Ionic polyplex-type nanocarriers consist of an ionic polymer chain conjugated with chemotherapeutic agents, and a counterionic polymer chain for gene loading. For instance, a self-assembled polyplex is prepared via an electrostatic interaction between protamine sulfate and doxorubicin-conjugated poly(L-aspartic acid) [175]. Subsequently, DNA was loaded into a polyplex that exhibited a comparable gene transfection efficiency to one of 25 kDa PEI. Consequently, the ionic polyplex could suppress the proliferative activity of HeLa cells via the combined delivery of a drug and a gene.

8.3.3.4 **Photothermal or Photodynamic Agents with Chemotherapeutic Drugs**

Recently, much effort has been directed towards the combination of photothermal or photodynamic agents with chemotherapeutic drugs [19, 176, 178, 179]. A localized therapy would be highly attractive for clinical cancer treatment, based on the pinpoint therapeutic accuracy in cancerous regions by using a laser-guided light source (NIR light). Unfortunately, the short penetration depth of a laser light (<1 cm) would require a rather invasive optical fiber-containing needle and/or an endoscope to be inserted into the tumor tissue, in order to access and treat any deep-seated cancerous tissues [19]. A noninvasive nanoparticle-based systemic chemotherapy would, therefore, be highly desirable, as drugs encapsulated in nanocarriers might diffuse over many cancerous regions, inducing the apoptosis of target cancer cells.

Some reports have been made of prototypical combinations of photothermal and chemotherapeutic agents. In this case, the typical nanocarrier design consists of a gold surface (the outer layer), and organic nanoparticles (such as liposomes and biodegradable polymers) which serve as a core and encapsulate the chemotherapeutic agents. The heat generated from gold nanostructures following irradiation with NIR light is sufficient to induce tumoricidal efficacy in localized areas. PLGA nanocarriers which contain doxorubicin covered by a gold over-layer, and with further conjugation of a cetuximab, anti-EGFR antibody, led to multiplexed therapeutic effects by providing: (i) targeting and inhibitory effects from the therapeutic antibody; (ii) a controlled "burst release" of drugs, accelerated by heat via irradiation with NIR light; and (iii) heat from the gold nanoshell (Figure 8.7) [176].

To date, two reports have been made of the nanoparticle-mediated combination of photodynamic and chemotherapy. The sequential combinations of anticancer drugs (doxorubicin) and photodynamic agents (meso-chlorin e_6 monoethylenediamine or 2,5-bis(5-hydroxymethyl-2-thienyl), bound to a *N*-(2-hydroxypropyl)methacrylamide (HPMA) copolymer and conjugated with a targeting moiety (Fab′), demonstrated a strong synergistic effect during the MTT assay against ovarian carcinoma OVCAR-3 cells [180, 181]. A similar strategy, using an aerosol OT™ (AOT)-alginate nanoparticle as a carrier for the simultaneous delivery of doxorubicin and methylene blue (a photodynamic agent) has been reported [178].

8.3.4 Activatable Smart Nanocomposites

Activatable – that is, stimuli-responsive – materials exhibit their conformational or phase changes corresponding to specific environmental variations in biological systems. Although these environmental variations, which include temperature, pH, specific enzymes and reactive oxygen species (ROS), might be small, they can trigger drastic changes in the structures of materials [24, 33, 182–208]. Activatable smart nanocomposites are mainly being developed for target-specific drug delivery systems for cancer treatment (Table 8.3) [209]. Whilst most studies have focused on temperature and pH stimuli, enzyme activity and ROS have been also investigated to stimulate targeted drug delivery carriers and molecular imaging nanoprobes [195, 196].

8.3.4.1 Temperature-Responsive Nanocomposites

A *hydrogel* is a network of polymer chains that are water-insoluble, but which can be chemically synthesized by the crosslinking of hydrophilic macromolecules. The main feature of a hydrogel is its high water-absorbent capability (up to 99%, w/w, of water). The water intake of most hydrogels is significantly influenced by temperature change and the degree of swelling. An increased swelling with increased

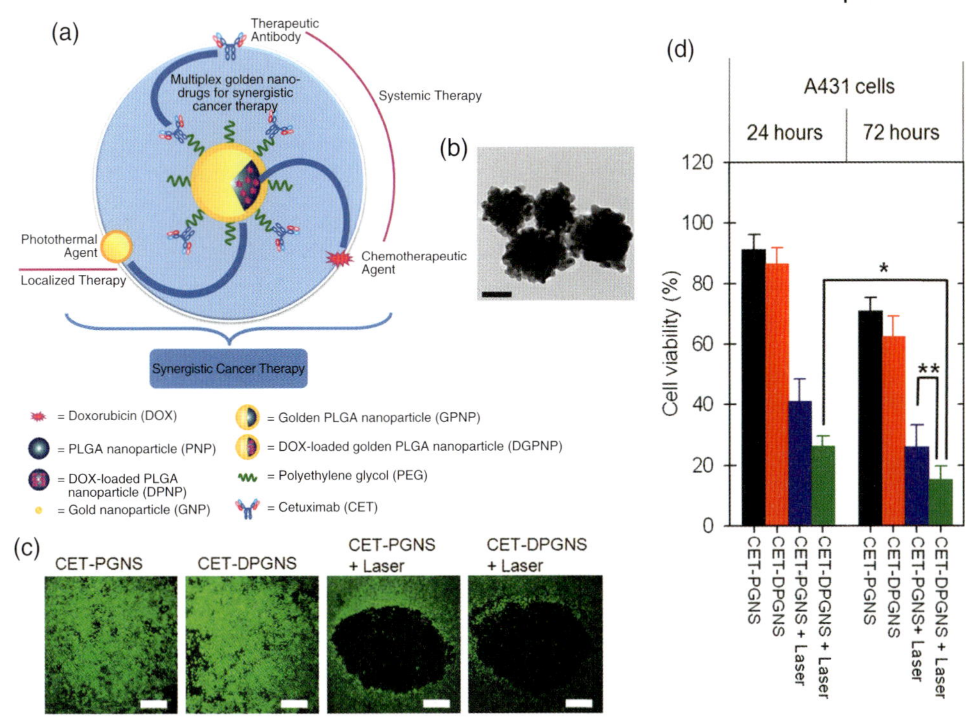

Figure 8.7 (a) Schematic illustration of multifunctional drug-loaded polymeric gold nanoshells (DPGNS) for synergistic cancer therapy; (b) TEM images of DPGNS. Scale bar = 50 nm; (c) Therapeutic efficacies of multifunctional smart polymer gold nanoshells: fluorescence images of A431 cells treated with cetuximab (CET)-PGNS and CET-DPGNS, respectively. After NIR laser exposure (820 nm, 15 W cm^{-2} for 10 min), the cells were further incubated for 48 h. All scale bars = 250 mm; (d) Cell viability of CET-PGNS, CET-DPGNS, IgG-PGNS, and IgG-DPGNS for A431 and MCF7 cells (* and ** both, $p < 0.05$). Reproduced with permission from Ref. [176].

temperature (positive thermosensitivity) is characteristic of polymers with a low critical solution temperature (LCST), whereas a decreased swelling with increased temperature (negative thermosensitivity) is characteristic of polymers having an upper critical solution temperature (UCST). The polymer properties are greatly affected by the temperature of the surroundings, due to the temperature-dependent changes in molecular interactions among polymer chains, or between the polymer and solvent molecules. In addition, the rate of swelling change is modulated by temperature change, as determined from the solute permeability of their membranes [202]. For example, PNIPAAm composed of hydrophobic comonomers (*n*-butylmethyacrylate) will shrink when the temperature is above the phase transition temperature (LCST) [204, 208]. Other PNIPAAm-based polymer networks demonstrate similar tendencies [57, 211].

A drug in the hydrophobic core will move out into the outer layer through diffusion, the process being accompanied by a swelling/deswelling behavior due to

Table 8.3 Activation mechanisms of stimuli-responsive nanocomposites and their applications.

Stimuli	Activation mechanism	Activatable materials	Application	Reference(s)
Temperature	Dehydration on heating	PNIPAAm-based copolymer PEG-Polyester Elastin-like polypeptide	Drug delivery	[57, 197–208]
pH	Acid-induced internal structural change (shrinking/ swelling or bond cleavage)	Polyacrylic acid-based copolymer	Drug delivery	[182–184, 193]
		Imidazole-based polymers	Drug delivery	[190, 191]
		Acetal-bond containing polymer	Drug delivery	[187, 192]
		Hydrazone-bond containing polymer	Drug delivery	[192, 194]
	Degradation	Calcium phosphate	Drug/gene delivery	[33, 86, 87]
Enzyme	Drug or imaging agents release due to specific peptide bond cleavage by enzyme (cathepsin or matrix metalloproteinase)	Enzymatic cleavable peptide	Drug delivery and imaging	[185, 186, 210]
Reactive oxygen species (ROS)	Oligomerization of monomers or chemical bond cleavage by ROS generated from myeloperoxidase	Monomers Blocked fluorophores	Drug delivery and imaging	[195, 196]

temperature changes. The drug can then be released from the temperature-responsive hydrogel-based outer layer because hydrogen bonding among the polymer network will become weakened with increasing temperature (Figure 8.8).

Various temperature-responsive nanocomposites which exhibit different activatable temperature ranges have been synthesized as core–shell-type micelles consisting of a thermosensitive PNIPAAm-based outer layer and a drug-loading hydrophobic core [199–201, 206, 207]. Biodegradable polymer-based drug carriers

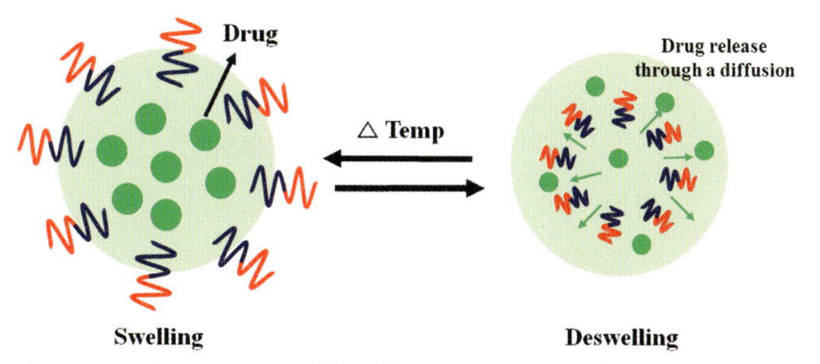

Figure 8.8 The squeezing effect of temperature-modulated drug release from temperature-responsive nanocomposites.

have also been synthesized using tri-block-copolymers (e.g., PLLA–PEG–PLLA), that can exhibit interior structural changes in the PLLA portion [197]. Additionally, *elastin-like polypeptides* are thermally responsive biopolymers that undergo an inverse phase transition (LCST) in response to an increase in temperature [198, 203, 205].

8.3.4.2 pH-Responsive Nanocomposites

The development of various pH-responsive polymers as cancer-targeted drug delivery systems has been prompted because the extracellular pH of the tumor tissue is lower (pH 5~6) than that of normal tissues (pH 7.4) (Figure 8.9) [33, 182–184, 187–194, 212–214]. pH-responsive polymers, employing acid-labile (degradable) bonds onto cationic gene-loaded polymer backbones, have also been employed for gene delivery [188, 189]. The activation mechanisms of pH-responsive nanocomposites are classified by: (i) a destabilization resulting from the ionization of pH-sensitive groups (e.g., imidazole) that bind to protons in the acidic condition; (ii) acid-induced internal structural changes due to a decrease in the permeability of the polymer membrane to solutes; and (iii) a low pH-induced bond cleavage, as in acetal and hydrazone bonds, through acid-catalyzed hydrolysis.

A unique pH-responsive inorganic nanocomposite using CNTs has been recently developed by the present authors, for targeted drug delivery to tumors. The π–π interaction of aromatic molecules to CNTs varied due to ionization by pH changes, and resulted in drug release. Moreover, the increased hydrophilicity of drugs at a lower pH stimulates the protonation of $-NH_2$ groups on the drug, thus reducing the hydrophobic interaction between the drug and the CNTs, consequently enhancing the rate of drug release. Another inorganic nanocarrier, a pH-responsive $CaPO_4$ nanoparticle, has been also used for cancer cell-specific siRNA delivery [33].

8.3.4.3 Enzyme-Responsive Nanocomposites

The specific enzymatic cleavage of a protein precursor (typically peptide cleavage) can be utilized for effective drug delivery or imaging. The release of therapeutic or imaging agents can be triggered by the cleavage of peptide chain, serving as

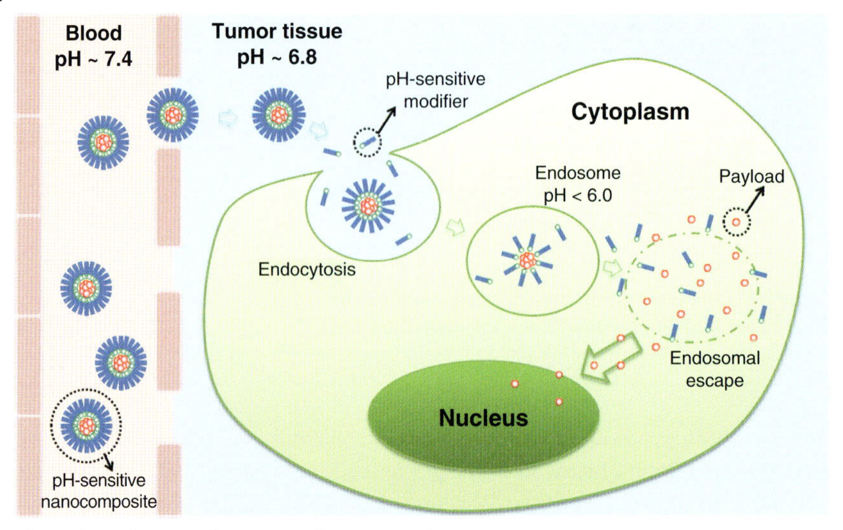

Figure 8.9 Schematic illustration of pH-responsive nanocomposites for tumor-targeting and intracellular release of payloads. Passive tissue targeting is achieved by a leaky tumor vasculature and ineffective lymphatic drainage (EPR effect). Active intracellular degradation can be initiated by functionalized surface modifier of nanocomposites [214].

a crosslinker between the agent and the carrier, in the presence of specific enzymes or protein antigens derived from specific disease cells. A notable example of this is the hydrolysis of an oligopeptide sequence glycyl-phenylalanyl-leucyl-glycine (GFLG) by cathepsin B, which has been implicated in the progression of various human tumors, including ovarian cancer [185, 186]. The MMPs have been recognized as another attractive enzyme system associated with cancer metastasis [210].

8.3.4.4 **Reactive Oxygen Species-Responsive Nanocomposites**

The ROS are highly reactive oxygen ions, free radicals and peroxides that are the natural byproducts of normal oxygen metabolism. However, ROS are also associated with the cellular activities for a variety of inflammatory responses, particularly in cardiovascular disease. Thus, the detection of ROS activity in the human body is very important for the diagnosis of numerous diseases. To date, myeloperoxidase (MPO) and hydrogen peroxide-responsive nanocomposites have been developed for *in vivo* imaging. Briefly, an HOCl-selective nanosized imaging probe, sulfonaphthoaminophenyl fluorescein (SNAPF), was used for *in vivo* imaging of the MPO-derived oxidant hypochlorous acid ($HOCl/OCl^-$) which is implicated in the pathogenesis of atherosclerosis and other inflammatory states [195, 196]. Nanoparticles formulated from peroxalate esters and fluorescent dyes can also be used to provide *in vivo* molecular imaging of hydrogen peroxide, with high specificity and sensitivity [215].

8.4 Conclusions and Outlook

The practical and widespread application of multifunctional nanocomposites in medical fields represents a major goal, mainly because many obstacles related to issues of toxicity and efficacy in the clinical application of nanocomposites must first be overcome. At present, the clear winners among a variety of components for nanocomposites appear to be the metal oxides, Fe_3O_4, $MnFe_2O_4$, and MnO, for MRI purposes, dendrimers and biodegradable polymers for drug delivery, and therapeutic biomolecules such as siRNA and antibodies, because they do not exhibit any salient toxicity, at least in short-term studies. Whilst it is quite unlikely that heavy metal-containing QDs will ever achieve practical *in vivo* applications, there is no doubt that these materials represent an excellent opportunity for *in vitro* molecular imaging and sensing. Yet, benefits will undoubtedly be acquired from the new knowledge regarding molecular events in biological systems, gathered from the use of QDs in animal models. Porous silicon nanoparticles, silica nanospheres and microspheres, CNTs, gold nanorods and nanoshells, lanthanide group metal oxides, with less toxicity issues in short-term studies, have yet to prove that they are safe for longer-term and frequent administration. Likewise, nanocomposites administered frequently therapeutic agents might have to undergo rigorous examinations, whereas nanocomposites used as image contrast agents, in small doses and less frequently administered, might be accepted with less resistance by the general medical fields.

In spite of the many discussions concerning toxicity and safety issues, it is believed that multifunctional nanocomposites will continue to attract interest well into the future. This is because: (i) new nanomaterials with new properties are emerging at a great rate; (ii) there is today a better understanding of cancer and other disease biology, and such knowledge can only be expanded; (iii) there will be a continuous quest for ways to make nanocomposites safer in the body; and – most importantly – (iv) nanocomposites will provide new properties never previously seen, to provide a better understanding of diseases, and more effective cures. The most notable new properties include multimodal imaging, theragnostics, multitherapy and activatability, all of which are based on the fact that nanocomposites may contain many different components. While conditions such as cancer, Parkinson's disease and Alzheimer's disease remain difficult to cure, with many past attempts at cure having failed, contemporary research groups, using conventional tools, have achieved only marginal success. Today, it seems that the only tools left are human ingenuity and a tool box containing artificial or naturally occurring nanomaterials that can be easily combined with existing materials. Yet, there remains among research groups a strong belief that multifunctional nanocomposites will soon find practical medical applications to provide cures for these most daunting diseases.

Acknowledgments

The present authors' studies were supported financially by MOST (KOSEF M10755020001-08N5502-00110) and MIHWAF (the Korea Health 21 R&D Project: A085136).

References

1 Ferrari, M. (2005) Cancer nanotechnology: opportunities and challenges. *Nature Reviews Cancer*, **5**, 161–71.

2 McNeil, S.E. (2009) Nanoparticle therapeutics: a personal perspective. *Wiley Interdisciplinary Reviews: Nanomedicine and Nanobiotechnology*, **1**, 264–71.

3 Mulder, W.J.M., Strijkers, G.J., van Tilborg, G.A.F., Cormode, D.P., Fayad, Z.A. and Nicolay, K. (2009) Nanoparticulate assemblies of amphiphiles and diagnostically active materials for multimodality imaging. *Accounts of Chemical Research*, **42**, 904–14.

4 Park, K., Lee, S., Kang, E., Kim, K., Choi, K. and Kwon, I.C. (2009) New generation of multifunctional nanoparticles for cancer imaging and therapy. *Advanced Functional Materials*, **19**, 1553–66.

5 Bawarski, W.E., Chidlowsky, E., Bharali, D.J. and Mousa, S.A. (2008) Emerging nanopharmaceuticals. *Nanomedicine*, **4**, 273–82.

6 Jun, Y.-W., Lee, J.-H. and Cheon, J. (2008) Chemical design of nanoparticle probes for high-performance magnetic resonance imaging. *Angewandte Chemie, International Edition*, **47**, 5122–35.

7 Gao, J., Gu, H. and Xu, B. (2009) Multifunctional magnetic nanoparticles: design, synthesis, and biomedical applications. *Accounts of Chemical Research*, **42**, 1097–107.

8 Pan, D., Lanza, G.M., Wickline, S.A. and Caruthers, S.D. (2009) Nanomedicine: perspective and promises with ligand-directed molecular imaging. *European Journal of Radiology*, **70**, 274–85.

9 Suh, W.H., Suslick, K.S., Stucky, G.D. and Suh, Y.-H. (2009) Nanotechnology, nanotoxicology, and neuroscience. *Progress in Neurobiology*, **87**, 133–70.

10 Michalet, X., Pinaud, F.F., Bentolila, L.A., Tsay, J.M., Doose, S., Li, J.J., Sundaresan, G., Wu, A.M., Gambhir, S.S. and Weiss, S. (2005) Quantum dots for live cells, *in vivo* imaging, and diagnostics. *Science*, **307**, 538–44.

11 Kostarelos, K., Bianco, A. and Prato, M. (2009) Promises, facts and challenges for carbon nanotubes in imaging and therapeutics. *Nature Nanotechnology*, **4**, 627–33.

12 Yang, J., Lee, C.-H., Ko, H.-J., Suh, J.-S., Yoon, H.-G., Lee, K., Huh, Y.-M. and Haam, S. (2007) Multifunctional magneto-polymeric nanohybrids for targeted detection and synergistic therapeutic effects on breast cancer. *Angewandte Chemie, International Edition*, **46**, 8836–9.

13 Lee, J.-H., Huh, Y.-M., Jun, Y.-W., Seo, J.-W., Jang, J.-T., Song, H.-T., Kim, S., Cho, E.-J., Yoon, H.-G., Suh, J.-S. and Cheon, J. (2007) Artificially engineered magnetic nanoparticles for ultra-sensitive molecular imaging. *Nature Medicine*, **13**, 95–9.

14 Yang, J., Lee, C.-H., Park, J., Seo, S., Lim, E.-K., Song, Y.J., Suh, J.-S., Yoon, H.-G., Huh, Y.-M. and Haam, S. (2007) Antibody conjugated magnetic PLGA nanoparticles for diagnosis and

treatment of breast cancer. *Journal of Materials Chemistry*, **17**, 2695–9.

15 Eaton, M. (2007) Nanomedicine: industry-wise research. *Nature Materials*, **6**, 251–3.

16 Wang, Y., Gao, S., Ye, W.-H., Yoon, H.S. and Yang, Y.-Y. (2006) Co-delivery of drugs and DNA from cationic core-shell nanoparticles self-assembled from a biodegradable copolymer. *Nature Materials*, **5**, 791–6.

17 Chen, J., Wang, D., Xi, J., Au, L., Siekkinen, A., Warsen, A., Li, Z.Y., Zhang, H., Xia, Y. and Li, X. (2007) Immuno gold nanocages with tailored optical properties for targeted photothermal destruction of cancer cells. *Nano Letters*, **7**, 1318–22.

18 Huang, X., El-Sayed, I.H., Qian, W. and El-Sayed, M.A. (2006) Cancer cell imaging and photothermal therapy in the near-infrared region by using gold nanorods. *Journal of the American Chemical Society*, **128**, 2115–20.

19 Hirsch, L.R., Stafford, R.J., Bankson, J.A., Sershen, S.R., Rivera, B., Price, R.E., Hazle, J.D., Halas, N.J. and West, J.L. (2003) Nanoshell-mediated near-infrared thermal therapy of tumors under magnetic resonance guidance. *Proceedings of the National Academy of Sciences of the United States of America*, **100**, 13549–54.

20 Hanahan, D. and Weinberg, R.A. (2000) The hallmarks of cancer. *Cell*, **100**, 57–70.

21 Phan, J.H., Moffitt, R.A., Stokes, T.H., Liu, J., Young, A.N., Nie, S. and Wang, M.D. (2009) Convergence of biomarkers, bioinformatics and nanotechnology for individualized cancer treatment. *Trends in Biotechnology*, **27**, 350–8.

22 Singh, R. and Lillard, J.W. Jr (2009) Nanoparticle-based targeted drug delivery. *Experimental and Molecular Pathology*, **86**, 215–23.

23 Peer, D., Karp, J.M., Hong, S., Farokhzad, O.C., Margalit, R. and Langer, R. (2007) Nanocarriers as an emerging platform for cancer therapy. *Nature Nanotechnology*, **2**, 751–60.

24 Hongrapipat, J., Kopeckov, P., Prakongpan, S. and Kopecek, J. (2008) Enhanced antitumor activity of combinations of free and HPMA copolymer-bound drugs. *International Journal of Pharmaceutics*, **351**, 259–70.

25 Ubr, V., Strohalm, J., Hirano, T., Ito, Y. and Ulbrich, K. (1997) Poly[*N*-(2-hydroxypropyl)methacrylamide] conjugates of methotrexate: synthesis and *in vitro* drug release. *Journal of Controlled Release*, **49**, 123–32.

26 Jung, Y., Park, H.-J., Kim, P.-H., Lee, J., Hyung, W., Yang, J., Ko, H., Sohn, J.-H., Kim, J.-H., Huh, Y.-M., Yun, C.-O. and Haam, S. (2007) Retargeting of adenoviral gene delivery via Herceptin-PEG-adenovirus conjugates to breast cancer cells. *Journal of Controlled Release*, **123**, 164–71.

27 Akinc, A., Thomas, M., Klibanov, A.M. and Langer, R. (2005) Exploring polyethylenimine-mediated DNA transfection and the proton sponge hypothesis. *Journal of Gene Medicine*, **7**, 657–63.

28 Zhang, C., Yadava, P. and Hughes, J. (2004) Polyethylenimine strategies for plasmid delivery to brain-derived cells. *Methods*, **33**, 144–50.

29 Yezhelyev, M.V., Qi, L., O'Regan, R.M., Nie, S. and Gao, X. (2008) Proton-sponge coated quantum dots for siRNA delivery and intracellular imaging. *Journal of the American Chemical Society*, **130**, 9006–12.

30 Hatakeyama, H., Ito, E., Akita, H., Oishi, M., Nagasaki, Y., Futaki, S. and Harashima, H. (2009) A pH-sensitive fusogenic peptide facilitates endosomal escape and greatly enhances the gene silencing of siRNA-containing nanoparticles *in vitro* and *in vivo*. *Journal of Controlled Release*, **139**, 127–32.

31 Liu, C., Zhao, G., Liu, J., Ma, N., Chivukula, P., Perelman, L., Okada, K., Chen, Z., Gough, D. and Yu, L. (2009) Novel biodegradable lipid nano complex for siRNA delivery significantly improving the chemosensitivity of human colon cancer stem cells to paclitaxel. *Journal of Controlled Release*. doi: 10.1016/j.jconrel.2009.08.013

32 Suh, M.S., Shim, G., Lee, H.Y., Han, S.-E., Yu, Y.-H., Choi, Y., Kim, K., Kwon, I.C., Weon, K.Y., Kim, Y.B. and Oh, Y.-K. Anionic amino acid-derived cationic lipid for siRNA delivery. *Journal of Controlled Release*, **140** (3), 268–76.

33 Zhang, M., Ishii, A., Nishiyama, N., Matsumoto, S., Ishii, T., Yamasaki, Y. and Kataoka, K. (2009) PEGylated calcium phosphate nanocomposites as smart environment-sensitive carriers for siRNA delivery. *Advanced Materials*, **21**, 3520–5.

34 Chia-Hung, L., Shih-Hsun, C., Yu-Jing, W., Yu-Ching, C., Nai-Tzu, C., Jeffrey, S., Chin-Tu, C., Chung-Yuan, M., Chung-Shi, Y. and Leu-Wei, L. (2009) Near-infrared mesoporous silica nanoparticles for optical imaging: characterization and *in vivo* biodistribution. *Advanced Functional Materials*, **19**, 215–22.

35 Vuu, K., Xie, J., McDonald, M.A., Bernardo, M., Hunter, F., Zhang, Y., Li, K., Bednarski, M. and Guccione, S. (2005) Gadolinium-rhodamine nanoparticles for cell labeling and tracking via magnetic resonance and optical imaging. *Bioconjugate Chemistry*, **16**, 995–9.

36 Kim, J.-H., Lee, S., Park, K., Nam, H.Y., Jang, S.Y., Youn, I., Kim, K., Jeon, H., Park, R.-W., Kim, I.-S., Choi, K. and Kwon, I.C. (2007) Protein-phosphorylation-responsive polymeric nanoparticles for imaging protein kinase activities in single living cells. *Angewandte Chemie, International Edition*, **46**, 5779–82.

37 Smith, A.M., Duan, H., Mohs, A.M. and Nie, S. (2008) Bioconjugated quantum dots for *in vivo* molecular and cellular imaging. *Advanced Drug Delivery Reviews*, **60**, 1226–40.

38 Gao, X., Yang, L., Petros, J.A., Marshall, F.F., Simons, J.W. and Nie, S. (2005) *In vivo* molecular and cellular imaging with quantum dots. *Current Opinion in Biotechnology*, **16**, 63–72.

39 Sharma, P., Brown, S., Walter, G., Santra, S. and Moudgil, B. (2006) Nanoparticles for bioimaging. *Advances in Colloid and Interface Science*, **123–126**, 471–85.

40 Jun, Y.-W., Huh, Y.-M., Choi, J.-S., Lee, J.-H., Song, H.-T., Kim, S., Yoon, S., Kim, K.-S., Shin, J.-S., Suh, J.-S. and Cheon, J. (2005) Nanoscale size effect of magnetic nanocrystals and their utilization for cancer diagnosis via magnetic resonance imaging. *Journal of the American Chemical Society*, **127**, 5732–3.

41 Sun, C., Lee, J.S.H. and Zhang, M. (2008) Magnetic nanoparticles in MR imaging and drug delivery. *Advanced Drug Delivery Reviews*, **60**, 1252–65.

42 Yang, J., Lee, T.-I., Lee, J., Lim, E.-K., Hyung, W., Lee, C.-H., Song, Y.J., Suh, J.-S., Yoon, H.-G., Huh, Y.-M. and Haam, S. (2007) Synthesis of ultrasensitive magnetic resonance contrast agents for cancer imaging using PEG-fatty acid. *Chemistry of Materials*, **19**, 3870–6.

43 Na, H.B., Lee, J.H., An, K., Park, Y.I., Park, M., Lee, I.S., Nam, D.-H., Kim, S., Kim, S.-H., Kim, S.-W., Lim, K.-H., Kim, K.-S., Kim, S.-O. and Hyeon, T. (2007) Development of a T1 contrast agent for magnetic resonance imaging using MnO nanoparticles. *Angewandte Chemie, International Edition*, **46**, 5397–401.

44 Kim, H., Lee, M., Kim, Y., Huh, J., Kim, H., Kim, M., Kim, T., Phan, V., Lee, Y.-B., Yi, G.-R., Haam, S. and Lee, K. (2009) Quantitative assessment of nanoparticle single crystallinity: palladium-catalyzed splitting of polycrystalline metal oxide nanoparticles. *Angewandte Chemie, International Edition*, **48**, 5129–33.

45 Xu, C., Tung, G.A. and Sun, S. (2008) Size and concentration effect of gold nanoparticles on X-ray attenuation as measured on computed tomography. *Chemistry of Materials*, **20**, 4167–9.

46 Sharma, P., Brown, S.C., Walter, G., Santra, S., Scott, E., Ichikawa, H., Fukumori, Y. and Moudgil, B.M. (2007) Gd nanoparticulates: from magnetic resonance imaging to neutron capture therapy. *Advanced Powder Technology*, **18**, 663–98.

47 Bridot, J.-L., Faure, A.-C., Laurent, S., Riviere, C., Billotey, C., Hiba, B., Janier, M., Josserand, V., Coll, J.-L., Vander Elst, L., Muller, R., Roux, S., Perriat, P. and Tillement, O. (2007) Hybrid gadolinium oxide nanoparticles: multimodal contrast agents for *in vivo* imaging. *Journal of the American Chemical Society*, **129**, 5076–84.

48 Carling, C.-J., Boyer, J.-C. and Branda, N.R. (2009) Remote-control photoswitching using NIR light. *Journal of the American Chemical Society*, **131**, 10838–9.

49 Soundararajan, A., Bao, A., Phillips, W.T., Perez Iii, R. and Goins, B.A. (2009) [186Re]Liposomal doxorubicin (Doxil): *in vitro* stability, pharmacokinetics, imaging and biodistribution in a head and neck squamous cell carcinoma xenograft model. *Nuclear Medicine and Biology*, **36**, 515–24.

50 Wang, S.X., Bao, A., Phillips, W.T., Goins, B., Herrera, S.J., Santoyo, C., Miller, F.R. and Otto, R.A. (2009) Intraoperative therapy with liposomal drug delivery: retention and distribution in human head and neck squamous cell carcinoma xenograft model. *International Journal of Pharmaceutics*, **373**, 156–64.

51 Levine, D.H., Ghoroghchian, P.P., Freudenberg, J., Zhang, G., Therien, M.J., Greene, M.I., Hammer, D.A. and Murali, R. (2008) Polymersomes: a new multi-functional tool for cancer diagnosis and therapy. *Methods*, **46**, 25–32.

52 Reynolds, A.R., Moein Moghimi, S. and Hodivala-Dilke, K. (2003) Nanoparticle-mediated gene delivery to tumour neovasculature. *Trends in Molecular Medicine*, **9**, 2–4.

53 Jin, S. and Ye, K. (2007) Nanoparticle-mediated drug delivery and gene therapy. *Biotechnology Progress*, **23**, 32–41.

54 Brannon-Peppas, L. and Blanchette, J.O. (2004) Nanoparticle and targeted systems for cancer therapy. *Advanced Drug Delivery Reviews*, **56**, 1649–59.

55 Wickline, S.A., Neubauer, A.M., Winter, P.M., Caruthers, S.D. and Lanza, G.M. (2007) Molecular imaging and therapy of atherosclerosis with targeted nanoparticles. *Journal of Magnetic Resonance Imaging*, **25**, 667–80.

56 Kim, J.-H., Park, K., Nam, H.Y., Lee, S., Kim, K. and Kwon, I.C. (2007) Polymers for bioimaging. *Progress in Polymer Science*, **32**, 1031–53.

57 Wang, Z.-C., Xu, X.-D., Chen, C.-S., Wang, G.-R., Wang, B., Zhang, X.-Z. and Zhuo, R.-X. (2008) Study on novel hydrogels based on thermosensitive PNIPAAm with pH sensitive PDMAEMA grafts. *Colloid and Surface B: Biointerfaces*, **67**, 245–52.

58 Etrych, T., Jelinkova, M.A., Blanka, R.O. and Ulbrich, K. (2001) New HPMA copolymers containing doxorubicin bound via pH-sensitive linkage: synthesis and preliminary *in vitro* and *in vivo* biological properties. *Journal of Controlled Release*, **73**, 89–102.

59 Soppimath, K.S., Liu, L.-H., Seow, W.Y., Liu, S.-Q., Powell, R., Chan, P. and Yang, Y.Y. (2007) Multifunctional core/shell nanoparticles self-assembled from pH-induced thermosensitive polymers for targeted intracellular anticancer drug delivery. *Advanced Functional Materials*, **17**, 355–62.

60 Xiao, K., Luo, J., Fowler, W.L., Li, Y., Lee, J.S., Xing, L., Cheng, R.H., Wang, L. and Lam, K.S. (2009) A self-assembling nanoparticle for paclitaxel delivery in ovarian cancer. *Biomaterials*, **30**, 6006–16.

61 Hrub, M., Kon, C. and Ulbrich, K. (2005) Polymeric micellar pH-sensitive drug delivery system for doxorubicin. *Journal of Controlled Release*, **103**, 137–48.

62 Roy, I., Ohulchanskyy, T.Y., Pudavar, H.E., Bergey, E.J., Oseroff, A.R., Morgan, J., Dougherty, T.J. and Prasad, P.N. (2003) Ceramic-based nanoparticles entrapping water-insoluble photosensitizing anticancer drugs: a novel drug carrier system for photodynamic therapy. *Journal of the American Chemical Society*, **125**, 7860–5.

63 Lee, H., Hu, M., Reilly, R.M. and Allen, C. (2007) Apoptotic epidermal growth factor (EGF)-conjugated block copolymer micelles as a nanotechnology platform for targeted combination therapy. *Molecular Pharmaceutics*, **4**, 769–81.

64 Sahoo, S.K. and Labhasetwar, V. (2003) Nanotech approaches to drug delivery and imaging. *Drug Discovery Today*, **8**, 1112–20.

65 Saad, M., Garbuzenko, O.B., Ber, E., Chandna, P., Khandare, J.J., Pozharov, V.P. and Minko, T. (2008) Receptor targeted polymers, dendrimers, liposomes: which nanocarrier is the most efficient for tumor-specific treatment and imaging? *Journal of Controlled Release*, **130**, 107–14.

66 Pope-Harman, A., Cheng, M.M.-C., Robertson, F., Sakamoto, J. and Ferrari, M. (2007) Biomedical nanotechnology for cancer. *The Medical Clinics of North America*, **91**, 899–927.

67 Chang, E.H. (2007) Nanomedicines: improving current cancer therapies and diagnosis. *Nanomedicine*, **3**, 339–339.

68 Sengupta, S., Eavarone, D., Capila, I., Zhao, G., Watson, N., Kiziltepe, T. and Sasisekharan, R. (2005) Temporal targeting of tumour cells and neovasculature with a nanoscale delivery system. *Nature*, **436**, 568–72.

69 Larpent, C., Cannizzo, C., Delgado, A., Gouanve, F., Sanghvi, P., Gaillard, C. and Bacquet, G. (2008) Convenient synthesis and properties of polypropyleneimine dendrimer-functionalized polymer nanoparticles. *Small*, **4**, 833–40.

70 King Heiden, T.C., Dengler, E., Kao, W.J., Heideman, W. and Peterson, R.E. (2007) Developmental toxicity of low generation PAMAM dendrimers in zebrafish. *Toxicology and Applied Pharmacology*, **225**, 70–9.

71 Kobayashi, H. and Brechbiel, M.W. (2005) Nano-sized MRI contrast agents with dendrimer cores. *Advanced Drug Delivery Reviews*, **57**, 2271–86.

72 Yang, W., Cheng, Y., Xu, T., Wang, X. and Wen, L.-P. (2009) Targeting cancer cells with biotin-dendrimer conjugates. *European Journal of Medicinal Chemistry*, **44**, 862–8.

73 Florence, A.T. and Hussain, N. (2001) Transcytosis of nanoparticle and dendrimer delivery systems: evolving vistas. *Advanced Drug Delivery Reviews*, **50**, S69–S89.

74 Lin, Y.-S. and Haynes, C.L. (2009) Synthesis and characterization of biocompatible and size-tunable multifunctional porous silica nanoparticles. *Chemistry of Materials*, **21**, 3979–86.

75 Kim, H.-J., Shin, K.-J., Han, M.K., An, K., Lee, J.-K., Honma, I. and Kim, H. (2009) One-pot synthesis of multifunctional mesoporous silica nanoparticle incorporated with zinc(II) phthalocyanine and iron oxide. *Scripta Materialia*, **61** (12), 1137–40.

76 Wu, S.-H., Lin, Y.-S., Hung, Y., Chou, Y.-H., Hsu, Y.-H., Chang, C. and Mou, C.-Y. (2008) Multifunctional mesoporous silica nanoparticles for intracellular labeling and animal magnetic resonance imaging studies. *ChemBioChem*, **9**, 53–7.

77 Qian, H.S., Guo, H.C., Ho, P.C.-L., Mahendran, R. and Zhang, Y. (2009) Mesoporous-silica-coated up-conversion fluorescent nanoparticles for photodynamic therapy. *Small*, **5** (20), 2285–90.

78 Moreno, M.S., Weyland, M., Midgley, P.A., Bengoa, J.F., Cagnoli, M.V., Gallegos, N.G., Alvarez, A.M. and Marchetti, S.G. (2006) Highly anisotropic distribution of iron nanoparticles within MCM-41 mesoporous silica. *Micron*, **37**, 52–6.

79 Trewyn, B.G., Nieweg, J.A., Zhao, Y. and Lin, V.S.Y. (2008) Biocompatible mesoporous silica nanoparticles with different morphologies for animal cell membrane penetration. *Chemical Engineering Journal*, **137**, 23–9.

80 Son, S.J., Bai, X. and Lee, S.B. (2007) Inorganic hollow nanoparticles and nanotubes in nanomedicine: Part 2: imaging, diagnostic, and therapeutic applications. *Drug Discovery Today*, **12**, 657–63.

81 Wu, G., Mikhailovsky, A., Khant, H.A., Fu, C., Chiu, W. and Zasadzinski, J.A. (2008) Remotely triggered liposome release by near-infrared light absorption via hollow gold nanoshells. *Journal of the American Chemical Society*, **130**, 8175–7.

82 Lim, Y.T., Cho, M.Y., Choi, B.S., Noh, Y.-W. and Chung, B.H. (2008) Diagnosis and therapy of macrophage cells using dextran-coated near-infrared responsive hollow-type gold nanoparticles. *Nanotechnology*, **37**, 375105.

83 Link, N., Brunner, T.J., Dreesen, I.A.J., Stark, W.J. and Fussenegger, M. (2007) Inorganic nanoparticles for transfection of mammalian cells and removal of viruses from aqueous solutions. *Biotechnology and Bioengineering*, **98**, 1083–93.

84 Hou, C.-H., Hou, S.-M., Hsueh, Y.-S., Lin, J., Wu, H.-C. and Lin, F.-H. (2009)

The *in vivo* performance of biomagnetic hydroxyapatite nanoparticles in cancer hyperthermia therapy. *Biomaterials*, **30**, 3956–60.

85 Pedraza, C.E., Bassett, D.C., McKee, M.D., Nelea, V., Gbureck, U. and Barralet, J.E. (2008) The importance of particle size and DNA condensation salt for calcium phosphate nanoparticle transfection. *Biomaterials*, **29**, 3384–92.

86 Hou, C.-H., Chen, C.-W., Hou, S.-M., Li, Y.-T. and Lin, F.-H. (2009) The fabrication and characterization of dicalcium phosphate dihydrate-modified magnetic nanoparticles and their performance in hyperthermia processes *in vitro*. *Biomaterials*, **30**, 4700–7.

87 Kester, M., Heakal, Y., Fox, T., Sharma, A., Robertson, G.P., Morgan, T.T., Altinoglu, E.I., Tabakovic, A., Parette, M.R., Rouse, S.M., Ruiz-Velasco, V. and Adair, J.H. (2008) Calcium phosphate nanocomposite particles for *in vitro* imaging and encapsulated chemotherapeutic drug delivery to cancer cells. *Nano Letters*, **8**, 4116–21.

88 Shin, J., Anisur, R.M., Ko, M.K., Im, G.H., Lee, J.H. and Lee, I.S. (2009) Hollow manganese oxide nanoparticles as multifunctional agents for magnetic resonance imaging and drug delivery. *Angewandte Chemie, International Edition*, **48**, 321–4.

89 Guo, S., Li, D., Zhang, L., Li, J. and Wang, E. (2009) Monodisperse mesoporous superparamagnetic single-crystal magnetite nanoparticles for drug delivery. *Biomaterials*, **30**, 1881–9.

90 Yang, P., Quan, Z., Hou, Z., Li, C., Kang, X., Cheng, Z. and Lin, J. (2009) A magnetic, luminescent and mesoporous core-shell structured composite material as drug carrier. *Biomaterials*, **30**, 4786–95.

91 Chen, J., Chen, S., Zhao, X., Kuznetsova, L.V., Wong, S.S. and Ojima, I. (2008) Functionalized single-walled carbon nanotubes as rationally designed vehicles for tumor-targeted drug delivery. *Journal of the American Chemical Society*, **130**, 16778–85.

92 Liu, Z., Cai, W., He, L., Nakayama, N., Chen, K., Sun, X., Chen, X. and Dai, H. (2006) *In vivo* biodistribution and highly efficient tumour targeting of carbon nanotubes in mice. *Nature Nanotechnology*, **2**, 47–52.

93 Bhirde, A.A., Patel, V., Gavard, J., Zhang, G., Sousa, A.A., Masedunskas, A., Leapman, R.D., Weigert, R., Gutkind, J.S. and Rusling, J.F. (2009) Targeted killing of cancer cells *in vivo* and *in vitro* with EGF-directed carbon nanotube-based drug delivery. *ACS Nano*, **3**, 307–16.

94 Piao, Y., Kim, J., Na, H.B., Kim, D., Baek, J.S., Ko, M.K., Lee, J.H., Shokouhimehr, M. and Hyeon, T. (2008) Wrap-bake-peel process for nanostructural transformation from [beta]-FeOOH nanorods to biocompatible iron oxide nanocapsules. *Nature Materials*, **7**, 242–7.

95 Kwak, E.L., Clark, J.W. and Chabner, B. (2007) Targeted agents: the rules of combination. *Clinical Cancer Research*, **13**, 5232–7.

96 Nimmagadda, S., Ford, E.C., Wong, J.W. and Pomper, M.G. (2008) Targeted molecular imaging in oncology: focus on radiation therapy. *Seminars in Radiation Oncology*, **18**, 136–48.

97 Pegram, M.D., Pietras, R., Bajamonde, A., Klein, P. and Fyfe, G. (2005) Targeted therapy: wave of the future. *Journal of Clinical Oncology*, **23**, 1776–81.

98 Merlin, J.-L., Barberi-Heyob, M. and Bachmann, N. (2002) *In vitro* comparative evaluation of trastuzumab (Herceptin®) combined with paclitaxel (Taxol®) or docetaxel (Taxotere®) in HER2-expressing human breast cancer cell lines. *Annals of Oncology*, **13**, 1743–8.

99 Ito, A., Kuga, Y., Honda, H., Kikkawa, H., Horiuchi, A., Watanabe, Y. and Kobayashi, T. (2004) Magnetite nanoparticle-loaded anti-HER2 immunoliposomes for combination of antibody therapy with hyperthermia. *Cancer Letters*, **212**, 167–75.

100 Ferrara, N. (2004) Vascular endothelial growth factor: basic science and clinical progress. *Endocrine Reviews*, **25**, 581–611.

101 Lee, J., Yang, J., Seo, S.-B., Ko, H.-J., Suh, J.-S., Huh, Y.-M. and Haam, S. (2008) Smart nanoprobes for ultrasensitive detection of breast cancer via magnetic resonance imaging. *Nanotechnology*, **19**, 485101.

102 Carter, P. (2001) Improving the efficacy of antibody-based cancer therapies. *Nature Reviews Cancer*, **1**, 118–29.

103 Allen, T.M. (2002) Ligand-targeted therapeutics in anticancer therapy. *Nature Reviews Cancer*, **2**, 750–63.

104 Phillips, J.A., Lopez-Colon, D., Zhu, Z., Xu, Y. and Tan, W. (2008) Applications of aptamers in cancer cell biology. *Analytica Chimica Acta*, **621**, 101–8.

105 Pestourie, C., Tavitian, B. and Duconge, F. (2005) Aptamers against extracellular targets for *in vivo* applications. *Biochimie*, **87**, 921–30.

106 Wang, A.Z., Bagalkot, V., Gu, F., Alexis, F., Vasilliou, C., Cima, M., Jon, S. and Farokhzad, O. (2007) Novel targeted aptamer-superparamagnetic iron oxide nanoparticle bioconjugates for combined prostate cancer imaging and therapy. *International Journal of Radiation Oncology, Biology, Physics*, **69**, S110–11.

107 Wang, L., Liu, B., Yin, H., Wei, J., Qian, X. and Yu, L. (2007) Selection of DNA aptamer that specific binding human carcinoembryonic antigen *in vitro*. *Journal of Nanjing Medical University*, **21**, 277–81.

108 Wang, A., Bagalkot, V., Vasilliou, C., Gu, F., Alexis, F., Zhang, L., Shaikh, M., Yuet, K., Cima, M., Langer, R., Kantoff, P., Bander, N., Jon, S. and Farokhzad, O. (2008) Superparamagnetic iron oxide nanoparticle-aptamer bioconjugates for combined prostate cancer imaging and therapy. *ChemMedChem*, **3**, 1311–15.

109 Farokhzad, O.C., Cheng, J., Teply, B.A., Sherifi, I., Jon, S., Kantoff, P.W., Richie, J.P. and Langer, R. (2006) Targeted nanoparticle-aptamer bioconjugates for cancer chemotherapy *in vivo*. *Proceedings of the National Academy of Sciences of the United States of America*, **103**, 6315–20.

110 Yoo, H.S. and Park, T.G. (2004) Folate-receptor-targeted delivery of doxorubicin nano-aggregates stabilized by doxorubicin-PEG-folate conjugate. *Journal of Controlled Release*, **100**, 247–56.

111 Lin, J.-J., Chen, J.-S., Huang, S.-J., Ko, J.-H., Wang, Y.-M., Chen, T.-L. and Wang, L.-F. (2009) Folic acid-Pluronic F127 magnetic nanoparticle clusters for combined targeting, diagnosis, and therapy applications. *Biomaterials*, **30**, 5114–24.

112 Flanagan, P.A., Duncan, R., Subr, V., Ulbrich, K., Kopeckov, P. and Kopecek, J. (1992) Evaluation of protein-*N*-(2-hydroxypropyl) methacrylamide copolymer conjugates as targetable drug-carriers. 2. Body distribution of conjugates containing transferrin, antitransferrin receptor antibody or anti-Thy 1.2 antibody and effectiveness of transferrin-containing daunomycin conjugates against mouse L1210 leukaemia *in vivo*. *Journal of Controlled Release*, **18**, 25–30.

113 Li, J.-L., Wang, L., Liu, X.-Y., Zhang, Z.-P., Guo, H.-C., Liu, W.-M. and Tang, S.-H. (2009) *In vitro* cancer cell imaging and therapy using transferrin-conjugated gold nanoparticles. *Cancer Letters*, **274**, 319–26.

114 Friedman, M., Orlova, A., Johansson, E., Eriksson, T.L.J., Höidén-Guthenberg, I., Tolmachev, V., Nilsson, F.Y. and Ståhl, S. (2008) Directed evolution to low nanomolar affinity of a tumor-targeting epidermal growth factor receptor-binding affibody molecule. *Journal of Molecular Biology*, **376**, 1388–402.

115 Alexis, F., Basto, P., Levy-Nissenbaum, E., Radovic-Moreno, A.F., Zhang, L., Pridgen, E., Wang, A.Z., Marein, S.L., Westerhof, K., Molnar, L.K. and Farokhzad, O.C. (2008) HER-2-targeted nanoparticle-affibody bioconjugates for cancer therapy. *ChemMedChem*, **3**, 1839–43.

116 Park, S. and Yoo, H.S. (2010) *In vivo* and *in vitro* anti-cancer activities and enhanced cellular uptakes of EGF fragment decorated doxorubicin nano-aggregates. *International Journal of Pharmaceutics*, **383**, 178–85.

117 Lily, Y., Hui, M., Wang, Y.A., Zehong, C., Xianghong, P., Xiaoxia, W., Hongwei, D., Chunchun, N., Qingan, Y., Gregory, A., Mark, Q.S., William, C.W., Xiaohu, G. and Shuming, N. (2009) Single chain epidermal growth factor receptor antibody conjugated nanoparticles for *in vivo* tumor targeting and imaging. *Small*, **5**, 235–43.

118 Acharya, S., Dilnawaz, F. and Sahoo, S.K. (2009) Targeted epidermal growth factor receptor nanoparticle bioconjugates for breast cancer therapy. *Biomaterials*, **30**, 5737–50.

119 Jae-Hyun, L., Kyuri, L., Seung Ho, M., Yuhan, L., Tae Gwan, P. and Jinwoo, C. (2009) All-in-one target-cell-specific magnetic nanoparticles for simultaneous molecular imaging and siRNA Delivery13. *Angewandte Chemie, International Edition*, **48**, 4174–9.

120 Smith, B.R., Cheng, Z., De, A., Koh, A.L., Sinclair, R. and Gambhir, S.S. (2008) Real-time intravital imaging of RGD-quantum dot binding to luminal endothelium in mouse tumor neovasculature. *Nano Letters*, **8**, 2599–606.

121 Temming, K., Schiffelers, R.M., Molema, G. and Kok, R.J. (2005) RGD-based strategies for selective delivery of therapeutics and imaging agents to the tumour vasculature. *Drug Resistance Updates*, **8**, 381–402.

122 Winter, P.M., Caruthers, S.D., Zhang, H., Williams, T.A., Wickline, S.A. and Lanza, G.M. (2008) Antiangiogenic synergism of integrin-targeted fumagillin nanoparticles and atorvastatin in atherosclerosis. *Journal of the American College of Cardiology: Cardiovascular Imaging*, **1**, 624–34.

123 Anne, M.N., Hoon, S., Patrick, M.W., Shelton, D.C., Todd, A.W., Robertson, J.D., David, S., Gregory, M.L. and Samuel, A.W. (2008) Nanoparticle pharmacokinetic profiling *in vivo* using magnetic resonance imaging. *Magnetic Resonance in Medicine*, **60**, 1353–61.

124 Lim, E.H., Danthi, N., Bednarski, M. and Li, K.C.P. (2005) A review: integrin $\alpha_v\beta_3$-targeted molecular imaging and therapy in angiogenesis. *Nanomedicine*, **1**, 110–14.

125 Roberts, M.J., Bentley, M.D. and Harris, J.M. (2002) Chemistry for peptide and protein PEGylation. *Advanced Drug Delivery Reviews*, **54**, 459–76.

126 Lutz, J.-F., Stiller, S., Hoth, A., Kaufner, L., Pison, U. and Cartier, R. (2006) One-pot synthesis of PEGylated ultrasmall iron-oxide nanoparticles and their *in vivo* evaluation as magnetic resonance imaging contrast agents. *Biomacromolecules*, **7**, 3132–8.

127 Moghimi, S.M., Hunter, A.C. and Murray, J.C. (2001) Long-circulating and target-specific nanoparticles: theory to practice. *Pharmacological Reviews*, **53**, 283–318.

128 Suh, W.H., Suh, Y.-H. and Stucky, G.D. (2009) Multifunctional nanosystems at the interface of physical and life sciences. *Nano Today*, **4**, 27–36.

129 Staufer, U., Akiyama, T., Gullo, M.R., Han, A., Imer, R., de Rooij, N.F., Aebi, U., Engel, A., Frederix, P.L.T.M., Stolz, M., Friederich, N.F. and Wirz, D. (2007) Micro- and nanosystems for biology and medicine. *Microelectronic Engineering*, **84**, 1681–4.

130 Seo, S.-B., Yang, J., Lee, E.-S., Jung, Y., Kim, K., Lee, S.-Y., Kim, D., Suh, J.-S., Huh, Y.-M. and Haam, S. (2008) Nanohybrids via a polycation-based nanoemulsion method for dual-mode detection of human mesenchymal stem cells. *Journal of Materials Chemistry*, **18**, 4402–7.

131 Seo, S.-B., Yang, J., Hyung, W., Cho, E.-J., Lee, T.-I., Song, Y.J., Yoon, H.-G., Suh, J.-S., Huh, Y.-M. and Haam, S. (2007) Novel multifunctional PHDCA/PEI nano-drug carriers for simultaneous magnetically targeted cancer therapy and diagnosis via magnetic resonance imaging. *Nanotechnology*, **18**, 475105.

132 Vijayanathan, V., Thomas, T. and Thomas, T.J. (2002) DNA nanoparticles and development of DNA delivery vehicles for gene therapy. *Biochemistry*, **41**, 14085–94.

133 Lynch, Z. (2009) The future of neurotechnology innovation. *Epilepsy and Behavior*, **15**, 120–2.

134 Liu, G., Garrett, M.R., Men, P., Zhu, X., Perry, G. and Smith, M.A. (2005) Nanoparticle and other metal chelation therapeutics in Alzheimer disease. *Biochimica et Biophysica Acta – Molecular Basis of Disease*, **1741**, 246–52.

135 Huynh, G.H., Deen, D.F. and Szoka, J.F.C. (2006) Barriers to carrier mediated drug and gene delivery to brain tumors. *Journal of Controlled Release*, **110**, 236–59.

136 Kobayashi, H., Koyama, Y., Barrett, T., Hama, Y., Regino, C.A.S., Shin, I.S.,

Jang, B.-S., Le, N., Paik, C.H., Choyke, P.L. and Urano, Y. (2007) Multimodal nanoprobes for radionuclide and five-color near-infrared optical lymphatic imaging. *ACS Nano*, **1**, 258–64.

137 Lee, S., Cha, E.-J., Park, K., Lee, S.-Y., Hong, J.-K., Sun, I.-C., Kim, S., Choi, K., Kwon, I.C., Kim, K. and Ahn, C.-H. (2008) A near-infrared-fluorescence-quenched gold-nanoparticle imaging probe for in-vivo drug screening and protease activity determination. *Angewandte Chemie, International Edition*, **47**, 2804–7.

138 Kim, J.-H., Lee, S., Kim, K., Jeon, H., Park, R.-W., Kim, I.-S., Choi, K. and Kwon, I.C. (2007) Polymeric nanoparticles for protein kinase activity. *Chemical Communications*, 1346–8.

139 Fernandez-Arguelles, M.T., Yakovlev, A., Sperling, R.A., Luccardini, C., Gaillard, S., Sanz Medel, A., Mallet, J.-M., Brochon, J.-C., Feltz, A., Oheim, M. and Parak, W.J. (2007) Synthesis and characterization of polymer-coated quantum dots with integrated acceptor dyes as fret-based nanoprobes. *Nano Letters*, **7**, 2613–17.

140 Zhu, L., Wu, W., Zhu, M.-Q., Han, J.J., Hurst, J.K. and Li, A.D.Q. (2007) Reversibly photoswitchable dual-color fluorescent nanoparticles as new tools for live-cell imaging. *Journal of the American Chemical Society*, **129**, 3524–6.

141 Saito, R., Krauze, M.T., Bringas, J.R., Noble, C., McKnight, T.R., Jackson, P., Wendland, M.F., Mamot, C., Drummond, D.C., Kirpotin, D.B., Hong, K., Berger, M.S., Park, J.W. and Bankiewicz, K.S. (2005) Gadolinium-loaded liposomes allow for real-time magnetic resonance imaging of convection-enhanced delivery in the primate brain. *Experimental Neurology*, **196**, 381–9.

142 Jackson, E.F., Esparza-Coss, E., Wen, X., Ng, C.S., Daniel, S.L., Price, R.E., Rivera, B., Charnsangavej, C., Gelovani, J.G. and Li, C. (2007) Magnetic resonance imaging of therapy-induced necrosis using gadolinium-chelated polyglutamic acids. *International Journal of Radiation Oncology, Biology, Physics*, **68**, 830–8.

143 Kim, J.S., Rieter, W.J., Taylor, K.M.L., An, H., Lin, W. and Lin, W. (2007) Self-assembled hybrid nanoparticles for cancer-specific multimodal imaging. *Journal of the American Chemical Society*, **129**, 8962–3.

144 Li, C., Winnard, P.T., Takagi, T., Artemov, D. and Bhujwalla, Z.M. (2006) Multimodal image-guided enzyme/prodrug cancer therapy. *Journal of the American Chemical Society*, **128**, 15072–3.

145 Talanov, V.S., Regino, C.A.S., Kobayashi, H., Bernardo, M., Choyke, P.L. and Brechbiel, M.W. (2006) Dendrimer-based nanoprobe for dual modality magnetic resonance and fluorescence imaging. *Nano Letters*, **6**, 1459–63.

146 Prinzen, L., Miserus, R.-J.J.H.M., Dirksen, A., Hackeng, T.M., Deckers, N., Bitsch, N.J., Megens, R.T., Douma, K., Heemskerk, J.W., Kooi, M.E., Frederik, P.M., Slaaf, D.W., van Zandvoort, M.A.M.J. and Reutelingsperger, C.P.M. (2006) Optical and magnetic resonance imaging of cell death and platelet activation using annexin A5-functionalized quantum dots. *Nano Letters*, **7**, 93–100.

147 Yang, H., Santra, S., Walter, G.A. and Holloway, P.H. (2006) GdIII-functionalized fluorescent quantum dots as multimodal imaging probes. *Advanced Materials*, **18**, 2890–4.

148 Mulder, W.J.M., Koole, R., Brandwijk, R.J., Storm, G., Chin, P.T.K., Strijkers, G.J., de Mello Donega, C., Nicolay, K. and Griffioen, A.W. (2005) Quantum dots with a paramagnetic coating as a bimodal molecular imaging probe. *Nano Letters*, **6**, 1–6.

149 Hong, R., Fischer, N.O., Emrick, T. and Rotello, V.M. (2005) Surface PEGylation and ligand exchange chemistry of FePt nanoparticles for biological applications. *Chemistry of Materials*, **17**, 4617–21.

150 Gao, J., Zhang, W., Huang, P., Zhang, B., Zhang, X. and Xu, B. (2008) Intracellular spatial control of fluorescent magnetic nanoparticles. *Journal of the American Chemical Society*, **130**, 3710–11.

151 Beaune, G., Dubertret, B., Clément, O., Vayssettes, C., Cabuil, V. and Ménager, C. (2007) Giant vesicles containing magnetic nanoparticles and quantum dots: feasibility and tracking by fiber confocal fluorescence microscopy. *Angewandte Chemie, International Edition*, **46**, 5421–4.

152 Xu, C., Xie, J., Ho, D., Wang, C., Kohler, N., Walsh, E.G., Morgan, J.R., Chin, Y.E. and Sun, S. (2008) Au–Fe_3O_4 dumbbell nanoparticles as dual-functional probes. *Angewandte Chemie, International Edition*, **47**, 173–6.

153 Lee, J.-H., Jun, Y.-W., Yeon, S.-I., Shin, J.-S. and Cheon, J. (2006) Dual-mode nanoparticle probes for high-performance magnetic resonance and fluorescence imaging of neuroblastoma13. *Angewandte Chemie, International Edition*, **45**, 8160–2.

154 Yang, J., Lim, E.-K., Lee, H.J., Park, J., Lee, S.C., Lee, K., Yoon, H.-G., Suh, J.-S., Huh, Y.-M. and Haam, S. (2008) Fluorescent magnetic nanohybrids as multimodal imaging agents for human epithelial cancer detection. *Biomaterials*, **29**, 2548–55.

155 Lee, H., Yu, M.K., Park, S., Moon, S., Min, J.J., Jeong, Y.Y., Kang, H.-W. and Jon, S. (2007) Thermally cross-linked superparamagnetic iron oxide nanoparticles: synthesis and application as a dual imaging probe for cancer *in vivo*. *Journal of the American Chemical Society*, **129**, 12739–45.

156 Choi, J.H., Nguyen, F.T., Barone, P.W., Heller, D.A., Moll, A.E., Patel, D., Boppart, S.A. and Strano, M.S. (2007) Multimodal biomedical imaging with asymmetric single-walled carbon nanotube/iron oxide nanoparticle complexes. *Nano Letters*, **7**, 861–7.

157 Min, K.H., Park, K., Kim, Y.-S., Bae, S.M., Lee, S., Jo, H.G., Park, R.-W., Kim, I.-S., Jeong, S.Y., Kim, K. and Kwon, I.C. (2008) Hydrophobically modified glycol chitosan nanoparticles-encapsulated camptothecin enhance the drug stability and tumor targeting in cancer therapy. *Journal of Controlled Release*, **127**, 208–18.

158 Bagalkot, V., Zhang, L., Levy-Nissenbaum, E., Jon, S., Kantoff, P.W., Langer, R. and Farokhzad, O.C. (2007) Quantum dot-aptamer conjugates for synchronous cancer imaging, therapy, and sensing of drug delivery based on bi-fluorescence resonance energy transfer. *Nano Letters*, **7**, 3065–70.

159 Bakalova, R., Ohba, H., Zhelev, Z., Ishikawa, M. and Baba, Y. (2004) Quantum dots as photosensitizers? *Nature Biotechnology*, **22**, 1360–1.

160 Kim, E., Yang, J., Choi, J., Suh, J.-S., Huh, Y.-M. and Haam, S. (2009) Synthesis of gold nanorod-embedded polymeric nanoparticles by a nanoprecipitation method for use as photothermal agents. *Nanotechnology*, **36**, 365602.

161 Kohler, N., Sun, C., Fichtenholtz, A., Gunn, J., Fang, C. and Zhang, M. (2006) Methotrexate-immobilized poly(ethylene glycol) magnetic nanoparticles for MR imaging and drug delivery. *Small*, **2**, 785–92.

162 Seo, S.-B., Yang, J., Lee, T.-I., Chung, C.-H., Song, Y.J., Suh, J.-S., Yoon, H.-G., Huh, Y.-M. and Haam, S. (2008) Enhancement of magnetic resonance contrast effect using ionic magnetic clusters. *Journal of Colloid and Interface Science*, **319**, 429–34.

163 Jon, S., Lee, H., Yu, M.K. and Jeong, Y.Y. (2007) Multifunctional superparamagnetic iron oxide nanoparticles (SPION) for cancer imaging and therapy. *Nanomedicine*, **3**, 348–348.

164 Lal, S., Clare, S.E. and Halas, N.J. (2008) Nanoshell-enabled photothermal cancer therapy: impending clinical impact. *Accounts of Chemical Research*, **41**, 1842–51.

165 Agrawal, A., Huang, S., Lin, A.W.H., Lee, M.-H., Barton, J.K., Drezek, R.A. and Pfefer, T.J. (2006) Quantitative evaluation of optical coherence tomography signal enhancement with gold nanoshells. *Journal of Biomedical Optics*, **11**, 041121.

166 Garber, K. (2009) Trial offers early test case for personalized medicine. *Journal of the National Cancer Institute*, **101**, 136–8.

167 Frei, E. (1972) Combination cancer therapy: presidential address. *Cancer Research*, **32**, 2593–607.

168 Walsh, C. (2000) Molecular mechanisms that confer antibacterial drug resistance. *Nature*, **406**, 775–81.

169 Hanahan, D., Bergers, G. and Bergsland, E. (2000) Less is more, regularly: metronomic dosing of cytotoxic drugs can target tumor angiogenesis in mice. *Journal of Clinical Investigation*, **105**, 1045–7.

170 Stinchcombe, T.E., Socinski, M.A., Lee, C.B., Hayes, D.N., Moore, D.T., Goldberg, R.M. and Dee, E.C. (2008) Phase I trial of nanoparticle albumin-bound paclitaxel in combination with gemcitabine in patients with thoracic malignancies. *Journal of Thoracic Oncology*, **3**, 521–6.

171 Bae, Y., Diezi, T.A., Zhao, A. and Kwon, G.S. (2007) Mixed polymeric micelles for combination cancer chemotherapy through the concurrent delivery of multiple chemotherapeutic agents. *Journal of Controlled Release*, **122**, 324–30.

172 El-Gendy, N. and Berkland, C. (2009) Combination chemotherapeutic dry powder aerosols via controlled nanoparticle agglomeration. *Pharmaceutical Research*, **26**, 1752–63.

173 Wiradharma, N., Tong, Y.W. and Yang, Y.-Y. (2009) Self-assembled oligopeptide nanostructures for co-delivery of drug and gene with synergistic therapeutic effect. *Biomaterials*, **30**, 3100–9.

174 Zhu, J.-L., Cheng, H., Jin, Y., Cheng, S.-X., Zhang, X.-Z. and Zhuo, R.-X. (2008) Novel polycationic micelles for drug delivery and gene transfer. *Journal of Material Chemistry*, **18**, 4433–41.

175 Cheng, H., Li, Y.-Y., Zeng, X., Sun, Y.-X., Zhang, X.-Z. and Zhuo, R.-X. (2009) Protamine sulfate/poly(l-aspartic acid) polyionic complexes self-assembled via electrostatic attractions for combined delivery of drug and gene. *Biomaterials*, **30**, 1246–53.

176 Yang, J., Lee, J., Kang, J., Oh, S.J., Ko, H.-J., Son, J.-H., Lee, K., Suh, J.-S., Huh, Y.-M. and Haam, S. (2009) Smart drug-loaded polymer gold nanoshells for systemic and localized therapy of human epithelial cancer. *Advanced Materials*, **21** (43), 4339–42.

177 MacDiarmid, J.A., Amaro-Mugridge, N.B., Madrid-Weiss, J., Sedliarou, I., Wetzel, S., Kochar, K., Brahmbhatt, V.N., Phillips, L., Pattison, S.T., Petti, C., Stillman, B., Graham, R.M. and Brahmbhatt, H. (2009) Sequential treatment of drug-resistant tumors with targeted minicells containing siRNA or a cytotoxic drug. *Nature Biotechnology*, **27**, 643–51.

178 Khdair, A., Handa, H., Mao, G. and Panyam, J. (2009) Nanoparticle-mediated combination chemotherapy and photodynamic therapy overcomes tumor drug resistance *in vitro*. *European Journal of Pharmaceutics and Biopharmaceutics*, **71**, 214–22.

179 Park, H., Yang, J., Seo, S., Kim, K., Suh, J., Kim, D., Haam, S. and Yoo, K.-H. (2008) Multifunctional nanoparticles for photothermally controlled drug delivery and magnetic resonance imaging enhancement. *Small*, **4**, 192–6.

180 Greco, F. and Vicent, M.J. (2009) Combination therapy: opportunities and challenges for polymer-drug conjugates as anticancer nanomedicines. *Advanced Drug Delivery Reviews*, **61** (13), 1203–13.

181 Shiah, J.G., Sun, Y., Kopeckov, P., Peterson, C.M., Straight, R.C. and Kopecek, J. (2001) Combination chemotherapy and photodynamic therapy of targetable *N*-(2-hydroxypropyl) methacrylamide copolymer-doxorubicin/ mesochlorin e6-OV-TL 16 antibody immunoconjugates. *Journal of Controlled Release*, **74**, 249–53.

182 Donini, C., Robinson, D.N., Colombo, P., Giordano, F. and Peppas, N.A. (2002) Preparation of poly(methacrylic acid-g-poly(ethylene glycol)) nanospheres from methacrylic monomers for pharmaceutical applications. *International Journal of Pharmaceutics*, **245**, 83–91.

183 Kyriakides, T.R., Cheung, C.Y., Murthy, N., Bornstein, P., Stayton, P.S. and Hoffman, A.S. (2002) pH-sensitive polymers that enhance intracellular drug delivery *in vivo*. *Journal of Controlled Release*, **78**, 295–303.

184 Kurkuri, M.D. and Aminabhavi, T.M. (2004) Poly(vinyl alcohol) and poly(acrylic acid) sequential interpenetrating network pH-sensitive microspheres for the delivery of diclofenac sodium to the intestine. *Journal of Controlled Release*, **96**, 9–20.

185 Veronese, F.M., Schiavon, O., Pasut, G., Mendichi, R., Andersson, L., Tsirk, A., Ford, J., Wu, G., Kneller, S., Davies, J. and Duncan, R. (2005) PEG doxorubicin conjugates: influence of polymer structure on drug release, *in vitro* cytotoxicity, biodistribution, and antitumor activity. *Bioconjugate Chemistry*, **16**, 775–84.

186 Satchi-Fainaro, R., Puder, M., Davies, J.W., Tran, H.T., Sampson, D.A., Greene, A.K., Corfas, G. and Folkman, J. (2004) Targeting angiogenesis with a conjugate of HPMA copolymer and TNP-470. *Nature Medicine*, **10**, 255–61.

187 Gillies, E.R., Goodwin, A.P. and Frechet, J.M.J. (2004) Acetals as pH-sensitive linkages for drug delivery. *Bioconjugate Chemistry*, **15**, 1254–63.

188 Murthy, N., Campbell, J., Fausto, N., Hoffman, A.S. and Stayton, P.S. (2003) Bioinspired pH-responsive polymers for the intracellular delivery of biomolecular drugs. *Bioconjugate Chemistry*, **14**, 412–19.

189 Murthy, N., Campbell, J., Fausto, N., Hoffman, A.S. and Stayton, P.S. (2003) Design and synthesis of pH-responsive polymeric carriers that target uptake and enhance the intracellular delivery of oligonucleotides. *Journal of Controlled Release*, **89**, 365–74.

190 Lee, E.S., Na, K. and Bae, Y.H. (2005) Doxorubicin loaded pH-sensitive polymeric micelles for reversal of resistant MCF-7 tumor. *Journal of Controlled Release*, **103**, 405–18.

191 Kim, G.M., Bae, Y.H. and Jo, W.H. (2005) pH-induced micelle formation of poly(histidine-*co*-phenylalanine)-block-poly(ethylene glycol) in aqueous media. *Macromolecular Bioscience*, **5**, 1118–24.

192 Gillies, E.R. and Frechet, J.M.J. (2005) pH-responsive copolymer assemblies for controlled release of doxorubicin. *Bioconjugate Chemistry*, **16**, 361–8.

193 Philippova, O.E., Hourdet, D., Audebert, R. and Khokhlov, A.R. (1997) pH-responsive gels of hydrophobically modified poly(acrylic acid). *Macromolecules*, **30**, 8278–85.

194 Ihre, H.R., Padilla De Jesus, O.L., Szoka, F.C. and Frechet, J.M.J. (2002) Polyester dendritic systems for drug delivery applications: design, synthesis, and characterization. *Bioconjugate Chemistry*, **13**, 443–52.

195 Shepherd, J., Hilderbrand, S.A., Waterman, P., Heinecke, J.W., Weissleder, R. and Libby, P. (2007) A fluorescent probe for the detection of myeloperoxidase activity in atherosclerosis-associated macrophages. *Chemistry and Biology*, **14**, 1221–31.

196 Chen, J.W., Sans, M.Q., Bogdanov, A. and Weissleder, R. (2006) Imaging of myeloperoxidase in mice by using novel amplifiable paramagnetic substrates. *Radiology*, **240**, 473–81.

197 Na, K., Lee, K.H., Lee, D.H. and Bae, Y.H. (2006) Biodegradable thermo-sensitive nanoparticles from poly(L-lactic acid)/poly(ethylene glycol) alternating multi-block copolymer for potential anti-cancer drug carrier. *European Journal of Pharmaceutical Sciences*, **27**, 115–22.

198 Meyer, D.E., Shin, B.C., Kong, G.A., Dewhirst, M.W. and Chilkoti, A. (2001) Drug targeting using thermally responsive polymers and local hyperthermia. *Journal of Controlled Release*, **74**, 213–24.

199 Chung, J.E., Yokoyama, M., Aoyagi, T., Sakurai, Y. and Okano, T. (1998) Effect of molecular architecture of hydrophobically modified poly(*N*-isopropylacrylamide) on the formation of thermoresponsive core-shell micellar drug carriers. *Journal of Controlled Release*, **53**, 119–30.

200 Liu, S.Q., Tong, Y.W. and Yang, Y.-Y. (2005) Incorporation and *in vitro* release of doxorubicin in thermally sensitive micelles made from poly(*N*-isopropylacrylamide-*co*-*N*,*N*-dimethylacrylamide)-*b*-poly(D,L-lactide-*co*-glycolide) with varying compositions. *Biomaterials*, **26**, 5064–74.

201 Chung, J.E., Yokoyama, M. and Okano, T. (2000) Inner core segment design for drug delivery control of thermo-responsive polymeric micelles. *Journal of Controlled Release*, **65**, 93–103.

202 Bae, Y.H., Okano, T. and Kim, S.W. (1989) Insulin permeation through thermo-sensitive hydrogels. *Journal of Controlled Release*, **9**, 271–9.

203 Furgeson, D.Y., Dreher, M.R. and Chilkoti, A. (2006) Structural optimization of a "smart" doxorubicin-polypeptide conjugate for thermally targeted delivery to solid tumors. *Journal of Controlled Release*, **110**, 362–9.

204 Emileh, A., Vasheghani-Farahani, E. and Imani, M. (2007) Swelling behavior, mechanical properties and network parameters of pH- and temperature-sensitive hydrogels of poly((2-dimethyl amino) ethyl methacrylate-co-butyl methacrylate. *European Polymer Journal*, **43**, 1986–95.

205 Chilkoti, A., Dreher, M.R., Meyer, D.E. and Raucher, D. (2002) Targeted drug delivery by thermally responsive polymers. *Advanced Drug Delivery Reviews*, **54**, 613–30.

206 Liu, S.Q., Tong, Y.W. and Yang, Y.Y. (2005) Thermally sensitive micelles self-assembled from poly(*N*-isopropylacrylamide-*co*-*N*,*N*-dimethylacrylamide)-*b*-poly(D,L-lactide-*co*-glycolide) for controlled delivery of paclitaxel. *Molecular BioSystems*, **1**, 158–65.

207 Chung, J.E., Yokoyama, M., Yamato, M., Aoyagi, T., Sakurai, Y. and Okano, T. (1999) Thermo-responsive drug delivery from polymeric micelles constructed using block copolymers of poly(*N*-isopropylacrylamide) and poly(butylmethacrylate). *Journal of Controlled Release*, **62**, 115–27.

208 Bae, Y.H., Okano, T., Hsu, R. and Kim, S.W. (1987) Thermo-sensitive polymers as on-off switches for drug release. *Macromolecular Rapid Communications*, **8**, 481–5.

209 Oh, K.T., Yin, H., Lee, E.S. and Bae, Y.H. (2007) Polymeric nanovehicles for anticancer drugs with triggering release mechanisms. *Journal of Materials Chemistry*, **17**, 3987–4001.

210 Bremer, C., Tung, C.-H. and Weissleder, R. (2001) *In vivo* molecular target assessment of matrix metalloproteinase inhibition. *Nature Medicine*, **7**, 743–8.

211 Park, T.G. (1999) Temperature modulated protein release from pH/temperature-sensitive hydrogels. *Biomaterials*, **20**, 517–21.

212 Newell, K., Franchi, A., Pouyssegur, J. and Tannock, I. (1993) Studies with glycolysis-deficient cells suggest that production of lactic acid is not the only cause of tumor acidity. *Proceedings of the National Academy of Sciences of the United States of America*, **90**, 1127–31.

213 Stubbs, M., McSheehy, P.M.J. and Griffiths, J.R. (1999) Causes and consequences of acidic pH in tumors: a magnetic resonance study. *Advances in Enzyme Regulation*, **39**, 13–30.

214 Ding, C., Gu, J., Qu, X. and Yang, Z. (2009) Preparation of multifunctional drug carrier for tumor-specific uptake and enhanced intracellular delivery through the conjugation of weak acid labile linker, *Bioconjugate Chemistry*, **20**, 1163–70.

215 Lee, D., Khaja, S., Velasquez-Castano, J.C., Dasari, M., Sun, C., Petros, J., Taylor, W.R. and Murthy, N. (2007) *In vivo* imaging of hydrogen peroxide with chemiluminescent nanoparticles. *Nature Materials*, **6**, 765–9.

9
Nanocomposites for Drug Delivery

Shoucang Shen, Yuan-Cai Dong, Wai Kiong Ng, Leonard Chia, and Reginald Beng Hee Tan

9.1
Introduction

In recent years, the application of nanocomposites as drug carriers for effective drug delivery has become a rapidly growing area in both academic and industrial research and development (R&D). Nanocomposites can be defined as multiphase solid materials, where at least one phase has one, two, or three dimensions at nanosize range (1~1000 nm), or which comprises structures with repeated nanoscale spacings of the different phases that make up the material. Due to the presence of nanostructured materials in the composite system, this new family of composite materials frequently exhibits remarkable improvements of material properties when compared to the matrix solid alone, or to conventional microcomposites and macrocomposites. These include mechanical properties, molecular permeability and the control of drug release, as well as properties related to engineering aspects, such as thermal stability, flame retardancy, chemical resistance, surface appearance, electrical conductivity, and optical clarity. Several reviews have been produced on the processing, characterization and applications of polymer-based nanocomposites [1, 2], among which polymer–clay nanocomposites have been extensively discussed [3, 4]. The layered clay materials possess a well-defined, ordered intra-lamellar space at nanoscale, which is potentially accessible to foreign species; such a nanostructure endows the composites with various improved properties and/or newly created novel functions. During the past decade, a variety of new nanostructured materials have been exploited to develop new applications for nanocomposites. The presence of nanostructured materials in composites plays a vital role in the creation of novel properties exhibiting multifunctional, high-performance characteristics beyond those of conventional macrophase materials, for various applications in the field of biomedicine. The application of nanocomposites in drug delivery, however, has been reviewed only minimally, and there is a clear lack of such information for those presently working in this field. Nanocomposites represent an excellent type of composite material, whereby interactions between the multi-phases may be maximized so as to provide a broad range

Nanomaterials for the Life Sciences Vol.8: Nanocomposites. Edited by Challa S. S. R. Kumar

ISBN: 978-3-527-32168-1

of novel properties for effective drug delivery. The controlled release of drugs from nanocomposite systems aims at their delivery at optimum dosage for extended periods, at increasing the efficacy of the drug, and at improving patient compliance.

In this chapter, the recent advances in nanocomposite systems for drug delivery will be reviewed. For clarity, the nanocomposites are classified as nanoparticles, nanofibers, two-dimensional (2-D)-nanostructure incorporated composites, and three-dimensional (3-D) nanocomposites. In addition to the extensively investigated polymer–inorganic nanocomposites, the application of polymer–polymer and inorganic–inorganic nanocomposites as drug carriers for effective drug delivery is also discussed.

9.2
Nanoparticle Composites for Drug Delivery

Nanoparticles made from organic or inorganic materials have attracted increasing attention during recent years for the delivery of drugs, peptides/proteins, genes, and vaccines [5–10]. In general, those nanoparticles used for drug delivery are less than 1000 nm in size, and the active agents can be dissolved, entrapped, or encapsulated within the particles, or adsorbed/conjugated onto the particles' surfaces. Based on the inner structure, nanoparticles may be classified as having a matrix or vesicular system (Figure 9.1). *Matrix-type nanoparticles* are homogeneous systems in which the active agents are dissolved/dispersed throughout the whole particles. The *vesicular-type nanoparticles* (which are also termed nanocapsules) are heterogeneous systems, where the active agents are confined within an aqueous or oily cavity enclosed by the organic or inorganic membrane [5, 6]. Depending on the fabrication process and the physico-chemical properties of the components which make up the nanoparticles, the active agents may exist as a solution, a solid solution, or as a crystalline or amorphous solid in the nanoparticles. Organic nanoparticles have been studied extensively due to their good biocompatibility, good biodegradability, and high loading efficiency. These include polymeric nano-

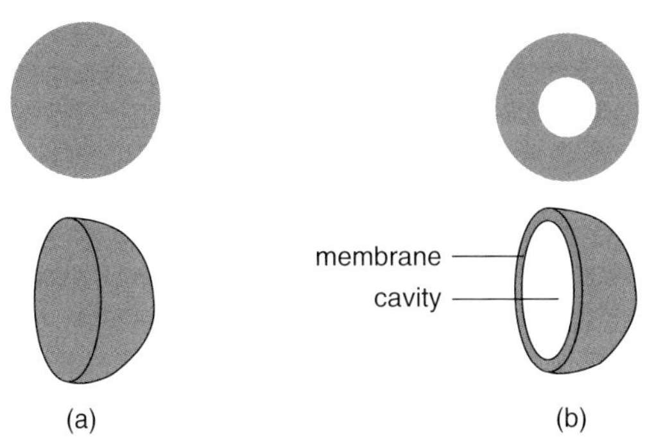

Figure 9.1 (a) Matrix-type and (b) vesicular-type nanoparticles.

particles, submicron liposomes, solid lipid nanoparticles (SLN), micelles, and dendrimers. By comparison, inorganic nanoparticles have been investigated to a lesser degree, with the majority of studies being performed with silica nanoparticles and iron oxide nanoparticles.

Both, *size* and *surface chemistry* play important roles in determining the performance of the administered drug-loaded nanoparticles and the eventual therapeutic efficiency. Due to their submicron size, intravenously administered nanoparticles possess an enhanced permeation and retention effect (EPR), and tend to accumulate within the sites of the pathological changes (this is known as "passive targeting") [11]. The extremely small nanoparticles (i.e., those <100 nm in size) are capable of avoiding uptake by the reticuloendothelial system (RES), and consequently exhibit an extended blood circulation time [12]. When orally and/or locally administered (e.g., topical, nasal, corneal), the nanoparticles – due to their large surface area – usually demonstrate a prolonged residence time and provide high local drug concentrations, leading to an increased bioavailability [13]. The importance of size is also reflected at the cellular level. It has been reported that the internalization of nanoparticles by cells (tumor cells, intestinal cells, etc.) is size-dependent – the smaller the particles, the greater the extent of internalization [14, 15]. Consequently, those nanoparticles loaded with the active agents may show a higher level of cytotoxicity against tumor cells due to their greater cellular uptake [16]. On the other hand, nanoparticles – on the basis of their small size – are able to cross physiological barriers (e.g., gastrointestinal, blood–brain), and this results in an improved bioavailability [17].

The *surface chemistry* of nanoparticles is equally important. For example, intravenously administered "naked" hydrophobic nanoparticles tend to be captured by the macrophages and to accumulate in the RES (liver, spleen, lung). This results in a short circulation time that is induced by the adsorption of blood proteins (opsonins) onto the particles' surface (the process of opsonization). Nanoparticles with a hydrophilic coating, such as poly(ethylene glycol) (PEG; often referred to as "stealth nanoparticles") have the ability to effectively suppress the opsonization process, which leads to an extended blood residence time [18].

Nanoparticle drug delivery systems possess many unique advantages that may lead to a reduced toxicity and increased therapeutic efficacy. The term "nanoparticles," when used in relation to drug delivery, should in fact be replaced by "composited nanoparticle," although this is not usually made clear in the relevant literature. This substitution should be made because:

- Nanoparticles, due to their high surface energy, have a natural tendency to aggregate. Stabilizers are generally required from the initial synthesis to the final *in vivo* applications.

- The surfaces of the drug-loaded nanoparticles are often decorated with specific molecules so as to impart novel functions, such as a long-circulation time, specific targeting, and stimulus-sensitivity. A surface layer is sometimes also required to reduce the toxicity of the nanoparticles and increase their biocompatibility. Iron oxide nanoparticles are an example of this.

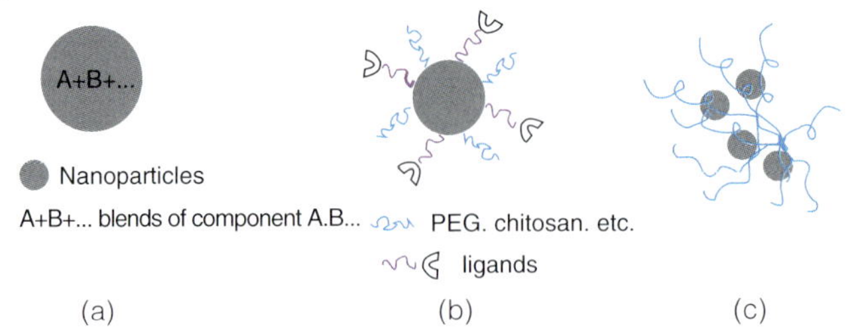

Figure 9.2 Morphologies of nanoparticles composite delivery systems. (a) The matrix or membrane (for nanocapsule) of nanoparticles is constructed from blends of several components (A + B + ...); (b) The core of nanoparticles is coated by a shell, which may be PEG, ligands, chitosan, etc., for functionalization; (c) Nanoparticles were mixed with or dispersed within other components, which may be in the submicron range or even larger. Note: Although the nanoparticles in the figure are shown only as the matrix system for simplicity, they may also be made from the vesicular system.

- The core of nanoparticles is, in some instances, composed of blends of several components rather than a single component, in order to achieve a desired loading efficiency, an increased stability, or an improved drug-release profile.

Thus, the term "nanoparticle composites" is more appropriate. In the following sections, "nanoparticle composites" may be defined as a drug delivery system in which at least one phase is made from nanoparticles although, excluding the active agents, there should be at least two or more components in the system. Based on this definition, a nanoparticle composite drug delivery system can be categorized under at least one of the morphologies illustrated in Figure 9.2.

Several excellent reviews have reported the development of nanoparticle drug delivery systems from different aspects, such as materials, preparation approaches, characterization, and applications [5–10, 19–21]. The following section will summarize the latest advances of these drug delivery systems, from the perspective of "nanoparticle composites." Details are provided of some of the most frequently studied materials, including organic nanoparticle composites (e.g., polymeric nanoparticles, liposomes, micelles, solid-lipid nanoparticles) and inorganic nanoparticle composites (e.g., iron oxide nanoparticles and silica nanoparticles).

9.2.1 Polymeric Nanoparticle Composites

Among the various biodegradable and nondegradable synthetic polymers, poly(lactide-*co*-glycolide) (PLGA) is used widely to fabricate nanoparticles, due to its excellent biocompatibility and biodegradability [22], while *chitosan* remains one of the most extensively used natural polymers for nanoparticle drug delivery. The recent advances in the use of PLGA and chitosan nanoparticles composites for drug delivery, which represent developments using synthetic and natural polymers, are reviewed in the following sections.

9.2.1.1 PLGA Nanoparticle Composites

PLGA with different molecular weights, lactide:glycolide ratios and crystallinity have been widely investigated for the fabrication of nanoparticles for drug delivery. In the human body, PLGA would undergo hydrolysis and be converted ultimately into carbon dioxide and water [22]. Consequently, in order to prepare PLGA nanoparticles with desired properties, it can be blended with other polymers or short-chain materials to produce nanoparticle composites, in a simple yet effective approach. The cationic PLGA nanoparticles composites reported by Basarkar *et al.* [23], which were prepared by blending PLGA and a methacrylate copolymer (Eudragit®100), were capable of efficiently and safely delivering plasmid DNA to prevent autoimmune diabetes. Ishihara *et al.* [24] developed long-circulating "stealth" PLGA nanoparticle composites from blends of PEG-block-PLGA and poly(lactic acid) (PLA) for the delivery of betamethasone disodium 21-phosphate (BP) to treat arthritis. The nanoparticle composites formed showed a prolonged blood circulation time, and a high therapeutic benefit was achieved. Similar stealth PLGA nanoparticle composites were also obtained by blending PLGA with poloxamer or poloxamine [25]. Likewise, thermosensitive nanoparticles composites were reported by Salehi *et al.* [26], that had been fabricated from blends of PLGA and poly(*N*-isopropylacrylamide-acrylamide-vinylpyrrolidone), using the emulsion–solvent–evaporation method.

The formation of core–shell-type PLGA nanoparticle composites is mostly adopted to endow the pure particles with certain novel functions, such as long circulating times, stimuli-sensitivity, targeting specificity, and bioadhesion. Both, Zhang *et al.* [27] and Chan *et al.* [28] used a nanoprecipitation method to obtain the long-circulating PLGA nanoparticles composites with a PEG shell; in this case, the PLGA nanoparticle core was surrounded by a lecithin monolayer, into which the 1,2-distearoyl-*sn*-glycero-3-phosphoethanolamine-N-carboxy(polyethylene glycol)2000 (DSPE-PEG2000-COOH) was inserted to form a PEG layer, as depicted in Figure 9.3. Bioadhesive cationic PLGA nanoparticle composites with a chitosan shell possess extended residence times, a characteristic which has been widely employed for the oral and localized (corneal, nasal, etc.) delivery of drugs,

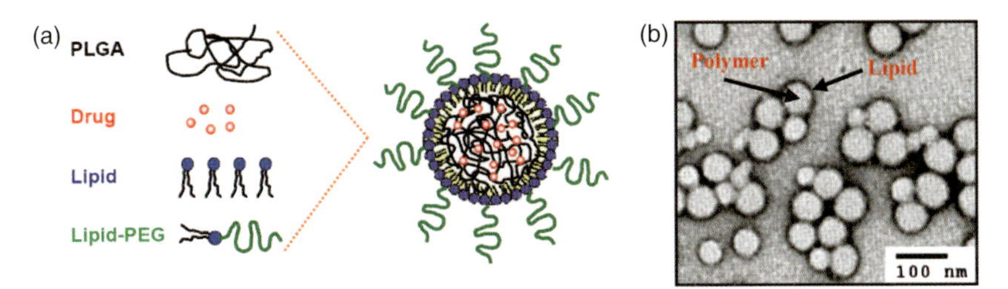

Figure 9.3 Core–shell-type PLGA nanoparticles composites. (a) Schematic representation; (b) Transmission electron microscopy (TEM) image. Reproduced with permission from Ref. [28]; © Elsevier Ltd.

and also for DNA/RNA delivery [29–35]. Nafee *et al.* [31] reported that this type of nanoparticle composite could improve the cellular uptake of the antisense oligonucleotides, 2′-methyl-RNA, while Yang *et al.* [32] demonstrated that paclitaxel-loaded PLGA nanoparticle composites with a chitosan coating showed a significantly higher uptake and cytotoxicity into A549 lung cancer cells, in comparison with the pure PLGA nanoparticles. A lung-specific increase in the activity of paclitaxel was also observed for nanoparticle composites.

The fabrication of targeted PLGA nanoparticle composites with a ligand shell (e.g., antibody, folate, peptide, protein) has been a very "hot" topic in recent years, as this strategy is considered to be most efficient approach for increasing the binding/uptake of drug-loaded pure nanoparticles into the cells of interest [36–41]. An improved bioavailability and therapeutic efficacy can thus be expected. Yin *et al.* [38] prepared PLGA nanoparticle composites with a lectin shell (wheat germ agglutinin) through conjugation for the oral delivery of thymopentin, and identified an enhanced bioadhesion and absorption of PLGA nanoparticle composites in the small intestine. Zhang *et al.* [41] described a PLGA nanoparticle composite with a conjugated peptide shell, that was rapidly endocytosed and trafficked to the lysosomes of human umbilical cord vascular endothelial cells (HUVECs) to a greater extent than were the pure nanoparticles. McCarron *et al.* [40] showed that camptothecin-loaded PLGA nanoparticle composites with antibody shell had a much lower half-maximal inhibitory concentration (IC_{50}) of 0.37 ng ml^{-1} in HCT116 cells after a 72 h treatment period, than did the camptothecin solution alone (21.8 ng ml^{-1}).

In addition to the core–shell- and blended matrix-type PLGA nanoparticle composites, a third form of nanoparticle composite has been envisaged as a possible drug carrier – that of PLGA nanoparticles interdispersed in another continuous matrix so as to form a composite drug delivery system. The desired properties of these composites – the dimensions of which are in the micron range or even larger – are acquired from the benefits of both the nanoparticles and the matrix materials. Dong *et al.* [42] described the development of PLGA nanoparticle/montmorillonite composites for the oral delivery of paclitaxel although, due to the viscous feature of the clay and the high surface areas of the nanoparticles, this type of nanoparticle composite had a prolonged residence time in the gastrointestinal tract, and a higher oral bioavailability of the loaded drug was expected. Composites created by entrapping PLGA nanoparticles in hydrogels (Figure 9.4) have also been developed for the intraperitoneal delivery of insulin and dexamethasone [43–45]. When insulin-loaded PLGA nanoparticles were administered intraperitoneally as a single dose (20 U kg^{-1}) to streptozotocin-induced diabetic mice, the blood glucose levels were decreased and maintained at low levels over a 24 h period. *In vitro* release further indicated that the entrapment of the nanoparticles into poly(vinyl alcohol) (PVA) hydrogels caused a reduction in both the release rate and total amount of insulin released. This suggested that the PLGA nanoparticles entrapped in the PVA hydrogels had shown a more suitable controlled release kinetics for protein delivery [43].

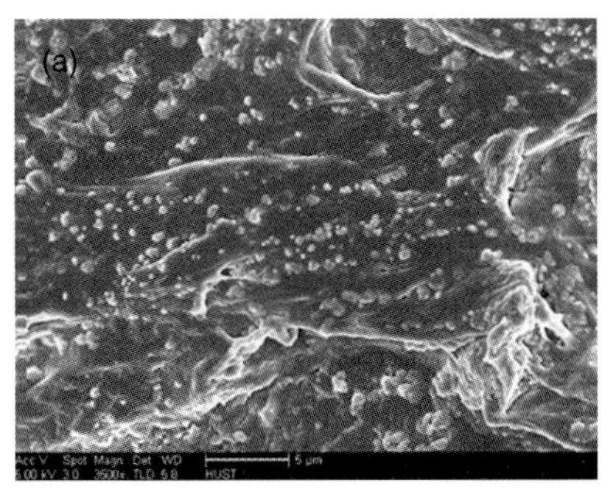
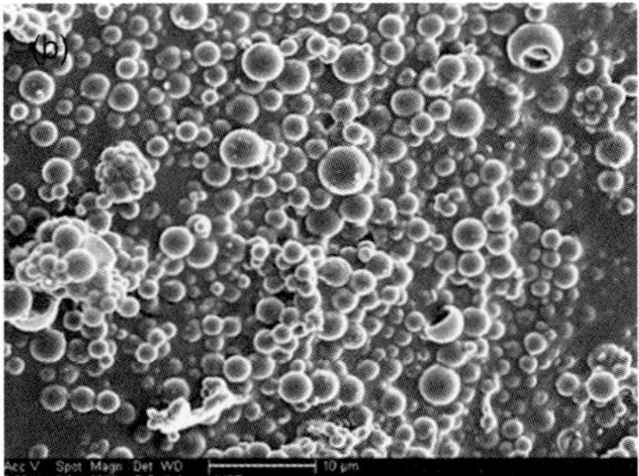

Figure 9.4 Scanning electron microscopy images of (a) insulin-loaded PLGA nanoparticles and (b) composites of insulin-loaded PLGA nanoparticles entrapped in hydrogels. Reproduced with permission from Ref. [43]; © Springer Science+Business Media, LLC.

9.2.1.2 Chitosan Nanoparticle Composites

Nanoparticles composed of natural polymers, such as chitosan, alginate, and hyaluronic acid, have been investigated for the delivery of drugs, peptide/proteins and genes. Chitosan nanoparticles, in particular, have undergone the widest examination due to their uniquely cationic and mucoadhesive properties [9, 46]. As an effective approach, a second natural polymer has often been incorporated into chitosan in order to stabilize the drug, to increase loading efficiency, to prolong the residence time at the site of drug absorption, and to achieve the desired drug release profiles [8, 47–55]. Chitosan/alginate nanoparticle composites have been shown as effective for the oral [47] and nasal [48] delivery of insulin, and for the ocular delivery of antibiotics [49]. Chitosan/hyaluronic acid nanoparticles composites have also been exploited for the ocular delivery of a model plasmid, pEGFP or pβ-gal [51, 52]. In transfection studies of gene-loaded chitosan/hyaluronic acid nanoparticle composites on human corneal epithelial (HCE) and a normal human conjunctival (IOBA-NHC) cell line, high transfection levels were identified (up to 15% of cells were transfected) without affecting cell viability [52]. The incorporation of poly(γ-glutamic acid) to chitosan/DNA nanoparticles was found to result in a significant increase in cellular uptake and transfection efficiency [50]. Chitosan/cyclodextrin nanoparticle composites have also been examined for macromolecular drug delivery, with the aim that cyclodextrin could increase the protein stability [53]. In addition to the aforementioned blended matrix-type composites, the development of core–shell-type chitosan nanoparticle composites with a hyaluronic acid shell has also been reported; these composites possessed a reduced toxicity and a prolonged circulation time in comparison to naked chitosan nanoparticles [54]. A third type of chitosan nanoparticle composite (see Figure 9.2c) has also been created by entrapping the chitosan/plasmid nanoparticles in a gene-activated matrix, and used for periodontal tissue engineering [55]. The plasmid exhibited a sustained release for over six weeks, and this novel nanoparticle composite was shown to promote the proliferation of periodontal ligament cells.

9.2.2 Liposomes

Liposomes are vesicular systems in which one or more lipid bilayer membranes surrounding discrete aqueous compartments. Due to their excellent biocompatibility and biodegradability, liposomes have found wide application in delivery systems for both hydrophilic and hydrophobic drugs. Typically, hydrophilic drugs are entrapped within the aqueous compartment of liposomes, while hydrophobic drugs are incorporated into the bilayer region [56–58]. Liposomes in the submicron range may serve as effective drug delivery vehicles, based mainly on the benefits of their small particle size. Indeed, submicron liposomes (1~1000 nm) used for drug delivery are seen as natural nanoparticle composite systems because: (i) cholesterol must be incorporated into the lipid bilayer to maintain membrane stability; and (ii) a mixture of lipids (rather than one type of lipid) is generally used to form the bilayer membrane. This is necessary in order to achieve the desired membrane fluidity and stability, both of which are important factors for the effectiveness of drug loading and the profile of drug release. As an example, whilst the encapsulation efficiency (EE) of paclitaxel into liposomes prepared from egg-yolk phosphatidylcholine (PC) was 60%, the incorporation of 10% phosphatidylglycerol (PG) into the liposomes increased the EE to 95% [59]. In addition, in order to achieve long-term circulatory functions, *N*-(carbonyl-methoxypolyethylene glycol 2000)-1,2-distearoyl-*sn*-glycero3-phosphoethanolamine sodium salt (MPEG-DSPE2000) must be incorporated into the bilayer membrane. An example of this is the commercial stealth Doxil®, the bilayer of which is composed of MPEG-DSPE2000, fully hydrogenated soy PC, and cholesterol [60–62]. Other PEG-containing molecules, such as methoxypolyethylene glycol-poly(D,L-lactide) (mPEG-PLA) have also been incorporated into the bilayer membrane to form stealth composites [63]. Core–shell-type mucoadhesive liposome composites were also produced by coating chitosan or poly(acrylic acid) for the delivery of drugs, either orally [64, 65] or via nebulization to the lungs [66].

In an effort to improve therapeutic efficacy, targeted liposome composites bearing ligands have been extensively studied during recent years. These ligands have included antibody [67–70], folate [71, 72], transferrin [73, 74], peptide [75–77], and affibody [78]. pH-sensitive or thermosensitive liposome composites have also been created by the incorporation of a pH-sensitive or thermosensitive material, such as *N*-isopropylacrylamide copolymer [79] or poly(*N*-(2-hydroxypropyl)methacrylamide mono/diacetate) [80], and these are regarded as the most efficient and safe drug carriers (Figure 9.5). The current trend, however, is to produce multifunctional liposomes. For example, long-circulating and pH-sensitive liposome composites were obtained by the incorporation of both DSPE-PEG2000 [81] or PEG-containing copolymers [82] and pH-sensitive molecules into the bilayer membrane. Multifunctional liposome composites bearing long-circulating, pH-

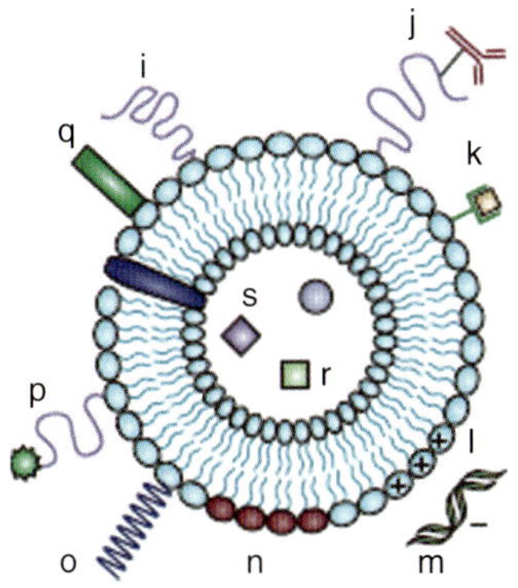

Figure 9.5 Schematic illustration of a multifunctional liposomes-a complex nanoparticle composite system. The liposomes may simultaneously possess protective polymer (i) for long-circulation times, an antibody (j) for targeting, a diagnostic label (k) for diagnosis, positively charged lipids (l) for complexation with DNA (m), stimuli-sensitive lipids (n) or polymer (o) for pH/thermosensitivity, a peptide (p) for cell penetration enhancement, and viral components (q) for inducing drug efflux into the cytoplasm. Drug, magnetic particles (r) for imaging and gold or silver nanoparticles (s) for electron microscopy may be loaded into such a composites system simultaneously. Reproduced with permission from Ref. [56]; © 2009, Nature Publishing Group.

sensitivity and targeting functions in one preparation have also been reported [83]. One particularly interesting liposome composite is that of LIM, where the liposomes are arranged in a microsphere, as reported by Feng and coworkers [84]. This novel composite system may serve to protect labile liposomes from degradation, as the drug-loaded liposomes are encapsulated inside the microspheres and released only gradually.

9.2.3 Solid Lipid Nanoparticle Composites

Solid lipid nanoparticles (SLNs) are submicron particles composed of lipids that exist in solid form at both room and body temperatures. Several reviews have provided the details of SLNs in different aspects [85–90]. As for polymeric nanoparticles, the SLNs are in fact nanoparticle composites, as the lipid core of the SLN is generally composed of a blend of solid and liquid lipids. Surfactants are also required in the system for stabilization purposes; for example, egg PC and PEGylated phospholipid were used as stabilizers for a trimyristin SLN [91]. In recent years, targeted core–shell-type SLN composites with ligand shells have also been examined with the aim of enhancing delivery efficiency. SLN composites bearing a transferring shell have been described for the delivery of quinine dihydrochloride to the brain [92], while those with a lectin-shell have been proposed for the oral administration of insulin [93]. SLNs were also incorporated into dextran hydrogels that were proposed as being suitable for oral drug delivery purposes [94].

9.2.4 Micelles

Polymeric micelles from amphiphilic copolymers are one of most widely studied drug carriers, due to their good stability, low critical micelle concentration (CMC), high loading efficiency, and feasibility of fabrication. Typically, PEG is used as the hydrophilic segment of the copolymer, which protrudes to form the shell during the preparation process; the hydrophobic segments (e.g., polylactide) form an aggregate as the core for drug loading. In most cases, when amphiphilic copolymers themselves are used they will self-assemble to form micelles which are capable of self-stabilization and will circulate for a long time due to the presence of the PEG shell. In recent years, however, *mixed micelles* have also been investigated to identify new functions. For example, targeted and/or pH-sensitive mixed micelles were described which had been fabricated from blends of poly(histidine)-*b*-polyethylene glycol (with or without ligands) and a conventional diblock copolymer, such as PLA-PEG [95–97]. Thermosensitive composite micelles have been prepared from blends of pluronic F-127 and Tween 80 [98]. A micelles/hydrogel composite was also reported in which polypeptide micelles were dispersed in PVA or chitosan hydrogels, for the simultaneous delivery of aspirin and doxorubicin [99].

9.2.5 Iron Oxide Nanoparticle Composites for Drug Delivery

Iron oxide nanoparticles (γ-Fe_2O_3 or Fe_3O_4) have long been studied for the purposes of bioimaging and drug delivery [100–102]. As magnetic nanoparticles alone were reported to be ineffective as drug carriers, due to limitations in drug loading, and to poor release rates and retention times in the bloodstream, the magnetic iron oxide nanoparticles must first be surface-modified to form a composite with increased stability and reduced toxicity, as well as imparting new functions. As shown in Figure 9.6, iron oxide nanoparticles composites can be produced by various means, including polymeric coating, functional ligands conjugation, silica coating, entrapment in liposomes, and micelles [100–102]. Beside bioimaging, for this purpose, gelatin [103], PVA [104], PEG fumarate [105], silica [106], and

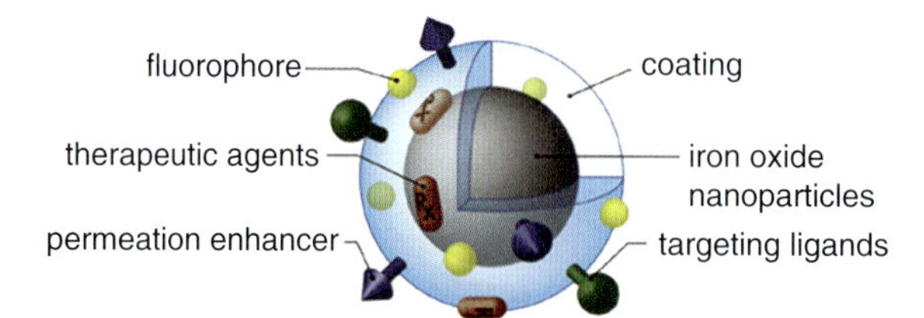

Figure 9.6 Schematic representation of iron oxide nanoparticle composites for drug delivery. Reproduced with permission from Ref. [100]; © Elsevier B.V.

poly(ethyleneimine) (PEI) [107] have also been used to form nanoparticle composites for drug and DNA delivery. Phanapavudhikul *et al.* [108] reported the details of an iron oxide nanoparticle composite which was achieved by encapsulating nanosized magnetite with an acrylate-based cationic copolymer made from methylmethacrylate (MMA), butyl acrylate (BA), and quinolinyl methacrylate (QMA), and modified with methoxy poly(ethylene glycol) methacrylate (MeOPEGMA) by using the water-replacement method. The composition of the co-polymer formulation was optimized based on zeta-potential measurements and freeze–thaw stability testing. The electrostatic interaction between the negatively charged model drug aspirin and the positively charged co-polymer played the most important role in drug loading and release. The drug release exhibited a biphasic profile, with an initial burst release followed by a prolonged slow release. Targeted iron oxide nanoparticle composites conjugated with ligands have also been reported, in which the adopted ligands included folate [109] and a urokinase plasminogen activator [110].

It should be noted here that a novel wrap–bake–peel process has been described for the preparation of biocompatible iron oxide nanocapsule composites [111], which involves silica coating, heat treatment, and final removal of the silica layer. The wrap–bake–peel process for the synthesis of water-dispersible magnetite nanocapsules for drug delivery is shown schematically in Figure 9.7. When the chemotherapeutic agent doxorubicin (DOX) was used as the model drug, the typical DOX loadings in the PEG-coated nanocapsules of hematite and magnetite were up to 17.8 wt% and 28.9 wt%, respectively. The PEG-coated magnetite nanocapsules (PEG–MNC) exhibited a very low cytotoxicity in SKBR-s cells, with iron concentrations up to $200\,\mu g\,ml^{-1}$. However, the DOX-loaded PEG-MNC exhibited

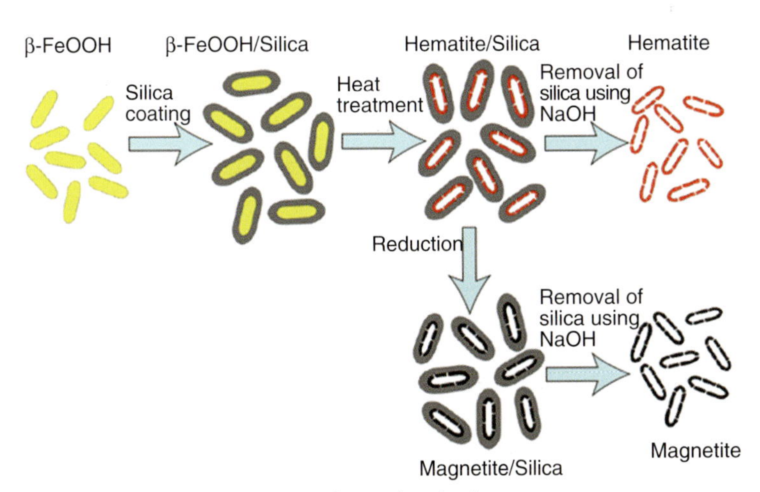

Figure 9.7 Schematic representation of procedure for the synthesis of uniform and water-dispersible iron oxide nanocapsules. Reproduced with permission from Ref. [111]; © Nature Publishing Group.

a higher cytotoxicity than free DOX in SKBR-3 cells, which indicated that the magnetite nanocapsules might represent an effective drug-delivery vehicle. In addition, the magnetic nanocapsules could be used as T_2 magnetic resonance imaging contrast agent.

9.2.6 Silica Nanoparticle Composites

Silica nanoparticles, especially in their mesoporous form, appear to be an excellent candidate for drug delivery, as their large surface area and pore volume serve as a natural reservoir for drug loading. In addition, they possess good biocompatibility, good biodegradability, and chemical and mechanical stability [112–114]. Similar to organic nanoparticles, the formation of silica nanoparticle composites by surface modification or matrix blending represents an effective approach to achieve the desired properties and to obtain a superior delivery efficiency. In order to reduce protein adsorption and enable long circulation times in the blood, stealth silica nanoparticles composites have been produced via surface modifications with different molecular weight PEGs. Hence, the reduction in protein adsorption was found to be constant when the PEG chain length was increased to above 3000 g mol^{-1} [115]. Schooneveld *et al.* [116] reported the production of stealth silica nanoparticles composites using DSPE-PEG; in this case, the blood circulation half-life was found to be 165 min, much longer than that of the bare silica nanoparticles (15 min). Lee *et al.* [117] produced positively charged silica nanoparticle composites by the incorporation of trimethylammonium, for the oral delivery of an anti-inflammatory drug, such that the drug was released only in the intestine, and not in the stomach. Antibody-functionalized silica nanoparticle composites were also prepared, and exhibited high targeting efficiency; in this case, the targeting capability was strongly affected by the antibody density [118]. Likewise, poly(*N*-isopropylacrylamide) (PNIPAM)-modified silica nanoparticle composites were shown to exhibit a temperature-dependent uptake and release of small molecules [119].

9.3 Nanofiber Composites for Drug Delivery

Among the various currently used nanostructures for drug delivery, nanofiber composites have attracted increased interest due to their easy fabrication, controllable size and/or shape, and sustained release properties. A wide variety of processing methods, including drawing, template synthesis [120], phase separation, self-assembly [121, 122], and electrospinning, have been examined for the synthesis of nanofiber composites [123]. Among these methods, electrospinning at room temperature is most commonly used to fabricate polymer-based nanofibers, as described elsewhere [124–130]. As the composite nanofibers were obtained under mild conditions, the activity of the drug was well preserved during the entire process. Electrospun nanofiber mats offer particular advantages for drug delivery, due to their large surface area and controllable adsorption/release properties. In order to achieve the desired chemical, physical and biological properties of nanofibrous scaffolds, combined multicomponent compositions have frequently been used. In addition, in order to produce nanofiber materials with controlled anisotropy and porosity, the electrospun scaffolds can be post-modified and functionalized for biomedical applications. The electrospinning of a polymer suspension with a preloaded drug has led to major opportunities for the application of nanofiber composite materials in drug delivery [131–133]. Although the release of drug molecules from the matrix of the polymer nanofibers is generally controlled by diffusion and scaffold degradation [134], the delivery properties (especially the rate of release) may also be tailored by the introduction of functionalized features. Functional nanofiber composites have also been applied in many areas of biomedicine, such as tissue engineering [135], wound dressing [136], and enzyme immobilization [125].

9.3.1 Delivery of Antibiotics

When biodegradable nanofiber materials have been incorporated with antibiotics, they have shown potential for wound dressing and healing, and for the prevention of post-surgical adhesions [137–139]. Kim *et al.* [137] reported the incorporation of a hydrophilic antibiotic (cefoxitin; Mefoxin®) into composite nanofibers containing PLGA and PLGA/PEG-*b*-PLA/PLA (80 : 15 : 5 wt%). For this, the antibiotic was first dissolved in an organic solvent, directly mixed with the polymers, and then electrospun such that the antibiotic was embedded within the composite nanofibers. An *in vitro* drug release study showed that the PLGA/PEG-*b*-PLA/PLA composites produced a more sustained release than did the PLGA mat with a single component. It was suggested that, due to the high ionic strength of the drug, and the minimal physical interactions between the polymer and drug, the drug had primarily located at the surface of the pure PLGA nanofibers. However, incorporating the amphiphilic PEG-*b*-PLA block copolymer to form composite nanofibers had allowed some of the hydrophilic antibiotic to become embedded in the nanofibers, and this had led to a higher drug loading and a sustained release of the antibiotic. The nanocomposite PLGA/PLA/PEG-*b*-PLA scaffold showed a sustained-release profile after the initial burst, with about 27% of the cefoxitin being released continuously over one week. The cefoxitin released from the electrospun scaffolds was shown to be structurally intact, and effective in inhibiting the growth of *Staphylococcus aureus* in both static (agar) and dynamic (liquid) environments. The ideal release profile of antibiotics displayed by this composite nanofiber scaffold indicated a potential role for the prevention of surgery-induced infections.

The use of both, PLA and poly(ε-caprolactone) (PCL) composite nanofibers was reported by Buschle-Diller *et al.* [140] for the delivery of antibiotics. When three different antibiotics – tetracycline, chlorotetracycline hydrochloride, and amphotericin B – were selected as model drugs, tetracycline was shown to be discharged from PCL at the highest rate, with amphotericin B the slowest. However, whereas the PCL nanofibers almost completely liberated all of the drugs over a time period, the PLA released only about 10% of the loaded drugs. However, a desirable drug release profile was achieved by forming a bicomponent PCL–PLA composite nanofiber. Subsequently, the feasibility of producing multicomponent, bioerodible polymeric nanofibers loaded with combinations of drugs for dual-controlled delivery, began to attract attention. Thus, Piras *et al.* [141] investigated the creation of a multicomponent, bioerodible polymer of poly(maleic anhydride-*alt*-2-methoxyethyl vinyl ether)-*n*-butyl hemiester, to prepare multicomponent drug-loaded nanofibers via electrospinning. For this, diclofenac sodium and human serum albumin were loaded into the formed nanofibers; the subsequent *in vitro* drug release evaluation showed that the composite nanofibers provided an independent delivery of the two active ingredients.

9.3.2 Delivery of Anticancer Drugs

Today, *chemotherapy* is an active research area, and the efficient delivery of anticancer drugs still faces a major challenge. Among various drug delivery vehicles, nanofibers have attracted increasing attention during recent years. For example, Song *et al.* [142] described a daunorubicin-loaded poly(*N*-isopropylacrylamide)-*co*-polystyrene (PNIPAM-*co*-PS) nanofiber produced via electrospinning, where the composite nanofibers enhanced the cell permeation and uptake of daunorubicin into drug-sensitive and drug-resistant cancer cells (e.g., leukemia K562). The results of subsequent MTT assays and electrochemical studies showed that the PNIPAM-*co*-PS nanofibers played an important role in facilitating cell tracking and drug delivery to the cancer cells. Meanwhile, the results of studies using atomic force microscopy (AFM) and confocal fluorescence microscopy indicated that an interaction of the PNIPAM-*co*-PS nanofibers with bioactive molecules on the membrane of leukemia cell lines could affect intracellular drug uptake in a positive manner, and lead to an efficient accumulation of daunorubicin in drug-sensitive and drug-resistant cancer cells. Chen *et al.* [143] also reported the incorporation of TiO_2 with PLA to form composite nanofibers, via electrospinning, to deliver daunorubicin. In this case, the drug molecules were shown readily to self-assemble on the surface of the nano-TiO_2–PLA composite; the nanofiber drug carrier then induced daunorubicin to permeate and accumulate in leukemia K562 cells.

The biodegradable polymer-based composite nanofibrous substrate appears to represent a promising approach for the local delivery of anticancer drugs [144], including postoperative local chemotherapy in a dual-controlled model. Xu *et al.* [145] reported that both hydrophobic and hydrophilic anticancer drugs (paclitaxel and doxorubicin, respectively) could be successfully loaded into PEG–PLA nanofiber mats by means of "emulsion-electrospinning," so as to achieve a multidrug delivery. As shown in Figure 9.8, the nanofibers exhibited a uniform diameter and smooth surface, without any evidence of drug crystals being present, which indicated that the drug had been well incorporated into the electrospun nanofibers.

Figure 9.8 Environmental scanning electron microscopy images of (a) (0.5 wt% doxorubicin + 0.5 wt% paclitaxel)/ PEG–PLA composite fibers and (b) (1.0 wt% doxorubicin + 1.0 wt% paclitaxel)/PEG–PLA composite fibers. Reproduced with permission from Ref. [145]; © 2008, Elsevier B.V.

In addition, neither morphology nor average diameter of the mediated nanofibers appeared to be affected by the variations in drug loading. In this case, the drug loading efficiency was 90.6% for doxorubicin and 97.0% for paclitaxel, in both drug-loaded fibers. As both drugs were loaded onto the same nanofibers, the release behaviors from the same fiber mats were to great extent dependent on their solubility properties and the distribution status in the fibers. Due to its high hydrophilicity, it was easier for doxorubicin to diffuse out from the fibers, and its release rate was always higher than that of the hydrophobic paclitaxel. Moreover, the release rate of paclitaxel was accelerated by the release of doxorubicin from the same drug-loaded fibers, as the rapid release of doxorubicin would clear more pathways and nanopores in the fibers as a result of doxorubicin diffusion and release through them, thus eventually facilitate the release of paclitaxel molecules to the buffer solution. Therefore, the release rate of paclitaxel was increased compared to paclitaxel-only-loaded nanofibers, as shown in Figure 9.9. An *in vitro* cytotoxicity assay showed that the co-electrospun medicated fiber mats showed the strongest cytotoxicity against rat Glioma C6 cancer cells. The cell growth inhibition rate, apoptosis rate, and percentage of cells in G_0/G_1 phase were higher than with a single drug-loaded system.

9.3.3 "Smart" Nanofiber Drug Carriers

"Smart" polymer-based devices can be manipulated to enable them to reversibly change their physico-chemical characteristics in response to their environment and drug-release profile. The drug release can be triggered by the environment or other external events, such as changes in pH [146, 147], temperature, or the presence of an analyte such as glucose [148, 149]. The smart composite polymer nanofibers may find wide application as temporary scaffolds in noninvasive procedures to deliver drugs or cells to particular parts of the body.

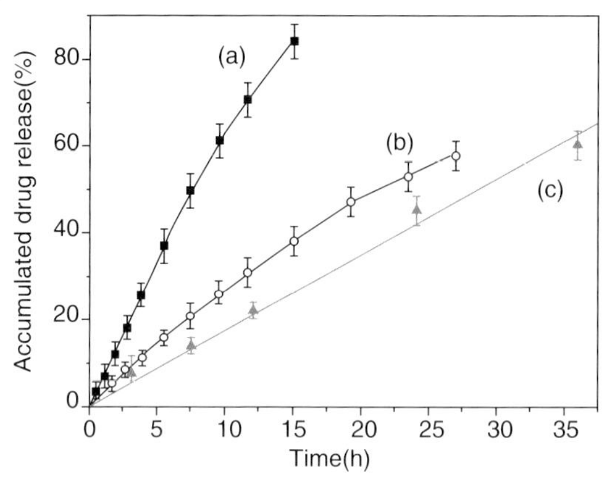

Figure 9.9 Release profiles of (a) doxorubicin (DOX) and (b) paclitaxel (PTX) from (1.0 wt% DOX + 1.0 wt% PTX)/PEG–PLA composite fibers, and of (c) paclitaxel from 1.0 wt% PTX/PEG–PLA fibers in 0.05 mol l^{-1} Tris–HCl buffer solutions in the presence of proteinase K (4 g ml^{-1}) at 37 °C. Reproduced with permission from Ref. [145]; © 2008, Elsevier B.V.

pH-sensitive drug carriers are of particular interest, as the pH environment clearly changes through the gastrointestinal tract. When a drug is administered orally, it passes through a low-pH zone (pH 1.2–1.5) in the stomach for 1–2 h, followed by near-neutral conditions in the intestine for dissolution and absorption into the body fluids. These variations in pH may affect the swelling of a nanofiber drug carrier, or alter the surface affinity with drug molecules, thus allowing a controlled drug delivery to a target region. Recently, Cao *et al.* [150] reported that poly[styrene-*co*-(maleic sodium anhydride)] and cellulose (SMA-Na/cellulose) hydrogel nanofibers had been prepared through the hydrolysis of precursor electrospun poly[styrene-*co*-(maleic anhydride)]/cellulose acetate (SMA/CA) nanofibers. In the presence of diethylene glycol, the SMA/CA composite nanofibers were crosslinked by esterification at 145 °C, and then hydrolyzed to yield crosslinked SMA-Na/cellulose hydrogel nanofibers. The water swelling ratio of the fabricated composite nanofibers was shown to depend on the pH condition; over the pH range of 2.5–5.5, the water absorption of SMA-Na-DEG/cellulose composite nanofiber was approximately 18.0 g g^{-1} (mainly because the cellulose nanofiber itself can absorb up to 17–19.0 g g^{-1} of water, due to the presence of three hydroxyl groups on each anhydroglucose unit). The swelling ratio then increased with the rise in pH, to reach about 20 g g^{-1} at pH 6. The maximum water swelling ratio of the SMA-Na-DEG/cellulose hydrogel nanofibers was 27.6 g g^{-1}, achieved in a buffer solution at pH 9.1. The characteristic two-step increase in the water swelling ratio results from a two-stage dissociation of the two protons in maleic acid, which have intrinsic dissociation constants (i.e., $pK_{a1} = 3.2$ and $pK_{a2} = 8.1$ at 25 °C). When the pH was increased, the –COOH group was ionized to a –COO– group, generating a higher negative charge density; this resulted in stronger electrostatic repulsion forces and led to a greater chain relaxation and more water absorption. The controllable swelling ratio of the nanofiber mat implied that the drug release could be controlled, since this was mainly under the effect of diffusion. Addition pH-sensitive composite nanofiber materials, fabricated by electrospinning, have been shown to exhibit a high potential for biomedical applications, especially for effective drug delivery [151].

A temperature-responsive composite nanofiber was recently reported by Jeong *et al.* [152]. In this case, a biodegradable and elastic polymer, PLA-*co*-ε-caprolactone) (PLCL), was electrospun to prepare nanofibers, after which *N*-isopropylacrylamide (NIPAAm) was grafted onto the surfaces under aqueous conditions, using ^{60}Co γ-irradiation. Subsequent scanning electron microscopy (SEM) studies confirmed that the PLCL nanofibers had maintained an interconnected pore structure after grafting with NIPAAm. The phase-separation characteristics of PNIPAAm in aqueous conditions possessed a unique temperature-responsive swelling behavior of PNIPAAm-g-PLCL nanofibers, and showed an ability to absorb a large amount of water below 32 °C, but to collapse abruptly when the temperature was increased to 40 °C. In accordance with the temperature-dependent changes in swelling behavior, the release rate of indomethacin and fluorescein isothiocyanate-labeled bovine serum albumin (FITC-BSA) loaded into PNIPAAm-g-PLCL nanofibers by a diffusion-mediated process, was regulated by the change in temperature. Both model drugs demonstrated a greater release rate at 40 °C relative to that at 25 °C. Such temperature-controlled release of drugs from PNIPAAm-g-PLCL nanofibers, by using γ-irradiation, may be used to design drugs and protein delivery carriers for various biomedical applications.

Chunder *et al.* [153] reported that ultrathin fibers comprising two weak polyelectrolytes, namely poly(acrylic acid) (PAA) and poly(allylamine hydrochloride) (PAH), could be fabricated using the electrospinning technique. The composite nanofibers exhibited both pH-sensitive and temperature-responsive properties for active ingredient delivery when the nanofiber surface was modified with a temperature-sensitive polymer. In this case, methylene blue (MB) was used as a model compound to evaluate the potential application of these fibers for drug delivery. MB release was controlled in a nonbuffered medium by changing the solution pH; typically, when MB-loaded fibers were immersed in solution at pH between 10 and 7, the PAA was fully charged but the negatively charged carboxylic groups remained bound to the positively charged MB molecules, preventing their release. However, when the solution pH was lowered to 6, the protons in solution combined with the carboxylic groups to replace the bound MB, which resulted in its release. At pH 6 the fibers remained quite blue, which suggested that some MB was still bound to the remaining carboxylic groups. When the pH of the solution was decreased further, to pH ~ 5, the MB was further released, and this continued until another equilibrium was reached. At pH 2, most of the carboxylic groups were protonated, so that most pre-bound MB was released. Such a pH-controlled release would be important if polyelectrolyte multilayers must be deposited while retaining the MB inside the fibers (see below). The sustained release of MB in a phosphate-buffered saline (PBS) solution was achieved by constructing perfluorosilane networks on the fiber surfaces as capping layers. A temperature-controlled release of MB was then achieved by depositing temperature-sensitive PAA/poly(*N*-isopropylacrylamide) (PNIPAAm) multilayers onto the fiber surfaces. The release of MB in PBS solution from PAA/PAH fibers coated with PAA/PNIPAAm multilayers was compared at room temperature (~25 °C) and at elevated temperature (40 °C). At room temperature, a greater time was required by the PAA/PNIPAAm multilayer-coated fibers to release all of the loaded MB, than by the bare fibers. This slow release was due to a hydrogen-bonded network on the coated fibers, which delayed the diffusion of MB. In contrast, the coated fibers released MB much faster at temperatures above PNIPAAm's lower critical solution temperature (LCST), of 32 °C; typically, at 40 °C the MB release took 50 min. It was suggested that PNIPAAm would form intramolecular hydrogen bonds above the LCST, which would in turn disrupt the PAA/PNIPAAm hydrogen-bond network and lead to an accelerated diffusion of MB. This pH-sensitive and temperature-responsive release of drugs from electrospun fibers might have application in drug carrier systems for the biomedical sciences.

9.3.4 Multifunctional Nanofibers

Polymer composites with multifunctions are of major importance for efficient drug delivery. For example, magnetic polymer composite nanofibers were exploited by the incorporation of magnetic Fe_3O_4 nanoparticles by electrospinning, and used for precise target delivery [154]. The composite nanofibers containing drugs and magnetite nanoparticles showed superparamagnetism at room temperature, and their saturation magnetization was shown to depend on the amount of Fe_3O_4 nanoparticles within the nanocomposites. The delivery could then be controlled by applying an external magnetic field. This type of composite, composed of biocompatible and biodegradable polymers, may prove to be of interest in a wide variety of applications in medicine, and especially for targeted drug delivery. Gold nanoparticle (AuNP)-loaded composite nanofibers were investigated for the high-sensitivity diagnosis of cancer cells, as well as drug delivery. For example, Song *et al.* [155] reported that poly(*N*-isopropylacrylamide)-*co*-polystyrene nanofibers with functionalized AuNPs would form a new nanocomposite that facilitated the accumulation of daunorubicin inside leukemia cells, and potentially could enhance the cure efficiency. These synergistic enhancement effects of nanocomposites on the uptake of daunorubicin in drug-resistant leukemia K562 cells were observed using electrochemical and confocal fluorescence microscopy. The results suggested that such nanocomposites might be used to facilitate drug delivery and provide an early diagnosis of cancer cells, with potential valuable applications in the related areas of biomedicine and bioanalysis.

The hydroxyapatite (HAp) nanoparticle-incorporated biodegradable nanofibrous scaffold has been used for tissue regeneration, as well as for drug/protein delivery

[156–158]. Nanocomposites containing HAp nanoparticles have been shown to elicit active bone growth, since HAp is a component of bone structure. It has been suggested that drug delivery and tissue engineering are closely related fields; indeed, tissue engineering may be viewed as a special case of drug delivery, where the target is to achieve a controlled delivery of mammalian cells, whereas the controlled release of therapeutic factors would enhance the efficacy of tissue engineering. Thus, biocompatible and biodegradable materials are required for multifunctions as both drug-delivery vehicles and tissue-engineering scaffolds. PLGA/HAp nanocomposites fit into this requirement. Fu *et al.* [159] investigated bone regeneration under the controlled release of bone morphogenetic protein-2 (BMP-2) from PLGA/HAp nanofibrous scaffold. As BMP-2 is easily inactivated enzymatically when exposed to serum *in vivo*, a sustained release would provide the best strategy to maintain high levels of BMP-2 in a local area, and this is the main reason for designing the release profile of scaffolds. Two methods were investigated for loading BMP-2 into nanofibrous scaffolds using an electrospinning method, namely encapsulation into the fibers, or coating on the fiber surface. The results revealed that BMP-2, when encapsulated into the fibers, retained its biological activity both *in vitro* and *in vivo*. The addition of HAp nanoparticles may alter the scaffold tensile strength and reduce the residual solvent content, and PLGA/HAp composite scaffolds developed in this way showed good morphological/mechanical strength, with the HAp nanoparticles being dispersed homogeneously within the PLGA matrix. Results obtained from animal experiments showed the bioactivity of BMP-2 released from a fibrous PLGA/HAp composite scaffold to be well maintained, and to lead to further improvements in new bone formation and the healing of segmental defects *in vivo*. In comparison, the BMP-2 loaded pure PLGA scaffold did not retain the bioactivity of BMP-2 *in vivo*, and had no effect on bone healing. These results reflected the advantages of composite nanofibers over single-component active ingredient carriers. The biocompatible and biodegradable HAp-containing nanocomposites would offer a platform for applications in bone regeneration, by providing a controlled release of BMP-2.

In general, composite nanofibers have been mainly fabricated by electrospinning, and investigated for application in drug delivery and tissue engineering. Nanocomposite fibrous carriers showed advantages in drug delivery applications as compared to traditional bulk materials and single-component nanomaterials in terms of effective targeting, sustained release and potential cytotoxicity. Functional composite nanofiber materials have promised a versatile nanoscale controlled or targeting drug delivery system, based on polymer–polymer and polymer–inorganic systems. These can be applied to the effective delivery of both small-molecule drugs and various classes of biomacromolecules, such as peptides, proteins and plasmid DNAs to desired regions of the body. The diffusional control mechanism associated with nanofibers may also provide the effective sustained release of encapsulated drug molecules from a nanofiber matrix.

9.4
Two-Dimensional Mesoporous Material-Incorporated Nanocomposites

Since the discovery of the M41S family of two-dimensional (2-D) nanostructured silica with ordered mesoporous silica materials in 1991 [160], silica and modified ordered mesoporous composite materials have been extensively explored, with attention mainly focused on the synthesis, properties and applications of these novel materials [161–163]. The pore size of these mesoporous materials (which are also referred to as "nanoporous" materials) is tunable over a range of 2 to 50 nm. These novel, ordered mesoporous silica materials with biocompatible amorphous pore walls have good potential for drug delivery and tissue regeneration [164–170]. The large pore volume allows a high drug loading into the ordered matrix, while the large surface area of the internal surface can be modified according to specific purposes for different drugs association on the surface of matrix. Since Vallet-Regi and coworkers [171] first reported that MCM-41-loaded ibuprofen could demonstrate a sustained release over a period of 80 h, mesoporous silica materials have attracted much attention as potential controlled drug-delivery systems. Organic-modified MCM-41 with functional groups on the material surface could further decrease and control the delivery rate of ibuprofen [172–175]. Moreover, the mesoporous materials could also be modified with various surface-functional species, while preserving the large pore channels and surface areas for drug adsorption and desorption [176–180]. These materials are also expected to provide a greater versatility for the delivery of drug molecules, where the release rate will depend to a large extent on the surface properties of the composite nanoporous drug carrier.

9.4.1
Inorganic 2-D Nanoporous Composites

9.4.1.1 Oxide-Modified Nanoporous Composites

Inorganic species incorporating mesoporous silica-based composites have exhibited good potential for effective drug delivery. For example, MgO-incorporated SBA-15 mesoporous silica was prepared by *in situ* coating [181, 182]. Shen *et al.* [183] reported that submicron particles with modified surfaces were synthesized via a simple, one-pot synthesis approach, and used as a drug carrier for controlled release. As seen in Figure 9.10, the ordered arranged nanoporous structure was well preserved with the incorporation of MgO. However, due to the alkalinity of the MgO species on the surface, the amount of a model drug (ibuprofen) that could be adsorbed onto the modified surface was increased when compared to the pure silica SBA-15, despite the surface area having been reduced by the surface modifications. The morphology of SBA-15 submicron particles was well preserved, with drug loadings of up to 30 wt%; this indicated that most of the drug molecules had been stored inside the nanoporous channels. The results obtained from *in vitro* experiments showed that the surface modification had greatly reduced the ibuprofen release rate, with only 63% of the adsorbed ibuprofen being released

Figure 9.10 Transmission electron microscopy images of MgO/SBA-15(Si/Mg = 5) submicron particles. (a) Taken along the parallel pore channels; (b) Taken vertical to the parallel pore channels. Reproduced with permission from Ref. [183]; © 2007, Pharmaceutical Society of Japan.

from the MgO/SBA-15 (Si/Mg = 20) during the first six hours. In contrast, ibuprofen release from the pure silica SBA-15 was complete within 1 h, under the same release conditions. The delayed release profile from the MgO-coated SBA-15 was attributed to an interaction of the ibuprofen molecules with the drug carrier surface, which was investigated using differential scanning calorimetry-thermogravimetric analysis (DSC-TGA) and Fourier transform infrared (FT-IR) spectroscopy. The TGA-DSC results indicated that ibuprofen adsorbed onto the SBA-15 surface was much different from the pure ibuprofen crystal (the pure ibuprofen crystal melted at 78 °C and vaporized at 220–250 °C, whereas the adsorbed ibuprofen was oxidized in air at 300–550 °C). The characteristic FT-IR band of the carboxyl groups of ibuprofen adsorbed onto the MgO-coated SBA-15 was shifted to 1591 cm^{-1}, due to a proton transfer of ibuprofen to the alkaline functional surface; this interaction was believed to be a contributory factor in the controlled release of ibuprofen from the carrier. The nanocomposites of MgO-coated SBA-15 showed an affinity with the acidic ibuprofen molecules, and retarded their release from the mesoporous matrix. The release rate of ibuprofen could also be modulated by varying the content of MgO, with the release rate decreasing when larger amounts of MgO-coated SBA-15 were present on the surface of the submicron particles.

Nanocomposites of mesoporous silica and calcium phosphate are of major interest in controlled local drug delivery, and have demonstrated excellent bioactivity and biodegradability. A nanoporous composite of silica containing calcium phosphate was shown to have a high porosity and to provide a sustained release of antibiotics [184]. The calcium phosphate species on the surface of drug carriers was shown to facilitate the development of a bioactive HAp layer following immersion in a simulated body fluid (SBF). In this way, silica–calcium phosphate nanocomposites can be used as carriers for antibiotics to treat osteomyelitis, while simultaneously helping to regenerate bone tissue. Recently, Sousa *et al.* [185] described a composite of nanoporous MCM-41 and HAp that could be fabricated via a hydrothermal synthesis process. The application of mesoporous silica MCM-41/HAp systems as matrices for the controlled delivery of drugs was investigated, to establish the influence of pore architecture and size on the release of the beta-blocker, atenolol. The results indicated that silica-ordered mesopores had the potential to encapsulate bioactive molecules, while the incorporation of an HAp phase into the mesoporous silica led to a significant change in the structural properties of the system. It appeared that the crystal growth of HAp had blocked the available spaces of the mesoporous silica, and that the original mesopores of MCM-41 were partially filled by HAp crystals. The incorporation of HAp, however, led to a decrease in textural characteristics such as surface area and pore volume. It was apparent, therefore, that MCM-41–HAp would act as a better drug release device than MCM-41, as the HAp phase would serve as a temporary barrier to prevent the rapid release of atenolol from the nanocomposite drug carrier.

9.4.1.2 **Mesoporous Bioactive Glasses**

Mesoporous bioactive glass (MBG) is a promising nanocomposite drug delivery system for controlled local drug delivery, notably in the treatment of bone infections. Xia *et al.* [177] reported that a well-ordered mesoporous bioactive glass with the composition of SiO_2–CaO–P_2O_5 (Si/Ca/P = 58:23:9, w/w) had been synthesized in aqueous solution, using a two-step, acid-catalyzed self-assembly process, followed by hydrothermal treatment. The bioactive glass thus obtained (M58S) exhibited an ordered mesoporous structure and a high specific surface area. An alternative material, M77S, was synthesized by increasing the silica content. When the antibiotic, gentamicin, was encapsulated into the MBG using an adsorption method, the extent of drug loading was threefold that of a conventional sol–gel 58S with the same composition. Drug release from the MBG occurred as an obvious two-step release process, with an initially rapid release followed by a relatively slow release. A comparison of gentamicin release in distilled water and in SBF showed that M58S effectively reduced the initial burst with 33 wt% of the total gentamicin being released into water, and 49 wt% into the SBF, during the first 24 h. The subsequent release rate fell notably with time, with cumulative release during the next 20 days reaching maximum values of 48 wt% into water and 61 wt% into SBF (Figure 9.11). With conventional 58S, drug release during the first 24 h was more extensive (but similar) in both water (ca. 53 wt%) and SBF (63 wt%), and after 20 days had reached 57 wt% and 85 wt%, respectively. In contrast to M58S and conventional 58S, drug release from M77S was slower into distilled water, but quicker into SBF. After soaking in distilled water and SBF, gentamicin was released much more slowly from M58S than from 58S. Drug release was also shown to be sensitive to the pH and the ionic concentration of the release medium, suggesting another possible means of controlling the release rate. In contrast to conventional sol–gel 58S, M58S showed a much greater ability to induce HAp formation; after 3 h in SBF only a particulate HAp growth was visible, but after 6 h the surface was covered with rod-like HAp crystals. No HAp precipitation was visible on the surface of 58S after soaking for 6 h, however. Clearly, as M58S possesses a greater ability to induce HAp formation than does

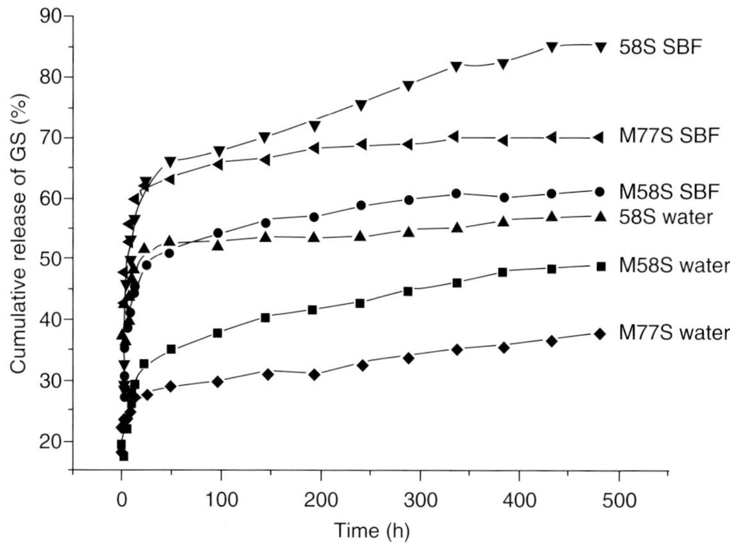

Figure 9.11 Cumulative gentamicin release from M77S, M58S and 58S in distilled water and simulated body fluid (SBF). Reproduced with permission from Ref. [177]; © 2005, Elsevier B.V.

58S, well-ordered MBGs might serve as bioactive drug release systems for the preparation of bone implant materials.

Recently, Cauda *et al.* [186] investigated the nanocomposites of an ordered silica SBA-15 mesostructure fabricated in a macroporous bioactive glass-ceramic scaffold of SiO_2–CaO–K_2O, for local drug delivery. The excellent bioactivity of the glass-ceramic, combined with the drug host/release properties of mesoporous silica, may lead to a role in local drug delivery from implants designed for tissue engineering. Due to the presence of an ordered mesoporous silica, the drug loading capacity was significantly increased, but bioactivity was retained. Similarly, Vitale-Brovarone *et al.* [187] studied a nanocomposite drug delivery system which contained glass-ceramic scaffolds of SiO_2–P_2O_5–CaO–MgO–Na_2O–K_2O–CaF_2 and nanoporous MCM-41 microspheres. In comparison with the original glass-ceramic system, these composite drug carriers were shown to adsorb and deliver greater amounts of a model drug (e.g., ibuprofen), with such benefits being attributed to the key role of the nanoporous silica matrix.

9.4.1.3 **Magnetic Functionalized Mesoporous Composites**

Magnetic functional nanocomposite materials composed of magnetic particles and a mesoporous silica matrix have attracted much attention due to their potential application in drug delivery, hyperthermic treatment of tumors, magnetic bioseparations, and magnetic resonance imaging. Pure magnetic nanoparticles tend to aggregate and have short circulating periods when directly exposed to a biological system, and therefore demonstrate less potential for biomedical applications [188, 189]. Magnetic composite materials, when used as drug delivery carriers, can permit the selective targeting of organs or tissues in the body, as well a means of accumulating nanoparticles along a therapeutic path by applying an external magnetic field. By embedding magnetic nanoparticles into nanoporous silica materials, it is possible to avoid problems of aggregation and to prolong the retention time within the circulatory system. Moreover, magnetic nanocomposites may play multifunctional roles as drug carriers for chemotherapy, and also in hyperthermic treatments. The latter approach can be used to eliminate bone-derived cancerous cells, as these are not viable at the high temperatures (>43 °C) created when external electromagnetic waves are applied to the magnetic nanocomposite.

Several methods have been reported for synthesis of nanocomposites with magnetic and mesoporous materials in different morphologies. Zhang *et al.* [190] fabricated magnetic hollow spheres with a periodic mesoporous organosilica shell and a Fe_3O_4 nanocrystal core. When monodispersed magnetite nanocrystals were embedded into the cavities of highly ordered periodic mesoporous organosilica hollow spheres, the particle size, thickness of the hollow shell and amount of magnetic particles encapsulated was controllable. Magnetic porous hollow silica drug carriers were also synthesized by using $CaCO_3/Fe_3O_4$ composite nanoparticles and cationic surfactant double templates [191]. In addition, Wang *et al.* [192] described a simple evaporation-induced self-assembly approach to synthesize ordered magnetic mesoporous γ-Fe_2O_3/SiO_2 nanocomposites with diverse mesostructures. The controllable release behaviors of lysozyme from these magnetic porous nanocomposites were investigated for the possible application of drug targeting and controlled release.

Recently, nanocomposites of magnetic functionalized and ordered mesoporous silica have attracted interest for controlled drug release and targeting [193]. Huang *et al.* [194] functionalized SBA-15 with Fe_xO_y nanoparticles embedded in the amorphous pore walls, using a sol–gel method. The magnetic composites obtained conserved the ordered mesoporous structure and showed a high surface area and pore volume after the formation of Fe_xO_y nanoparticles. The latter were mainly composed of γ-Fe_2O_3, which contributed to the superparamagnetic properties of the Fe_xO_y/SBA-15 composites. The resultant nanocomposites showed sufficient response to an external magnetic field, and could be used as carriers in drug delivery systems to reach targeted locations under the control of an external magnetic field. The magnetic composites demonstrated sustained release profiles with ibuprofen as a model drug; in this case, the release rate was affected by the pore size, particle morphology, and surface property of these systems. Magnetic rice grain-like SBA-15 composites which possessed a regular morphology showed a higher release rate compared to Fe_xO_y-loaded classical SBA-15 blocks with a larger particle size.

Uniform, rattle-like, hollow magnetic mesoporous spheres were prepared and used as drug carriers to achieve a targeted sustained release [195]. For this, Fe_3O_4 nanoparticles were encapsulated in the core of mesoporous silica, using a sol–gel reaction, to form rattle-type hollow magnetic mesoporous microspheres. A ring-

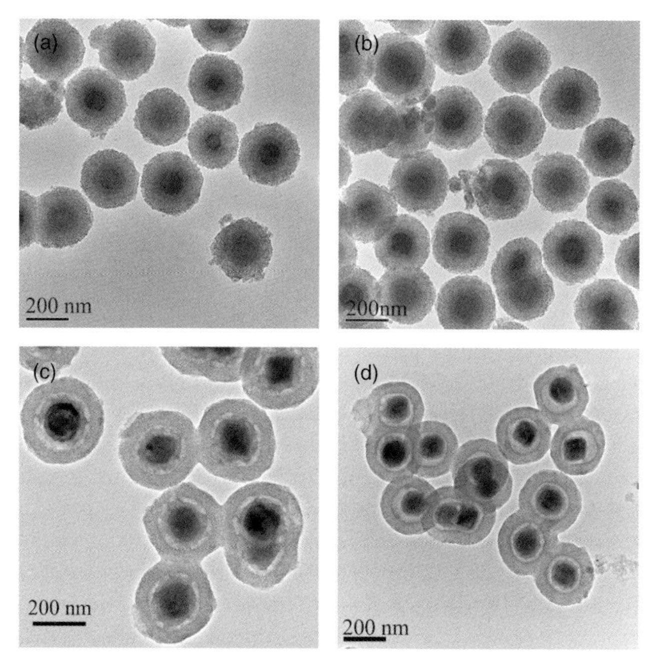

Figure 9.12 (a–d) Transmission electron microscopy images of hollow magnetic mesoporous nanocomposites (H1–H4) obtained by different hydrothermal conditions (H_2O vol% = 25, 50, 75, 100, respectively). Reproduced with permission from Ref. [195]; © Wiley-VCH Verlag GmbH & Co. KGaA, Weinheim.

shaped cavity between the core and the mesoporous silica shell was generated by two-step coating and a simple hydrothermal treatment in a water/ethanol medium. The TEM images of the hollow magnetic mesoporous nanocomposites are shown in Figure 9.12. The nanocomposite hollow magnetic mesoporous spheres showed a relatively high magnetization strength (>20 emu g^{-1}), while the middle layer cavity contributed to a high drug loading capacity (302 mg g^{-1}). The release investigation identified a sustained-release behavior of a model drug (e.g., ibuprofen) from the nanocomposites of a magnetic and mesoporous silica system, and also indicated that the release process followed Fick's law. These hollow magnetic mesoporous spheres with a sophisticated middle cavity layer represent promising candidates for targeted drug-delivery carriers.

In addition to silica-based mesoporous substrates, other biocompatible mesoporous materials – such as mesoporous bioactive glass and carbon – have also been functionalized with magnetic nanoparticles for potential application in drug targeting. Li *et al.* [196] described the preparation of nanocomposites composed of magnetic nanoparticles and mesoporous bioactive glass, by using by a facile one-pot synthesis. Nanocomposites in which Fe_3O_4 nanoparticles had been confined and dispersed in ordered mesoporous glass matrices were prepared by a simultaneous evaporation-induced self-assembly of Ca, P, Si and Fe sources, with subsequent reduction in an H_2 atmosphere. Experiments to demonstrate ibuprofen storage and release with these composites identified adjustable loading amounts ranging from 199 to 420 mg g^{-1}, with a sustained release that extended for 100 h. A superparamagnetic behavior was determined, and the saturation magnetization of the bioactive glass composites was shown to increase with a higher content of the Fe species. This type of magnetic and mesoporous bioactive glass nanocomposite might (potentially) be applicable for selective targeted drug delivery and the hyperthermia treatment of bone tumors. Recently, Yuan *et al.* [197] described a co-casting method for the synthesis of mesoporous composites of α-Fe and ordered mesoporous carbon, in which the α-Fe nanoparticles were embedded into the wall of the carbon. The composites showed superparamagnetic behavior, and were monitored as carriers for the delivery of tetracycline hydrochloride. The mesopore surface area and mesopore volume of the magnetic-carbon composites were found to be crucial factors for drug adsorption. In contrast, desorption kinetic studies showed that the desorption rate of tetracycline hydrochloride from the nanocomposite was related to the average pore diameter of the magnetic-mesoporous carbon, with the sample having the largest pore size exhibiting the highest desorption rate. The drug desorption kinetics are well-depicted by a pseudo-second-order kinetic equation. Furthermore, Cao *et al.* [198] reported that novel silica-coated iron–carbon nanocomposite particles had been prepared for use in targeting therapy as a drug carrier, and tested in pigs. The results of *in vivo* experiments, using $^{99m}TcO_4$-adsorbed composite particles, showed a prominent biodistribution in the left hepatic lobe of pigs under the control of an external magnetic field (Figure 9.13). Typically, greater quantities of drugs could be delivered to the target areas; when doxorubicin-adsorbed composite particles were

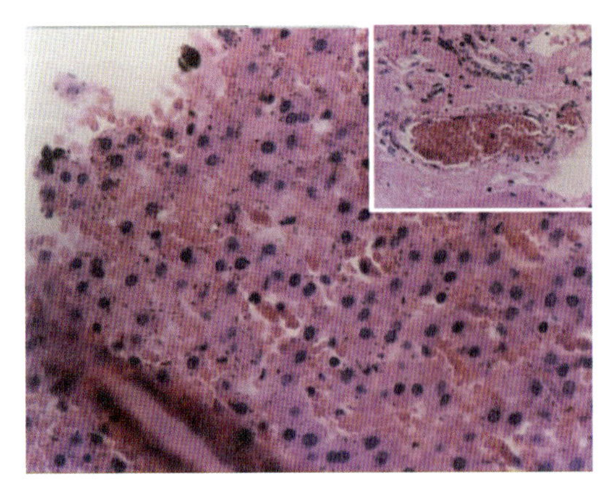

Figure 9.13 Photomicrograph showing a large amount of tiny composite particles scattered in hepatic cells and the tissue interface of the targeted liver area (hematoxylin and eosin (HE) staining; original magnification ×200). The inset shows extravasation of the composite particles outside the hepatic vessel (HE staining, original magnification ×400). Reproduced with permission from Ref. [198]; © 2007, Wiley Periodicals, Inc.

infused intra-arterially, the doxorubicin content of hepatic tissue was 23.8-fold higher in the targeted area of the left lobe than in the nontargeted area of the right lobe. These results also suggested that the composite particles could penetrate through the capillary wall around the tissue interstitium and hepatic cells, under the driving force of an external magnetic force in the targeted area. The indication here was that these novel silica-coated iron–carbon composite particles might be applied as targeted treatment for certain types of tumor, as an effective drug carrier.

9.4.1.4 Luminescent Mesoporous Composites

The nanocomposites of mesoporous silica and fluorescent particles may have potential application in the fields of disease diagnosis, drug delivery, and therapy. In addition to the high pore volume for storage and delivery, the presence of fluorescent properties enables the tracking and evaluation of the efficiency of the drug release. Fluorescent nanocomposites may be used to track the path of delivery, to provide information regarding the mechanism of drug delivery, and can also be used in the qualitative and quantitative detection of disease location and drug release efficiency. Consequently, the design, synthesis and application of fluorescent mesoporous nanocomposites for drug delivery have recently attracted much attention [199].

Yang *et al.* [200] suggested that a luminescence functionalization of ordered mesoporous SBA-15 silica might be achieved by depositing a $YVO_4:Eu^{3+}$ phosphorescent layer on its surface, via the Pechini sol–gel process, and that this would result in the formation of a $YVO_4:Eu^{3+}$/SBA-15 nanocomposite material. The obtained nanocomposites, by combining the advantages of the mesoporous structure of SBA-15 and the strong red luminescence property of $YVO_4:Eu^{3+}$, could be used as a novel functional drug carrier for "smart" drug delivery. As might be expected, the pore volume, surface area and pore size of SBA-15 was decreased following deposition of the $YVO_4:Eu^{3+}$ layer and the adsorption of ibuprofen. The ibuprofen-loaded $YVO_4:Eu^{3+}$/SBA-15 system continued to exhibit the red emission of Eu^{3+} (617 nm, $^5D_0$-$^7F_2$) under ultraviolet (UV) irradiation, and also controlled the drug release. In addition, as shown in Figure 9.14, the photoluminescence (PL) intensity was increased with the increasing release of ibuprofen, and reached a maximum when ibuprofen had been completely released from the drug storage system. It is well known that the emission of Eu^{3+} will be suppressed to some extent in an environment which contains a high phonon frequency. The organic groups in ibuprofen, with vibration frequencies between 1000 and 3250 cm^{-1}, will quench the emission of Eu^{3+} to a great extent in the ibuprofen–$YVO_4:Eu^{3+}$/SBA-15 system. With the release of ibuprofen, however, its quenching effect on the emission of Eu^{3+} will be weakened, which will result in an increased emission intensity. The emission intensity of Eu^{3+} was seen to increased in line with the cumulative released of ibuprofen, which led to the extent of drug release being easily identifiable, trackable, and monitorable via changes of luminescence. This correlation between emission intensity and extent of drug release might, potentially, be used to monitor the drug release process and efficiency during the course of disease therapy. Yang *et al.* [201, 202] also functionalized mesoporous MCM-41 with

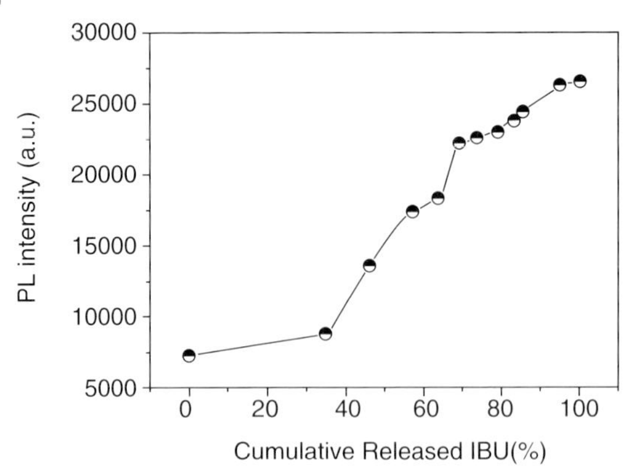

Figure 9.14 Photoluminescence emission intensity of Eu^{3+} in IBU–YVO_4:Eu^{3+}/SBA-15 as a function of cumulative release amount of ibuprofen (IBU). Reproduced with permission from Ref. [200]; © American Chemical Society.

fluorescent properties, and used it for drug delivery purposes. Tests of drug release indicated that the ibuprofen release rate could be controlled by regulating the morphology of the materials. The photoluminescence intensity also increased in line with cumulative release of ibuprofen from the Eu^{3+} functionalized MCM-41 system. Similarly, fluorescent properties were created on porous HAp [203] and bioactive glass [204], the aim being to create nanocomposites that could serve as drug carriers for trackable local drug delivery systems.

9.4.2 Polymer–Silica Nanocomposites

When 2-D ordered nanoporous silica or bioactive glass are utilized as carriers, drug release from the ceramic carriers is controlled by diffusion, which occurs immediately when the ceramic carriers are dispersed into water, especially if the drug is very water-soluble. Although the above-mentioned drug release profile can be mediated by modifying the pore size and particle morphology, the drug release patterns of most ceramic drug delivery systems are typically characterized by a very sharp initial burst, followed by slower release in the later stages; however, this is disadvantageous for long-term sustained drug delivery. In order to control drug release from uniformly arranged nanoporous channels, a biodegradable polymer was used to cap the opening of the mesoporous drug carrier. The drug-loaded mesoporous silica particles were then encapsulated in biodegradable polymers, such that the release profiles could be further controlled by the polymer caps and by degradation of the polymer. Xue and Shi [205] reported that PLGA/mesoporous silica hybrid nanocomposites can be synthesized via a novel sol–gel route, assisted by a single emulsion solvent evaporation. In this case, the gentamicin-loaded mesoporous silica showed a sharp initial burst during the first day, followed by a rather constant and low-level release over the subsequent three weeks. In comparison to the mesoporous silica with biodegradable polymer encapsulation, the nanocomposite structure achieved a reduced initial burst, while the composite microspheres demonstrated three distinct drug-release stages: an initial burst; a plateau stage with a slow release rate; and a sustained-release stage, which was controlled by degradation of the PLGA encapsulant. The initial burst of the low drug-loaded hybrid structure was 27.4wt%, and the nanocomposite structure delivered over a release period of up to five weeks. More importantly, the initial release burst of the hybrid structure was reduced significantly compared to that of the bare mesoporous silica. Similar polymer–silica xerogel composite microspheres were also found to control the release of gentamicin for nine weeks [206]. These distinct behaviors confirm the promise of this hybrid structure as a new drug release material for bone-filling purposes. Recently, Huang *et al.* [207] described a nanocomposite drug carrier system which combined the features of nanoporous structures and magnetic function with a biodegradable polymer coating. For this, magnetically functionalized mesoporous silica spheres with average diameters ranging from 150nm to 2μm were synthesized by depositing magnetic Fe_xO_y nanoparticles onto MCM-41 hosts, using a sol–gel method. After having loaded a model drug (e.g., ibuprofen) onto the functionalized mesoporous matrix by adsorption, the drug-containing spheres were coated with a biodegradable PLGA, using the solid-in-oil-in-water (S/O/W) single-emulsion solvent evaporation method, so as to form PLGA hybrid Fe_xO_y-loaded silica spheres, as shown in Figure 9.15, such that large spheres with diameters of tens of micrometers were formed. The drug-loaded silica particles were encapsulated in the polymer matrix, while silica particles were also seen to be dispersed on the surface of spheres. The hollow form of the composite spheres is shown clearly in Figure 9.15e and f, and this was confirmed using N_2 adsorption/desorption analysis. Although the typical size of the PLGA hybrid spheres obtained using the S/O/W method was greater than the uncoated spheres, the achievement of a reduced initial burst with a sustained release stage that lasted for 20 days ensured that the PLGA hybrid magnetic spheres should be considered as vectors for a long-term release system. The PLGA-coated magnetic composites showed a response to an external magnetic field that was similar to that of Fe_xO_y-loaded samples. Hence, such nanocomposites of PLGA and mesoporous silica may serve as potential vectors for drug targeting and prolonged controlled-release systems.

Polymer materials were also incorporated into mesoporous silica using an *in situ* synthesis to achieve the controlled release of drug molecules. Lin *et al.* [208] directly synthesized a nanocomposite of polymer–mesoporous silica nanoparticles via a dual-template technology. For this, the cationic polymer-quaternized poly[bis(2-chloroethyl)ether-*alt*-1,3-bis[3-(dimethylamino)propyl]urea] (PEPU) and the anionic surfactant sodium dodecyl sulfate (SDS) were used to form a homogeneous co-micelle system to fabricate mesoporous silica spherical nanoparticles with diameters of 50 to 180nm. After removal of the free surfactants by extraction, a mesoporous silica was obtained in which the cationic polymer (PEPU) was retained in the pore. The use of these materials as hosts allowed investigations to

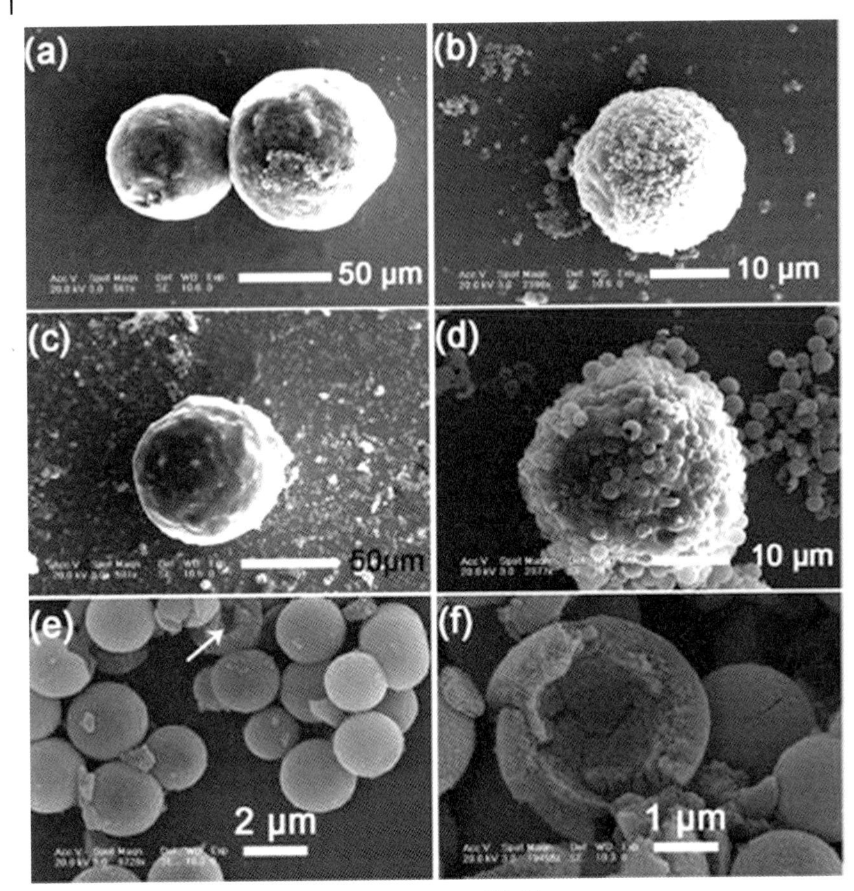

Figure 9.15 Scanning electron microscopy images of PLGA hybrid composites containing ibuprofen and Fe_xO_y-MCM-41 particles with different sizes: (a) 150 nm, (b) 500 nm, (c) 800 nm; (d–f) Bimodal silica spheres containing a MCM-41 mesoporous system. Reproduced with permission from Ref. [207]; © 2009, American Chemical Society.

be made of the drug delivery properties. As a result of the strong interaction between the cationic polymer and the drug molecules, these nanocomposite drug delivery systems demonstrated well-sustained release profiles, such that the burst release of ibuprofen in 30 min was reduced to 10 wt%, while releases of 35 wt% and 50 wt% were attained in 10 h and 20 h, respectively. Overall, the delivery process was well controlled, and took 80 h to achieve total release. This effect was attributed to the stronger interaction between the ibuprofen molecule and the cationic polymer remaining in the pore of the nanocomposite drug carriers. Thus, cationic polymer-functionalized mesoporous silicas would represent good candidates for controlled drug delivery. In addition to nanocomposites which included the incorporation of a polymer, mesoporous silica was also modified with short-chain organic functional groups so as to control the release of drug molecules. Lin *et al.* [209] described mesoporous composites with silica and hybrid organic silica that were synthesized by co-condensation, in which the surface-functional organic species played important roles in controlling both the adsorption capacity and the release rate of the drug molecules. When a mesoporous composite was incorporated with a hydrophobic ethylene group on the wall framework, a higher loading capacity and a significantly lower release rate of tetracycline were observed. These controlled release properties were attributed to the stronger hydrophobic–hydrophobic van der Waals attractions with drug molecules. The incorporation of organic compounds into composites of mesoporous materials may also create stimuli-responsive properties for smart drug delivery, the recent progress of which is discussed in the following section.

9.4.3
Stimuli-Responsive Nanocomposites

In addition to a sustained drug release, stimuli-responsive nanocomposite mesoporous drug carriers possess other smart features of high drug loading, site-selection and controlled release profiles, which improved the targeted therapeutic efficacy. With desired polymers incorporated with mesoporous materials, stimuli-responsive properties can be obtained to aid the controlled drug release as the polymer is highly dispersed on the large surface of mesoporous silica and has an intimate interaction with the loaded drug molecules. Drug carriers with stimuli-responsive polymers are sensitive to either chemical signals (such as pH, metabolites and ionic factors or solutes present in a system) or physical stimuli (such as temperature, ultrasound or electrical potential). These responses may alter the interaction between the polymer chains and thus affect the solubility, swelling behavior and crystalline/amorphous transition, such that the drug release rate can be controlled.

In the case of mesoporous silica and polymer composite drug delivery systems, several stimuli-responsive strategies have attracted attention [210, 211]. Recently, Xie *et al.* [212] described a pH-controlled dual-drug release system that involved the use of nanocomposites of MBG grafted with a co-polymer. For this, a poly(c-benzyl-L-glutamate)-poly(ethylene glycol) (PBLG-g-PEG) graft copolymer was first incorporated into a MBG, after which water-soluble gentamicin and fat-soluble naproxen were used as model drugs to study this composite system. Initially, a pH-controlled release of the individual drugs was achieved, predominantly via the release of gentamicin from the MBG in an acidic environment, and the rapid release of naproxen in an alkaline environment from the polypeptide nanomicelles. This variation in release behavior was mainly attributed to the different release rates of gentamicin from MBG and naproxen from PBLG-g-PEG, at different pH values. In this composite delivery system, the rate of gentamicin release from MBG was controlled not only by the diffusion mechanism, but also by rate at which gentamicin was desorbed from the pore surface, where the drug interacted with the Si–OH groups of the MBG [177]. Subsequently, when the pH of the release medium was lowered, the gentamicin desorption rate was rapidly increased such that the drug was quickly released. However, when the H^+ concentration falls, the gentamicin molecules become very difficult to desorb from the MBG surface, and this results in a slow rate of release. In contrast, in the case of a hydrophobic drug, which has limited water-solubility, the use of an amphiphilic polymeric core–shell nanoparticles can provide a sustained and controlled drug release. At acidic pH, the naproxen entrapped in the cores of the nanomicelles is minimally ionized, which leads to a slow release. However, as the pH is increased the carboxyl groups of the naproxen molecules in the PBGL-g-PEG core–shell nanomicelles will gradually become ionized and migrate from the inner core to the outer surface. Consequently, the rate of naproxen release will increase in line with the increase in ionization. The release profiles of both naproxen and gentamicin from the dual-drug delivery system suggested that the release of these two quite different drugs can be controlled by altering the pH of the environment. In a similar study, Xu *et al.* [213] demonstrated a pH-controlled drug release from mesoporous silica tablets that had been coated with hydroxypropyl methylcellulose phthalate. As shown in Figure 9.16, a simple pH-controlled drug release system could be installed by coating a pH-sensitive polymer, hydroxypropyl methylcellulose phthalate (HPMCP), onto a drug-loaded mesoporous SBA-15 tablet. The use of famotidine as a model drug allowed the effects of coating times and drying temperature on drug release to be investigated. In simulated gastric fluid (SGF, pH 1.2), famotidine was completely released from mesoporous silica tablets without HPMCP coating in only 2h, but by increasing the coating times and drying temperature its release was greatly delayed, despite the experiments being performed in the same medium. For a silica tablet with two layers of HPMCP, only 4.0wt% of the famotidine was released within 4h. However, in a simulated

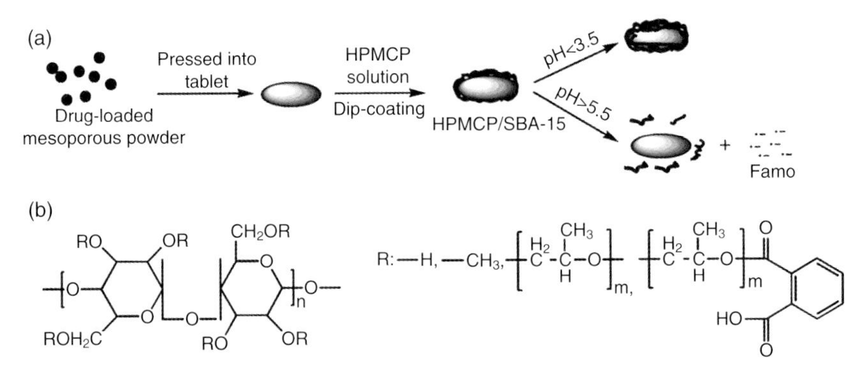

Figure 9.16 (a) Schematic representation of the pH-sensitive drug release system HPMCP/SBA-15; (b) Chemical structure of hydroxypropyl methylcellulose phthalate (HPMCP). Reproduced with permission from Ref. [213]; © 2008, Elsevier Ltd.

intestinal fluid (SIF, pH 7.4), the HPMCP coating had no obvious effect on famotidine release. Clearly, although this composite delivery system delayed famotidine release in SGF, it had no influence on its release in SIF. This so-called "intelligence" of the HPMCP/SBA-15 system not only matched the requirements for an intestinal drug release system, but also avoided any possible gastric instability of the drug. It may, therefore, represent a potential candidate for the targeted treatment of intestinal diseases. Additional pH-responsive, controlled release investigations have been conducted to monitor controlled drug delivery with mesoporous composite drug carriers [214, 215].

The presence of a thermoresponsive polymer can control the release profile of a drug that has been loaded into the pore channels of mesoporous silica, by varying the environmental temperature. Similar to the nanoparticle and nanofiber composite systems, the most intensively investigated temperature-sensitive polymer, poly(*N*-isopropylacrylamide) (PNIPAM), which is capable of undergoing a hydrophilic–hydrophobic transition at the LCST, can be coated onto mesoporous silica particles via a radical co-polymerization [216]. Zhou *et al.* [217] grafted a thermoresponsive PNIPAM polymer inside a mesoporous silica, by using an atom transfer radical polymerization (ATRP), and subsequently investigated the control of drug release in response to the environmental temperature. This nanocomposite delivery system demonstrated a high ibuprofen storage capacity of 58 wt% (ibuprofen/silica). The thermoresponsive properties of drug release from a PNIPAM- grafted mesoporous silica loaded with ibuprofen is shown in Figure 9.17. At low temperatures (<30 °C), the drug was confined within the pores, owing to an expansion of the PNIPAM molecular chains and hydrogen bonding between the PNIPA and ibuprofen. With increasing temperature, however, the polymer chains became hydrophobic, which resulted in a collapse of the hydrogen bonds and a loss of affinity with the drug molecules inside the pores. This thermoresponsive swelling of the polymer chains, as a result of a change in temperature, could be used to control drug release. Surprisingly, however, when the PNIPAM polymer

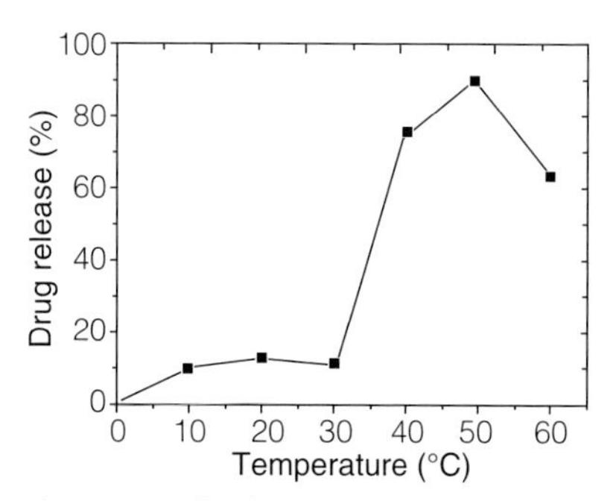

Figure 9.17 The dependence of ibuprofen release on temperature in a solution of simulated body fluid, with pH maintained at 7.4. Reproduced with permission from Ref. [217]; © 2007, The Royal Society of Chemistry 2007.

was coated onto the surface of mesoporous silica nanoparticles, the drug delivery system showed a reverse of the drug release profile [119]. At a low temperature (25 °C), the soluble PNIPAM chains caused the entrances of the pores to open below the phase transition temperature, and this led to a high drug release rate. In contrast, as the PNIPAM chains were insoluble in PBS at 38 °C, the entrances of the pores became blocked by the collapsed PNIPAM chains, such that release of the drug from the voids of the mesoporous silica nanoparticles was significantly obstructed. Consequently, the drug delivery could, again, be controlled by manipulating the environmental temperature. Moreover, Zhu *et al.* [218] reported a site-selective controlled delivery system that had been fabricated via the *in situ* assembly of stimuli-responsive ordered SBA-15 and magnetic particles. In this case, the nanocomposite drug carrier combined the advantages of mesoporous silica and PNIPAM multilayers with a stimuli-responsive property, and used the magnetic particles as site-selective labels. Such a drug delivery system, based on magnetic SBA-15/PNIPAM composites, would have potential applications in targeted and temperature-responsive controlled drug release [219].

In addition to the pH-sensitive and thermoresponsive features created on mesoporous silica particles incorporated with functional polymers, other stimuli-responsive characters have been investigated. For example, Cho *et al.* [220] investigated the efficacy of electrical stimulation on the delivery of nerve growth factors by coating an electrically conducting polymer onto mesoporous silica nanoparticles that could then be used as drug carriers. Subsequently, the controlled release of nerve growth factors in the presence of an electrical stimulation caused a significant promotion in neurite extension. Kim *et al.* [221] showed that ultrasound could also be used to externally trigger a smart drug release from a poly(dimethylsiloxane)-mesoporous silica composite. In fact, this ultrasound-induced pulsatile release system could be used as a device for controlled, targeted and on-demand drug delivery. The pressure-sensitive adhesive performance of a poly(2-ethyl hexylacrylate-*co*-acrylic acid)/silicate nanocomposite was also investigated for transdermal drug delivery [222]. Mal *et al.* [223] described the photocontrolled reversible release of guest molecules from coumarin-modified mesoporous silicas, in which the uptake, storage and release of organic molecules in MCM-41 could be regulated through the photocontrolled and reversible intermolecular dimerization of coumarin derivatives attached to the pore outlets as UV-sensitive caps.

Although a stimuli-responsive controlled release can be achieved by incorporating functional polymers with mesoporous silica, by varying the environmental conditions such as pH, temperature, magnetic, chemical and light, the application of these smart techniques is limited in certain circumstances. The temperature of the human body is normally maintained at 37 ± 0.5 °C, but may reach up to 40 °C during a fever; consequently a temperature-responsive drug delivery system might not easily control the release rate of an orally administered drug. A pH-responsive situation is applicable only in the environment of the stomach and intestine; once the drug-loaded particles have reached other parts of the body the pH will be held constant so that the drug release rate will not vary. In addition, as UV light may cause harm to human tissues, it is not plausible to include a UV-light-controlled drug release once the drug-loaded composite carriers are inside the body. Although, for transdermal drug delivery, most stimuli-responsive techniques can be applied, the drug absorption will be limited due to highly effective barrier properties of the skin.

9.5 Three-Dimensional Nanostructured Nanocomposites

Previously, three-dimensional (3-D) nanostructured composites have been developed for efficient drug delivery and, in general, have been integrated with tissue regeneration scaffolds. To date, a number of attempts have been made to fabricate 3-D nanocomposite scaffolds for the purpose of both of tissue engineering and local drug delivery. The nanostructured and microstructured 3-D scaffold enable the controlled delivery of highly specific biomolecular moieties, which have been proven to be effective in bone regeneration and in the guidance of functional angiogenesis.

Wei *et al.* [224] investigated the application of nanocomposites of 3-D scaffolds comprised of PLGA microspheres and poly(L-lactic acid) (PLLA) nanofibers, for the controlled delivery of platelet-derived growth factor-BB (PDGF-BB). The macroporous PLLA-based nanofibrous scaffolds were fabricated through a combination of phase separation and sugar-leaching techniques [225]. As shown in Figure 9.18, the 3-D nanofibrous scaffold thus created was shown to have a high porosity, of 98%. The scaffolds were characterized as multilevel porous structures with regular spherical macropores of 250–425 μm diameter, micro-interpore openings of ~100 μm, and nanofibers with diameters of 50–500 nm, which were similar in size to those of native collagen fibers. In the presence of PLGA microspheres, the tissue engineering scaffolds demonstrated good mechanical properties, with well-interconnected macroporous and nanofibrous structures, and were capable of providing a controlled delivery of the growth factor. The incorporation of protein-containing PLGA microspheres into the PLLA scaffold offered an excellent opportunity to control the release kinetics of proteins from the scaffold matrices. The 3-D porous scaffold system was capable of controlling the release of bioactive PDGF-BB in a temporal manner, with a low burst release on the first day, although subsequently the release profiles were fairly constant, with a sustained release of up to 42 days. The release kinetics from the scaffolds were shown to be governed by a degradation of the incorporated microspheres, which caused a significant reduction of the burst effect. When this type of PLLA scaffold was incorporated with nanosized HAp, a significant bone-like apatite deposition was facilitated throughout the scaffold in a SBF [225]. The pre-incorporation of nanosized HAp eliminated the induction period and facilitated apatite growth in the SBF. The compressive modulus was increased substantially when a continuous apatite layer was formed on the pore walls of the scaffold. The resultant composite scaffold was shown to mimic the natural bone matrix, with a combination of an organic phase

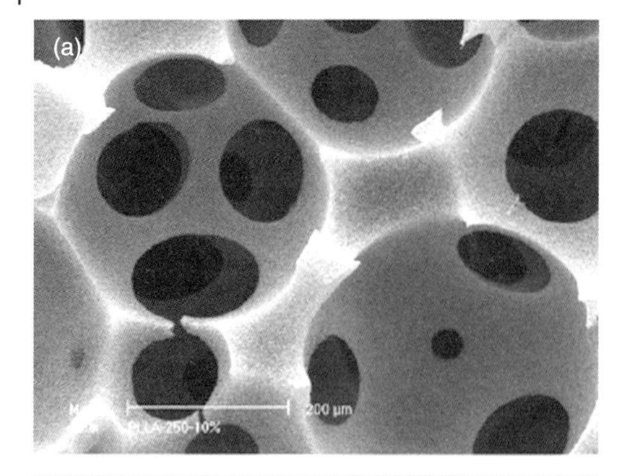

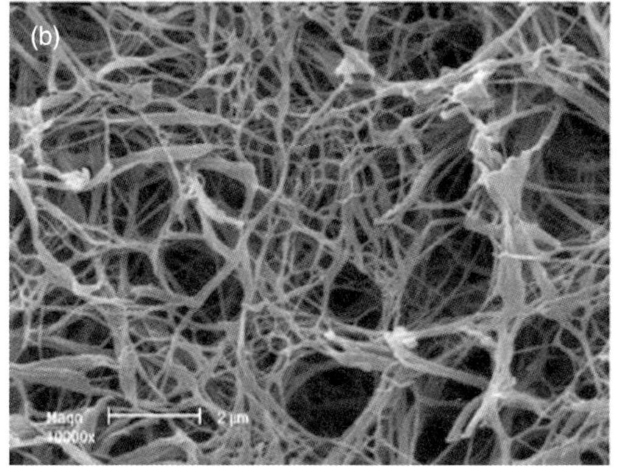

Figure 9.18 Scanning electron microscopy images of PLLA nanofibrous scaffolds before microsphere incorporation. (a) Low magnification, ×200; (b) High magnification, ×10 000. Reproduced with permission from Ref. [224]; © 2006, Elsevier B.V.

(a polymer such as PLLA) and an inorganic apatite phase. The demonstrated bioactivity of the apatite layer, together with well-controlled 3-D macroporous and nanofibrous structures, means that these novel nanocomposite scaffolds would be desirable for bone tissue engineering and the controlled delivery of growth factors.

A 3-D scaffold of PLGA/HAp has also been fabricated by electrospinning, and used for protein delivery and bone tissue regeneration [226]. The PLGA/HAp nanocomposite 3-D scaffolds exhibited a good morphology, with the HAp nanoparticles being dispersed homogeneously inside the PLGA matrix within the scaffold. It was also shown that human BMP-2 could successfully maintain its integrity and natural conformations after undergoing the process of electrospinning. The nanocomposite scaffolds allowed a sustained (2–8 weeks) release of BMP-2, with the release rate being accelerated in line with increasing the HAp content. Cell-culture experiments have shown that the encapsulation of HAp might enhance the attachment of cells to scaffolds, and also lead to a reduced cytotoxicity. Clearly, these integrated 3-D scaffolds would be advantageous in a bone tissue engineering role.

Recently, 3-D porous composites of MBG scaffolds were shown to exhibit activity for the crystal growth of HAp on their surfaces [227–229]. Zhu *et al.* [230] prepared a 3-D porous MBG scaffold with an interconnected macroporous network and pore diameters of 300–500 μm, for use as a drug carrier to provide a controlled delivery. The walls of the macroporous network were composed of mesoporous nanostructures with a pore size of 4.8 nm, and exhibited a higher surface area and pore volume. The mesoporous structure was shown to play an important role in drug loading and release rate, with the drug uptake capacity of the MBG scaffold being more than twofold that of the scaffold without a nanoporous structure. During the entire period in the SBF, gentamicin was released from the MBG scaffold at a much lower rate than from the bioactive glasses. Hence, this type of 3-D porous MBG scaffold might serve as a local drug delivery system for bone tissue regeneration.

A number of 3-D nanocomposites have also been investigated for drug delivery purposes. For example, microspheres of PLGA with 3-D microporous network structures were prepared using a simple solvent evaporation method, and used to monitor controlled protein delivery [231]. After having entrapped the proteins into the microporous structure, calcium alginate was used to block the micropores of the protein-loaded porous microspheres, and to control the diffusion of active components. When compared to pure PLGA microspheres, the protein release rate from the calcium alginate-coated porous carriers was shown to be well controlled, with the initial burst having been suppressed. Consequently, a sustained protein release from the nanocomposite with 3-D porous structures was achieved over 25 days. In addition, other 3-D structured materials have been prepared and used for drug delivery. For example, grid-like 3-D scaffolds of PLLA fabricated using a porogen leaching technique enabled an improved delivery of erythropoetin proteins [232]. Likewise, 3-D nanocontainers with nanoporous walls were prepared by the self-assembly of lithographically patterned two-dimensional (2-D) cruciforms with solder hinges, and showed promising roles in drug delivery, molecular separations, and cell-based therapy [233]. Three-dimensional, flower-like porous brushite crystals were also created, via an internal phase emulsion process, and used as drug carriers for the controlled release of antibiotics (e.g., sodium ampicillin) over a 14-day period [234]. Whilst these 3-D scaffolds have undergone intensive investigation regarding their use in tissue engineering [235–237], they have also attracted much attention in relation to their possible roles in the sustained release of proteins and antibiotics, in facilitating tissue regeneration, and protecting the surrounding tissues from infection.

9.6 Hydrogel Nanocomposite Drug Carriers

Hydrogels are hydrophilic, crosslinked polymer chain networks that can provide "stealth" properties, and also exert control over their stability, mechanical properties, and degradation profiles; hence, they have great potential in a variety of medical applications [238]. Due to their high water content and elastic structure, hydrogels tend to demonstrate excellent biocompatibility, such that their nanocomposites have recently attracted considerable attention for use in drug delivery systems, due mainly to their enhanced response to the environment. Moreover, the properties of these hydrogel nanocomposites can easily be tailored, simply by manipulating the individual properties of the biopolymer and the nanostructured material. Nanocomposite hydrogels of clays and biopolymers have been proposed as drug carriers for both effective controlled release and site-specific delivery, because these composite systems can provide improvements to mechanical and rheological properties, they can control drug permeability and water uptake, and are also responsive to stimuli [239, 240]. Generally, these improved properties for controlled drug delivery cannot be achieved by a single component of pure phase biopolymers or clays. In general, clay particles with at least one dimension on the nanoscale (1–1000 nm) are used as fillers for biopolymers to form nanocomposite hydrogels in which there is a strong interaction between the clay particles and the polymer matrix.

Several clay–polymer hydrogel nanocomposites have been fabricated and investigated for applications in controlled drug release. With the presence of clay fillers, the swelling response to stimuli was improved and drug delivery effectively controlled. Liu *et al.* [241] investigated the drug release behavior of chitosan–montmorillonite (CS–MMT) nanocomposite hydrogels, following electrostimulation. The exfoliated silica nanosheets were shown to be capable of acting as crosslinkers, so as to form a network structure between the CS and the MMT, and this difference in crosslinking density strongly affected the release of vitamin B_{12} under electrostimulation. At a lower MMT concentration (1 wt%), the release kinetics of vitamin B_{12} from the nanohydrogel showed pseudo-zero-order behavior, and the release mechanism was altered, under electrostimulation, from a diffusion-controlled to a swelling-controlled mode. A further increase in the MMT content reduced both the diffusion exponent and the responsiveness of the nanohydrogel to electrostimulation. In addition, a consecutively repeated "on" and "off" operation showed that, whilst the electroresponsiveness of the nanohydrogel with higher MMT concentrations was reduced, its anti-fatigue behavior was considerably improved.

In the case of nanocomposite hydrogels, the swelling ratio can be used to mediate the drug release profile, and was affected by the incorporation of clay [242, 243]. The swelling ratio was shown to decrease in line with increased amounts of MMT incorporated into a NIPAAm hydrogel; consequently, although drug release from the hydrogel was suppressed by a high loading of MMT, its mechanical strength was improved [244]. In contrast, the swelling ratio of nanocomposite hydrogels based on poly[acrylic acid-*co*-poly(ethylene glycol) methyl ether acrylate] (PEGMEA) was increased by the incorporation of hydrotalcite to control the drug release rate [245]. An *in situ* polymerization of PEGMEA with hydrotalcite, as well as an exfoliation of the intercalated hydrotalcite, was achieved in xerogels and swollen gels. The swelling ratio was dramatically increased when the hydrotalcite content was increased from 0 to 10 mol%, and the drug release rate was enhanced.

The drug-release behavior was also affected by different release factors, including the electrostatic attraction and repulsion between the gel and the drug, and between the hydrogel networks. If the electrostatic charges of the drug solute and hydrogel were different, then an electrostatic attraction would exist between them, and the drug would become strongly bound in the nanocomposite gels, and the release ratios would be lower. In contrast, if the electrostatic charges of the drug solutes and hydrogel were the same, the release rate of drug from the hydrogel would be higher. Those nanocomposite hydrogels with a higher swelling ratio could cause uncharged drug solutes to be released through a gel more easily.

Magnetic nanocomposites of temperature-sensitive hydrogels have been developed and shown to be responsive to alternating magnetic fields. Satarkar *et al.* [246] developed "intelligent" nanocomposites by incorporating superparamagnetic Fe_3O_4 particles into negative temperature-sensitive PNIPAM hydrogels. The systems were characterized as having properties of temperature-responsive swelling, remote heating on the application of an alternating magnetic field, and remote-controlled drug delivery. The rise in temperature caused by the application of an external alternating magnetic field depended on the Fe_3O_4 particle loading in the system. This class of biomaterials holds promise for use in remote-controlled drug delivery systems, notably for the pulsatile release of drug molecules on demand. When Hawkins *et al.* [247] investigated the controlled degradation and drug release from magnetic nanocomposite hydrogels by applying an alternating magnetic field, the heating effect caused an increase in the degradation rate of a degradable hydrogel and, in turn, an increase in the rate at which the drug was released from the system.

9.7 Summary and Outlook

Today, research into nanocomposite carriers for effective drug delivery is very much on the "fast track," as modified drug carriers with nanostructured features deliver active ingredients in an enhanced and controlled manner, providing major benefits to the temporal and/or spatial control of drug release. Composite nanoparticles with core–shell and hollow structures continue to undergo intensive investigation as drug carriers, the main advantage being that their small particle size facilitates their internalization by cells and the crossing of physiological barriers. Currently, one-dimensional (1-D) composite nanofibers and 3-D composite nanoscaffolds are fabricated by electrospinning, and play important roles in both tissue regeneration and drug delivery (sometimes in combination), whereas 2-D nanoporous silica-incorporated composites increase drug loading on the basis of their large surface areas and pore volumes. The ability to combine a functional polymer with a nanocomposite has led to the development of "smart" drug carriers that are stimulus-responsive, allowing drug delivery to be controlled simply by varying the environment. Today, with many nanocomposite drug carriers being at a developmental stage, further investigations must be conducted before any of their benefits can be translated to practical applications in pharmaceuticals and biomedical engineering.

An ideal nanocomposite smart drug carrier should possess the basic properties of biocompatibility and low cytotoxicity, in addition to intelligent features such as stimuli-responsive and/or targeting abilities. Nanostructured composites should also provide a full protection for the loaded drug molecules, especially in the case of environment-sensitive proteins and genes; this is especially important for the transfection of genes, as some are prone to degradation in the endosome, lysosome, and cytoplasm. Recently, much attention has been devoted to the design of nanocomposite carriers with proteins to control tissue regeneration and gene delivery.

As with most mature technologies, the substantial benefits derived during the early stages of development have been partly offset by the considerable debate regarding the safety of nanotechnology. Unfortunately, the rapid pace of research into the use of nanomaterials for drug delivery has exceeded that of any pharmacological and toxicological research into the effects of nanomaterials within the biological environment [248]. The toxicity of nanostructured materials, including nondimensional nanoparticles, 1-D nanofibers and 2-D nanoporous materials incorporated into nanocomposites, must first be monitored in animals, under more realistic conditions. Likewise, when considering possible adverse effects to health, studies must be addressed towards any possible carcinogenic potential of nanostructured materials. To date, research into the potential toxicity of nanomaterials lags behind the development of new nanostructured composites. Clearly, with insufficient information available regarding the pharmacology and toxicology of nanostructured materials, it becomes difficult to assess the risks of the associated nanotechnology. Fortunately, however, safety concerns continue to attract attention, and strict tests and validation procedures must now be carried out on drug delivery applications and large-scale productions. Today, steady progress is being made to provide a more comprehensive understanding of nanostructured drug carriers for drug loading, and of the release mechanisms of these hybrid materials as compared to single-component systems.

References

1 Koo, J.H. (2006) *Polymer Nanocomposites: Processing, Characterization, and Applications*, McGraw-Hill, New York.

2 Thomas, S. and Zaikov, G.E. (2008) *Polymer Nanocomposite Research Advances*, Nova Science Publishers.

3 Manias, E., Touny, A., Wu, L., Strawhecker, K., Lu, B. and Chung, T.C. (2001) Polypropylene/montmorillonite nanocomposites. Review of the synthetic routes and materials properties. *Chemistry of Materials*, **13**, 3516–23.

4 Pinnavaia, T.J. and Beall, G.W. (2001) *Polymer-Clay Nanocomposites*, John Wiley & Sons, New York.

5 Soppimathmath, K.S., Aminabhavi, T.M., Kulkarni, R.A. and Rudzinski, W.E. (2001) Biodegradable polymeric nanoparticles as drug delivery devices. *Journal of Controlled Release*, **70**, 1–20.

6 Brigger, I., Dubernet, C. and Couvreur, P. (2001) Nanoparticles in cancer therapy and diagnosis. *Advanced Drug Delivery Reviews*, **54**, 631–51.

7 Brannon-Peppas, L. and Blanchette, J.O. (2004) Nanoparticle and targeted system for cancer therapy. *Advanced Drug Delivery Reviews*, **56**, 1649–59.

8 Hamidi, M., Azadi, A. and Rafiei, P. (2008) Hydrogel nanoparticles in drug delivery. *Advanced Drug Delivery Reviews*, **60**, 1638–49.

9 Liu, Z., Jiao, Y., Wang, Y., Zhou, C. and Zhang, Z. (2008) Polysaccharide-based nanoparticles as drug delivery systems. *Advanced Drug Delivery Reviews*, **60**, 1650–62.

10 Husseini, G.A. and Pitt, W.G. (2008) Micelles and nanoparticles for ultrasonic drug and gene delivery. *Advanced Drug Delivery Reviews*, **60**, 1137–52.

11 Yuan, F., Leuning, M., Huang, S.K., Berk, D.A., Papahajopoulos, D. and Jain, R.K. (1994) Microvascular permeability and interstitial penetration of sterically stabilized (stealth) liposomes in human tumor xenograft. *Cancer Research*, **54**, 3352–6.

12 Storm, G., Belliot, S.O., Daemen, T. and Lasic, D.D. (1995) Surface modification of nanoparticles to oppose uptake by the mononuclear phagocyte system. *Advanced Drug Delivery Reviews*, **17**, 31–48.

13 Chen, H. and Langer, R. (1998) Oral particulate delivery: status and future trends. *Advanced Drug Delivery Reviews*, **34**, 339–50.

14 Desai, M.P., Labhasetwar, V., Walter, E., Levy, R.J. and Amidon, G.L. (1997) The mechanism of uptake of biodegradable microparticles in Caco-2 cells is size dependant. *Pharmaceutical Research*, **14**, 1568–73.

15 Win, K.Y. and Feng, S.S. (2005) Effect of particle size and surface coating on cellular uptake of polymeric nanoparticles for oral delivery of anticancer drugs. *Biomaterials*, **26**, 2713–22.

16 Jin, C., Bai, L., Wu, H., Song, W., Guo, G. and Dou, K. (2009) Cytotoxicity of paclitaxel incorporated in PLGA nanoparticles on hypoxic human tumor cells. *Pharmaceutical Research*, **26**, 1776–84.

17 Bhardwaj, V., Hariharan, S., Bala, I., Lamprecht, A., Kumar, N., Panchagnula, R. and Kumar, M.N.V.R. (2005) Pharmaceutical aspects of polymeric nanoparticles for oral drug delivery. *Journal of Biomedical Nanotechnology*, **1**, 235–58.

18 Gref, R., Domb, A., Quellec, P., Blunk, T., Müller, R.H., Verbavatz, J.M. and Langer, R. (1995) The controlled intravenous delivery of drugs using PEG-coated sterically stabilized nanospheres. *Advanced Drug Delivery Reviews*, **16**, 215–33.

19 Abdelwahed, W., Degobert, G., Stainmesse, S. and Fessi, H. (2006) Freeze-drying of nanoparticles: formulation, process and storage considerations. *Advanced Drug Delivery Reviews*, **58**, 1688–713.

20 Vauthier, M. and Bouchemal, K. (2009) Methods for the preparation and manufacturing of polymeric nanoparticles. *Pharmaceutical Research*, **26**, 1025–58.

21 Hawkins, M.J., Soon-shiong, P. and Desai, N. (2008) Protein nanoparticles as drug carriers in clinical medicine. *Advanced Drug Delivery Reviews*, **60**, 876–85.

22 Jain, R.A. (2000) The manufacturing techniques of various drug loaded biodegradable poly(lactide-*co*-glycolide) (PLGA) devices. *Biomaterials*, **21**, 2475–90.

23 Basarkar, A. and Singh, J. (2009) Poly(lactide-*co*-glycolide)-polymethacrylate nanoparticles for intramuscular delivery of plasmid encoding interleukin-10 to prevent autoimmune diabetes in mice. *Pharmaceutical Research*, **26**, 72–81.

24 Ishihara, T., Kubota, T., Choi, T. and Higaki, M. (2009) Treatment of experimental arthritis with stealth-type polymeric nanoparticles encapsulating betamethasone phosphate. *Journal of Pharmacology and Experimental Therapeutics*, **329**, 412–17.

25 Santander-Ortega, M.J., Csaba, N., Alonso, M.J., Ortega-Vinuesa, J.L. and Bastos-González, D. (2007) Stability and physicochemical characteristics of PLGA, PLGA:poloxamer and PLGA:poloxamine blend nanoparticles: a comparative study. *Colloids and Surfaces A:*

Physicochemical and Engineering Aspects, **296**, 132–40.

26 Salehi, R., Davaran, S., Rashidi, M.R. and Entezami, A.A. (2009) Thermosensitive nanoparticles prepared from poly(*N*-isopropylacrylamide-acrylamide-vinylpyrrolidone) and its blend with poly(lactide-*co*-glycolide) for efficient drug delivery system. *Journal of Applied Polymer Science*, **111**, 1905–10.

27 Zhang, L., Chan, J., Gu, F., Rhee, J.-W., Wang, A., Radovic-Moreno, A., Alexis, F., Langer, R. and Farokhzad, O.C. (2008) Self-assembled lipid-polymer hybrid nanoparticles: a robust drug delivery platform. *ACS Nano*, **2**, 1696–702.

28 Chan, J.M., Zhang, L., Yuet, K.P., Liao, G., Rhee, J.-W., Langer, R. and Farokhzad, O.C. (2009) PLGA-lecithin-PEG core-shell nanoparticles for controlled drug delivery. *Biomaterials*, **30**, 1627–34.

29 Kim, B.-S., Kim, C.-S. and Lee, K.-M. (2008) The intracellular uptake ability of chitosan-coated poly(D,L-lactide-*co*-glycolide) nanoparticles. *Archives of Pharmacal Research*, **31**, 1050–4.

30 Yang, R., Shim, W.S., Cui, F.D., Cheng, G., Han, X., Jin, Q.R., Kim, D.D., Chung, S.J. and Shim, C.K. (2009) Enhanced electronic interaction between chitosan-modified PLGA nanoparticles and tumor. *International Journal of Pharmaceutics*, **371**, 1–2.

31 Nafee, N., Taetz, S. and Schneider, M. (2007) Chitosan-coated PLGA nanoparticles for DNA/RNA delivery: effect of the formulation parameters. *Nanomedicine: Nanotechnology, Biology and Medicine*, **3**, 173–83.

32 Yang, R., Yang, S.-G., Shim, W.-S., Cui, F., Cheng, G., Kim, I.-W., Kim, D.-D., Chung, S.-J. and Shim, C.-K. (2009) Lung specific delivery of paclitaxel by chitosan-modified PLGA nanoparticles via transient formation of microaggregates. *Journal of Pharmaceutical Sciences*, **98**, 970–84.

33 Grabovac, V. and Bernkop-Schnürch, A. (2007) Development and *in vitro* evaluation of surface modified poly(lactide-*co*-glycolide) nanoparticles with chitosan-4-thiobutylamidine. *Drug Development and Industrial Pharmacy*, **33**, 767–74.

34 Nafee, N., Schneider, M., Schaefer, U.F. and Lehr, C.-M. (2009) Relevance of the colloidal stability of chitosan/PLGA nanoparticles on their cytotoxicity profile. *International Journal of Pharmaceutics*, **381**, 130–9.

35 Taetz, S., Nafee, N., Beisner, J., Piotrowska, K., Baldes, C., Mürdter, T.E., Huwer, H., Schneider, M., Schaefer, U.F., Klotz, U. and Lehr, C.-M. (2009) The influence of chitosan content in cationic chitosan/PLGA nanoparticles on the delivery of antisense 2′-O-methyl-RNA directed against telomerase in lung cancer cells. *European Journal of Pharmaceutics and Biopharmaceutics*, **72**, 358–69.

36 Yang, J., Lee, C.-H., Park, J., Seo, S., Lim, E.-K., Song, Y.J., Suh, J.-S., Yoon, H., Hun, Y.-M. and Haam, S. (2007) Antibody conjugated magnetic PLGA nanoparticles for diagnosis and treatment of breast cancer. *Journal of Materials Chemistry*, **17**, 2695–9.

37 Kocbek, P., Obermajer, N., Cegnar, M., Kos, J. and Kristl, J. (2007) Targeting cancer cells using PLGA nanoparticles surface modified with monoclonal antibody. *Journal of Controlled Release*, **120**, 18–26.

38 Yin, Y., Chen, D., Qiao, M., Wei, X. and Hu, H. (2007) Lectin-conjugated PLGA nanoparticles loaded with thymopentin: *ex vivo* bioadhesion and *in vivo* biodistribution. *Journal of Controlled Release*, **123**, 27–38.

39 Kou, G., Gao, J., Wang, H., Chen, H., Li, B., Zhang, D., Wang, S., Hou, S., Qian, W., Dai, J., Zhong, Y. and Guo, Y. (2007) Preparation and characterization of paclitaxel-loaded PLGA nanoparticles coated with cationic SM5-1 single-chain antibody. *Journal of Biochemistry and Molecular Biology*, **40**, 731–9.

40 McCarron, P.A., Marouf, W.M., Quinn, D.J., Fay, F., Burden, R.E., Olwill, S.A. and Scott, C.J. (2008) Antibody targeting of camptothecin-loaded PLGA nanoparticles to tumor cells. *Bioconjugate Chemistry*, **19**, 1561–9.

41 Zhang, N., Chittapuso, C., Ampassavate, C., Siahaan, T.J. and Berkland, C. (2008)

PLGA nanoparticle-peptide conjugate effectively targets intercellular molecule-1. *Bioconjugate Chemistry*, **19**, 145–52.

42 Dong, Y.C. and Feng, S.S. (2005) Poly(D,L-lactide-*co*-glycolide)/montmorillonite nanoparticles for oral delivery of anticancer drugs. *Biomaterials*, **26**, 6068–76.

43 Liu, J., Zhang, S.M., Chen, P.P., Cheng, L., Zhou, W., Tang, W.X., Chen, Z.W. and Ke, C.M. (2007) Controlled release of insulin from PLGA nanoparticles embedded within PVA hydrogels. *Journal of Materials Science: Materials in Medicine*, **18**, 2205–10.

44 Yeo, Y., Ito, T., Bellas, E., Highley, C.B., Marini, R. and Kohane, D.S. (2007) In-situ cross-linkable hyaluronan hydrogels containing polymeric nanoparticles for preventing postsurgical adhesions. *Annals of Surgery*, **245**, 819–24.

45 Cascone, M.G., Pot, P.M., Lazzeri, L. and Zhu, Z. (2002) Release of dexamethasone from PLGA nanoparticles entrapped into dextran/poly(vinyl alcohol) hydrogels. *Journal of Materials Science: Materials in Medicine*, **13**, 265–9.

46 Agnihotri, S.A., Mallikarjuna, N.N. and Aminabhavi, T.M. (2004) Recent advances on chitosan-based micro- and nanoparticles in drug delivery. *Journal of Controlled Release*, **100**, 5–25.

47 Sarmento, B., Ribeiro, A., Veiga, F., Sampaio, P., Neufeld, R. and Ferreira, D. (2007) Alginate/chitosan nanoparticles are effective for oral insulin delivery. *Pharmaceutical Research*, **24**, 2198–206.

48 Goycoolea, F.M., Lollo, G., Remunan-Lopez, C., Quaglia, F. and Alonso, M.J. (2009) Chitosan-alginate blended nanoparticles as carriers for the transmucosal delivery of macromolecules. *Biomacromolecules*, **10**, 1736–43.

49 Motwani, S.K., Chopra, S., Talegaonkar, S., Kohl, K., Ahmad, F.J. and Khar, R.K. (2008) Chitosan-sodium alginate nanoparticles as submicroscopic reservoirs for ocular delivery: formulation, optimization and *in vitro* characterization. *European Journal of Pharmaceutics and Biopharmaceutics*, **68**, 513–25.

50 Peng, S.F., Yang, M.J., Su, C.J., Chen, H.L., Lee, P.W., Wei, M.C. and Sung, H.W. (2009) Effects of incorporation of poly(gamma-glutamic acid) in chitosan/DNA complex nanoparticles on cellular uptake and transfection efficiency. *Biomaterials*, **30**, 1797–808.

51 de la Fuente, M., Seijo, B. and Alonso, M.J. (2008) Bioadhesive hyaluronan-chitosan nanoparticles can transport genes across the ocular mucosa and transfect ocular tissue. *Gene Therapy*, **15**, 668–76.

52 de la Fuente, M., Seijo, B. and Alonso, M.J. (2008) Novel hyaluronic acid-chitosan nanoparticles for ocular gene therapy. *Investigative Ophthalmology and Visual Science*, **49**, 2016–24.

53 Krauland, A.H. and Alonso, M.J. (2007) Chitosan/cyclodextrin nanoparticles as macromolecular drug delivery system. *International Journal of Pharmaceutics*, **340**, 134–42.

54 Nasti, A., Zaki, N.M., Leonardis, P., Ungphaiboon, S., Sansongsak, P., Rimoli, M.G. and Tirelli, N. (2009) Chitosan/TPP and chitosan/TPP-hyaluronic acid nanoparticles: systematic optimization of the preparative process and preliminary biological evaluation. *Pharmaceutical Research*, **28**, 1918–30.

55 Peng, L., Cheng, X., Zhuo, R., Lan, J., Wang, Y., Shi, B. and Li, S. (2009) Novel gene-activated matrix with embedded chitosan/plasmid DNA nanoparticles encoding PDGF for periodontal tissue engineering. *Journal of Biomedical Materials Research, Part A*, **90**, 564–76.

56 Torchilin, V.P. (2005) Recent advances with liposomes as pharmaceutical carriers. *Nature Reviews Drug Discovery*, **4**, 145–60.

57 Torchilin, V.P. and Weissig, V. (2007) *Liposomes*, 2nd edn, Oxford University Press.

58 Huang, S.-L. (2008) Liposomes in ultrasonic drug and gene delivery. *Advanced Drug Delivery Reviews*, **60**, 1167–76.

59 Crosasso, P., Ceruti, M., Brusa, P., Arpicco, S., Dosio, F. and Cattel, L. (2000) Preparation, characterization and properties of sterically stabilized paclitaxel-containing liposomes. *Journal of Controlled Release*, **63**, 19–30.

60 Yang, T., Cui, F.D., Choi, M.K., Cho, J.W., Chung, S.J., Shim, C.K. and Kim, D.D. (2007) Enhanced solubility and stability of PEGylated liposomal paclitaxel: *in vitro* and *in vivo* evaluation. *International Journal of Pharmaceutics*, **338**, 317–26.

61 Dadashzadeh, S., Vali, A.M. and Rezaie, M. (2008) The effect of PEG coating on *in vitro* cytotoxicity and *in vivo* disposition of topotecan loaded liposomes in rats. *International Journal of Pharmaceutics*, **353**, 251–9.

62 Abu Lila, A.S., Kizuki, S., Doi, Y., Suzuki, T., Ishida, T. and Kiwada, H. (2009) Oxaliplatin encapsulated in PEG-coated cationic liposomes induces significant tumor growth suppression via a dual-targeting approach in a murine solid tumor model. *Journal of Controlled Release*, **137**, 8–14.

63 Lu, Y., Li, J. and Wang, G. (2008) *In vitro* and *in vivo* evaluation of mPEG-PLA modified liposomes loaded glycyrrhetinic acid. *International Journal of Pharmaceutics*, **356**, 274–81.

64 Werle, M. and Takeuchi, H. (2009) Chitosan-aprotinin coated liposomes for oral peptide delivery: development, characterization and *in vivo* evaluation. *International Journal of Pharmaceutics*, **370**, 26–32.

65 Werle, M., Hironaka, K., Takeuchi, H. and Hoyer, H. (2009) Development and *in vitro* characterization of liposomes coated with thiolated poly(acrylic acid) for oral drug delivery. *Drug Development and Industrial Pharmacy*, **35**, 209–15.

66 Zaru, M., Manca, M.-L., Fadda, A.M. and Antimisiaris, S.G. (2009) Chitosan-coated liposomes for delivery to lungs by nebulization. *Colloids and Surfaces B*, **71**, 88–95.

67 Gupta, B. and Torchilin, V.P. (2007) Monoclonal antibody 2C5-modified doxorubicin-loaded liposomes with significantly enhanced therapeutic activity against intracranial human brain U-87 MG tumor xenografts in nude mice. *Cancer Immunity*, **56**, 1215–23.

68 Pan, H., Han, L., Chen, W., Yao, M. and Lu, W. (2007) Targeting to tumor necrotic regions with biotinylated antibody and streptavidin modified liposomes. *Journal of Controlled Release*, **125**, 228–35.

69 Hatakeyama, H., Akita, H., Ishida, E., Hashimoto, K., Kobayashi, H., Aoki, T., Yasuda, J., Obata, K., Kikuchi, H., Ishida, T., Kiwada, H. and Harashima, H. (2007) Tumor targeting of doxorubicin by anti-MT1-MMP antibody-modified PEG liposomes. *International Journal of Pharmaceutics*, **342**, 194–200.

70 Elbayoumi, T.A. and Torchilin, V.P. (2007) Enhanced cytotoxicity of monoclonal anticancer antibody 2G5-modified doxorubicin-loaded PEGylated liposomes against various tumor cell lines. *European Journal of Pharmaceutical Sciences*, **32**, 159–68.

71 Elbayoumi, T.A. and Torchilin, V.P. (2009) Tumor-targeted nanomedicines: enhanced antitumor efficacy *in vivo* of doxorubicin-loaded, long-circulating liposomes with cancer-specific monoclonal antibody. *Clinical Cancer Research*, **15**, 1973–80.

72 Xiang, G., Wu, J., Lu, Y., Liu, Z. and Lee, R.J. (2008) Synthesis and evaluation of a novel ligand for folate-mediated targeting liposomes. *International Journal of Pharmaceutics*, **356**, 29–36.

73 Kamaly, N., Kalber, T., Thanou, M., Bell, J.D. and Miller, A.D. (2009) Folate receptor targeted bimodal liposomes for tumor magnetic resonance imaging. *Bioconjugate Chemistry*, **20**, 648–55.

74 Li, X., Ding, L., Xu, Y., Wang, Y. and Ping, Q. (2009) Targeted delivery of doxorubicin using stealth liposomes modified with transferrin. *International Journal of Pharmaceutics*, **373**, 116–23.

75 Doi, A., Kawabata, S., Iida, K., Yokoyama, K., Kajimoto, Y., Kuroiwa, T., Shirakawa, T., Kirihata, M., Kasaoka, S., Muruyama, K., Kumada, H., Sakurai, Y., Masunaga, S., Ono, K. and Miyatake, S. (2008) Tumor-specific targeting of sodium borocaptate (BSH) to malignant glioma by transferrin-PEG liposomes: a modality for boron neutron capture therapy. *Journal of Neuro-oncology*, **87**, 287–94.

76 Gupta, B., Levchenko, T.S. and Torchilin, V.P. (2007) TAT peptide-modified liposomes provide enhanced gene delivery to intracranial human brain tumor xenografts in nude mice. *Oncology Research*, **16**, 351–9.

77 Garg, A., Tisdale, A.W., Haidari, E. and Kokkoli, E. (2009) Targeting colon cancer cells using PEGylated liposomes modified with a fibronectin-mimetic peptide. *International Journal of Pharmaceutics*, **366**, 201–10.

78 Demirgoz, D., Garg, A. and Kokkoli, E. (2008) PR_b-targeted PEGylated liposomes for prostate cancer therapy. *Langmuir*, **24**, 13518–24.

79 Beuttler, J., Rothdiener, M., Muller, D., Frejd, F.Y. and Kontermann, R.E. (2009) Targeting of epidermal growth factor receptor (EGFR)-expressing tumor cells with sterically stabilized affibody liposomes (SAL). *Bioconjugate Chemistry*, **20**, 1201–8.

80 Bertrand, N., Fleischer, J.G., Wasan, K.M. and Leroux, J.-C. (2009) Pharmacokinetics and biodistribution of *N*-isopropylacrylamide copolymers for the design of pH-sensitive liposomes. *Biomaterials*, **30**, 2598–605.

81 Paasonen, L., Romberg, B., Storm, G., Yliperttula, M., Urtti, A. and Hennink, W.E. (2007) Temperature-sensitive poly(*N*-(2-hydroxypropyl)methacrylamide mono/dilactate)-coated liposomes for triggered contents release. *Bioconjugate Chemistry*, **18**, 2131–6.

82 Leite, E.A., dos Santos Giuberti, C., Wainstein, A.J.A., Wainstein, A.P.D.L., Coelho, L.G.V., Lana, A.M.Q., Savassi-Rocha, P.R. and De Oliveira, M.C. (2009) Acute toxicity of long-circulating and pH-sensitive liposomes containing cisplatin in mice after intraperitoneal administration. *Life Sciences*, **84**, 641–9.

83 Momekova, D., Rangelov, S., Yanev, S., Nikolova, E., Konstantinov, S., Romberg, B., Storm, G. and Lambov, N. (2007) Long-circulating, pH-sensitive liposomes sterically stabilized by copolymers bearing short blocks of lipid-mimetic units. *European Journal of Pharmaceutical Sciences*, **32**, 308–17.

84 Feng, S.S., Ruan, G. and Li, Q.-T. (2004) Fabrication and characterization of a novel drug delivery device liposomes-in-microsphere (LIM). *Biomaterials*, **25**, 5181–9.

85 Pardeike, J., Hommoss, A. and Müller, R.H. (2009) Lipid nanoparticles (SLN, NLC) in cosmetic and pharmaceutical dermal products. *International Journal of Pharmaceutics*, **366**, 170–84.

86 Kaur, I.P., Bhandari, R., Bhandari, S. and Kakkar, V. (2008) Potential of solid lipid nanoparticles in brain targeting. *Journal of Controlled Release*, **127**, 97–109.

87 Wong, H.L., Bendayan, R., Rauth, A.M., Li, Y. and Wu, X.Y. (2007) Chemotherapy with anticancer drugs encapsulated in solid lipid nanoparticles. *Advanced Drug Delivery Reviews*, **59**, 491–504.

88 Almeida, A.J. and Souto, E. (2007) Solid lipid nanoparticles as a drug delivery system for peptides and proteins. *Advanced Drug Delivery Reviews*, **59**, 478–90.

89 Date, A.A., Joshi, M.D. and Patravale, V.B. (2007) Parasitic disease: liposomes and polymeric nanoparticles versus lipid nanoparticles. *Advanced Drug Delivery Reviews*, **59**, 505–21.

90 Wissing, S.A., Kayser, O. and Müller, R.H. (2004) Solid lipid nanoparticles for parenteral drug delivery. *Advanced Drug Delivery Reviews*, **56**, 1257–72.

91 Lee, M.-K., Lim, S.-J. and Kim, C.-K. (2007) Preparation, characterization and *in vitro* cytotoxicity of paclitaxel-loaded sterically stabilized solid lipid nanoparticles. *Biomaterials*, **28**, 2137–46.

92 Gupta, Y., Jain, A. and Jain, S.K. (2007) Transferrin-conjugated solid lipid nanoparticles for enhanced delivery of quinine dihydrochloride to the brain. *Journal of Pharmacy and Pharmacology*, **59**, 935–40.

93 Zhang, N., Ping, Q., Huang, G., Xu, W., Cheng, Y. and Han, X. (2006) Lectin-modified solid lipid nanoparticles as carriers for oral administration of insulin. *International Journal of Pharmaceutics*, **327**, 153–9.

94 Casadei, M.A., Cerreto, F., Cesa, S., Giannuzzo, M., Feeney, M., Marianecci, C. and Paolicelli, P. (2006) Solid lipid nanoparticles incorporated in dextran hydrogels: a new drug delivery system for oral formulations. *International Journal of Pharmaceutics*, **325**, 140–6.

95 Oh, K.T., Lee, E.S., Kim, D. and Bae, Y.H. (2008) L-Histidine-based pH-sensitive anticancer drug carrier micelle: reconstitution and brief evaluation of its systemic toxicity. *International Journal of Pharmaceutics*, **358**, 177–83.

96 Kim, D., Lee, E.S., Oh, K.T., Gao, Z.G. and Bae, Y.H. (2008) Doxorubicin-loaded polymeric micelle overcomes multidrug resistance of cancer by double-targeting folate and early endosomal pH. *Small*, **4**, 2043–50.

97 Yin, H.Q. and Bae, Y.H. (2009) Physicochemical aspects of doxorubicin-loaded pH-sensitive polymeric formulations from a mixture of poly(L-histidine)-*b*-poly(L-lactide)-*b*-poly(ethylene glycol). *European Journal of Pharmaceutics and Biopharmaceutics*, **71**, 223–30.

98 Yang, Y., Wang, J., Zhang, X., Lu, W. and Zhang, Q. (2009) A novel mixed micelle gel with thermo-sensitive property for the local delivery of docetaxel. *Journal of Controlled Release*, **135**, 175–82.

99 Wei, L., Cai, C., Lin, J. and Chen, T. (2009) Dual-drug delivery system based on hydrogel/micelle composites. *Biomaterials*, **30**, 2606–13.

100 Sun, C., Lee, J.S.H. and Zhang, M. (2008) Magnetic nanoparticles in MR imaging and drug delivery. *Advanced Drug Delivery Reviews*, **60**, 1252–65.

101 McCarthy, J.R. and Weissleder, R. (2008) Multifunctional magnetic nanoparticles for targeted imaging and therapy. *Advanced Drug Delivery Reviews*, **60**, 1241–51.

102 Gupta, A.K. and Gupta, M. (2005) Synthesis and surface engineering of iron oxide nanoparticles for biomedical applications. *Biomaterials*, **26**, 3995–4021.

103 Gaihre, B., Khil, M.S., Lee, D.R. and Kim, H.Y. (2009) Gelatin-coated magnetic iron oxide nanoparticles as carrier system: drug loading and *in vitro* drug release study. *International Journal of Pharmaceutics*, **365**, 180–9.

104 Mahmoudi, M., Simchi, A., Imani, M., Milani, A.S. and Stroeve, P. (2008) Optimal design and characterization of superparamagnetic iron oxide nanoparticles coated with polyvinyl alcohol for targeted delivery and imaging. *Journal of Physical Chemistry B*, **112**, 14470–81.

105 Mahmoudi, M., Simchi, A., Imani, M. and Hafeli, U.O. (2009) Superparamagnetic iron oxide nanoparticles with rigid cross-linked polyethylene glycol fumarate coating for application in imaging and drug delivery. *Journal of Physical Chemistry C*, **113**, 8124–31.

106 Maver, U., Bele, M., Makovec, D., Campelj, S., Jamnik, J. and Gaberšček, M. (2009) Incorporation and release of drug into/from superparamagnetic iron oxide nanoparticles. *Journal of Magnetism and Magnetic Materials*, **321**, 3187–92.

107 McBain, S.C., Yiu, H.H.P., Ei Haj, A. and Dobson, J. (2007) Polyethyleneimine functionalized iron oxide nanoparticles as agents for DNA delivery and transfection. *Journal of Materials Chemistry*, **17**, 2561–5.

108 Phanapavudhikul, P., Shen, S., Ng, W.K. and Tan, R.B. (2008) Formulation of Fe_3O_4/acrylate co-polymer nanocomposites as potential drug carriers. *Drug Delivery*, **15**, 177–83.

109 Woo, K., Moon, J., Choi, K.-S., Seong, T.Y. and Yoon, K.-H. (2009) Cellular uptake of folate-conjugated lipophilic superparamagnetic iron oxide nanoparticles. *Journal of Magnetism and Magnetic Materials*, **321**, 1610–12.

110 Yang, L., Cao, Z., Sajja, H.K., Mao, H., Wang, L., Geng, H., Xu, H., Jiang, T., Wood, W.C., Nie, S. and Wang, Y.A. (2008) Development of receptor targeted magnetic iron oxide nanoparticles for efficient drug delivery and tumor imaging. *Journal of Biomedical Nanotechnology*, **4**, 439–49.

111 Piao, Y.Z., Kim, J., Na, H., Kim, D., Baek, J.S., Ko, M.K., Lee, J.H., Shokouhimehr, M. and Hyeo, T. (2008) Wrap–bake–peel process for nanostructural transformation from β-FeOOH nanorods to biocompatible iron oxide nanocapsules. *Nature Materials*, **7**, 242–7.

112 Slowing, I.I., Trewyn, B.G., Giri, S. and Lin, V.S.-Y. (2007) Mesoporous silica nanoparticles for drug delivery and biosensing applications. *Advanced Functional Materials*, **17**, 1125–236.

113 Slowing, I.I., Vivero-Escoto, J.L., Wu, C.-W. and Lin, V.S.-Y. (2008) Mesoporous silica nanoparticles as controlled release drug delivery and gene transfection carriers. *Advanced Drug Delivery Reviews*, **60**, 1278–88.

114 Lu, J., Liong, M., Zink, J.I. and Tamanoi, F. (2007) Mesoporous silica nanoparticles as a delivery system for hydrophobic anticancer drugs. *Small*, **3**, 1341–6.

115 Yagüe, C., Moros, M., Grazú, V., Arruebo, M. and Santamaria, J. (2008) Synthesis and stealthing study of bare and PEGylated silica micro- and nanoparticles as potential drug-delivery vectors. *Chemical Engineering Journal*, **137**, 45–53.

116 van Schooneveld, M.M., Vucic, E., Koole, R., Zhou, Y., Stocks, J., Cormode, D.P., Tang, C.Y., Gordon, R.E., Nicolay, K., Meijerink, A., Fayad, Z.A. and Mulder, W.J.M. (2008) Improved biocompatibility and pharmacokinetics of silica nanoparticles by means of a lipid coating: a multimodality investigation. *Nano Letters*, **8**, 2517–25.

117 Lee, C.-H., Lo, L.-W., Mou, C.-Y. and Yang, C.-S. (2008) Synthesis and characterization of positive-charge functionalized mesoporous silica nanoparticles for oral delivery of an anti-inflammatory drug. *Advanced Functional Materials*, **18**, 3283–92.

118 Tsai, C.P., Chen, C.Y., Hung, Y., Chang, F.H. and Mou, C.-Y. (2009) Monoclonal antibody-functionalized mesoporous silica nanoparticles for selective targeting breast cancer cells. *Journal of Materials Chemistry*, **19**, 5737–43.

119 You, Y.Z., Kalebaila, K.K., Brock, S.L. and Oupicky, D. (2008) Temperature-controlled uptake and release in PNIPAM-modified porous silica nanoparticles. *Chemistry of Materials*, **20**, 3354–9.

120 Niu, Z., Liu, J., Lee, L.A., Bruckman, M.A., Zhao, D., Koley, G. and Wang, Q. (2007) Biological templated synthesis of water-soluble conductive polymeric nanowires. *Nano Letters*, **7**, 3729–33.

121 Zhang, L.J., Peng, H., Hsu, C.F., Kilmartin, P.A. and Travas-Sejdic, J. (2007) Self-assembled polyaniline nanotubes grown from a polymeric acid solution. *Nanotechnology*, **18**, 115607.

122 Kong, L.R., Lu, X.F. and Zhang, W.J. (2008) Facile synthesis of multifunctional multiwalled carbon nanotubes/Fe_3O_4 nanoparticles/polyaniline composite nanotubes. *Journal of Solid State Chemistry*, **181**, 628–36.

123 Jayaraman, K., Kotaki, M., Zhang, Y.Z., Mo, X.M. and Ramakrishna, S. (2004) Recent advances in polymer nanofibers. *Journal of Nanoscience and Nanotechnology*, **4**, 52–65.

124 Sill, T.J. and von Recum, H.A. (2008) Electrospinning: Applications in drug delivery and tissue engineering. *Biomaterials*, **29**, 1989–2006.

125 Liang, D., Hsiao, B.S. and Chu, B. (2007) Functional electrospun nanofibrous scaffolds for biomedical applications. *Advanced Drug Delivery Reviews*, **59** (14), 1392–412.

126 Martins, A., Reis, R.L. and Neves, N.M. (2008) Electrospinning: processing technique for tissue engineering scaffolding. *International Materials Reviews*, **53**, 257–74.

127 Kriegel, C., Arrechi, A., Kit, K., McClements, D.J. and Weiss, J. (2008) Fabrication, functionalization, and application of electrospun biopolymer nanofibers. *Critical Reviews in Food Science and Nutrition*, **48**, 775–97.

128 Zhang, B.F., Li, C.J. and Chang, M. (2009) Curled poly(ethylene glycol terephthalate)/poly(ethylene propanediol terephthalate) nanofibers produced by side-by-side electrospinning. *Polymer Journal*, **41**, 252–3.

129 Li, X.Q., Su, Y., Chen, R., He, C.L., Wang, H.S. and Mo, X.M. (2009) Fabrication and properties of core-shell structure P(LLA-CL) nanofibers by coaxial electrospinning. *Journal of Applied Polymer Science*, **111**, 1564–70.

130 Shah, P.N., Parth, N., Manthe, R.L., Lopina, S.T. and Yun, Y.H. (2009) Electrospinning of L-tyrosine polyurethanes for potential biomedical applications. *Polymer*, **50**, 2281–9.

131 Ashammakhi, N., Wimpenny, I., Nikkola, L. and Yang, Y. (2009) Electrospinning: methods and development of biodegradable nanofibres for drug release. *Journal of Biomedical Nanotechnology*, **5**, 1–19.

132 Venugopal, J., Prabhakaran, M.P., Molamma, P., Low, S., Choon, A.T., Deepika, G., Dev, V.R.G. and Ramakrishna, S. (2009) Continuous nanostructures for the controlled release of drugs. *Current Pharmaceutical Design*, **15**, 1799–808.

133 Hadjiagyrou, M. and Chiu, J.B. (2008) Enhanced composite electrospun nanofiber scaffolds for use in drug delivery. *Expert Opinion on Drug Delivery*, **5**, 1093–106.

134 Peng, H.S., Zhou, S.B., Guo, T., Li, Y.S., Li, X.H., Wang, J.X. and Weng, J. (2008) *In vitro* degradation and release profiles for electrospun polymeric fibers containing paracetamol. *Colloids and Surfaces B*, **66**, 206–12.

135 Jia, X.Q. and Kiick, K.L. (2009) Hybrid multicomponent hydrogels for tissue engineering. *Macromolecular Bioscience*, **9**, 140–56.

136 Ignatova, M., Manolova, N., Markova, N. and Rashkov, I. (2009) Electrospun non-woven nanofibrous hybrid mats based on chitosan and PLA for wound-dressing applications. *Macromolecular Bioscience*, **9**, 102–11.

137 Kim, K., Luu, Y.K., Chang, C., Fang, D., Hsiao, B.S., Chua, B. and Hadjiargyrou, M. (2004) Incorporation and controlled release of a hydrophilic antibiotic using poly(lactide-*co*-glycolide)-based electrospun nanofibrous scaffolds. *Journal of Controlled Release*, **98**, 47–56.

138 Katti, D.S., Robinson, K.W., Ko, F.K. and Laurencin, C.T. (2004) Bioresorbable nanofiber-based systems for wound healing and drug delivery: optimization of fabrication parameters. *Journal of Biomedical Materials Research, Part B*, **70B**, 286–96.

139 Zong, X., Li, S., Chen, E., Garlick, B., Kim, K., Fang, D., Chiu, J., Zimmerman, T., Brathwaite, C., Hsiao, B.S. and Chu, B. (2004) Prevention of postsurgery-induced abdominal adhesions by electrospun bioabsorbable nanofibrous poly(lactide-*co*-glycolide)-based membranes. *Annals of Surgery*, **240**, 910–15.

140 Buschle-Diller, G., Cooper, J., Xie, Z.W., Wu, Y., Waldrup, J. and Ren, X.H. (2007) Release of antibiotics from electrospun bicomponent fibers. *Cellulose*, **14**, 553–62.

141 Piras, A.M., Chiellini, F., Chiellini, E., Nikkola, L. and Ashammakhi, N. (2008) New multicomponent bioerodible electrospun nanofibers for dual-controlled drug release. *Journal of Bioactive and Compatible Polymers*, **23**, 423–43.

142 Song, M., Guo, D.D., Pan, C., Jiang, H., Chen, C., Zhang, R.Y., Gu, Z.Z. and Wang, X.M. (2008) The application of poly(*N*-isopropylacrylamide)-*co*-polystyrene nanofibers as an additive agent to facilitate the cellular uptake of an anticancer drug. *Nanotechnology*, **19**, 165102.

143 Chen, C., Lv, G., Pan, C., Song, M., Wu, C., Guo, D., Wang, X., Chen, B. and Gu, Z. (2007) Poly(lactic acid) (PLA) based nanocomposites – a novel way of drug releasing. *Biomedical Materials*, **2**, L1–4.

144 Xu, X.L., Chen, X.S., Ma, P.A., Wang, X.R. and Jing, X.B. (2008) The release behavior of doxorubicin hydrochloride from medicated fibers prepared by emulsion-electrospinning. *European Journal of Pharmaceutics and Biopharmaceutics*, **70**, 165–70.

145 Xu, X., Chen, X., Wang, Z. and Jing, X. (2009) Ultrafine PEG–PLA fibers loaded with both paclitaxel and doxorubicin hydrochloride and their *in vitro* cytotoxicity. *European Journal of Pharmaceutics and Biopharmaceutics*, **72**, 18–25.

146 Gupta, P., Vermani, K. and Garg, S. (2002) Hydrogels: from controlled release to pH-responsive drug delivery. *Drug Discovery Today*, **7**, 569–79.

147 Sheng, D., Palaniswamy, R. and Kam, C.T. (2008) pH-Responsive polymers:

synthesis, properties and applications. *Soft Matter*, **4**, 435–49.

148 Panda, J.J., Mishra, A., Basu, A. and Chauhan, V.S. (2008) Stimuli responsive self-assembled hydrogel of a low molecular weight free dipeptide with potential for tunable drug delivery. *Biomacromolecules*, **9**, 2244–50.

149 Mano, J.F. (2008) Stimuli-responsive polymeric systems for biomedical applications. *Advanced Engineering Materials*, **10**, 515–27.

150 Cao, S., Hua, B. and Liu, H. (2009) Synthesis of pH-responsive crosslinked poly[styrene-*co*-(maleic sodium anhydride)] and cellulose composite hydrogel nanofibers by electrospinning. *Polymer International*, **58**, 545–51.

151 Zhao, Y., Yokoi, H., Tanaka, M., Kinoshita, T. and Tan, T.W. (2008) Self-assembled pH-responsive hydrogels composed of the RATEA16 peptide. *Biomacromolecules*, **9**, 1511–18.

152 Jeong, S.I., Lee, Y.M., Lee, J., Shin, Y.M., Shin, H., Lim, Y.M. and Nho, Y.C. (2008) Preparation and characterization of temperature-sensitive poly(*N*-isopropylacrylamide)-g-poly(L-lactide-*co*-epsilon-caprolactone) nanofibers. *Macromolecular Research*, **16**, 139–48.

153 Chunder, A., Sarkar, S., Yu, Y.B. and Zhai, L. (2007) Fabrication of ultrathin polyelectrolyte fibers and their controlled release properties. *Colloids and Surfaces B*, **58**, 172–9.

154 Tan, S.T., Wendorff, J.H., Pietzonka, C., Jia, Z.H. and Wang, G.Q. (2005) Biocompatible and biodegradable polymer nanofibers displaying superparamagnetic properties. *ChemPhysChem*, **6**, 1461–5.

155 Song, M., Wang, X.M., Wang, C.X., Pan, C., Fu, D.G. and Gu, Z.Z. (2009) Application of the blending of PNIPAM-*co*-PS nanofibers with functionalized Au nanoparticles for the high-sensitive diagnosis of cancer cells. *Journal of Nanoscience and Nanotechnology*, **9**, 876–9.

156 Kim, H.-W., Song, J.-W. and Kim, H.-E. (2005) Nanofiber generation of gelatin-hydroxyapatite biomimetics for guided tissue regeneration. *Advanced Functional Materials*, **15**, 1988–94.

157 Hosseinkhani, H., Hosseinkhani, M., Khademhosseini, A. and Kobayashi, H. (2007) Bone regeneration through controlled release of bone morphogenetic protein-2 from 3-D tissue engineered nano-scaffold. *Journal of Controlled Release*, **117**, 380–6.

158 Goldberg, M., Langer, R. and Jia, X.Q. (2007) Nanostructured materials for applications in drug delivery and tissue engineering. *Journal of Biomaterials Science, Polymer Edition*, **18**, 241–68.

159 Fu, Y.-C., Nie, H., Ho, M.-L., Wang, C.-K. and Wang, C.-H. (2008) Optimized bone regeneration based on sustained release from three-dimensional fibrous PLGA/HAp composite scaffolds loaded with BMP-2. *Biotechnology and Bioengineering*, **99**, 996–1006.

160 Kresge, C.T., Leonowicz, M.E., Roth, W.J., Vartuli, J.C. and Beck, J.S. (1992) Ordered mesoporous molecular sieves synthesized by a liquid-crystal template mechanism. *Nature*, **359**, 710–12.

161 Ying, J.Y., Mehnert, C.P. and Wong, M.S. (1999) Synthesis and applications of supramolecular-templated mesoporous materials. *Angewandte Chemie, International Edition*, **38**, 56–77.

162 Corma, A. (1997) From microporous to mesoporous molecular sieve materials and their use in catalysis. *Chemical Reviews*, **97**, 2373–420.

163 Linden, M., Schacht, S., Schüth, F., Steel, A. and Unger, K.K. (1998) Recent advance in nano- and macroscale control of hexagonal, mesoporous materials. *Journal of Porous Materials*, **5**, 177–93.

164 Vallet-Regí, M.A., Ruiz-Gonzalez, L.I. and Gonzalez-Calbet, J.M. (2006) Revisiting silica based ordered mesoporous materials: medical applications. *Journal of Materials Chemistry*, **16**, 26–31.

165 Izquierdo-Barba, I., Ruiz-González, L., Doadrio, J.C., González-Calbet, J.M. and Vallet-Regí, M. (2005) Tissue regeneration: a new property of mesoporous materials. *Solid State Science*, **7**, 983–9.

166 Balas, F., Manzano, M., Horcajada, P. and Vallet-Regí, M. (2006) Confinement and controlled release of bisphosphonates on ordered mesoporous silica-based materials. *Journal of the American Chemical Society*, **128**, 8116–17.

167 Giri, S., Trewyn, B.G. and Lin, V.S. (2007) Mesoporous silica nanomaterial-based biotechnological and biomedical delivery systems. *Nanomedicine*, **2**, 99–111.

168 Rigby, S.P., Fairhead, M. and van der Walle, C.F. (2008) Engineering silica particles as oral drug delivery vehicles. *Current Pharmaceutical Design*, **14**, 1821–31.

169 Anglin, E.J., Cheng, L., Freeman, W.R. and Sailor, M.J. (2008) Porous silicon in drug delivery devices and materials. *Advanced Drug Delivery Reviews*, **60**, 1266–77.

170 Wang, S. (2009) Ordered mesoporous materials for drug delivery. *Microporous and Mesoporous Materials*, **117**, 1–9.

171 Vallet-Regí, M., Ramila, A., del Real, R.P. and Pariente, J.P. (2001) A new property of MCM-41: drug delivery system. *Chemistry of Materials*, **13**, 308–11.

172 Munoz, B., Ramila, A., Pariente, J.P., Diaz, I. and Vallet-Regí, M. (2003) MCM-41 organic modification as drug delivery rate regulator. *Chemistry of Materials*, **15**, 500–3.

173 Qu, F.Y., Zhu, G.S., Huang, S.Y., Li, S.G. and Qiu, S.L. (2006) Effective controlled release of captopril by silylation of mesoporous MCM-41. *ChemPhysChem*, **7**, 400–6.

174 Tang, Q.L., Xu, Y., Wu, D., Sun, Y.H., Wang, J.Q., Xu, J. and Deng, F. (2006) Studies on a new carrier of trimethylsilyl-modified mesoporous material for controlled drug delivery. *Journal of Controlled Release*, **114**, 41–6.

175 Zeng, W., Qian, X.F., Zhang, Y.B., Yin, J. and Zhu, Z.K. (2005) Organic modified mesoporous MCM-41 through solvothermal process as drug delivery system. *Materials Research Bulletin*, **40**, 766–72.

176 Doadrio, J.C., Sousa, E.M.B., Izquierdo-Barba, I., Doadrio, A.L., Perez-Pariente, J. and Vallet-Regí, M. (2006) Functionalization of mesoporous materials with long alkyl chains as a strategy for controlling drug delivery pattern. *Journal of Materials Chemistry*, **16**, 462–6.

177 Xia, W. and Chang, J. (2006) Well-ordered mesoporous bioactive glasses (MBG): a promising bioactive drug delivery system. *Journal of Controlled Release*, **110**, 522–30.

178 Doadrio, A.L., Sousa, E.M.B., Doadrio, J.C., Pariente, J.P., Izquierdo-Barba, I. and Vallet-Regí, M. (2004) Mesoporous SBA-15 HPLC evaluation for controlled gentamicin drug delivery. *Journal of Controlled Release*, **97**, 125–32.

179 Yang, Q., Wang, S.H., Fan, P.W., Wang, L.F., Di, Y., Lin, K.F. and Xiao, F.S. (2005) pH-responsive carrier system based on carboxylic acid modified mesoporous silica and polyelectrolyte for drug delivery. *Chemistry of Materials*, **17**, 5999–6003.

180 Zhang, L., Qiao, S., Jin, Y., Cheng, L., Yan, Z. and Lu, G.Q. (2008) Hydrophobic functional group initiated helical mesostructured silica for controlled drug release. *Advanced Functional Materials*, **18**, 3834–42.

181 Wei, Y.L., Wang, Y.M., Zhu, J.H. and Wu, Z.Y. (2003) In-situ coating of SBA-15 with MgO: direct synthesis of mesoporous solid bases from strong acidic systems. *Advanced Materials*, **15**, 1943–5.

182 Wang, Y.M., Wu, Z.Y., Wei, Y.L. and Zhu, J.H. (2005) In situ coating metal oxide on SBA-15 in one-pot synthesis. *Microporous and Mesoporous Materials*, **84**, 127–36.

183 Shen, S., Chow, P.S., Chen, F. and Tan, R.B.H. (2007) Submicron particles of SBA-15 modified with MgO as carriers for controlled drug delivery. *Chemical and Pharmaceutical Bulletin*, **55**, 985–91.

184 El-Ghannam, A., Ahmed, K. and Omran, M. (2005) Nanoporous delivery system to treat osteomyelitis and regenerate bone: gentamicin release kinetics and bactericidal effect. *Journal of Biomedical Materials Research, Part B*, **73B**, 277–84.

185 Sousa, A., Souza, K.C. and Sousa, E.M.B. (2008) Mesoporous silica/apatite nanocomposite: special synthesis route to control local drug delivery. *Acta Biomaterialia*, **4**, 671–9.

186 Cauda, V., Fiorilli, S., Onida, B., Vernè, E., Brovarone, C.V., Viterbo, D., Croce, G., Milanesio, M. and Garrone, E. (2008) SBA-15 ordered mesoporous silica inside a bioactive glass–ceramic scaffold for local drug delivery. *Journal of Materials Science: Materials in Medicine*, **19**, 3303–10.

187 Brovarone, C.V., Baino, F., Miola, M., Mortera, R., Onida, B. and Vernè, E. (2009) Glass–ceramic scaffolds containing silica mesophases for bone grafting and drug delivery. *Journal of Materials Science: Materials in Medicine*, **20**, 809–20.

188 Ruiz-Hernández, E., López-Noriega, A., Arcos, D., Izquierdo-Barba, I., Terasaki, O. and Vallet-Regí, M. (2007) Aerosol-assisted synthesis of magnetic mesoporous silica spheres for drug targeting. *Chemistry of Materials*, **19**, 3455–63.

189 Zhou, J., Wu, W., Caruntu, D., Yu, M.H., Martin, A., Chen, J.F., O'Connor, C.J. and Zhou, W.L. (2007) Synthesis of porous magnetic hollow silica nanospheres for nanomedicine application. *Journal of Physical Chemistry C*, **111**, 17473–7.

190 Zhang, L., Qiao, S., Jin, Y., Chen, Z., Gu, H. and Lu, G.Q. (2008) Magnetic hollow spheres of periodic mesoporous organosilica and Fe_3O_4 nanocrystals: fabrication and structure control. *Advanced Materials*, **20**, 805–9.

191 Ma, H., Zhou, J., Caruntu, D., Yu, M.H., Chen, J.F., O'Connor, C.J. and Zhou, W.L. (2008) Fabrication of magnetic porous hollow silica drug carriers using $CaCO_3/Fe_3O_4$ composite nanoparticles and cationic surfactant double templates. *Journal of Applied Physiology*, **103**, 07A320–3.

192 Wang, Y., Ren, J., Liu, X., Wang, Y., Guo, Y., Guo, Y. and Lu, G. (2008) Facile synthesis of ordered magnetic mesoporous γ-Fe_2O_3/SiO_2 nanocomposites with diverse mesostructures. *Journal of Colloid and Interface Science*, **326**, 158–65.

193 Sen, T. and Bruce, I.J. (2009) Mesoporous silica–magnetite nanocomposites: fabrication, characterization and applications in biosciences. *Microporous and Mesoporous Materials*, **120**, 246–51.

194 Huang, S., Yang, P., Cheng, Z., Li, C., Fan, Y., Kong, D. and Lin, J. (2008) Synthesis and characterization of magnetic Fe_xO_y@SBA-15 composites with different morphologies for controlled drug release and targeting. *Journal of Physical Chemistry C*, **112**, 7130–7.

195 Zhao, W., Chen, H., Li, Y., Li, L., Lang, M. and Shi, J. (2008) Uniform rattle-like hollow magnetic mesoporous sphere as drug delivery carriers and their sustained –release property. *Advanced Functional Materials*, **18**, 2780–8.

196 Li, X., Wang, X., Hua, Z. and Shi, J. (2008) One-pot synthesis of magnetic and mesoporous bioactive glass composites and their sustained drug release property. *Acta Materialia*, **56**, 3260–5.

197 Yuan, X., Xing, W., Zhuo, S.-P., Han, Z., Wang, G., Gao, X. and Yan, Z.-F. (2009) Preparation and application of mesoporous Fe/carbon composites as a drug carrier. *Microporous and Mesoporous Materials*, **117**, 678–84.

198 Cao, H., Gan, J., Wang, S., Xuan, S., Wu, Q., Li, C., Wu, C., Hu, C. and Huang, G. (2008) Novel silica-coated iron–carbon composite particles and their targeting effect as a drug carrier. *Journal of Biomedical Materials Research*, **86A**, 671–7.

199 Kong, D., Yang, P., Wang, Z., Chai, P., Huang, S., Lian, H. and Lin, J. (2008) Mesoporous silica coated CeF_3:Tb3+ particles for drug release. *Journal of Nanomaterials*, **2008**, 312792.

200 Yang, P., Huang, S., Kong, D. and Lin, J. (2007) Luminescence functionalization of SBA-15 by $YVO_4:Eu^{3+}$ as a novel drug delivery system. *Inorganic Chemistry*, **46**, 3203–11.

201 Yang, P., Quan, Z., Li, C., Lian, H., Huang, S. and Lin, J. (2008) Fabrication, characterization of spherical $CaWO_4$:Ln @MCM-41(Ln = Eu^{3+}, Dy^{3+}, Sm^{3+}, Er3+) composites and their applications as drug release systems. *Microporous and Mesoporous Materials*, **116**, 524–31.

202 Yang, P., Quan, Z., Lu, L., Huang, S. and Lin, J. (2008) Luminescence functionalization of mesoporous silica with different morphologies and applications as drug delivery systems. *Biomaterials*, **29**, 692–702.

203 Yang, P., Quan, Z., Li, C., Kang, X., Lian, H. and Lin, J. (2008) Bioactive, luminescent and mesoporous europium-doped hydroxyapatite as a drug carrier. *Biomaterials*, **29**, 4341–7.

204 Fan, Y., Yang, P., Huang, S., Jiang, J., Lian, H. and Lin, J. (2009) Luminescent and mesoporous europium-doped bioactive glasses (MBG) as a drug carrier. *Journal of Physical Chemistry C*, **113**, 7826–30.

205 Xue, J.M. and Shi, M. (2004) PLGA/mesoporous silica hybrid structure for controlled drug release. *Journal of Controlled Release*, **98**, 209–17.

206 Xue, J.M., Tan, C.H. and Lukito, D. (2006) Biodegradable polymer–silica xerogel composite microspheres for controlled release of gentamicin. *Journal of Biomedical Materials Research, Part B*, **78B**, 417–22.

207 Huang, S., Fan, Y., Cheng, Z., Kong, D., Yang, P., Quan, Z., Zhang, C. and Lin, J. (2009) Magnetic mesoporous silica spheres for drug targeting and controlled release. *Journal of Physical Chemistry C*, **113**, 1775–84.

208 Lin, H., Zhu, G., Xing, J., Gao, B. and Qiu, S. (2009) Polymer-mesoporous silica materials templated with an oppositely charged surfactant/polymer system for drug delivery. *Langmuir*, **25**, 10159–64.

209 Lin, C.X.C., Qiao, S.Z., Yu, C.Z., Ismadji, S. and Lu, G.Q.M. (2009) Periodic mesoporous silica and organosilica with controlled morphologies as carriers for drug release. *Microporous and Mesoporous Materials*, **117**, 213–19.

210 Sorenson, M.H., Samoshina, Y., Claesson, P.M. and Alberius, P. (2009) Sustained release of ibuprofen from polyelectrolyte encapsulated mesoporous carriers. *Journal of Dispersion Science and Technology*, **30**, 892–902.

211 Liu, C.Y., Hu, J.H., Yang, D. and Yang, W.L. (2009) Preparation of multi-responsive mesoporous silica microspheres and its application in controlled drug release. *Acta Chimica Sinica*, **67**, 843–9.

212 Xia, W., Chang, J., Lin, J. and Zhu, J. (2008) The pH-controlled dual-drug release from mesoporous bioactive glass/polypeptide graft copolymer nanomicelle composites. *European Journal of Pharmaceutics and Biopharmaceutics*, **69**, 546–52.

213 Wu, W., Gao, Q., Xu, Y., Wu, D. and Sun, Y. (2009) pH-controlled drug release from mesoporous silica tablets coated with hydroxypropyl methylcellulose phthalate. *Materials Research Bulletin*, **44**, 606–12.

214 Zhu, Y., Shi, J., Shen, W., Dong, X., Feng, J., Ruan, M. and Li, Y. (2005) Stimuli-responsive controlled drug release from a hollow mesoporous silica sphere/polyelectrolyte multilayer core-shell structure. *Angewandte Chemie, International Edition*, **44**, 5083–7.

215 Sun, G.Y., Chang, Y.P., Li, S.H., Li, Q.Y., Xu, R., Gu, J.M. and Wang, E.B. (2009) pH-responsive controlled release of antitumour-active polyoxometalate from mesoporous silica materials. *Dalton Transactions*, **23**, 4481–7.

216 Park, J.-H., Lee, Y.-H. and Oh, S.-G. (2007) Preparation of thermosensitive PNIPAm-grafted mesoporous silica particles. *Macromolecular Chemical Physics*, **208**, 2419–27.

217 Zhou, Z., Zhu, S. and Zhang, D. (2007) Grafting of thermo-responsive polymer inside mesoporous silica with large pore size using ATRP and investigation of its use in drug release. *Journal of Materials Chemistry*, **17**, 2428–33.

218 Zhu, S., Zhou, Z. and Zhang, D. (2007) Control of drug release through the in situ assembly of stimuli-responsive ordered mesoporous silica with magnetic particles. *ChemPhysChem*, **8**, 2478–83.

219 Zhu, Y., Kaskel, S., Ikoma, T. and Hanagata, N. (2009) Magnetic SBA-15/poly(*N*-isopropylacrylamide) composite: preparation, characterization and temperature-responsive drug release property. *Microporous and Mesoporous Materials*, **123**, 107–12.

220 Cho, Y., Shi, R., Ivanisevic, A. and Borgens, R.B. (2009) A mesoporous silica nanosphere-based drug delivery system using an electrically conducting polymer. *Nanotechnology*, **20**, 275102.

221 Kim, H.-J., Matsuda, H., Zhou, H. and Hon, I. (2006) Ultrasound-triggered smart drug release from a poly(dimethylsiloxane)–mesoporous silica composite. *Advanced Materials*, **18**, 3083–8.

222 Rana, P.K. and Sahoo, P.K. (2007) Synthesis and pressure sensitive adhesive performance of poly(EHA-*co*-AA)/silicate nanocomposite used in transdermal drug delivery. *Applied Polymer Science*, **106**, 3915–21.

223 Mal, N.K., Fujiwara, M. and Tanaka, Y. (2003) Photocontrolled reversible release of guest molecules from coumarin-modified mesoporous silica. *Nature*, **421**, 350–3.

224 Wei, G., Jin, Q., Giannobile, W.V. and Ma, P.X. (2006) Nano-fibrous scaffold for controlled delivery of recombinant human PDGF-BB. *Journal of Controlled Release*, **112**, 103–10.

225 Wei, G.B. and Ma, P.X. (2006) Macroporous and nano-fibrous polymer scaffolds and polymer/bone-like apatite composite scaffolds generated by sugar spheres. *Journal of Biomedical Materials Research*, **78A**, 306–15.

226 Nie, H., Soh, B.W., Fu, Y.-C. and Wang, C.-H. (2007) Three-dimensional fibrous PLGA/HAp composite scaffold for BMP-2 delivery. *Biotechnology and Bioengineering*, **99**, 223–34.

227 Yun, H.-S., Kim, S.-E. and Hyeon, Y.-T. (2007) Design and preparation of bioactive glasses with hierarchical pore networks. *Chemical Communications*, **2**, 2139–42.

228 Zhu, Y., Wu, C., Ramaswamy, Y., Kockrick, E., Simon, P., Kaskel, S. and Zreiqat, H. (2008) Preparation, characterization, and *in vitro*, bioactivity of mesoporous bioactive glasses (MBGs) scaffolds for bone tissue engineering. *Microporous and Mesoporous Materials*, **112**, 494–503.

229 Li, X., Wang, X., Chen, H., Jiang, P., Dong, X. and Shi, J. (2007) Hierarchically porous bioactive glass scaffolds synthesized with a PUF and P123 cotemplated approach. *Chemistry of Materials*, **19**, 4322–6.

230 Zhu, Y. and Kaskel, S. (2009) Comparison of the *in vitro* bioactivity and drug release property of mesoporous bioactive glasses (MBGs) and bioactive glasses (BGs) scaffolds. *Microporous and Mesoporous Materials*, **118**, 176–82.

231 Sun, L., Zhou, S., Wang, W., Li, X., Wang, J. and Weng, J. (2009) Preparation and characterization of porous biodegradable microspheres used for controlled protein delivery. *Colloids and Surfaces A*, **345**, 173–81.

232 El-Ayoubi, R., Eliopoulos, N., Diraddo, R., Galipeau, J. and Yousefi, A.-M. (2008) Design and fabrication of 3D porous scaffolds to facilitate cell-based gene therapy. *Tissue Engineering Part A*, **14**, 1037–48.

233 Wang, J.H., Patel, M. and Garcias, D.H. (2009) Self assembly of three-dimensional nanoporous containers. *Nano*, **4**, 1–5.

234 Lim, H.N., Kassim, A., Huang, N.M., Khiew, P.S. and Chiu, W.S. (2009) Three-dimensional flower-like brushite crystals prepared from high internal phase emulsion for drug delivery application. *Colloids and Surfaces A*, **345**, 211–18.

235 Tsang, V.L. and Bhatia, S.N. (2004) Three-dimensional tissue fabrication. *Advanced Drug Delivery Reviews*, **56**, 1635–47.

236 Borenstein, J.T., Weinberg, E.J., Orrick, B.K., Sundback, C., Kaazempur-Mofrad, M.R. and Vacanti, J.P. (2007) Microfabrication of three-dimensional engineered scaffolds. *Tissue Engineering*, **13**, 1837–44.

237 Schaefermeier, P.K., Szymanski, D., Weiss, F., Fu, P., Lueth, T., Schmitz, C., Meiser, B.M., Reichart, B. and Sodian, R. (2009) Design and fabrication of three-dimensional scaffolds for tissue engineering of human heart valves. *European Surgical Research*, **42**, 49–53.

238 Kopeček, J. (2007) Hydrogel biomaterials: a smart future? *Biomaterials*, **28**, 5185–92.

239 Hussain, F., Hojjati, M., Okamoto, M. and Gorga, R.E. (2006) Review article: polymer-matrix nanocomposites, processing, manufacturing, and application: an overview. *Journal of Composite Materials*, **40**, 1511–75.

240 Viseras, C., Aguzzi, C., Cerezo, P. and Bedmar, M.C. (2008) Biopolymer-clay nanocomposites for controlled drug delivery. *Materials Science Technology*, **24**, 1020–6.

241 Liu, K.-H., Liu, T.-Y., Chen, S.-Y. and Liu, D.M. (2008) Drug release behavior of chitosan–montmorillonite nanocomposite hydrogels following electrostimulation. *Acta Biomaterialia*, **4**, 1038–45.

242 Xiang, Y., Peng, Z. and Chen, D. (2006) A new polymer/clay nano-composite hydrogel with improved response rate and tensile mechanical properties. *European Polymer Journal*, **42**, 2125–32.

243 Ma, J., Xu, Y., Fan, B. and Liang, B. (2007) Preparation and characterization of sodium carboxymethylcellulose/ poly(*N*-isopropylacrylamide)/clay semi-IPN nanocomposite hydrogels. *European Polymer Journal*, **43**, 2221–8.

244 Lee, W.-F. and Fu, Y.-T. (2003) Effect of montmorillonite on the swelling behavior and drug-release behavior of nanocomposite hydrogels. *Journal of Applied Polymer Science*, **89**, 3652–60.

245 Lee, W.-F. and Chen, Y.-C. (2004) Effect of hydrotalcite on the physical properties and drug-release behavior of nanocomposite hydrogels based on poly[acrylic acid-*co*-poly(ethylene glycol) methyl ether acrylate] gels. *Journal of Applied Polymer Science*, **94**, 692–9.

246 Starkar, N.S. and Hilt, J.Z. (2008) Hydrogel nanocomposites as remote-controlled biomaterials. *Acta Biomaterialia*, **4**, 11–16.

247 Hawkins, A.S., Starkar, N.S. and Hilt, J.Z. (2009) Nanocomposite degradable hydrogels: demonstration of remote controlled degradation and drug release. *Pharmaceutical Research*, **26**, 667–73.

248 Rzigalinski, B.A. and Strobl, J.S. (2009) Cadmium-containing nanoparticles: perspectives on pharmacology and toxicology of quantum dots. *Toxicology and Applied Pharmacology*, **238**, 280–8.

10
Nanocomposites for Bone Tissue Engineering

Kalpana S. Katti, Dinesh R. Katti, and Avinash H. Ambre

10.1
Introduction

Bone, as a connective tissue, provides structural support, protects vital internal organs, assists movement by coordinating with muscles, and stores calcium- and phosphate-based minerals [1, 2]. Whilst bone is a dynamic tissue that undergoes continuous remodeling [3], it is incapable of self-regeneration, especially in the case of major defects that, until now, have been rectified using methods such as grafting and metallic implants. Whereas, *autografts* and *allografts* have limited applications, owing to the potential for immune rejection and pathogen transfer, metallic implants are limited by their poor integration with the host tissue, corrosion, and failure due to fatigue [4]. In recent years, tissue engineering has evolved as a significant field presenting viable solutions for hard tissue replacement. As stated by Langer and Vacanti, in their pioneering studies:

> "Tissue engineering is an interdisciplinary field that applies the principles of engineering and life sciences towards the development of biological substitutes that restore, maintain or improve tissue function" [5].

In the past, many investigations have been conducted–and, indeed are ongoing–to rectify bone defects by applying the principles of tissue engineering. In this chapter, current advances in the use of novel nanocomposite systems, as applied to bone tissue engineering, are reviewed. Despite many reports having been made describing the use of nanocomposites in bone tissue engineering, in terms of material, clinical, and process parameter optimization, no such comprehensive reviews have yet been provided. Today, many fabrication methods are utilized in both laboratory and industrial scenarios, and an overview of these is provided, together with details of the relevant interface characterization and modeling methodologies. A comprehensive overview of the material components of bone tissue engineering scaffolds, as well as their fabrication and characterization routes, is

Nanomaterials for the Life Sciences Vol.8: Nanocomposites. Edited by Challa S. S. R. Kumar

ISBN: 978-3-527-32168-1

also provided. The major components of tissue engineering are scaffolds, cells, and growth factors. The material choices for the design of scaffolds using nanocomposite systems, and the fabrication routes to their design are detailed in Sections 10.2, 10.3, and 10.4, respectively, while methods used to verify the biocompatibility and bioactivity of scaffolds are outlined in Sections 10.5 and 10.6. The engineering of bone–ligament and bone–cartilage interfaces represents an important aspect of successful tissue engineering (see Section 10.7), and details of the modeling techniques used to evaluate interfaces in tissue engineering nanocomposite scaffolds are provided in Section 10.8. Finally, some ideas on the future of novel nanocomposite scaffolds in bone tissue engineering are provided in Section 10.9.

10.2 Tissue Engineering Scaffolds

Bone exhibits an hierarchical structure, with structural units that range from the microscale to the nanoscale [1]. In addition, bone consists of cells that reside in an extracellular matrix (ECM) that consist of structural (collagen and elastin) and adhesive (fibronectin and vitronectin) protein fibers, in the nanometer range [6]. An environment which consists of nanoscale features is more conducive for an initial cell attachment and proliferation, due to the increased sensitivity of the cells via the filopodia and microspikes [7]. This type of cellular behavior has also been attributed to an increase in the number of atoms and crystal grains, along with an increase in surface area in the case of nanostructures [8, 9]. It has also been shown that the interaction of proteins (such as fibronectin, vitronectin, laminin and collagen) that affect the behavior of osteoblasts occurs to a greater extent on nanoscale materials [9]. Cells that belong to the osteogenic lineage (e.g., osteoblasts, osteocytes, osteoclasts), in addition to scaffold and growth factors, represent the key components for bone tissue engineering (Figure 10.1) that are used to mimic the *in vivo* environment for bone tissue regeneration in order to cure bone defects (although it should be noted that the osteoblasts, osteocytes and osteoclasts may also be derived from stem cells).

As noted above, the *scaffold* serves as one of the key components of bone tissue engineering, and should be designed to mimic the hierarchical structure of the ECM in order to replicate the intracellular and intercellular responses required in cell differentiation and proliferation [6, 10]. Scaffolds are porous structures (see Figure 10.2) that support cell growth, proliferation, and differentiation. The cell-surface receptors react to the mechanical properties of the ECM by converting mechanical signals to chemical signals [6]. These receptors also interact with chemical ligands present in a nanostructured ECM, thus affecting the cell behavior, while the nanofibrous structure of the ECM is also responsible for the clustering of chemical ligands that affect cell behavior. Consequently, it is possible to modulate cell behavior via both the mechanical and chemical properties of the surrounding three-dimensional (3-D) environment, with such modulation being

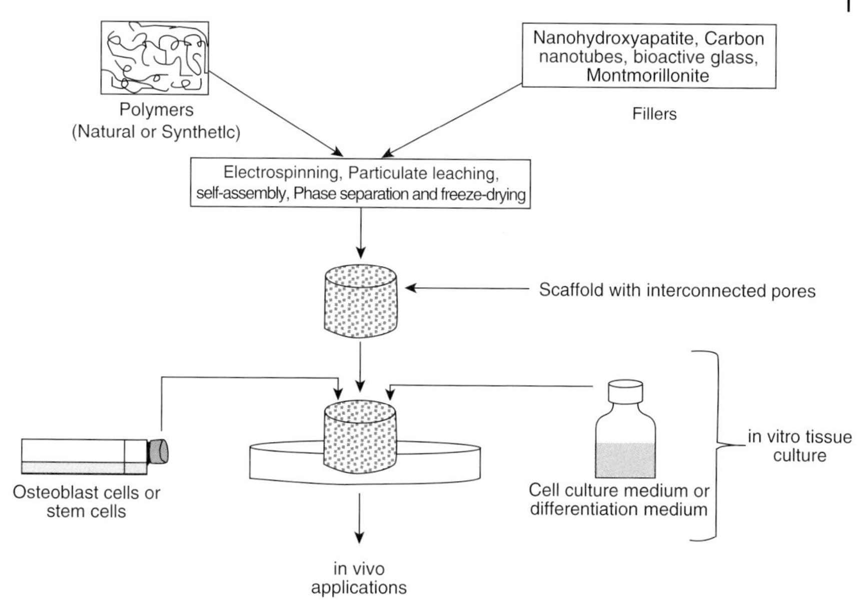

Figure 10.1 Bone tissue engineering: the general scheme.

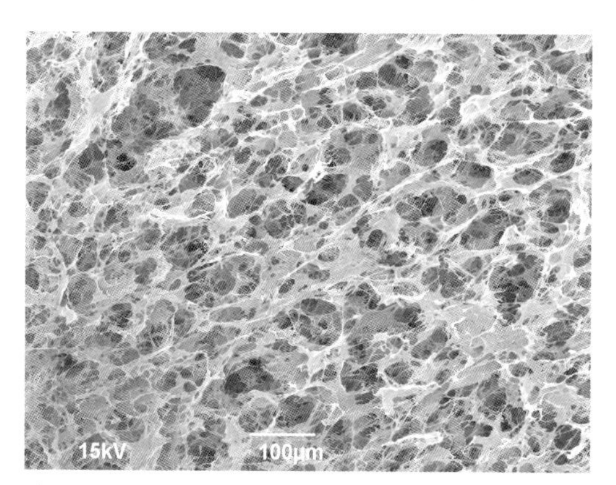

Figure 10.2 Scanning electron microscopy image of a bone tissue engineering scaffold structure. The scaffold is a nanocomposite material composed of a polyelectrolyte complex of polygalacturonic acid and chitosan with nanoscale hydroxyapatite particles.

both space- and time-dependent. The cells will be influenced by the spatial arrangement of the surrounding environment, and by the time-dependent changes in the molecules involved in adhesion between the cell and the ECM [6, 11]. Thus, the design of a scaffold can be described as an initial critical step for successful bone tissue regeneration. Other major challenges in scaffold design pertain to the regeneration of a large volume of tissue that contains different types of cells, and to the creation of an effective distribution of blood vessels around the regenerating tissue [12]. The organization of these different cell types in the regenerating tissue is essential for its correct functioning, and is heavily dependent on cell–cell interactions that are mediated through cadherins, a group of type-1 transmembrane proteins that play an important role in cell adhesion. This, in turn, provides the challenge of engineering the cell surfaces so as to initiate effective cellular interactions that will lead to an organization of cells in a regenerating tissue. Additional strategies, such as the coculture of different types of cell and the use of vascular endothelial growth factor (VEGF) have been followed in attempts to overcome the problems of developing an effective distribution of blood vessels within a regenerating tissue, although at present the problems of stabilizing the developing blood vessels and delivering accurate concentrations of VEGF are inadequately resolved [12]. The major challenges associated with bone tissue engineering include the development of a scaffold with appropriate porosity and pore size, effective macroscale mechanical properties, and appropriate rates of biodegradation. While designing the scaffolds, it is also important to consider that, although scaffold systems are expected to contain complicated information that will favor the regeneration of a tissue, the over-design of a scaffold may place limitations on any transition to clinical practice [12]. Nonetheless, nanocomposites based on organic–inorganic systems have the potential to overcome these challenges associated with bone tissue engineering.

The development of nanocomposite scaffolds which possess the required characteristics depends on factors such as the correct selection of organic and inorganic components, and the processing routes used to create the scaffolds. These factors may further affect the biocompatibility and mechanical properties (via interfacial interactions) of the nanocomposite scaffolds. These factors, and the current position in terms of materials design for bone tissue engineering scaffolds, are described in the following sections.

10.3 Materials

A wide variety of organic and inorganic materials have been used in the design of bone tissue engineering scaffolds. Typically, the organic materials used to create nanocomposite scaffolds are either natural or synthetic polymers, while the most commonly used inorganic materials are nanohydroxyapatite (nano-HAp), silicates, and carbon nanotubes (CNTs). The success of these materials depends on whether they are able to mimic the nuances of the natural ECM.

10.3.1 Hydroxyapatite

Hydroxyapatite (HAp) is the major component of the inorganic phase of bone, and its natural or synthetic forms are known to be bioactive, osteoconductive, nontoxic, and nonimmunogenic, with a crystal structure that is similar to the HAp present in bone tissue. It has been shown that HAp at the nanoscale (nano-HAp) is functionally more effective than HAp at the microscale, due to improved surface properties such as a greater surface roughness, lower contact angles, and a reduced pore size. These characteristics of nano-HAp are known to improve cell adhesion, differentiation and growth via an increased adsorption of specific proteins [1, 13]. Nano-HAp is more active in influencing the unfolding of adhesive proteins to a greater extent, and in increasing the number of available arginine-glycine-aspartic (RGD) acid sequences present in adhesive proteins for the cell-surface receptors [13]. Among the many methods used to synthesize nano-HAp can be included solid-state, wet-chemical, hydrothermal, mechanochemical, pH shockwave and microwave techniques. Nano-HAp has been used in a variety of studies to prepare nanocomposites for bone tissue engineering but, more recently, it has been incorporated into polymers such as chitosan [14–16], poly-2-hydroxyethylmethacrylate-polycaprolactone (PHEMA/PCL) [17], poly(lactic acid) (PLA) [18], poly(3-hydroxybutyrate-*co*-3-hydroxyvalerate) [19], poly(ε-caprolactone) (PCL) [20], and poly(propylene fumarate) [21]. Each of these studies has demonstrated the value of nano-HAP in the development of nanocomposites for bone tissue engineering.

10.3.2 Silicates

Very few studies have been conducted regarding the use of silicates, such as montmorillonite (MMT) clays, for the purpose of bone tissue engineering. In the past, MMT clays have been used to prepare nanocomposites with improved mechanical properties, decreased gas permeabilities and flammabilities, and improved biodegradability [22]. The development of scaffolds with adequate mechanical properties capable of withstanding the rigors of the *in vivo* environment after implantation, without adversely affecting the functions of the surrounding tissues, represents one of the major challenges of bone tissue engineering. The potential for MMT clays to improve the mechanical properties of the composites can be exploited in the case of the scaffolds used for bone tissue engineering. Recently, it has been confirmed that MMT clays possess medicinal properties [23–25] and also show potential to improve the mechanical properties of composites used in bone tissue engineering applications [15, 26–28].

10.3.3 Carbon Nanotubes

Carbon nanotubes (CNTs), which are members of the fullerene family of carbon allotropes, have mechanical properties in the gigapascal range, possess a high surface area, and are also internalized by the cells. Functionalized CNTs have been used in studies related to cancer treatment [29], as scaffolds for neurite growth [30], as antiseptic bandages [31], and as biosensors to detect insulin [32]. The exceptional mechanical properties and surface area of CNTs may prove to be useful for bone tissue engineering applications and, indeed, several studies have been conducted in this respect. For example, Shi *et al.* [33] prepared nanocomposite scaffolds using both functionalized and nonfunctionalized ultra-short single-walled carbon nanotubes (SWCNTs) and poly(propylene fumarate) (PPF). The functionalized ultra-short SWCNTs were shown to be more effective at increasing the compressive properties of the scaffolds, on which rat bone mesenchymal stem cells (MSC) were shown to attach and proliferate. In another study, injectable nanocomposites of PPF and SWCNTs were prepared that not only had improved compressive and flexural properties, but also showed that the dispersion of the CNTS was affected by their concentration in the nanocomposite [34]. A similar type of nanocomposite was prepared by Sitharam *et al.* [35], using PPF, SWCNTs, and ultra-short SWCNTs. Notably, the rheological properties of the nanocomposites that affect their injectability were seen to depend on the *size* of the CNTs, whereas the mechanical properties depended on their surface area. Likewise, when Sitharam *et al.* [36] assessed the *in vivo* biocompatibility of scaffolds composed of ultra-short CNTs and PPF/propylene fumarate diacrylate, bone growth was shown to occur more rapidly on the PPF/propylene fumarate diacrylate/CNT scaffold than on a scaffold not containing CNTs. Clearly, composite scaffolds containing CNTs are osteoconductive and also promote osteogenesis. A number of other studies have involved the use of CNTs based on composite systems such as chitosan/multiwalled CNTs [37], HAp/CNTs [38, 39], poly(carbonate)urethane/multiwall CNTs [40], and poly(L-lactide)/magnetic multiwall CNTs [41], and all have emphasized the role of CNTs as useful bone tissue-engineering materials.

10.3.4
Bioactive Glass

Bioactive glass, as an inorganic material, can be used in the preparation of nanocomposites for bone tissue engineering applications. These glasses are characterized by the rapid formation of an hydroxycarbonate apatite layer on their surfaces, that promotes their adhesion to the bone apatite. Apatite layer formation on bioactive glasses occurs when they are in contact with fluids that are identical or similar to human plasma [42–45]. In addition, they have a faster bonding rate with bone, and also stimulate osteogenic cell production upon dissolution, due to the release of Si, Ca, P, and Na ions. Notably, bioactive glasses prepared using the sol–gel method are nanostructured, as opposed to those prepared by the melt method [43, 46]. Bioactive glasses prepared via the sol–gel method also showed a greater bioactivity than those prepared using the melt method [47–49]. Recently, much interest has been shown in using bioactive glass nanoparticles to prepare composites that can be used in bone tissue engineering. For example, when Hong *et al.* [50] prepared bioactive glass ceramic nanoparticles by combining sol–gel and coprecipitation methods, the bioactivity of the nanoparticles was confirmed by the surface formation of HAp, as demonstrated using Fourier transform infrared (FT-IR) spectroscopy and X-ray diffraction (XRD). Likewise, nanocomposite scaffolds prepared by Hong *et al.* [51], by incorporating bioactive glass nanoparticles into

PLA, exhibited increases in compressive strength and compressive modulus, and also a higher content of bioactive glass nanoparticles. These scaffolds proved to be bioactive, with *in vitro* degradation rates in phosphate-buffered saline (PBS) faster than those of scaffolds prepared without bioactive glass nanoparticles; this difference was considered due to the hydrophilic nature of the bioactive glass. Liu *et al.* [52] modified the bioactive glass nanoparticles with low-molecular-weight poly(L-lactide), and used these modified nanoparticles to prepare poly(L-lactide) composite scaffolds that had an improved tensile strength of up to 6 wt% loading. Biocompatibility studies performed with these scaffolds, using rabbit bone marrow stromal cells, indicated that composites with unmodified bioactive glass nanoparticles had a better initial cell adhesive and proliferative properties than scaffolds with modified particles, but this trend was reversed after seven days. Couto *et al.* [53] developed hydrogel composites using chitosan and bioactive glass nanoparticles, with an aim to prepare injectable nanocomposites useful for tissue engineering. Both, rheological and *in vitro* studies aimed at determining the bioactivity of these hydrogel composites suggested that they might prove useful for bone tissue regeneration. Mansur *et al.* [54] prepared nanocomposite scaffolds based on poly(vinyl alcohol) (PVA) and bioactive glass that had an interconnected macroporous structure, with pores in the range of 10 to 500 μm, and a mesoporous structure in the nanometer range. The compressive properties of these scaffolds were shown to be affected by the PVA content and its degree of hydrolysis.

10.3.5 Polymers

As bone is considered to be a nanocomposite, with the inorganic phase as its major component, it might be argued that in order to develop nanocomposites with appropriate mechanical properties that mimic those of bone, the inorganic phase should be relatively significant. However, it is important to note that the mechanical properties of bone are based on an hierarchical structure that is very difficult to mimic when preparing nanocomposites for bone tissue engineering. Consequently, the correct selection of an organic phase becomes vitally important when preparing nanocomposites for bone tissue engineering. In addition to mimicking the features of the ECM, the organic phase must have a biodegradative rate that matches the bone tissue regeneration rate. Likewise, the organic phase should have biodegradation products that cause changes in the local pH, but which can easily be excreted from the body without affecting any of the pH-dependent enzymic reactions that occur simultaneously within the cells and their environment. The organic phase should also possess adequate mechanical properties to support the growing tissue in the dynamic *in vivo* environment, and not trigger any immune responses. Morphologically, the organic phase may be in the form of either natural or synthetic polymers.

Natural polymers, obtained from natural sources such as animals and plants, possess biomolecules or sequences of biomolecules that are favorable for cell attachment and differentiation [55]. These polymers can be degraded by enzymes or by hydrolysis, are similar to biological macromolecules, and may not trigger an immune response due to their similarity to the ECM [56]. Some of their key disadvantages, however, include an uncontrolled biodegradation rate, batch-to-batch variations in the mechanical properties, difficulties with scale-up [55], and the possibility of infections when polymers are derived from an animal source, such as collagen. It is also difficult to predict or estimate the *in vivo* biodegradation rate in these polymers, because of the different extents of enzymic activity in different individuals. In the case of synthetic polymers, it is possible to control biodegradation rates by the appropriate introduction of functional groups into the polymeric chain (by grafting), or by selecting appropriate monomers when preparing the polymers. The mechanical properties of synthetic polymers depend on their molecular weight, and can be controlled by using an effective polymer synthesis method such as addition-polymerization, condensation-polymerization, and/or ring-opening polymerization. Unfortunately, these polymers may pose certain problems such as chronic inflammation, and even a lack of desired cell response as they do not possess the natural biomolecule sequences that promote cell attachment. Moreover, uncertainties involved in the complete removal of solvents during the creation of some synthetic polymers may have an adverse effect on the cell response. The natural and synthetic polymers used for bone tissue engineering are discussed in the following sections, together with the details of some recent studies involving their use in this area.

10.3.5.1 Natural Polymers

10.3.5.1.1 Collagen Collagen (type I) is a protein that is known to be the major component of the ECM of most tissues, and so can be used extensively for tissue engineering applications. Collagen exhibits good biocompatibility, a low immune response, good mechanical strength, and can also be crosslinked, which helps to control its mechanical and degradation properties [56]. Kikuchi *et al.* [57] used glutaraldehyde to prepare crosslinked collagen/HAp nanocomposites, and showed the mechanical strength of the composites to be improved with the glutaraldehyde content. *In vivo* studies involving the implantation of these crosslinked nanocomposites into rabbit tibia suggested that the osteoclastic resorption of these composites decreased with the increasing glutaraldehyde content, without any adverse reactions. Kim *et al.* [58] developed microspheres of collagen–apatite nanocomposites, and studied the *in vitro* response of rat bone marrow stem cells on these microspheres. The results showed that the rat bone marrow stem cells were able to attach and grow on these nanocomposite microspheres, which indicated that they could be used for bone tissue regeneration. Several other studies involving use of collagen have also emphasized its importance in bone tissue engineering [59–63].

10.3.5.1.2 Chitosan Chitosan, another important biopolymer, is an incompletely deacetylated form of chitin, and has been studied extensively for bone tissue engineering applications. Chitosan is known to be biocompatible, to show antimicrobial activity, and to be capable of activating macrophages and promoting the proliferation of fibroblasts [64]. In recent years there has been a sustained interest in chitosan-based nanocomposites for bone tissue engineering. For example, Thein-Han *et al.* [14] prepared biomimetic chitosan–nano-HAp scaffolds which showed the mechanical properties and response of pre-osteoblasts to be improved, and the degradation rate to be decreased, by the addition of nano-HAP. These studies also suggested that the mechanical properties of the scaffolds were affected by the molecular weight and degree of deacetylation of chitosan. Zhang *et al.* [65] studied nanofibrous scaffolds made from chitosan and HAp with attributes similar to the ECM, found that such scaffolds favored the formation of bone tissue. Kong *et al.* [66] used chitosan to prepare chitosan–nano-HAp scaffolds, and found the biocompatibility and bioactivity of chitosan to be further enhanced by the addition of nano-HAp. Li *et al.* [67] prepared hybrid scaffolds of chitosan–alginate that had an improved Young's modulus, an improved shape retention in solutions over a wide pH range, and a better cell proliferation as compared to pure chitosan scaffolds. In a more recent study, Cai *et al.* [16] developed nanocomposites based on chitosan, PLA and HAp with a highly improved elastic modulus and compressive strength that were dependent on the poly(lactic acid) content. The latter material, when used in these composites, was responsible for the rod-like shape of HAp. Although many studies been conducted to investigate its bone tissue engineering applications, chitosan remains the major component due to the presence of amino groups, its tunable degree of deacetylation that may affect its ability to form complexes with other polyelectrolyte molecules, and its mechanical properties. The amino groups of chitosan are protonated in acidic media, and can form complexes with other polymeric molecules where the functional groups have a complementary charge. Indeed, this has increased the chances of developing novel nanocomposite systems involving the use of chitosan. For example, Verma *et al.* [68] developed a nanofibrous polyelectrolyte scaffold based on chitosan, poly(galacturonic acid) and HAp that had a much superior osteoblast cell adhesion compared to a scaffold without HAp.

10.3.5.1.3 **Cellulose** Cellulose is a natural polymer that is found in plants such as cotton, wood, and straw, and is produced by microbes such as *Acetobacter xylinum*. The plasma membrane of *A. xylinum* consists of terminal enzyme complexes that are responsible for the formation of sub-elementary cellulose fibrils on the bacterial cell surface. These fibrils are extruded into the incubation medium, where they accrue to create microfibrils that subsequently form a tight ribbon structure. The ribbon structures from many bacterial cells then form cellulose, which has a sheet-like structure. Cellulose is convertible into many derivatives, including carboxymethylcellulose (CMC), cellulose nitrate, cellulose acetate and cellulose xanthate, and it can also be molded or processed into fibers. Although cellulose cannot be degraded enzymatically (this is one of its disadvantages), it can be rendered degradable by altering its higher-order structure [56]. Bacterial cellulose is superior to plant cellulose in aspects such as mechanical properties (tensile strength and modulus), surface area, water-holding capacity and crystallinity [69]. Wan *et al.* [69] studied the prospects of cellulose as a useful bone tissue engineering material by preparing nanocomposites based on HAp and bacterial cellulose. The prepared nanocomposites had a 3-D network structure with the formation of carbonated HAp after soaking in a simulated body fluid (SBF). Li *et al.* [70] oxidized bacterial cellulose with sodium periodate to obtain 2,3-dialdehyde bacterial cellulose, a biodegradable and biocompatible form of bacterial cellulose. Whilst these studies did not involve the preparation of composites, it was interesting to note that the 2,3-dialdehyde bacterial cellulose scaffolds were able to degrade both in SBF and PBS, and were able to form a microporous structure that was in a range useful for cell growth. Jiang *et al.* [71] found that the microstructure, mechanical properties, degradation and bioactivity of nano-HAp/chitosan/CMC scaffolds was suitable for bone tissue engineering applications. Subsequent studies conducted by Muller *et al.* [72] and Svensson *et al.* [73] have also shown that cellulose can be a useful material for tissue engineering applications.

10.3.5.1.4 **Starch** Starch is a natural polymer found in plants such as corn, rice, potato, and wheat, and is capable of being crosslinked, acetylated, converted into thermoplastics, and blended with synthetic polymers to improve its mechanical properties [56]. When Salgado *et al.* [74] developed scaffolds based on starch–cellulose acetate blends, the scaffolds showed compressive properties that were adequate for bone tissue engineering. Moreover, the fact that the scaffolds favored cell proliferation, and the formation of ECM and mineralization, suggested a use for starch-based materials in bone tissue engineering. Later, when Marques *et al.* [75] investigated composites of HAp and corn starch blended with polymers such as ethylene vinyl alcohol, cellulose acetate, and PCL, their results not only highlighted the suitability of starch for bone tissue engineering but also detailed the importance of surface roughness, the miscibility of the blend systems used, and the HAp content on osteoblast cell behavior. The results of other studies [76–78] have implied that starch might serve as a useful bone tissue engineering material.

10.3.5.1.5 **Hyaluronic Acid** Hyaluronic acid, a natural polymer that consists of disaccharide units of α-1,4-D-glucuronic acid and β-1,3-*N*-acetyl-D-glucosamine, is found in connective tissues such as umbilical cord, synovial fluid, and vitreous [79] It also known to be present in the ECM, and to participate in maintaining the structure of the ECM, morphogenesis, wound repair, and metastasis [80]. Previously, hyaluronic acid has been used in studies related to bone tissue engineering. For example, Lisignoli *et al.* [81] investigated an esterified hyaluronic acid-based scaffold with non-woven, micron-sized fibrous structures to study the differentiation and mineralization of rat bone marrow stromal cells, in the presence and absence of basic fibroblast growth factor (bFGF). Subsequently, bFGF was found to enhance the mineralization rate and expression of differentiation markers such as osteocalcin and osteopontin. The conclusion of these studies was that hyaluronic acid-based scaffolds were capable of supporting bone growth before implantation. As an extension of these *in vitro* studies, when Lisignoli *et al.* [82] implanted hyaluronic acid-based scaffolds that had been cultured with rat bone marrow stromal cells, into rats with defects in their forearm bones, the scaffolds showed bone growth trends that were similar to those found in the *in vitro* studies. Ji *et al.* [79] prepared hyaluronic acid-based scaffolds by electrospinning and found them suitable for cell attachment and spreading. Likewise, Kim *et al.* [80] used hyaluronic acid-based scaffolds that had been seeded with human MSC for curing bone defects in rats; support for the new bone tissue formation by the scaffolds was shown to be more enhanced in the presence of bone morphogenic protein (BMP-2).

10.3.5.1.6 **Silk Fibroin Polymers** Silk fibroin polymers, which are obtained from animal sources such as spiders (e.g., *Nephila clavipes*) and worms (e.g., *Bombyx mori*), consist of secondary β-sheet structures formed due to the specific arrangement of hydrophobic and hydrophilic domains composed of specific amino acid sequences, have also been investigated with regards to nanocomposite design. The hydrophobic domains form the β-sheet structure (ordered), while their combination with the less-ordered hydrophilic structures is responsible for the mechanical properties of silk. Silk fibroin obtained from *B. mori* consists of light and heavy protein chains that are linked by disulfide bonds and coated with sericins. Silk fibroin is biocompatible, enzymatically degradable, and has good mechanical properties. One major advantage is that the properties of silk fibroin can be adjusted according to requirements, by using different processing routes.

10.3.5.2 Synthetic Polymers

10.3.5.2.1 **Aliphatic Polyesters** Aliphatic polyesters, including poly(glycolic acid) (PGA), PLA, poly(lactic-*co*-glycolic acid) (PLGA) and PCL represent a class of synthetic polymers that can be used for bone tissue engineering applications. The differences in the properties of the polymers in this category depend on the monomer units from which they are constructed.

PGA is characterized by a moderate crystallinity, a high melting point, poor solubility in organic solvents, and an ability to undergo hydrolytic degradation. It has been suggested that PGA can be used in situations in which an initially tough and faster degrading material is needed, although the possibility of local pH variations due to its degradation products necessitates the consideration of its use in regions that have a buffering capacity or a mechanism for the rapid removal of degradation products.

PLA can be prepared as either D or L isomeric forms, denoted as PLLA or PDLA, respectively. As the degradation product of PLLA (i.e., L-lactic acid) is found in human metabolism, its use may be preferable. The higher solubility of PLA as compared to PGA in organic solvents may prove to be advantageous for electrospinning. The properties of PLGA can be controlled by adjusting the lactic acid/glycolic acid ratio.

PCL is semicrystalline, has good solubility in organic solvents, and a low melting point with a degradation rate less than that of PGA and PLGA. Aliphatic polyesters have been used in numerous studies related to bone tissue engineering. For example, Lee *et al.* [83] prepared a PLLA/MMT scaffold with improved mechanical properties and porosity that was suitable for bone tissue engineering. Liu *et al.* [84] showed that the fall in pH due to the degradation of PLGA can be reduced by incorporating titania nanoparticles into the PLGA. These nanoparticles served as a buffer to control the fall in pH that further enhanced the significance of these studies [85] in relation to the behavior of osteoblasts on PLGA/titania nanocomposites. Nejati *et al.* [86] prepared PLLA/HAp nanocomposites with a microstructure and a compressive strength that was close to that of cancellous bone. When Serrano *et al.* [87] studied the biocompatibility of PCL using murine fibroblasts, PCL was shown to serve as a good substrate for fibroblast adhesion, and also increased the mitochondrial activity of the fibroblasts. Shor *et al.* [88] developed PCL/HAp scaffolds with a controlled microstructure (by using an extrusion technique) that showed an ability for mineralization upon seeding with fetal bovine osteoblasts. Erisken *et al.* [89] prepared graded nanocomposites scaffolds using PCL and β-tricalcium phosphate with a spatial variation in structure and composition, and an ability to form ECM and create mineralization. Likewise, several other studies have demonstrated the use of aliphatic polyesters as useful bone tissue engineering materials [18, 90–95].

10.3.5.2.2 **Polyphosphazene** Polyphosphazene is a synthetic polymer with a backbone of alternating phosphorus and nitrogen atoms, with each phosphorus atom linked to two organic side groups. The degradation of polyphosphazene can be modulated by making changes in the side groups; this provides it with an advantage over aliphatic polyesters, which require changes to be made to their backbone structures in order to control their degradation rates [96, 97]. The incorporation of side groups such as imidazolyl, amino acid esters, glycolate esters and lactate esters, causes polyphosphazene to become sensitive to hydrolysis and to release nontoxic products. Bhattacharya *et al.* [98] prepared polyphosphazene–nano-HAp composite nanofibers as a potential material system for bone tissue engineering application. Likewise, Nukavarapu *et al.* [99] produced three different types of polyphosphazene with different side groups, and further used a phenylalanine ethyl ester-substituted polyphosphazene with a glass transition temperature (T_g) higher than 37 °C (physiological temperature) to develop scaffolds with nano-HAp. These scaffolds showed demonstrated a compressive strength close to human cancellous bone, and also proved favorable for osteoblast cell adhesion and proliferation.

10.3.5.2.3 **Poly(propylene fumarate)** Poly(propylene fumarate) (PPF) is an unsaturated linear polyester which is capable of being crosslinked by using agents such as methyl methacrylate (MMA), *N*-vinyl pyrrolidinone, PPF-diacrylate and poly(ethyleneglycol)–diacrylate [21, 100]. The fact that the degradation products of PPF are nontoxic is important from a tissue engineering perspective. Horch *et al.* [101] used alumoxane nanoparticles to prepare PPF composites with improved flexural properties for bone tissue engineering applications. Lee *et al.* [21] found that a PPF–HAp nanocomposite system was suitable for the attachment and proliferation of murine pre-osteoblast cells, and could serve as suitable system for bone tissue engineering. Several other studies [102–105] have also reported the suitability of PPF for bone tissue engineering applications. One such study demonstrated an interesting possibility of developing injectable nanocomposites made from PPF and SWCNTs [103], whereby the functionalization of the CNTs played an important role in the rheological behavior and degree of crosslinking of these nanocomposites that, in turn, proved to be important for the development of injectable nanocomposites. Studies related to injectable nanocomposites are important for tissue engineering, because these composites provide the option of minimizing complex surgical procedures, of developing scaffolds for curing defects having unusual shapes, and of delivering cells and bioactive molecules to the defect sites, along with the nanocomposite solution.

Currently, a wide variety of synthetic polymers is available that could be developed and used to assist in the design of nanocomposites for tissue engineering applications. Over the years, improvements in the understanding of polymerization mechanisms has increased the possibilities of developing synthetic polymer systems, according to the requirements for bone tissue engineering. Other than some of the synthetic polymers discussed above, the main synthetic polymer systems to have been developed and studied for bone tissue engineering are poly(ester urethane) [106], poly(urethane) [107–109], and PVA [54, 110].

10.4 Processing Methods/Routes for Nanocomposites in Bone Tissue Engineering

The fact that nanoscale features have an important effect on cell behavior emphasizes the importance of nanocomposites in bone tissue engineering. In order to obtain appropriate nanoscale features or nanostructures in the final nanocomposite, it is necessary to use appropriate processing methods. Today, a number of challenges are associated with the preparation of these nanocomposites, one of which is the need to conserve the intrinsic properties of the materials used. The processing method used might affect the conformation of the polymer chains in the case of polymers, or the distribution of nanosized inorganic materials in the form of particles while preparing nanocomposites. The changes in conformation of the polymer chains in the case of natural polymers may affect the available biomolecule sequences and cell behavior. In addition, the properties of polymeric materials, such as melting point, T_g, viscosity and resistance to the solvents used during processing, may also limit the number of routes by which such processing can be effected. In the case of bone regeneration, it is necessary first to develop nanocomposites with structural features that mimic, or at least resemble, the structural traits of the ECM. In this context, there are at present several routes available for the development of nanocomposites for bone tissue engineering, and these are discussed in the following sections.

10.4.1 Electrospinning

Electrospinning is a simple technique that can be used to produce fibers having diameters in the nanometer range, which makes it useful when mimicking the structure of the ECM proteins. In addition, fibers produced by electrospinning have a high surface area-to-mass ratio (40 to $100\,m^2\,g^{-1}$), as well as superior stiffness and tensile strength [111, 112]. The possibility to spin fibers at room temperature from a variety of polymers, thus producing aligned nanofibers, and/or to spin fibers simultaneously from separate solutions so as to generate layered scaffold structures, emphasized the attraction of electrospinning in bone tissue engineering applications [10, 111]. The electrospinning apparatus consists of a high-voltage supply, a capillary tube containing a polymer solution connected to a needle or a pipette, and a grounded metallic collector; the process is shown schematically in Figure 10.3. The needle and the metallic collector act as two electrodes between which a high voltage (several tens of KV) is applied. The application of a high voltage supply generates an electric field between the two electrodes that produces electrostatic forces (generated by electrostatic repulsions between the surface charges on the droplet and coulombic force exerted by the external electric field) in the spherical droplet liquid (polymer solution) at the needle tip, and opposes the surface tension forces in the liquid droplet. This results in a deformation of the spherical droplet to a conical shape, known as the "Taylor cone." When the electric field surpasses a critical value, the electrostatic force

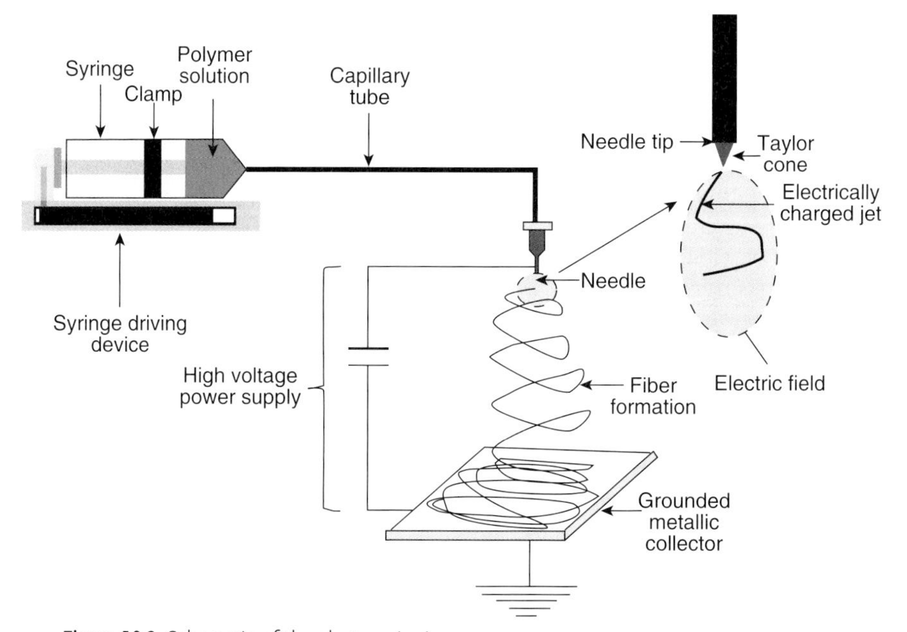

Figure 10.3 Schematic of the electrospinning process.

overcomes the surface tension forces, and this results in the ejection of an electrically charged jet from the tip of the Taylor cone. The electrically charged jet interacts with the externally applied electric field; this makes the jet unstable, and causes it to bend as it moves towards the grounded collector, thus producing long, ultrafine fibers (these are produced by splitting of the jet due to repulsive forces) in the form of a nonwoven structure [113]. The evaporation of solvent from the charged jet takes place as the jet moves towards the collector. Different types of collector, such as plate, cylinder, disc and frame, can be used to control the alignment of the electrospun fibers [8]. The electrospinning process–and thus the dimensions of the electrospun fibers–are affected by several factors, including viscosity, conductivity, surface tension, the magnitude of the applied voltage, the distance between the needle and the collector, the solution flow rate, and the solution temperature. Unfortunately, the electrospinning process is associated with certain limitations, such as the use of organic solvents and a poor generation of controlled, 3-D pore structure. In the case of natural polymers categorized as proteins, the electrospinning process may alter their structures, further affecting the cell behavior on the scaffolds. This use of chemical crosslinking may also be necessary for some electrospun polymers, to prevent them from dissolving in aqueous media [6]. With regards to bone tissue engineering, the electrospinning process has been used widely to develop polymer or polymer nanocomposite scaffolds. Some of the polymer and polymer nanocomposite systems (in the form of electrospun scaffolds) that have been investigated for bone tissue regeneration applications include PCL/nano-HAp/collagen [114], polyphosphazene/nano-HAp [98], carboxymethyl chitin/PVA [115], PLGA/nano-HAp [92], silk fibroin/BMP/nano-HAp [116], PLGA/amorphous tricalcium phosphate [117], and a thiolated hyaluronic acid derivative [79].

10.4.2 Phase Separation

Phase separation is a thermodynamic process that is used for preparing interwoven, nanofibrous scaffolds in bone tissue engineering. Whilst the thermally induced phase separation method is most often used for phase separation, it is also possible to use a nonsolvent for the polymer in order to induce phase separation. This process is shown schematically in Figure 10.4.

Thermally induced phase separation is of two types, namely "solid–liquid phase separation" and "liquid–liquid phase separation," depending on the crystallization temperature (freezing point) of the solvent used. In case of solid–liquid phase separation, the crystallization temperature of the solvent used is higher than the liquid–liquid phase separation temperature; this causes the solvent to crystallize and the polymer to separate when the temperature of the polymer solution is lowered. The crystallized solvent is further removed by freeze-drying (sublimation), which leaves behind pores with a morphology similar to that of the solvent crystallites. Thus, it is possible to control the pore structure and type of phase separation by using solvents having different crystallization (freezing) properties.

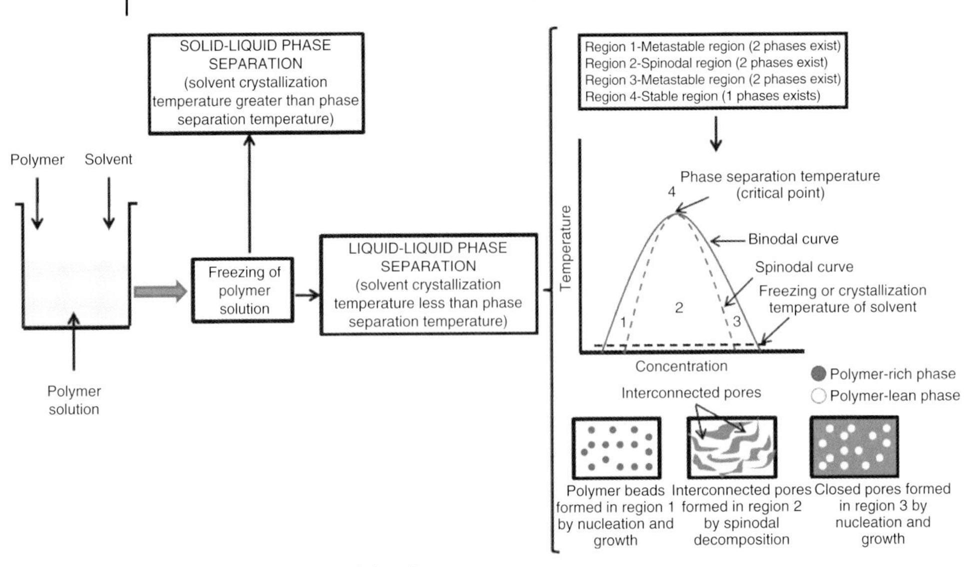

Figure 10.4 Schematic of the phase separation process.

In order for liquid–liquid phase separation to take place, the crystallization temperature of the solvent used must be lower than the liquid–liquid phase separation temperature; this causes the polymer to separate into polymer-rich and polymer-lean phases. This type of phase separation is affected by the upper or a lower critical solution temperature of the polymers involved, and is posited to take place either due to a spinodal liquid–liquid phase separation, or to a nucleation and growth mechanism. Phase separation by nucleation and growth is thought to occur in the metastable region of the temperature–composition phase diagram, where the solution is stable with respect to the small fluctuations in composition. The polymer concentration under the metastable region controls the structure of the polymer solid that is obtained after removal of the solvent with a low polymer concentration; this results in a powder-like polymer solid and a high concentration that causes a closed-pore structure of the polymer solid. The spinodal region is unstable, and small variations in polymer concentration will cause a decrease in free energy that results in a wave of concentration fluctuations in the polymer solution. These further result in the formation of interconnected polymer-rich and polymer-lean (mostly solvent) phases, to produce a scaffold with a continuous-pore network, following removal of the polymer-lean phase.

It is also possible to induce phase separation at room temperature by adding a nonsolvent to the polymer solution. so as to obtain a gel from which the solvent can then be extracted by using water. The gel is further cooled below the T_g of the polymer, and freeze-dried to produce a nanofibrous scaffold.

One advantages of the phase separation technique is that the morphology of the scaffold can be controlled by changing parameters such as the polymer type and concentration, the freezing temperature, and the use of different types of porogen.

This technique can also be useful for preparing scaffolds of different shapes, according to requirements, and for maintaining batch-to-batch consistency. In spite of being a simple technique, phase separation remains a laboratory-scale procedure that is limited to a few polymers. The phase separation technique has been used to prepare scaffolds based on polymer systems such as PEG/PLLA [118], PLA–dextran blend [119], PLLA [120], PLGA [121], HAp/poly(hydroxybutyrate-*co*-hydroxyvalerate) [122], and HAp/chitosan-gelatin [123].

10.4.3 Freeze-Drying

The technique of freeze-drying is used to remove the solvent following phase separation of the polymer solution, either by lowering the temperature or by adding a nonsolvent to the polymer. During freeze-drying, the temperature is maintained sufficiently low that any remixing of the phase-separated polymer solution is prevented. The freeze-drying technique, which is shown schematically in Figure 10.5, has been used in several studies related to bone tissue regeneration, for developing scaffolds based on polymers such as PLLA [124], chitosan [68, 125], gelatin [126], CMC [127], poly(ether ester) [128], and silk fibroin/hyaluronan [129, 130]. Other studies have also involved the use of nanocomposite scaffolds prepared via the phase separation technique for bone tissue engineering applications. Typical examples of polymer nanocomposite scaffolds fabricated via freeze-drying include PLA/nano-HAp [131], collagen/HAp [132], chitosan/HAp [14, 133], gelatin/HAp [134, 135], and PDLLA/PLGA/bioglass [136]. Polyelectrolyte complex (PEC) fibrous scaffolds for tissue engineering have also been synthesized, and scaffolds fabricated via a freeze-drying methodology [68]. Typically, the design parameters to be optimized are the temperature ranges and time, as well as the concentrations of the polymer solutions.

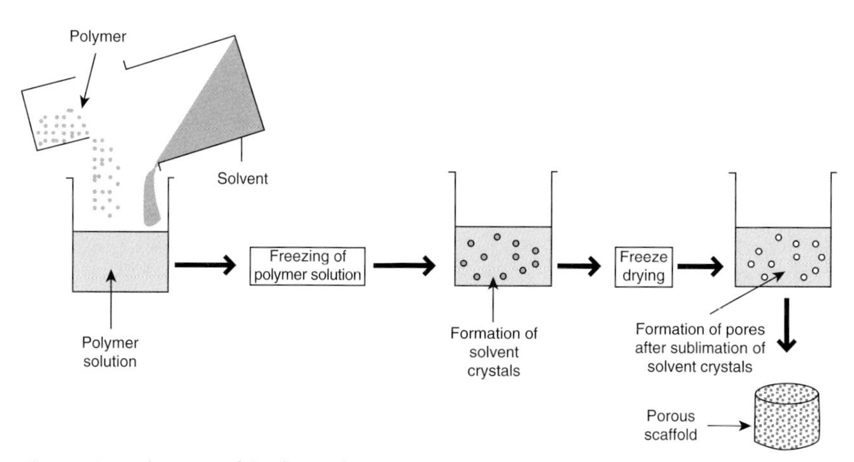

Figure 10.5 Schematic of the freeze-drying process.

10.4.4 Particulate Leaching

Particulate leaching is a relatively simple technique that is used to prepare porous scaffolds by using porogens that are soluble in water or nontoxic solvents; typical examples include sugar, sodium chloride and saccharose. A polymer solution into which the porogens are dispersed is cast into a mold; the solvent is then removed by evaporation, after which leaching of the porogen produces a porous scaffold. The process is shown schematically in Figure 10.6. The main advantage of particulate leaching is that it provides an effective control of pore size and porosity, simply by varying the size and amount of the porogens. However, it has certain limitations: (i) that the solvent removal by evaporation may be incomplete; (ii) that there may be a lack of interconnectivity and open-pore structure in scaffolds requiring a low porosity (due to too-few contact points between the porogens); and (iii) it is more suited to producing thin scaffolds. Despite particulate leaching having been used in combination with other techniques, the lack of interconnectivity represents the major limitation. When Liu *et al.* [137] prepared gelatin/apatite nanofibrous scaffolds by combining thermally induced phase separation and particulate (porogen) leaching techniques, the mold was pre-heated to develop interconnectivity between the particulates (and thus between the scaffold pores), which in turn helped to distribute the cells throughout the scaffold. Several studies related to bone tissue generation have been conducted using scaffolds prepared via particulate leaching, with recent trends seeking to further improve this method [119, 138–142].

10.4.5 Self-Assembly

The self-assembly technique, which is so useful for preparing nanofibrous scaffolds, is present everywhere in Nature, from microscopic to macroscopic levels, and is "… the spontaneous association and organization of numerous individual entities into coherent and well-defined structures without external intervention [143]." Molecular self-assembly involves diffusion followed by the association of

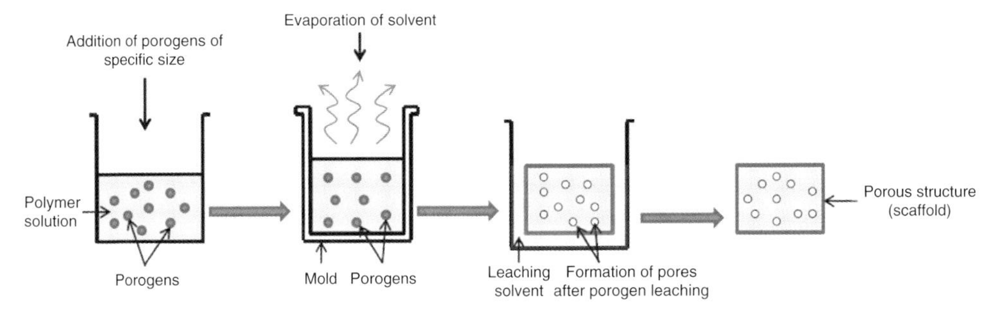

Figure 10.6 Schematic of the particulate leaching process.

molecules through noncovalent interactions, such as hydrogen-bonding, ionic bonding, hydrophobic interactions, and van der Waals interactions. These interactions, although weak, are capable of forming the higher-ordered structures seen in biomacromolecules, because of their large numbers [143, 144]. Furthermore, the self-assembly process can be affected by changes in pH and exposure to cations. Some common examples include the self-assembly of phospholipids into vesicles and tubules, and of polypeptide chains into the triple-helix structure of collagen. Synthesized peptide amphiphiles may self-assemble into a triple-helix structure of collagen in an aqueous solvent [145]. A peptide amphiphile is composed of a collagen sequence Gly-Val-Lys-Gly-Asp-Lys-Gly-Asn-Pro-Gly-Trp-Pro-Gly-Ala-Pro that forms the head group, which is connected in turn to a long-chain mono- or di-alkyl ester lipid that forms the lipophilic tail. The head group forms the triple-helix structure, while the tail region associates through hydrophobic interactions to stabilize the 3-D structure of the head group. Self-assembly has been used to create self-assembled structures such as β-sheets, hydrogels with interwoven nanofibers using ionic self-complementary oligopeptides, nanofibers from dendrimers, and nanotubes from polyglutamines [144]. Although self-assembly can be used to produce nanofibers having lesser diameters compared to those prepared by electrospinning, the process is limited by a lack of control when producing macrosized pores, and an inability to produce mechanically stable 3-D scaffolds. Other limitations include the complicated nature of self-assembly, and the low productivity.

10.5 Biocompatibility

Biocompatibility is defined as “the ability of a material to perform with an appropriate host response in a specific application” [146]. Rickert *et al.* [147] have described biocompatibility not only in terms of lack of cytotoxicity of a material, but also in terms of its biofunctionality that allows local and organospecific interactions between the cell and the biomaterial. Further, they have described biomaterial and biofunctionality as biological, biophysical, and biochemical interactions between the material and the body. Biocompatible materials are expected to fulfill the following criteria in that they [147]:

- should not instigate allergies, immunological and foreign-body reactions;
- should not produce cytotoxic, teratogenic, mutagenic, and/or oncogenic effects;
- should be able to degrade in a controlled manner, and the degradation products should also fulfill the above mentioned criteria; and
- should be capable of enduring the procedures used for their sterilization. them

The success and utility of a newly developed material as a useful tissue engineering biomaterial depends on its biocompatibility. Although the mechanical properties of a material also play an important role in cellular response, it is necessary to determine the biocompatibility of a newly developed material before considering the mechanical properties of that material. The two most common methods used to determine the biocompatibility of a material are briefly described in the following sections.

10.5.1 Counting Cell Densities

This basic method can be used to determine the growth of cells over a particular cell culture period, and is valuable when studying cell growth over 2-D substrates in tissue culture. The cells are first seeded over a 2-D substrate in the form of polymer or polymer composite film, in the presence of a cell culture medium; the number of cells attached to the 2-D substrate is then recorded. The cells are counted after having recorded images of different portions of the 2-D substrate, using an inverted microscope. In order to increase the visibility of cells under the microscope, they may be stained with a dye; the cell density is then calculated by counting cells over a specific area of the image. Although useful, this method cannot be applied to 3-D scaffolds; moreover, great care must be taken when counting the cells.

10.5.2 Spectroscopic Techniques

The MTT assay is commonly used to determine cell viability or cell proliferation. In this method, the dye MTT [3-(4,5-dimethylthiazol-2-yl)-2,5-diphenyltetrazolium bromide] is taken up only by viable cells, which convert it (via mitochondrial succinate dehydrogenase) to a purple-colored formazan derivative; the formazan is dissolved in dimethyl sulfoxide and the absorbance of the solution monitored at 540–570 nm. The absorbance value is proportional to the number of viable cells. The MTT assay can be used for both 2-D substrates and scaffolds in well-plates, to obtain quantitative information with regards to cell viability or cell proliferation. The assay may also be used to monitor cytotoxicity.

10.6 Bioactvity

One of the earliest definitions of a bioactive material to have emerged was "...one which has been designed to induce specific biological activity [148]." Yet, over time, bioactive materials have come to be known as bone-bonding materials, and have to satisfy requirements for the formation of bone-like apatite on their surface when implanted in a living body. It was further suggested that *in vivo* apatite formation should be reproducible in a SBF with ionic concentrations equal to those of human blood plasma. Further, bioactivity is described as the ability of a material to form a bonding layer having a bone-like apatite structure, which speeds up the formation of bone [149]. Bioactivity studies are carried out to predict the bone bioactivity of a material by assessing the formation of apatite on its surface, when immersed in a SBF. The SBF contains salts such as NaCl, $NaHCO_3$, KCl, Na_2SO_4, and $CaCl_2$, at appropriate concentration, in deionized water. The materials to be assessed are soaked in the SBF (the pH of which has been adjusted to a physiological value) for the desired time, after which surface apatite formation is monitored using scanning electron microscopy (SEM) coupled to energy-dispersive spectroscopy (EDS), XRD analysis, and FT-IR spectroscopy. The soaking of tissue engineering biomaterials in SBF has been carried out in several studies to assess their bioactivity [66, 150–153]. Both, *in vivo* and *in vitro* studies may be carried out to further assess the biological response of appropriate cellular systems.

10.7 Engineering the Bone–Cartilage/Bone–Ligament Interface

The thrust in most studies related to tissue engineering has been towards developing novel nanocomposite systems with properties that favor the formation of an individual tissue. For this purpose, a persistent effort has been made over the years to improve or alter the chemistry and mechanical properties of the materials used in such systems, and this has been complemented by use of improved processing techniques. In addition, considerable progress has been made in tissue engineering with regards to the regeneration of individual tissues. It is clear that, for successful clinical applications, it is necessary for the regenerated tissue to be capable of integrating with the surrounding tissue. Consequently, the interface between a regenerated tissue and the surrounding tissue(s) are important, and the engineering of the interface is regarded as the next critical step. With regards to bone tissue engineering, the bone–cartilage and bone–ligament interfaces play significant roles in the successful implantation of a bone tissue-engineered construct. The bone–cartilage and/or bone–ligament interfaces serve to reduce the stress concentration between dissimilar tissues (e.g., bone–cartilage) by controlling stress transfer between the tissues [154, 155]. The interfaces are able to facilitate stress transfer due to the presence of various cell types, in addition to a spatially graded biochemical composition with an hierarchical arrangement of the biochemical components [156]. For example, the bone–ligament interface consists of three different regions, depending on the type of cell and composition:

- The first region, known as the "ligament proper," is composed of aligned collagens I and III, and fibroblasts.
- The second region consists of two sub-regions that represent the uncalcified and the calcified fibrocartilage. The uncalcified region contains collagens I and II along with fibrochondrocytes, whereas the calcified region consists of collagens I, II, and X along with hypertrophic chondrocytes.
- The third region is the subchondral bone that is adjacent to the calcified fibrocartilage.

The bone–cartilage interface has similar zones referred to as the superficial zone, the middle zone, deep zone, calcified cartilage, and the subchondral bone. The compressive moduli values increase from the superficial zone to the subchondral zone, which helps to reduce stress concentration. In addition, the typical alignment of the collagen fibers in each zone is considered to contribute to an effective stress transfer across the interface.

Studies related to engineering of the interface represent emerging trends in tissue engineering, and are concentrated towards the anterior cruciate ligament (ACL)–bone interface, mainly as the ACL is very prone to injury. The ACL has a limited vascularization and a poor healing capacity, such that a surgical approach is indispensable. Consequently, studies conducted to date have focused on developing strategies by which to engineer the highly complex ACL–bone interface. One approach has been to develop scaffolds with different phases, and to co-culture different cell types in one such scaffold. Splazzi *et al.* [157] developed triphasic scaffolds based on PLGA, in which the phases were composed of a PLGA mesh, PLGA microspheres, and PLGA-bioactive glass composite microspheres, with each phase having a different surface area. When fibroblasts and osteoblasts were seeded separately in these phases, both formed an ECM containing collagen I, while the osteoblasts also formed a mineralized matrix. The unseeded area between the two phases showed the presence of both osteoblasts and fibroblasts. Thus, the designed scaffold was successful in co-culturing different cell types; moreover, the fibrocartilage-like region in between demonstrated the potential to form an interfacial phase. As a continuation of these studies, Splazzi *et al.* [158] developed similar triphasic scaffolds, the difference being that one phase was composed of a polygalactin mesh, rather than a PLGA mesh. The three phases were fabricated separately, and then joined using an organic solvent; subsequent sintering led to the creation of an integrated triphasic structure. Different cells (e.g., fibroblasts, osteoblasts, and chondrocytes) were seeded separately in the three phases of the scaffold, the aim being to create a cellular composition similar to a fibrocartilage tissue in an *in vivo* environment. A considerable growth of vascularized collagenous tissue was observed in all phases of the tricultured scaffold after four weeks. The presence of collagens II and X, associated with fibrocartilage, which was detected after eight weeks, indicated the formation of a fibrocartilage-like tissue. One other finding was that those scaffolds pre-seeded with cells showed a better tissue growth and vascularization *in vivo* when compared to the unseeded scaffolds. The results obtained from these studies showed a good potential for the regeneration of interfaces between soft tissues and bone (e.g., fibrocartilage interface region), for "seamless" integration with the surrounding *in vivo* environment. In another interesting study, Phillips *et al.* [156] showed how the spatial distribution of a retrovirus encoding an osteogenic transcription factor could be used to control the organization of fibroblasts and osteoblasts in a 3-D scaffold. The aim here was to avoid the discontinuities observed in those interfacial regions engineered by using biphasic or triphasic scaffolds. In this case, a spatial distribution of the osteogenic transcription factor in the collagen scaffold was achieved by immobilizing the transcription factors on the poly(L-lysine) coating present on the scaffold. The coating thickness was varied by using a dip-coating method that further affected the spatial distribution of the osteogenic transcription factor. On seeding fibroblasts onto this collagen scaffold, both mineralized and nonmineralized zones of the fibroblastic ECM were observed, depending on the spatial distribution of the osteogenic transcription factor. These scaffolds were also shown capable of maintaining the spatial distribution of the osteogenic transcription factor and, consequently, a zonal organization *in vivo*. Although the formation of an interfacial fibrocartilage region was not seen in these scaffolds, the results demonstrated a potential for developing continuous complex interfacial structures, as seen at the bone–ligament interface. The possible development of materials with gradations in composition to create interfaces was proposed by Li *et al.* [159], who employed a combination of electrospinning and soaking in SBF to generate a variation in the deposition of calcium phosphate-based minerals along the length of the electrospun scaffold. The proliferation and attachment of cells was varied along the scaffold length, with the greatest attachment and proliferation seen in those areas of maximum mineral deposition. The mechanical properties also showed a variation along the length of the scaffold, that appeared to depend on the mineral content.

The results obtained from the above-described studies in relation to interface tissue engineering, have provided further insight into the requirements for a scaffold when engineering the bone–ligament interface. The scaffolds fabricated for bone–ligament interface tissue engineering should exhibit variations in their structure, composition, and mechanical properties. In addition, they should be able to support the growth and differentiation of different cell types, and be conducive to so-called "heterotypic" and "homotypic" cellular communication. One important factor when designing scaffolds for bone–ligament interface tissue engineering is the formation of different, but properly integrated, phases without or with minimum discontinuities between them. However, other issues must be considered, perhaps the most important of which is a need to improve the present understanding of "structure–function" relationships at the bone–ligament interface. In particular, regeneration of the interface due to interactions between the different cell types at the interface have been proposed to control the "phenotypic changes" or the "trans-differentiation" of these cell types. Overall, an optimal engineering of the bone–ligament interface will play an important role in the effective transition of an *in vivo* tissue-engineered construct to its clinical application.

10.8
Interfacial Interactions in Nanocomposites for Bone Tissue Engineering

Since cellular responses are affected by the mechanical properties of the ECM, it is important to consider the mechanical properties of the nanocomposites fabricated for bone tissue engineering. Further, the mechanical properties and cellular responses of polymer nanocomposite materials are acutely influenced by the interface between the polymer and the filler (particle) [160–163], and may be studied by using characterization techniques such as FT-IR spectroscopy and atomic force microscopy (AFM), in addition to simulation studies.

10.8.1 Fourier Transform-Infrared (FT-IR) Spectroscopy

A typical FT-IR infrared spectrum is obtained by passing infrared radiation through a sample, and then determining what fraction of the radiation has been absorbed by the sample, at a certain energy. The band positions seen in the FT-IR spectrum of a sample represent the molecular signature of the vibrations of the bonds in that sample. The frequency of IR radiation that matches with the vibrational frequency of the atom's bonds in a sample, and the change in the dipole moment of a molecule during the vibration of bonds, represent the basic requirements for the absorption bands of a molecule to appear in the FT-IR spectrum. The nanocomposites used for bone tissue engineering are multicomponent mixtures of polymers and modified fillers. In the case of a nanocomposite, the shifts in band positions observed in the FT-IR spectrum are indicative of the interactions between the ions and functional groups at the polymer–filler interface. Many studies related to bone tissue engineering have involved the use of natural polymers with sequences similar to biological molecules, in which hydrogen bonding is extensive. Although, individually, hydrogen bonds are weak, their presence in large numbers (as in these types of polymer) will have a significant effect on their interactions with the fillers used to prepare the nanocomposites from the polymeric base.

As well as controlling mechanical properties, the interfacial interactions may also affect the rate of biodegradation, and this is an important issue for bone tissue engineering applications. Consequently, FT-IR studies may prove to be a valuable to obtain insights into these types of interaction, by using different sampling methods such as transmission, photoacoustic, reflectance, and microspectroscopy. The added advantage of using FT-IR spectroscopy to characterize the nanocomposites is an ability to examine surfaces by using reflectance techniques, such as attenuated total reflectance and specular reflectance. This is especially important when studying the role of surface chemistry in cellular responses. Previously, FT-IR spectroscopy has been used to characterize the interfaces in composites prepared for bone tissue engineering applications [15, 16, 94, 164, 165]. In one such study with chitosan nanocomposites containing MMT clay and HAp [15], FT-IR spectroscopy was used to examine the interfacial interactions that affected the nanomechanical properties of the nanocomposites. Likewise, in the case of chitosan/polygalacturonic acid/HAp nanocomposites, FT-IR spectroscopy was used to monitor the interfacial interactions that affected the lattice parameters of HAp [165]. Besides having the capability to characterize interfaces in polymer composites, FT-IR spectroscopy can also be used to characterize tissues grown on engineered scaffolds, in order to evaluate any differences between the molecular components of tissues grown naturally or *in vitro*. This may also be useful for developing an understanding of the bone mineralization process.

10.8.2 Atomic Force Microscopy

Atomic force microscopy is a nondestructive surface characterization technique that can be used for high-resolution surface imaging and, in combination with indentation, to determine the mechanical properties through nanoindentation, modulus mapping and nanodynamic mechanical analysis (nano-DMA). As discussed previously, cells respond favorably to nanoscale features, and therefore it is necessary to determine the topographic changes and mechanical properties at the nanoscale. AFM is capable of both assessments, and hence can be used to characterize the nanocomposites used for bone tissue engineering applications. The AFM nanoindentation technique is used to estimate mechanical properties such as elasticity and hardness at the nanoscale, whereas modulus mapping provides quantitative information regarding the variations in elastic modulus values on a surface at high spatial resolution. The use of AFM in studying the interfaces in nanocomposites has been demonstrated in some important studies [163, 166–168], from which quantitative data relating to nanomechanical properties were acquired. Such nanomechanical properties were attributed to interfacial interactions between the organic (polymer) and inorganic (HAp, MMT clay) components.

10.8.3 Modeling and Simulation Studies

The study of interfacial interactions in polymer nanocomposites or naturally occurring composite materials through modeling and simulation represents an emerging trend to improve the present understanding of the properties exhibited by this type of material, as well as aiding in materials design and selection. Modeling techniques such as molecular dynamics, steered molecular dynamics, discrete element methods and finite element methods can be used with biological systems such as bone and bone biomaterials. The molecular interfaces in bone biomaterials exhibit a variety of molecular interactions that have significant impacts on the mechanical response, degradation and bioactivity of the biomaterial (Figure 10.7). Molecular dynamics (MD) is a computational method used to calculate the time-dependent behavior of a molecular system; that is, the future positions and velocities of atoms based on their current positions and velocities. MD employs Newton's second law of motion to simulate the movements of individual atoms, and alters the intermolecular degrees of freedom that represent the changes in atomic position with time in small steps, so as to establish a minimum energy configuration. In steered MD, a molecular system is steered in a pathway of interest using an external pull through a spring of given stiffness. An external force is applied to the molecules to investigate their mechanical properties, and also to speed up those processes that are too slow to model; the atoms are also pulled at a constant force or a constant velocity. These techniques have been used to study the interactions between tropocollagen and HAp [169], polymers (polyacrylic acid, PCL) and HAp [170], protein–mineral in nacre [171, 172], polyacrylic acid and HAp

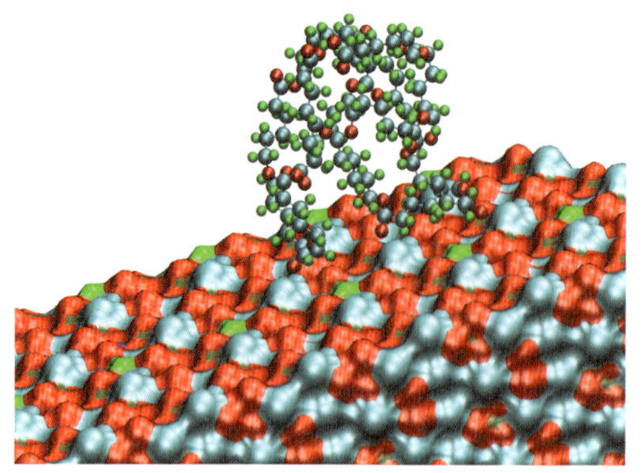

Figure 10.7 Molecular model showing the hydroxyapatite–polycaprolactone interface in hydroxyapatite–polycaprolactone nanocomposites used in bone tissue engineering.

[173–175], and collagen–HAp [176]. Other additional techniques used to study the interactions between collagen and HAp in bone are based on continuum micromechanics [177]. These studies endorsed the importance of interactions between the organic and inorganic phases of a composite material, based on its mechanical properties. Some of the studies that have involved the investigation of interactions between HAP and collagen (protein) in bone might prove useful when designing nanocomposites with adequate mechanical properties for bone tissue engineering applications.

10.9
Summary and Conclusions

This chapter has provided a comprehensive overview of the materials, fabrication methods and characterization techniques used in nanocomposite systems for bone tissue engineering. The preparation of scaffolds that mimic the structure of the ECM, and have satisfactory mechanical properties, biodegradation rates and porosity, represent the main challenges of bone tissue engineering. Bearing this in mind, it would appear that nanocomposite systems possess the potential to overcome these challenges; consequently, factors such as materials selection, processing routes and biocompatibility requirements associated with the preparation of nanocomposite scaffolds are at present the subject of many investigations. The importance of interfacial interactions in the nanocomposites produced for bone tissue engineering applications, and the methods used to study such interactions, have also been addressed. Indeed, the evaluation of interfacial interactions between the polymer–mineral and/or the polymer–filler in nanocomposites may help to improve the present understanding with regards to the design of scaffolds with effective mechanical properties. These interfacial interactions may also affect the conformation of the polymer chains in nanocomposites, as well as their properties, while the functional groups at the surface may affect the cellular response. Clearly, modeling techniques, and the information that they provide, will enhance bone tissue engineering efforts.

Although most challenges in bone tissue engineering relate to the design of appropriate scaffolds, other issues must be addressed. An example of this is the current limited understanding of the mechanisms involved in bone formation, and the growth rate or rate of formation of natural bone following a minor bone injury or the application of stress. Although most current studies involve the development of scaffolds based on new material systems and their biocompatibility, they do not address the importance of bone formation rate on synthetic scaffolds. Any improvement in the fundamental understanding of natural bone formation, and its formation rate, will help to predict bone formation rates on engineered scaffolds, thus increasing practical utility. Biodegradation is an important issue in bone tissue engineering, and in general the rate of bone regeneration should match that of scaffold biodegradation. Thus, any information regarding bone formation rates will aid in the preparation of scaffolds with appropriate biodegradation rates. In addition, the development of scaffolds with pore sizes sufficient for nutrient flows that are favorable for tissue growth has also been reported. However, it should be stressed that the *in vivo* environment is *dynamic*, and the flow rates of nutrient-containing fluids may vary in a dynamic environment. As the forces exerted by the fluids on the pore walls may affect cell behavior, it is important to quantitate such fluid flow through the pores of a scaffold system, using modeling techniques. Current advances in materials processing through the application of advanced techniques, combined with improved tools used for characterization and simulation purposes, will further aid the development of novel nanocomposite systems for bone tissue engineering.

References

1 Murugan, R. and Ramakrishna, S. (2005) Development of nanocomposites for bone grafting. *Composites Science and Technology*, **65** (15-16), 2385–406.

2 Dorozhkin, S.V. (2007) Calcium orthophosphates. *Journal of Materials Science*, **42** (4), 1061–95.

3 Boyle, W.J., Simonet, W.S. and Lacey, D.L. (2003) Osteoclast differentiation and activation. *Nature*, **423** (6937), 337–42.

4 Salgado, A.J., Coutinho, O.P. and Reis, R.L. (2004) Bone tissue engineering: state of the art and future trends. *Macromolecular Bioscience*, **4** (8), 743–65.

5 Langer, R. and Vacanti, J.P. (1993) Tissue engineering. *Science*, **260** (5110), 920–6.

6 Goldberg, M., Langer, R. and Jia, X.Q. (2007) Nanostructured materials for applications in drug delivery and tissue engineering. *Journal of Biomaterials Science – Polymer Edition*, **18** (3), 241–68.

7 Stevens, M.M. and George, J.H. (2005) Exploring and engineering the cell surface interface. *Science*, **310** (5751), 1135–8.

8 Murugan, R., Huang, Z.M., Yang, F. and Ramakrishna, S. (2007) Nanofibrous scaffold engineering using

electrospinning. *Journal of Nanoscience and Nanotechnology*, **7** (12), 4595–603.

9 Christenson, E.M., Anseth, K.S., van den Beucken, L., Chan, C.K., Ercan, B., Jansen, J.A. *et al.* (2007) Nanobiomaterial applications in orthopedics. *Journal of Orthopaedic Research*, **25** (1), 11–22.

10 Barnes, C.P., Sell, S.A., Boland, E.D., Simpson, D.G. and Bowlin, G.L. (2007) Nanofiber technology: designing the next generation of tissue engineering scaffolds. *Advanced Drug Delivery Reviews*, **59** (14), 1413–33.

11 Toh, Y.-C., Ng, S., Khong, Y.M., Zhang, X., Zhu, Y., Lin, P.-C. *et al.* (2006) Cellular responses to a nanofibrous environment. *Nano Today*, **1** (3), 34–43.

12 Place, E.S., Evans, N.D. and Stevens, M.M. (2009) Complexity in biomaterials for tissue engineering. *Nature Materials*, **8** (6), 457–70.

13 Chan, C.K., Kumar, T.S.S., Liao, S., Murugan, R., Ngiam, M. and Ramakrishman, S. (2006) Biomimetic nanocomposites for bone graft applications. *Nanomedicine*, **1** (2), 177–88.

14 Thein-Han, W.W. and Misra, R.D.K. (2009) Biomimetic chitosan-nanohydroxyapatite composite scaffolds for bone tissue engineering. *Acta Biomaterialia*, **5** (4), 1182–97.

15 Katti, K.S., Katti, D.R. and Dash, R. (2008) Synthesis and characterization of a novel chitosan/montmorillonite/hydroxyapatite nanocomposite for bone tissue engineering. *Biomedical Materials*, **3** (3), 12.

16 Cai, X., Tong, H., Shen, X., Chen, W., Yan, J. and Hu, J. Preparation and characterization of homogeneous chitosan-polylactic acid/hydroxyapatite nanocomposite for bone tissue engineering and evaluation of its mechanical properties. *Acta Biomaterialia*, **5** (7), 2693–703.

17 Huang, J., Lin, Y.W., Fu, X.W., Best, S.M., Brooks, R.A., Rushton, N. *et al.* (2007) Development of nano-sized hydroxyapatite reinforced composites for tissue engineering scaffolds. *Journal of Materials Science: Materials in Medicine*, **8** (11), 2151–7.

18 Kothapalli, C.R., Shaw, M.T. and Wei, M. (2005) Biodegradable HA-PLA 3-D porous scaffolds: effect of nano-sized filler content on scaffold properties. *Acta Biomaterialia*, **1** (6), 653–62.

19 Cool, S.M., Kenny, B., Wu, A., Nurcombe, V., Trau, M., Cassady, A.I. *et al.* (2007) Poly(3-hydroxybutyrate-*co*-3-hydroxyvalerate) composite biomaterials for bone tissue regeneration: *in vitro* performance assessed by osteoblast proliferation, osteoclast adhesion and resorption, and macrophage proinflammatory response. *Journal of Biomedical Materials Research, Part A*, **82A** (3), 599–610.

20 Lee, H.J., Kim, S.E., Choi, H.W., Kim, C.W., Kim, K.J. and Lee, S.C. (2007) The effect of surface-modified nano-hydroxyapatite on biocompatibility of poly([epsilon]-caprolactone)/hydroxyapatite nanocomposites. *European Polymer Journal*, **43** (5), 1602–8.

21 Lee, K.-W., Wang, S., Yaszemski, M.J. and Lu, L. (2008) Physical properties and cellular responses to crosslinkable poly(propylene fumarate)/hydroxyapatite nanocomposites. *Biomaterials*, **29** (19), 2839–48.

22 Sinha Ray, S. and Okamoto, M. (2003) Polymer/layered silicate nanocomposites: a review from preparation to processing. *Progress in Polymer Science*, **28** (11), 1539–641.

23 Viseras, C., Aguzzi, C., Cerezo, P. and Lopez-Galindo, A. (2007) Uses of clay minerals in semisolid health care and therapeutic products. *Applied Clay Science*, **36** (1–3), 37–50.

24 Lee, W.-F. and Fu, Y.-T. (2003) Effect of montmorillonite on the swelling behavior and drug-release behavior of nanocomposite hydrogels. *Journal of Applied Polymer Science*, **89** (13), 3652–60.

25 Dong, Y. and Feng, S.-S. (2005) Poly(D,L-lactide-*co*-glycolide)/montmorillonite nanoparticles for oral delivery of anticancer drugs. *Biomaterials*, **26** (30), 6068–76.

26 Zheng, J.P., Wang, C.Z., Wang, X.X., Wang, H.Y., Zhuang, H. and Yao, K.D. (2007) Preparation of biomimetic three-dimensional gelatin/montmorillonite-chitosan scaffold for tissue engineering. *Reactive and Functional Polymers*, **67** (9), 780–8.

27 Depan, D., Kumar, A.P. and Singh, R.P. (2009) Cell proliferation and controlled drug release studies of nanohybrids based on chitosan-g-lactic acid and montmorillonite. *Acta Biomaterialia*, **5** (1), 93–100.

28 Marras, S.I., Kladi, K.P., Tsivintzelis, I., Zuburtikudis, I. and Panayiotou, C. (2008) Biodegradable polymer nanocomposites: the role of nanoclays on the thermomechanical characteristics and the electrospun fibrous structure. *Acta Biomaterialia*, **4** (3), 756–65.

29 Shvedova, A.A., Kisin, E.R., Porter, D., Schulte, P., Kagan, V.E., Fadeel, B. *et al.* (2009) Mechanisms of pulmonary toxicity and medical applications of carbon nanotubes: two faces of Janus? *Pharmacology and Therapeutics*, **121** (2), 192–204.

30 Marianna, F. and Mukasa, B. (2008) Carbon nanotubes as functional excipients for nanomedicines: I. Pharmaceutical properties. *Nanomedicine: The Official Journal of the American Academy of Nanomedicine*, **4** (3), 173–82.

31 Simmons, T.J., Lee, S.H., Park, T.J., Hashim, D.P., Ajayan, P.M. and Linhardt, R.J. (2009) Antiseptic single wall carbon nanotube bandages. *Carbon*, **47** (6), 1561–4.

32 Wang, Y. and Li, J. A carbon nanotubes assisted strategy for insulin detection and insulin proteolysis assay. *Analytica Chimica Acta*, **650** (1), 49–53.

33 Shi, X.F., Sitharaman, B., Pham, Q.P., Liang, F., Wu, K., Billups, W.E. *et al.* (2007) Fabrication of porous ultra-short single-walled carbon nanotube nanocomposite scaffolds for bone tissue engineering. *Biomaterials*, **28** (28), 4078–90.

34 Shi, X.F., Hudson, J.L., Spicer, P.P., Tour, J.M., Krishnamoorti, R. and Mikos, A.G. (2005) Rheological behaviour and mechanical characterization of injectable poly(propylene fumarate)/single-walled carbon nanotube composites for bone tissue engineering. *Nanotechnology*, **16** (7), S531–8.

35 Sitharaman, B., Shi, X.F., Tran, L.A., Spicer, P.P., Rusakova, I., Wilson, L.J. *et al.* (2007) Injectable in situ cross-linkable nanocomposites of biodegradable polymers and carbon nanostructures for bone tissue engineering. *Journal of Biomaterials Science – Polymer Edition*, **18** (6), 655–71.

36 Sitharaman, B., Shi, X., Walboomers, X.F., Liao, H., Cuijpers, V., Wilson, L.J. *et al.* (2008) *In vivo* biocompatibility of ultra-short single-walled carbon nanotube/biodegradable polymer nanocomposites for bone tissue engineering. *Bone*, **43** (2), 362–70.

37 Abarrategi, A., Gutiérrez, M.C., Moreno-Vicente, C., Hortigüela, M.J., Ramos, V., López-Lacomba, J.L. *et al.* (2008) Multiwall carbon nanotube scaffolds for tissue engineering purposes. *Biomaterials*, **29** (1), 94–102.

38 Xu, J.L., Khor, K.A., Sui, J.J. and Chen, W.N. (2009) Preparation and characterization of a novel hydroxyapatite/carbon nanotubes composite and its interaction with osteoblast-like cells. *Materials Science and Engineering: C*, **29** (1), 44–9.

39 Balani, K., Anderson, R., Laha, T., Andara, M., Tercero, J., Crumpler, E. *et al.* (2007) Plasma-sprayed carbon nanotube reinforced hydroxyapatite coatings and their interaction with human osteoblasts *in vitro*. *Biomaterials*, **28** (4), 618–24.

40 Khang, D., Kim, S.Y., Liu-Snyder, P., Palmore, G.T.R., Durbin, S.M. and Webster, T.J. (2007) Enhanced fibronectin adsorption on carbon nanotube/poly(carbonate) urethane: independent role of surface nano-roughness and associated surface energy. *Biomaterials*, **28** (32), 4756–68.

41 Feng, J., Sui, J., Cai, W., Wan, J., Chakoli, A.N. and Gao, Z. (2008) Preparation and characterization of magnetic multi-walled carbon nanotubes-poly(L-lactide) composite. *Materials Science and Engineering: B*, **150** (3), 208–12.

42 Burg, K.J.L., Porter, S. and Kellam, J.F. (2000) Biomaterial developments for bone tissue engineering. *Biomaterials*, **21** (23), 2347–59.

43 Rezwan, K., Chen, Q.Z., Blaker, J.J. and Boccaccini, A.R. (2006) Biodegradable and bioactive porous polymer/inorganic composite scaffolds for bone tissue engineering. *Biomaterials*, **27** (18), 3413–31.

44 Padilla, S., Sánchez-Salcedo, S. and Vallet-Regí, M. (2007) Bioactive glass as precursor of designed-architecture scaffolds for tissue engineering. *Journal of Biomedical Materials Research, Part A*, **81A** (1), 224–32.

45 Jones, J.R., Ehrenfried, L.M. and Hench, L.L. (2006) Optimising bioactive glass scaffolds for bone tissue engineering. *Biomaterials*, **27** (7), 964–73.

46 Mahony, O. and Jones, J.R. (2008) Porous bioactive nanostructured scaffolds for bone regeneration: a sol-gel solution. *Nanomedicine*, **3** (2), 233–45.

47 Pereira, M.M. and Hench, L.L. (1996) Mechanisms of hydroxyapatite formation on porous gel-silica substrates. *Journal of Sol-Gel Science and Technology*, **7** (1), 59–68.

48 Li, R., Clark, A.E. and Hench, L.L. (1991) An investigation of bioactive glass powders by sol-gel processing. *Journal of Applied Biomaterials*, **2** (4), 231–9.

49 Sepulveda, P., Jones, J.R. and Hench, L.L. (2001) Characterization of melt-derived 45S5 and sol-gel-derived 58S bioactive glasses. *Journal of Biomedical Materials Research*, **58** (6), 734–40.

50 Hong, Z., Liu, A., Chen, L., Chen, X. and Jing, X. (2009) Preparation of bioactive glass ceramic nanoparticles by combination of sol-gel and coprecipitation method. *Journal of Non-Crystalline Solids*, **355** (6), 368–72.

51 Hong, Z., Reis, R.L. and Mano, J.F. (2008) Preparation and *in vitro* characterization of scaffolds of poly(L-lactic acid) containing bioactive glass ceramic nanoparticles. *Acta Biomaterialia*, **4** (5), 1297–306.

52 Liu, A., Hong, Z., Zhuang, X., Chen, X., Cui, Y., Liu, Y. *et al.* (2008) Surface modification of bioactive glass nanoparticles and the mechanical and biological properties of poly(L-lactide) composites. *Acta Biomaterialia*, **4** (4), 1005–15.

53 Couto, D.S., Hong, Z. and Mano, J.F. (2009) Development of bioactive and biodegradable chitosan-based injectable systems containing bioactive glass nanoparticles. *Acta Biomaterialia*, **5** (1), 115–23.

54 Mansur, H.S. and Costa, H.S. (2008) Nanostructured poly(vinyl alcohol)/bioactive glass and poly (vinyl alcohol)/chitosan/bioactive glass hybrid scaffolds for biomedical applications. *Chemical Engineering Journal*, **137** (1), 72–83.

55 Langer, R. (1997) Tissue engineering: a new field and its challenges. *Pharmaceutical Research*, **14** (7), 840–1.

56 Mano, J.F., Silva, G.A., Azevedo, H.S., Malafaya, P.B., Sousa, R.A., Silva, S.S. *et al.* (2007) Natural origin biodegradable systems in tissue engineering and regenerative medicine: present status and some moving trends. *Journal of The Royal Society Interface*, **4** (17), 999–1030.

57 Kikuchi, M., Matsumoto, H.N., Yamada, T., Koyama, Y., Takakuda, K. and Tanaka, J. (2004) Glutaraldehyde cross-linked hydroxyapatite/collagen self-organized nanocomposites. *Biomaterials*, **25** (1), 63–9.

58 Kim, H.W., Gu, H.J. and Lee, H.H. (2007) Microspheres of collagen-apatite nanocomposites with osteogenic potential for tissue engineering. *Tissue Engineering*, **13** (5), 965–73.

59 Venugopal, J., Low, S., Choon, A.T., Kumar, T.S.S. and Ramakrishna, S. (2008) Mineralization of osteoblasts with electrospun collagen/hydroxyapatite nanofibers. *Journal of Materials Science: Materials in Medicine*, **19** (5), 2039–46.

60 Lode, A., Bernhardt, A. and Gelinsky, M. (2008) Cultivation of human bone marrow stromal cells on three-dimensional scaffolds of mineralized collagen: influence of seeding density on colonization, proliferation and osteogenic differentiation. *Journal of Tissue Engineering and Regenerative Medicine*, **2** (7), 400–7.

61 Ngiam, M., Liao, S., Patil, A.J., Cheng, Z., Chan, C.K. and Ramakrishna, S. (2009) The fabrication of nano-hydroxyapatite on PLGA and PLGA/collagen nanofibrous composite scaffolds and their effects in osteoblastic behavior for bone tissue engineering. *Bone*, **45** (1), 4–16.

62 Sotome, S., Uemura, T., Kikuchi, M., Chen, J., Itoh, S., Tanaka, J. *et al* (2004) Synthesis and *in vivo* evaluation of a novel hydroxyapatite/collagen-alginate as a bone filler and a drug delivery carrier of bone morphogenetic protein. *Materials Science and Engineering C*, **24** (3), 341–7.

63 Tsai, S.W., Hsu, F.Y. and Chen, P.L. (2008) Beads of collagen-nanohydroxyapatite composites prepared by a biomimetic process and the effects of their surface texture on cellular behavior in MG63 osteoblast-like cells. *Acta Biomaterialia*, **4** (5), 1332–41.

64 Krayukhina, M.A. *et al.* (2008) Polyelectrolyte complexes of chitosan: formation, properties and applications. *Russian Chemical Reviews*, **77** (9), 799.

65 Zhang, Y., Venugopal, J.R., El-Turki, A., Ramakrishna, S., Su, B. and Lim, C.T. (2008) Electrospun biomimetic nanocomposite nanofibers of hydroxyapatite/chitosan for bone tissue engineering. *Biomaterials*, **29** (32), 4314–22.

66 Kong, L., Gao, Y., Lu, G., Gong, Y., Zhao, N. and Zhang, X. (2006) A study on the bioactivity of chitosan/nano-hydroxyapatite composite scaffolds for bone tissue engineering. *European Polymer Journal*, **42** (12), 3171–9.

67 Li, Z., Ramay, H.R., Hauch, K.D., Xiao, D. and Zhang, M. (2005) Chitosan-alginate hybrid scaffolds for bone tissue engineering. *Biomaterials*, **26** (18), 3919–28.

68 Verma, D., Katti, K.S. and Katti, D.R. (2009) Polyelectrolyte-complex nanostructured fibrous scaffolds for tissue engineering. *Materials Science and Engineering: C*, **29** (7), 2079–84.

69 Wan, Y.Z., Hong, L., Jia, S.R., Huang, Y., Zhu, Y., Wang, Y.L. *et al.* (2006) Synthesis and characterization of hydroxyapatite-bacterial cellulose nanocomposites. *Composites Science and Technology*, **66** (11–12), 1825–32.

70 Li, J., Wan, Y., Li, L., Liang, H. and Wang, J. (2009) Preparation and characterization of 2,3-dialdehyde bacterial cellulose for potential biodegradable tissue engineering scaffolds. *Materials Science and Engineering: C*, **29** (5), 1635–42.

71 Jiang, L., Li, Y., Wang, X., Zhang, L., Wen, J. and Gong, M. (2008) Preparation and properties of nano-hydroxyapatite/chitosan/carboxymethyl cellulose composite scaffold. *Carbohydrate Polymers*, **74** (3), 680–4.

72 Müller, F.A., Müller, L., Hofmann, I., Greil, P., Wenzel, M.M. and Staudenmaier, R. (2006) Cellulose-based scaffold materials for cartilage tissue engineering. *Biomaterials*, **27** (21), 3955–63.

73 Svensson, A., Nicklasson, E., Harrah, T., Panilaitis, B., Kaplan, D.L., Brittberg, M. *et al.* (2005) Bacterial cellulose as a potential scaffold for tissue engineering of cartilage. *Biomaterials*, **26** (4), 419–31.

74 Salgado, A.J., Gomes, M.E., Chou, A., Coutinho, O.P., Reis, R.L. and Hutmacher, D.W. (2002) Preliminary study on the adhesion and proliferation of human osteoblasts on starch-based scaffolds. *Materials Science and Engineering: C*, **20** (1–2), 27–33.

75 Marques, A.P. and Reis, R.L. (2005) Hydroxyapatite reinforcement of different starch-based polymers affects osteoblast-like cells adhesion/spreading and proliferation. *Materials Science and Engineering: C*, **25** (2), 215–29.

76 Santos, M.I., Fuchs, S., Gomes, M.E., Unger, R.E., Reis, R.L. and Kirkpatrick, C.J. (2007) Response of micro- and macrovascular endothelial cells to starch-based fiber meshes for bone tissue engineering. *Biomaterials*, **28** (2), 240–8.

77 Gomes, M.E., Godinho, J.S., Tchalamov, D., Cunha, A.M. and Reis, R.L. (2002) Alternative tissue engineering scaffolds based on starch: processing methodologies, morphology, degradation and mechanical properties. *Materials Science and Engineering: C*, **20** (1–2), 19–26.

78 Neves, N.M., Kouyumdzhiev, A. and Reis, R.L. (2005) The morphology, mechanical properties and ageing behavior of porous injection molded starch-based blends for tissue engineering scaffolding. *Materials Science and Engineering: C*, **25** (2), 195–200.

79 Ji, Y., Ghosh, K., Shu, X.Z., Li, B., Sokolov, J.C., Prestwich, G.D. *et al.* (2006) Electrospun three-dimensional hyaluronic acid nanofibrous scaffolds. *Biomaterials*, **27** (20), 3782–92.

80 Kim, J., Kim, I.S., Cho, T.H., Lee, K.B., Hwang, S.J., Tae, G. *et al.* (2007) Bone regeneration using hyaluronic acid-based hydrogel with bone morphogenic protein-2 and human mesenchymal stem cells. *Biomaterials*, **28** (10), 1830–7.

81 Lisignoli, G., Zini, N., Remiddi, G., Piacentini, A., Puggioli, A., Trimarchi, C. *et al.* (2001) Basic fibroblast growth factor enhances *in vitro* mineralization of rat bone marrow stromal cells grown on non-woven hyaluronic acid based polymer scaffold. *Biomaterials*, **22** (15), 2095–105.

82 Lisignoli, G., Fini, M., Giavaresi, G., Nicoli Aldini, N., Toneguzzi, S. and Facchini, A. (2002) Osteogenesis of large segmental radius defects enhanced by basic fibroblast growth factor activated bone marrow stromal cells grown on non-woven hyaluronic acid-based polymer scaffold. *Biomaterials*, **23** (4), 1043–51.

83 Lee, J.H., Park, T.G., Park, H.S., Lee, D.S., Lee, Y.K., Yoon, S.C. *et al.* (2003) Thermal and mechanical characteristics of poly(-lactic acid) nanocomposite scaffold. *Biomaterials*, **24** (16), 2773–8.

84 Liu, H., Slamovich, E.B. and Webster, T.J. (2006) Less harmful acidic degradation of poly(lactic-*co*-glycolic acid) bone tissue engineering scaffolds through titania nanoparticle addition. *International Journal of Nanomedicine*, **1** (4), 541–5.

85 Liu, H. *et al.* (2005) Increased osteoblast functions on nanophase titania dispersed in poly-lactic-*co*-glycolic acid composites. *Nanotechnology*, **16** (7), S601.

86 Nejati, E., Mirzadeh, H. and Zandi, M. (2008) Synthesis and characterization of nano-hydroxyapatite rods/poly(L-lactide acid) composite scaffolds for bone tissue engineering. *Composites Part A: Applied Science and Manufacturing*, **39** (10), 1589–96.

87 Serrano, M.C., Pagani, R., Vallet-Regí, M., Peña, J., Rámila, A., Izquierdo, I. *et al.* (2004) *In vitro* biocompatibility assessment of poly([var epsilon]-caprolactone) films using L929 mouse fibroblasts. *Biomaterials*, **25** (25), 5603–11.

88 Shor, L., Güçeri, S., Wen, X., Gandhi, M. and Sun, W. (2007) Fabrication of three-dimensional polycaprolactone/hydroxyapatite tissue scaffolds and osteoblast-scaffold interactions *in vitro*. *Biomaterials*, **28** (35), 5291–7.

89 Erisken, C., Kalyon, D.M. and Wang, H.J. (2008) Functionally graded electrospun polycaprolactone and beta-tricalcium phosphate nanocomposites for tissue engineering applications. *Biomaterials*, **29** (30), 4065–73.

90 Yang, Y., Yan, S.F., Li, X.X., Yin, J.B. and Chen, X.S. (2008) Preparation of poly(L-lactide)/surface-grafting silica nanocomposites and their surface induction of bone-like apatite in simulated body fluid. *Chemical Journal of Chinese Universities – Chinese*, **29** (11), 2294–8.

91 Karp, J., Shoicet, M. and Davies, J. (2003) Bone formation on two-dimensional poly(DL-lactide- co -glycolide) (PLGA) films and three-dimensional PLGA tissue engineering scaffolds *in vitro*. *Journal of Biomedical Materials Research*, **64a** (2), 388.

92 Jose, M., Thomas, V., Johnson, K., Dean, D. and Nyairo, E. (2009) Aligned PLGA/HA nanofibrous nanocomposite scaffolds for bone tissue engineering. *Acta Biomaterialia*, **5**, 305–15.

93 Chen, V.J. and Ma, P.X. (2004) Nano-fibrous poly(L-lactic acid) scaffolds with interconnected spherical macropores. *Biomaterials*, **25** (11), 2065.

94 Verma, D., Katti, K. and Katti, D. (2006) Bioactivity in in situ hydroxyapatite-polycaprolactone composites. *Journal of Biomedical Materials Research Part A*, **78A** (4), 772–80.

95 Lei, Y., Rai, B., Ho, K.H. and Teoh, S.H. (2007) *In vitro* degradation of novel bioactive polycaprolactone–20% tricalcium phosphate composite scaffolds for bone engineering. *Materials Science and Engineering: C*, **27** (2), 293–8.

96 Krogman, N.R., Weikel, A.L., Kristhart, K.A., Nukavarapu, S.P., Deng, M., Nair, L.S. *et al.* (2009) The influence of side group modification in polyphosphazenes on hydrolysis and cell adhesion of blends with PLGA. *Biomaterials*, **30** (17), 3035–41.

97 Lee, Y.K. (2005) Design parameters of polymers for tissue engineering applications. *Macromolecular Research*, **13** (4), 277–84.

98 Bhattacharyya, S., Kumbar, S.G., Khan, Y.M., Nair, L.S., Singh, A., Krogman, N.R. *et al.* (2009) Biodegradable polyphosphazene-nanohydroxyapatite composite nanofibers: scaffolds for bone tissue engineering. *Journal of Biomedical Nanotechnology*, **5** (1), 69–75.

99 Nukavarapu, S.P., Kumbar, S.G., Brown, J.L., Krogman, N.R., Weikel, A.L., Hindenlang, M.D. *et al.* (2008) Polyphosphazene/nano-hydroxyapatite composite microsphere scaffolds for bone tissue engineering. *Biomacromolecules*, **9** (7), 1818–25.

100 Wang, S., Lu, L. and Yaszemski, M.J. (2006) Bone-tissue-engineering material poly(propylene fumarate): correlation between molecular weight, chain dimensions, and physical properties. *Biomacromolecules*, **7** (6), 1976–82.

101 Horch, R.A., Shahid, N., Mistry, A.S., Timmer, M.D., Mikos, A.G. and Barron, A.R. (2004) Nanoreinforcernent of poly(propylene fumarate)-based networks with surface modified alumoxane nanoparticles for bone tissue engineering. *Biomacromolecules*, **5** (5), 1990–8.

102 Mistry, A.S., Cheng, S.H., Yeh, T., Christenson, E., Jansen, J.A. and Mikos, A.G. (2009) Fabrication and *in vitro* degradation of porous fumarate-based polymer/alumoxane nanocomposite scaffolds for bone tissue engineering. *Journal of Biomedical Materials Research, Part A*, **89A** (1), 68–79.

103 Shi, X.F., Hudson, J.L., Spicer, P.P., Tour, J.M., Krishnamoorti, R. and Mikos, A.G. (2006) Injectable nanocomposites of single-walled carbon nanotubes and biodegradable polymers for bone tissue engineering. *Biomacromolecules*, **7** (7), 2237–42.

104 Wang, S., Kempen, D.H.R., Yaszemski, M.J. and Lu, L. (2009) The roles of matrix polymer crystallinity and hydroxyapatite nanoparticles in modulating material properties of photo-crosslinked composites and bone marrow stromal cell responses. *Biomaterials*, **30** (20), 3359–70.

105 Cai, Z.-Y., Yang, D.-A., Zhang, N., Ji, C.-G., Zhu, L. and Zhang, T. (2009) Poly(propylene fumarate)/(calcium sulphate/[beta]-tricalcium phosphate) composites: preparation, characterization and *in vitro* degradation. *Acta Biomaterialia*, **5** (2), 628–35.

106 Boissard, C.I.R., Bourban, P.E., Tami, A.E., Alini, M. and Eglin, D. (2009) Nanohydroxyapatite/poly(ester urethane) scaffold for bone tissue engineering. *Acta Biomaterialia*, **5** (9), 3316–27.

107 Zhao, C.-X. and Zhang, W.-D. (2008) Preparation of waterborne polyurethane nanocomposites: polymerization from functionalized hydroxyapatite. *European Polymer Journal*, **44** (7), 1988–95.

108 Dong, Z., Li, Y. and Zou, Q. (2009) Degradation and biocompatibility of porous nano-hydroxyapatite/polyurethane composite scaffold for bone tissue engineering. *Applied Surface Science*, **255** (12), 6087–91.

109 Zanetta, M., Quirici, N., Demarosi, F., Tanzi, M.C., Rimondini, L. and Farè, S. (2009) Ability of polyurethane foams to support cell proliferation and the differentiation of MSCs into osteoblasts. *Acta Biomaterialia*, **5** (4), 1126–36.

110 Srivastava, V.K., Rastogi, A., Goel, S.C. and Chukowry, S.K. (2007) Implantation of tricalcium phosphate-polyvinyl alcohol-filled carbon fibre reinforced polyester resin composites into bone marrow of rabbits. *Materials Science and Engineering: A*, **448** (1–2), 335–9.

111 Huang, Z.-M., Zhang, Y.Z., Kotaki, M. and Ramakrishna, S. (2003) A review on polymer nanofibers by electrospinning and their applications in nanocomposites. *Composites Science and Technology*, **63** (15), 2223–53.

112 Desai, K., Kit, K., Li, J. and Zivanovic, S. (2008) Morphological and surface properties of electrospun chitosan nanofibers. *Biomacromolecules*, **9**, 1000–6.

113 Geng, X., Kwon, O.-H. and Jang, J. (2005) Electrospinning of chitosan dissolved in concentrated acetic acid solution. *Biomaterials*, **26** (27), 5427–32.

114 Venugopal, J., Vadgama, P., Kumar, T.S.S. and Ramakrishna, S. (2007) Biocomposite nanofibres and osteoblasts for bone tissue engineering. *Nanotechnology*, **18** (5), Article no. 055101.

115 Shalumon, K.T., Binulal, N.S., Selvamurugan, N., Nair, S.V., Menon, D., Furuike, T. *et al.* (2009) Electrospinning of carboxymethyl chitin/poly(vinyl alcohol) nanofibrous scaffolds for tissue engineering applications. *Carbohydrate Polymers*, **77** (4), 863–9.

116 Li, C., Vepari, C., Jin, H.-J., Kim, H.J. and Kaplan, D.L. (2006) Electrospun silk-BMP-2 scaffolds for bone tissue engineering. *Biomaterials*, **27** (16), 3115–24.

117 Schneider, O.D., Weber, F., Brunner, T.J., Loher, S., Ehrbar, M., Schmidlin, P.R. *et al.* (2009) *In vivo* and *in vitro* evaluation of flexible, cottonwool-like nanocomposites as bone substitute material for complex defects. *Acta Biomaterialia*, **5** (5), 1775–84.

118 Kim, H.D., Bae, E.H., Kwon, I.C., Pal, R.R., Nam, J.D. and Lee, D.S. (2004) Effect of PEG-PLLA diblock copolymer on macroporous PLLA scaffolds by thermally induced phase separation. *Biomaterials*, **25** (12), 2319–29.

119 Cai, Q., Yang, J., Bei, J. and Wang, S. (2002) A novel porous cells scaffold made of polylactide-dextran blend by combining phase-separation and particle-leaching techniques. *Biomaterials*, **23** (23), 4483–92.

120 He, L., Zhang, Y., Zeng, X., Quan, D., Liao, S., Zeng, Y. *et al.* (2009) Fabrication and characterization of poly(L-lactic acid) 3D nanofibrous scaffolds with controlled architecture by liquid-liquid phase separation from a ternary polymer-solvent system. *Polymer*, **50** (16), 4128–38.

121 Hua, F.J., Park, T.G. and Lee, D.S. (2003) A facile preparation of highly interconnected macroporous poly(L-lactic acid-*co*-glycolic acid) (PLGA) scaffolds by liquid-liquid phase separation of a PLGA-dioxane-water ternary system. *Polymer*, **44** (6), 1911–20.

122 Jack, K.S., Velayudhan, S., Luckman, P., Trau, M., Grøndahl, L. and Cooper-White, J. (2009) The fabrication and characterization of biodegradable HA/PHBV nanoparticle-polymer composite scaffolds. *Acta Biomaterialia*, **5** (7), 2657–67.

123 Zhao, F., Yin, Y., Lu, W.W., Leong, J.C., Zhang, W., Zhang, J. *et al.* (2002) Preparation and histological evaluation of biomimetic three-dimensional hydroxyapatite/chitosan-gelatin network composite scaffolds. *Biomaterials*, **23** (15), 3227–34.

124 Kim, J.-W., Taki, K., Nagamine, S. and Ohshima, M. (2008) Preparation of poly(L-lactic acid) honeycomb monolith structure by unidirectional freezing and freeze-drying. *Chemical Engineering Science*, **63** (15), 3858–63.

125 Ho, M.-H., Kuo, P.-Y., Hsieh, H.-J., Hsien T-Y, Hou L.-T., Lai, J.-Y. *et al.* (2004) Preparation of porous scaffolds by using freeze-extraction and freeze-gelation methods. *Biomaterials*, **25** (1), 129–38.

126 Wu, X., Liu, Y., Li, X., Wen, P., Zhang, Y., Long, Y. *et al.* (2010) Preparation of aligned porous gelatin scaffolds by unidirectional freeze-drying method. *Acta Biomaterialia*, **6** (3), 1167–77.

127 Yuan, N.-Y., Lin, Y.-A., Ho, M.-H., Wang, D.-M., Lai, J.-Y. and Hsieh, H.-J. (2009) Effects of the cooling mode on the structure and strength of porous scaffolds made of chitosan, alginate, and carboxymethyl cellulose by the freeze-gelation method. *Carbohydrate Polymers*, **78** (2), 349–56.

128 Deschamps, A.A., Claase, M.B., Sleijster, W.J., de Bruijn, J.D., Grijpma, D.W. and Feijen, J. (2002) Design of segmented poly(ether ester) materials and structures for the tissue engineering of bone. *Journal of Controlled Release*, **78** (1–3), 175–86.

129 Garcia-Fuentes, M., Meinel, A.J., Hilbe, M., Meinel, L. and Merkle, H.P. (2009)

Silk fibroin/hyaluronan scaffolds for human mesenchymal stem cell culture in tissue engineering. *Biomaterials*, **30** (28), 5068–76.

130 Mandal, B.B. and Kundu, S.C. (2009) Cell proliferation and migration in silk fibroin 3D scaffolds. *Biomaterials*, **30** (15), 2956–65.

131 Wei, G. and Ma, P.X. (2004) Structure and properties of nano-hydroxyapatite/polymer composite scaffolds for bone tissue engineering. *Biomaterials*, **25** (19), 4749–57.

132 Yunoki, S., Marukawa, E., Ikoma, T., Sotome, S., Fan, H.S., Zhang, X.D. *et al* (2007) Effect of collagen fibril formation on bioresorbability of hydroxyapatite/collagen composites. *Journal of Materials Science: Materials in Medicine*, **18** (11), 2179–83.

133 Oliveira, J.M., Rodrigues, M.T., Silva, S.S., Malafaya, P.B., Gomes, M.E., Viegas, C.A. *et al.* (2006) Novel hydroxyapatite/chitosan bilayered scaffold for osteochondral tissue-engineering applications: scaffold design and its performance when seeded with goat bone marrow stromal cells. *Biomaterials*, **27** (36), 6123–37.

134 Kim, H.W., Kim, H.E. and Salih, V. (2005) Stimulation of osteoblast responses to biomimetic nanocomposites of gelatin-hydroxyapatite for tissue engineering scaffolds. *Biomaterials*, **26** (25), 5221–30.

135 Landi, E., Valentini, F. and Tampieri, A. (2008) Porous hydroxyapatite/gelatin scaffolds with ice-designed channel-like porosity for biomedical applications. *Acta Biomaterialia*, **4** (6), 1620–6.

136 Maquet, V., Boccaccini, A.R., Pravata, L., Notingher, I. and Jérôme, R. (2004) Porous poly([alpha]-hydroxyacid)/Bioglass® composite scaffolds for bone tissue engineering. I: preparation and *in vitro* characterisation. *Biomaterials*, **25** (18), 4185–94.

137 Liu, X., Smith, L.A., Hu, J. and Ma, P.X. (2009) Biomimetic nanofibrous gelatin/apatite composite scaffolds for bone tissue engineering. *Biomaterials*, **30** (12), 2252–8.

138 Gong, Y., Zhou, Q., Gao, C. and Shen, J. (2007) *In vitro* and *in vivo* degradability and cytocompatibility of poly(L-lactic acid) scaffold fabricated by a gelatin particle leaching method. *Acta Biomaterialia*, **3** (4), 531–40.

139 Hou, Q., Grijpma, D.W. and Feijen, J. (2003) Porous polymeric structures for tissue engineering prepared by a coagulation, compression moulding and salt leaching technique. *Biomaterials*, **24** (11), 1937–47.

140 Oh, S.H., Kang, S.G., Kim, E.S., Cho, S.H. and Lee, J.H. (2003) Fabrication and characterization of hydrophilic poly(lactic-*co*-glycolic acid)/poly(vinyl alcohol) blend cell scaffolds by melt-molding particulate-leaching method. *Biomaterials*, **24** (22), 4011–21.

141 Reignier, J. and Huneault, M.A. (2006) Preparation of interconnected poly([epsilon]-caprolactone) porous scaffolds by a combination of polymer and salt particulate leaching. *Polymer*, **47** (13), 4703–17.

142 Kim, S.-S., Sun Park, M., Jeon, O., Yong Choi, C. and Kim, B.-S. (2006) Poly(lactide-*co*-glycolide)/hydroxyapatite composite scaffolds for bone tissue engineering. *Biomaterials*, **27** (8), 1399–409.

143 Kyle, S., Aggeli, A., Ingham, E. and McPherson, M.J. (2009) Production of self-assembling biomaterials for tissue engineering. *Trends in Biotechnology*, **27** (7), 423–33.

144 Ma, P.X. (2008) Biomimetic materials for tissue engineering. *Advanced Drug Delivery Reviews*, **60** (2), 184–98.

145 Smith, L.A. and Ma, P.X. (2004) Nano-fibrous scaffolds for tissue engineering. *Colloids and Surfaces B: Biointerfaces*, **39** (3), 125–31.

146 Ratner, B., Hoffmann, A., Schoen, F. and Lemons, J. (2004) *An Introduction to Materials in Medicine*, 2nd edn, Elsevier, Academic Press.

147 Rickert, D., Lendlein, A., Peters, I., Moses, M. and Franke, R.-P. (2006) Biocompatibility testing of novel multifunctional polymeric biomaterials for tissue engineering applications in head and neck surgery: an overview. *European Archives of Oto-Rhino-Laryngology*, **263** (3), 215–22.

148 Bohner, M. and Lemaitre, J. (2009) Can bioactivity be tested *in vitro* with SBF solution? *Biomaterials*, **30** (12), 2175–9.

149 Ratner, B., Hoffmann, A., Schoen, F. and Lemons, J. (1996) *An Introduction to Materials in Medicine*, 1st edn, Academic Press, San Diego, CA.

150 Zhu, Y., Wu, C., Ramaswamy, Y., Kockrick, E., Simon, P., Kaskel, S. *et al.* (2008) Preparation, characterization and *in vitro* bioactivity of mesoporous bioactive glasses (MBGs) scaffolds for bone tissue engineering. *Microporous and Mesoporous Materials*, **112** (1–3), 494–503.

151 Maeda, Y., Jayakumar, R., Nagahama, H., Furuike, T. and Tamura, H. (2008) Synthesis, characterization and bioactivity studies of novel [beta]-chitin scaffolds for tissue-engineering applications. *International Journal of Biological Macromolecules*, **42** (5), 463–7.

152 Torres, F.G., Nazhat, S.N., Sheikh, Md, Fadzullah, S.H., Maquet, V. and Boccaccini, A.R. (2007) Mechanical properties and bioactivity of porous PLGA/TiO_2 nanoparticle-filled composites for tissue engineering scaffolds. *Composites Science and Technology*, **67** (6), 1139–47.

153 Shen, H., Hu, X., Yang, F., Bei, J. and Wang, S. (2009) The bioactivity of rhBMP-2 immobilized poly(lactide-*co*-glycolide) scaffolds. *Biomaterials*, **30** (18), 3150–7.

154 Yang, P. and Temenoff, J. (2009) Engineering orthopedic tissue interfaces. *Tissue Engineering: Part B*, **15** (2), 127–41.

155 Moffat, K., Sun, W., Pena, P., Chahine, N., Doty, S., Ateshian, G. *et al.* (2008) Characterization of the structure–function relationship at the ligament-to-bone interface. *Proceedings of the National Academy of Sciences of the United States of America*, **105** (23), 7947–52.

156 Phillips, J., Burns, K., Doux, J., Guldberg, R. and Garcia, A. (2008) Engineering graded tissue interfaces. *Proceedings of the National Academy of Sciences of the United States of America*, **105** (34), 12170–5.

157 Splazzi, J., Doty, S., Moffat, K., Levine, W. and Lu, H. (2008) Development of controlled matrix heterogeneity on a triphasic scaffold for orthopedic interface tissue engineering. *Tissue Engineering*, **12**, 3947–508.

158 Spalazzi, J., Dagher, E., Doty, S., Guo, E., Rodeo, S. and Lu, H. (2008) *In vivo* evaluation of a multiphased scaffold designed for orthopaedic interface tissue engineering and soft tissue-to-bone integration. *Journal of Biomedical Materials Research, Part A*, **86A**, 1–12.

159 Li, X., Xie, J., Lipner, J., Yuan, X., Thomopoulos, S. and Xia, Y. (2009) Nanofiber scaffolds with gradations in mineral content for mimicking the tendon-to-bone insertion site. *Nano Letters*, **9**, 2763–8.

160 Khanna, R., Katti, K. and Katti, D. (2009) *Journal of Engineering Mechanics*, **35**, 468–78.

161 Misra, S., Valappil, S., Roy, I. and Boccaccini, A. (2006) Polyhydroxyalkanoate (PHA)/inorganic phase composites for tissue engineering applications. *Biomacromolecules*, **7** (8), 2249–58.

162 Ribeiro, R., Ganguly, P., Darensbourg, D., Usta, M., Ucisik, A., Liang *et al.* (2007) Biomimetic study of a polymeric composite material for joint repair applications. *Journal of Materials Research*, **22**, 1632–9.

163 Verma, D., Katti, K.S., Katti, D.R. and Mohanty, B. (2008) Mechanical response and multilevel structure of biomimetic hydroxyapatite/polygalacturonic/chitosan nanocomposites. *Materials Science and Engineering: C*, **28** (3), 399–405.

164 Verma, D., Katti, K.S. and Katti, D.R. (2008) Effect of biopolymers on structure of hydroxyapatite and interfacial interactions in biomimetically synthesized hydroxyapatite/biopolymer nanocomposites. *Annals of Biomedical Engineering*, **36** (6), 1024–32.

165 Verma, D., Katti, K. and Katti, D. (2006) Experimental investigation of interfaces in hydroxyapatite/polyacrylic acid/polycaprolactone composites using photoacoustic FTIR spectroscopy. *Journal of Biomedical Materials Research, Part A*, **77A** (1), 59–66.

166 Sikdar, D., Katti, D., Katti, K. and Mohanty, B. (2007) Effect of organic modifiers on dynamic and static nanomechanical properties and crystallinity of intercalated clay–polycaprolactam nanocomposites.

Journal of Applied Polymer Science, **105**, 790–802.

167 Sikdar, D., Katti, D., Katti, K. and Mohnaty, B. (2009) Influence of backbone chain length and functional groups of organic modifiers on crystallinity and nanomechanical properties of intercalated clay-polycaprolactam nanocomposites. *International Journal of Nanotechnology*, **6** (5/6), 468–92.

168 Mohanty, B., Katti, K.S. and Katti, D.R. Experimental investigation of nanomechanics of the mineral-protein interface in nacre. *Mechanics Research Communications*, **35** (1–2), 17–23.

169 Bhowmick, R., Katti, K. and Katti, D. (2009) Mechanisms of load-deformation behavior of molecular collagen in hydroxyapatite-tropocollagen molecular system: steered molecular dynamics study. *Journal of Engineering Mechanics*, **135** (5), 413–21.

170 Bhowmick, R., Katti, K. and Katti, D. (2009) Molecular interactions of degradable and non-degradable polymers with hydroxyapatite influence mechanics of polymer-hydroxyapatite nanocomposite biomaterials. *International Journal of Nanotechnology*, **6** (5/6), 511–29.

171 Ghosh, P., Katti, D. and Katti, K. (2008) Mineral and protein-bound water and latching action control mechanical behavior at protein-mineral interfaces in biological nanocomposites. *Journal of Nanomaterials*, **2008**, Article no. 562973.

172 Ghosh, P., Katti, D. and Katti, K. (2007) Mineral proximity influences mechanical response of proteins in biological mineral–protein hybrid systems. *Biomacromolecules*, **8**, 851–6.

173 Bhowmick, R., Katti, K. and Katti, D. (2008) Influence of mineral on the load deformation behavior of polymer in hydroxyapatite-polyacrylic acid nanocomposite biomaterials: a steered molecular dynamics study. *Journal of Nanoscience and Nanotechnology*, **8**, 2075–84.

174 Bhowmik, R., Katti, K.S., Verma, D. and Katti, D.R. (2007) Probing molecular interactions in bone biomaterials: through molecular dynamics and Fourier transform infrared spectroscopy. *Materials Science and Engineering: C*, **27** (3), 352–71.

175 Bhowmik, R., Katti, K.S. and Katti, D. (2007) Molecular dynamics simulation of hydroxyapatite–polyacrylic acid interfaces. *Polymer*, **48** (2), 664–74.

176 Bhowmik, R., Katti, K. and Katti, D. (2007) Mechanics of molecular collagen is influenced by hydroxyapatite in natural bone. *Journal of Materials Science*, **42** (21), 8795–803.

177 Fritsch, A. and Hellmich, C. (2007) 'Universal' microstructural patterns in cortical and trabecular, extracellular and extravascular bone materials: micromechanics-based prediction of anisotropic elasticity. *Journal of Theoretical Biology*, **244** (4), 597–620.

11
Nanocomposites for Tissue Engineering

Michael Gelinsky and Sascha Heinemann

11.1
Introduction

The healing of a tissue defect can, in principle, be achieved in five possible ways: (i) by spontaneous (self-) healing; (ii) by the transplantation of autologous tissue from another site; (iii) by the implantation of a cell-free biomaterial; (iv) through cell therapy; and (v) by using a tissue engineering approach. Selection of the optimal method depends on the tissue involved, the defect site, and the size and specific self-healing capacity, which might be reduced due to older age, to the presence of systemic disease (such as osteoporosis in the case of a bone defect), or to the physical state after radiotherapy. Whereas, a small bone defect in an otherwise healthy and young patient will require only suitable mechanical stabilization for easy and rapid self-healing, a large lesion of the articular cartilage will hardly heal without the use of autologous cells, that have been preferentially pre-seeded and pre-cultivated in a scaffold material. Although the transplantation of autologous tissues (e.g., bone from the iliac crest) is still seen as the "gold standard" for critical size defects [1], the availability of such materials is very limited, and the grafting process always involves an additional surgical procedure.

The term tissue engineering is defined as the development of artificial constructs, consisting of a matrix (scaffold) and living cells for the regeneration of a tissue by combining methods from materials science and life sciences [2]. To meet these requirements, an interdisciplinary field has developed during the past few decades which combines methods from engineering, biology, and medicine. To discriminate the tissue engineering approach from cell therapies, it is necessary to focus on the tissue differentiation processes. The term "tissue engineering" should be used only when at least the first steps towards a tissue-specific differentiation take place in the cell culture laboratory (*in vitro*) – when the cells have begun to communicate and interact with each other, and to synthesize an extracellular matrix (ECM). However, when the suspended cells are implanted into a defect site, or seeded onto a scaffold material directly before implantation, tissue differentiation will occur only after implantation (*in vivo*); such an approach can

Nanomaterials for the Life Sciences Vol.8: Nanocomposites. Edited by Challa S. S. R. Kumar

ISBN: 978-3-527-32168-1

be termed cell therapy. An example of the latter is the well-known autologous chondrocyte transplantation (ACT), in which suspended cells are implanted into a defect of articular cartilage [3]. The expression "*in vivo* tissue engineering" is in fact a contradiction in terms, and should not be used because it cannot be distinguished clearly from normal healing processes which start after the implantation of a biocompatible, porous material. A novel approach aims to create functionalized biomaterials that are intended to attract cells (which are advantageous to the healing process) after implantation, by generating concentration gradients of chemoattractants. An example of this strategy, based on the chemoattractive effect of stroma-derived factor 1α (SDF-1α) towards mesenchymal stem cells (MSC), was recently demonstrated by Thieme and coworkers [4]. The general aim here is to concentrate the cells necessary for the healing process directly into the defect site, but without any pre-seeding and cultivation of cells on the scaffold material *in vitro*. This interesting approach can be termed "*in situ* tissue engineering."

In this chapter, attention will be focused on scaffolds for tissue engineering in the original sense, although most of the materials described could also be used as implant materials or matrices for cell therapies or, with a suitable functionalization, for *in situ* tissue engineering applications. It is becoming clear, in this rapidly developing area of research, that nanocomposites are better suited to the development of scaffolds than monophasic materials. In contrast to the many recent reviews on nanocomposite scaffolds, this chapter will provide an exhaustive overview of various types of nanocomposite used for tissue engineering scaffolds, rather than of those biomaterials that cannot serve as matrices for cell seeding and cultivation. Details of scaffold requirements for successful tissue engineering are provided, together with an explanation of natural nanocomposites such as ECMs and their use as scaffolds, and a description of currently available synthetic nanocomposite scaffolds.

11.1.1
Requirements for Successful Tissue Engineering

The most important property of any biomaterial that is to be used as a scaffold in tissue engineering is its biodegradability (including physiological degradation and cellular resorption), so that a new healthy tissue can ultimately be formed, leading to the complete healing of a defect. Natural (healthy) tissues are always better adapted to the local situation than any artificial biomaterial or cell–matrix construct, and it is for this reason that any synthetic material must, finally, totally disappear so as not to disturb biological tissue formation and remodeling. Hence, details will be provided only of biodegradable nanocomposites, despite some non-degradable materials (e.g., metal alloys, ceramics) often being required for mechanical stabilization or fixation purposes.

The requirements for a material to be applied as a scaffold in tissue engineering are not only manifold but also challenging, due to the complex and sensitive host system and differences between tissues. The basic requirements, independent of the host tissue and implantation site, include:

- **Biocompatibility:** This fundamental requirement characterizes the ability of a material to perform with an appropriate host response in a specific *in vivo* application [5]. The scaffold material and the surrounding host tissue should coexist, without having any undesirable effect on each other: the material must not elicit any unresolved inflammatory response, immunogenicity, cytotoxicity, irritation, allergenic, or carcinogenic effects.
- **Biodegradability:** The material must degrade, to be replaced by newly formed tissue, during the healing process [6]. The degradation may be due to a dissolution process, an enzymatic digestion or active resorption, and carried out by cells (such as osteoclasts in the case of a material replacing bone). During biodegradation, no toxic or otherwise harmful degradation products should be produced, and the scaffold material should disappear within a similar time span to the new tissue being formed [7].
- **Porosity:** During the *in vitro* cultivation time the scaffold must allow effective and homogeneous cell seeding, in addition to supplies of oxygen and nutrients. After implantation, a rapid vascularization (ingrowth of blood capillaries) of the whole construct must be possible. Although the most suitable parameters remain the subject of much controversy, it is generally accepted that the pore system must be interconnecting, that the overall porosity should be about 90%, and that the pore diameters should be at least 100 μm [8]. Exceptions to this are the scaffold used to engineer epithelial or endothelial tissues, which may be membrane-like but must *not* be porous.
- **Mechanical properties:** The scaffold material must be biomechanically compatible with the host tissue, and must not collapse during the patient's normal activities. In the case of load-bearing implantation sites (e.g., in large bone defects), this requirement can hardly be fulfilled, especially in combination with the necessary high porosity. Consequently, at such sites the tissue engineering scaffolds must be implanted together with nondegradable devices such as metal plates, nails or screws, which can be removed when the defect has healed sufficiently [9].
- **Biointegration:** A bioadhesive contact between scaffold and host tissue must be established, allowing a rapid integration and cellular, as well as vascular, ingrowth [10].
- **Easy manufacture and handling:** The manufacture of the scaffold should be simple, and sterilization of the surface and the volume of the material both possible and safe. In addition, handling of the material, with regards not only to cell seeding but also to implantation, must be easy [11].
- **Cost-effective production:** The production of a tissue engineering scaffold must be possible in a cost-effective manner. The up-scaling of all processes should be facile, and allow adaptation to a defect- or even patient-specific size and shape [9].

The reasons why it is difficult for monophasic materials to fulfill all of these conditions, and why composites (especially nanocomposites) are increasingly being investigated for use as scaffolds for tissue engineering, are outlined in the following sections. Mother Nature often makes use of highly organized nanocomposites, an example being the bone ECM, a nanocomposite that consists of fibrous protein collagen type I and the nanocrystalline calcium phosphate mineral phase, hydroxyapatite (HAp).

The methods used to prepare porous three-dimensional (3-D) scaffolds do not differ between conventional monophasic materials and nanocomposites, and consequently will not be reviewed in this chapter. Whilst the basic methods have been described by Chevalier *et al.* [12], more general considerations regarding the preferred morphology and architecture of tissue engineering scaffolds have been provided by Bonfield [13].

11.1.2 Composites and Nanocomposites

11.1.2.1 Composites

It is well-recognized that, whilst many demanding applications cannot be satisfied by monophasic materials, defined mixtures can demonstrate useful combinations of properties, and this has led to the development of *composites*. Originally, composites were created for use as constructional and technical material, but subsequently found their way into biomedical materials research and scaffold developments for tissue engineering applications.

Although a universal definition of the term "composite" does not exist, in the field of biomaterials composites are generally described as follows [14, 15]:

- A composite is a combination of two or more heterogeneous materials.
- It is a multiphase material, built up by combining materials in which the phases differ in both composition and form.
- The phases of the composite retain their identities and properties, and are bonded, which is why an interface is maintained between them.
- A composite provides improved specific or synergistic characteristics that are not obtainable by any of the original phases alone.

In general, four types of composite can be distinguished: (i) fibrous composites, where the fibers are in a matrix; (ii) laminar composites, in which the phases are in layers; (iii) particulate composites, where the particles or flakes are in a matrix); and (iv) hybrid composites, which are combinations of any of the above.

Most composites applicable for the manufacture of tissue engineering scaffolds consist of a structural phase (mostly inorganic particles, whiskers, fibers, lamellae or meshes), embedded in a continuous matrix phase (mostly an organic substance). The function of the structural phase is usually to improve the strength of the composite, whereas the matrix phase acts as a binder for the inorganic building blocks and can also provide elasticity and ductility. Decreasing the size of the basic units of the incorporated phase to the level of the basic units of the matrix phase will enhance the homogeneity of the composite and its adjustability.

11.1.2.2 Nanocomposites

A composite becomes a *nanocomposite* when at least one of the components is of nanoscale size [7], where the dimension of a basic unit is ≤100 nm in at least one direction. Decreasing the size leads to an increase in the surface area and, in turn, to an enhanced cohesion between the components. The mechanical properties such as compressive strength are also generally improved by nanosizing [16].

11.1.2.3 Hybrids

In biomaterials research, the term *hybrid* or *hybrid material* is often used when two phases are blended on the molecular scale. The characterization of a hybrid material can be based on nature of the phases, the interactions that occur between the phases, and the resulting structure. With regards to the matrix, combinations of crystalline/amorphous or organic/inorganic are common, while the building units of the phases can be molecules, macromolecules, particles, or fibers. The inorganic phases are commonly formed *in situ* by molecular precursors, which often tend to form clusters or particles potentially may be templated by an organic phase. Previously, two classes of hybrid have been distinguished, based on the bonding characteristics [15]:

- Class I hybrids are those with weak interactions between the phases, due to the presence of van der Waals forces, hydrogen bonds, or weak electrostatic interactions. Typical examples are organic polymers with an entrapped inorganic phase lacking a strong interaction between the phases. Weak crosslinking can occur by the inorganic moieties, or by the polymer matrix. If the inorganic phase forms its own network, in addition to the organic network, the term interpenetrating networks (IPN) is used. Typically, the formation of a sol–gel inorganic phase in the presence of an organic network-forming phase will result in an IPN.

- Class II hybrids exhibit strong chemical interactions between the phases; they mostly include discrete building blocks or polymers that are covalently bound to each other.

The terms *nanocomposites* and *nanohybrids* are not clearly discriminated, and often are used synonymously in biomedical publications. Even the term "hybrid nanocomposite" has been used [17], which can be defined as a combination of several nanocomposites.

11.2 Biological Nanocomposites

During the process of evolution, a variety of nanocomposites has arisen among the species. For example, invertebrates often produce composites of calcium carbonate mineral particles and organic phases (e.g., nacre), or those consisting of amorphous silicon dioxide and organic components (e.g., in sponge spicules) [18, 19]. In contrast, vertebrates have developed nanocomposites with calcium phosphate mineral phases, such as the ECM of bone (including fish bone), dentin, and antler. In the field of biomineralization research, current attention is largely focused on investigations into biological organic–inorganic composites [20], with increasing attempts made to mimic biological nanocomposites by following a *biomimetic* approach [21, 22]; examples of these are provided in the following sections. Despite a lack of space preventing these fascinating biological nanocomposites [23, 24] from being described in detail at this point, some aspects of bone will be introduced, on the basis of its important role in tissue engineering applications.

11.2.1 Bone

The ECM of bone tissue is a highly organized and hierarchically structured nanocomposite that serves as the natural matrix for the attachment and growth of bone cells in the living organism. As the ECM is considered the optimal scaffold material for bone tissue engineering, many attempts have been made to mimic its unique properties. The hierarchical organization of (cortical) bone tissue, from the macroscopic level down to the nanometer scale, is shown in Figure 11.1.

In mammals, several types of bone are present, or are formed temporarily during development or healing; these are mainly compact (cortical), trabecular (spongy), and woven bone. The major components of bone are hydroxyapatite (HAp, a calcium phosphate mineral phase) nanocrystals and collagen type I (a fibrous protein) fibrils, which together form a true nanocomposite at the first level of organization. The third major component, water, is located within the fibril, in the gaps between the triple-helical collagen molecules, and also between fibrils

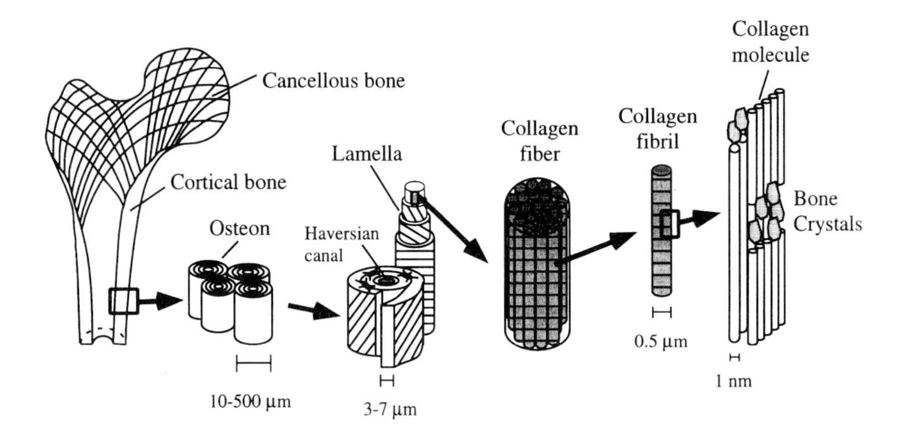

Figure 11.1 Hierarchical organization of cortical bone tissue. Reproduced with permission from Ref. [25] with permission.

and between fibers [26]. The bone apatite crystals are plate-shaped, with average lengths and widths of 50 nm and 25 nm, respectively; the thickness may vary, however, from around 1.5 nm for mineralized tendon up to 4 nm for some types of mature bone. The organic component of the bone ECM is mainly represented by type I collagen and a low content of noncollagenous proteins such as osteocalcin and osteopontin. Despite their low concentrations, the noncollagenous proteins – together with other organic substances such as glycosaminoglycans – seem to play an important role in controlling HAp crystallization and, therefore, the formation of the nanocomposite.

Interestingly, the 3-D packing of the triple helical collagen molecules in the fibril appears to be unclear, despite a variety of models having been proposed during the past decades [27]. Consequently, the details of the interaction between the mineral and the organic phase on the nanometer scale also remain somewhat vague. The experimental data obtained, mostly using transmission electron microscopy (TEM), have revealed the nanoscopic dimensions of the HAp crystals, together with a partly observable patterned alignment that fits with the 67 nm spacing of the gap regions in the collagen fibrils. This finding was interpreted as a strong suggestion that nucleation of the HAp platelets begins in the gap regions of the fibril. A typical TEM image of a nonstained, calcified mammalian bone sample in which the mineral particles are visible as dark areas, is shown in Figure 11.2.

The nanoscopic size of the mineral phase, combined with the plate-like morphology, results in a very high surface area. The interaction between the inorganic (mineral) components and the fibrous collagen matrix is therefore very strong. The collagen matrix (together with noncollagenous proteins and glycosaminoglycans, etc.) controls not only the size and shape of the HAp crystals, but also their crystallographic orientation. It has been shown that the *c*-axes of the hexagonal

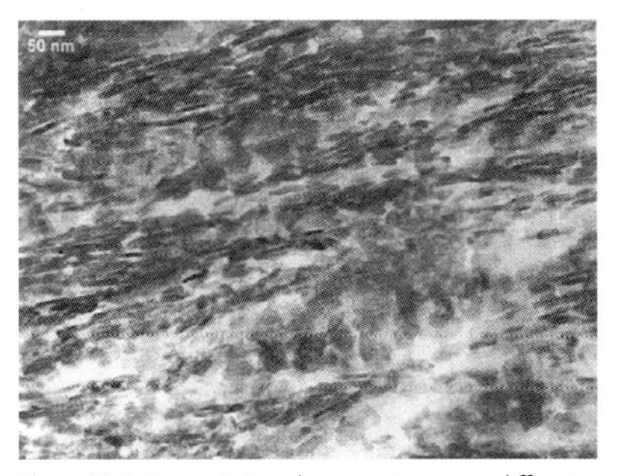

Figure 11.2 Transmission electron microscopy diffraction contrast image of inorganic bovine trabecular bone; an ultra thin section of a calcified sample. Reproduced with permission from Ref. [28].

HAp platelets are well aligned with the collagen fibril long axis [26]. Collagen is synthesized by osteoblasts, which also express the enzyme alkaline phosphatase (ALP) that is mainly responsible for calcium phosphate mineral formation. With the exception of woven bone, the collagen fibrils are deposited in a sheet-like manner, and with a parallel fiber alignment (called "lamellae") into the free space, which is created by the resorbing activity of the osteoclasts during bone remodeling. The lamellae form osteons in compact bone, and trabecules in spongy bone; both of these structural elements are responsible for the outstanding mechanical properties of bone tissue and its perfect adaptation to local force distribution [25].

The optimization of mechanical strength is one motivation for Nature to develop hierarchical structures, and to take advantage of the combination of nanometer-sized inorganic phases glued by organic phases. Gao *et al.* discussed the influence of the shape, size, and distribution of a mineral phase embedded in an organic matrix, as present in many biological materials, with respect to the optimized mechanical strength [29]. A large volume concentration, as well as a large aspect ratio of the mineral building blocks, results in the extreme stiffness of such composites. The optimal size of a single mineral crystal is limited to the nanometer scale, due to the destabilizing effects of flaws.

11.2.2 Biological Nanocomposites as Scaffolds

It is clear that the outstanding properties of natural nanocomposites such as bone ECM have led to their direct use as scaffold materials for tissue engineering applications. Although, for many decades, autologous bone (mostly harvested from the iliac crest) has been used as an implant material in surgery [1], it cannot be used in other patients without causing problems of immune rejection. Nonetheless, autologous bone contains living cells and hence is considered not only as a scaffold but also as an osteoinductive graft that can be used directly to fill defect sites.

Before using the bone ECM as a matrix for seeding and cultivating cells (and hence as a scaffold in the true sense of tissue engineering), it must first be decellularized and processed in order to remove all immunogenic components. In general, mostly bovine (thus, xenogeneic) cancellous bone is used for this purpose, with such materials having been approved and available commercially for several years (e.g., Tutobone®) [30–32]. Unfortunately, this processing may weaken the bone's mechanical properties to some extent and cells, when seeded onto such scaffold material, do not become embedded in the matrix (as would occur before decellularization) but rather adhere to the surface of the trabecular structure. In addition, the properties of each specimen will differ, and come to depend on the pore structure and the direction of loading according to the local situation of the bone segment used to prepare the implant material. As a consequence, attention has been focused on the development of artificial scaffold materials, in an attempt to mimic the composition and structure of the natural bone ECM, rather than to use the ECM directly as a scaffold for tissue engineering.

Badylak and coworkers recently reviewed the use of other ECMs as implant materials and scaffolds for tissue engineering, focusing on the skin, mucosa, and pericardium [33, 34]. These biological materials can be interpreted as solely organic nanocomposites that consist of interpenetrating networks of mostly nanoscopic fibrous biopolymers (proteins such as collagen or elastin, polysaccharides, etc.). Due to their sheet-like morphology, these materials can be used preferably as scaffolds for tissues of the intestinal tract, skin and mucosa, or as patches in the case of dura defects or hernia therapies. Several of these materials have already been approved and are available commercially [34] for medical therapies. The main advantages of such processed natural ECMs are their biological nanotopography, which supports cell attachment, proliferation and differentiation. In addition, most of these materials are easily degradable *in vivo*, and the degradation products can be metabolized and used to generate a new biological ECM during defect healing. Their main disadvantage is an insufficient pore size, which prevents 3-D cell seeding and limits their application to flat tissue constructs.

11.3 Organic–Organic Nanocomposites

Pure organic nanocomposites for tissue engineering make use of natural or synthetic polymers. Typically, the natural biopolymers include proteins such as collagen, gelatin, fibroin (derived from silk), and fibrin, and polysaccharides such as chitin, chitosan, alginate, hyaluronic acid, starch, and cellulose derivatives. The synthetic polymers include poly(lactic acid) (PLA), poly(glycolic acid) (PGA), poly(vinyl alcohol) (PVA), and poly(caprolactone) (PCL).

In the case of material combinations that consist only of organic substances, it is often difficult to distinguish between true nanocomposites and polymer blends or mixtures. On returning to the definitions provided in Section 11.1.2, the phases that comprise a composite must retain their identities and properties, and must be bonded – which is why an interface between them is maintained. Therefore, in most pure organic nanocomposites at least one of the components will form a fibrous network which is embedded in a second phase and serves as a matrix – which is why, in a nanocomposite, the fibrous component must have diameters less than 100 nm, by definition. Some examples of synthetic organic–organic nanocomposite scaffolds are listed in Table 11.1.

Table 11.1 Examples of organic–organic nanocomposite scaffolds.

Component 1	Component 2	Reference(s)
Collagen	Silk fibroin	[35]
Chitosan	Polycaprolactone (PCL)	[36]
Gelatin	Elastin	[37]
Cellulose	Polyvinyl alcohol (PVA)	[38, 39]

The fabrication of nanoscopic biopolymer fibers is straightforward for members of the collagen family, because they are able to form fibrils with diameters less than 100nm, via a self-assembly processes that occurs not only in organisms (*in vivo*) but also in the laboratory (*in vitro*). In addition, preformed polymeric networks can be isolated from natural ECMs (see Section 11.2.1), and combined with other synthetic polymers or biopolymers to result in novel nanocomposite materials. The production of nanofibers from synthetic polymers, or generally from polymer solutions, requires the use of methods such as electrospinning, a rapidly developing area of research field (though not only in biomedical engineering) [40, 41]. Due to their nanoscale topographies, electrospun fibers mimic the corresponding properties of natural ECMs and are, therefore, currently being investigated as scaffolds for soft tissues, especially for skin [42, 43]. Unfortunately, scaffolds that consist solely of electrospun nanofibers lack mechanical stability, and must be embedded in (or combined with) other components, at least before a porous 3-D scaffold can be manufactured. In many reports, the electrospun fibers have been combined with inorganic phases in order to produce composites with sufficient mechanical strength. These types of nanocomposite are further discussed in Section 11.5.3, while nanocomposites containing carbon nanotubes (CNTs) are described in Section 11.6.

11.4 Inorganic–Inorganic Nanocomposites

Nanocomposite scaffolds that consist solely of inorganic constituents are of relevance only to the tissue engineering of bone. By combining different mineral phases, the solubility – and therefore also the degradability – of the resultant composite can be altered and adapted to clinical needs. These materials can be formed either directly from two (or more) (nano)crystalline phases, or by embedding a crystalline into an amorphous phase (e.g., silica). Both approaches have used been successfully to manufacture nanocomposites with interesting properties. Stoichiometric HAp is known to be the most stable and least-soluble calcium phosphate phase in aqueous solutions under physiological conditions [23]. However, in order to overcome this problem, HAp was combined with other more-soluble calcium phosphate phases, such as β-tricalcium phosphate (β-TCP). When at least one of the two phases is nanoscopic, a nanocomposite can be achieved. The combination in specific ratios results in biphasic calcium phosphates (BCPs) with controllable degradation behavior and bioactivity [44, 45], while the greater release of calcium and phosphate ions from BCPs after implantation also promotes the formation of new bone. It has been shown that a composition consisting of 20% HAp and 80% β-TCP is able to stimulate the osteogenic differentiation of human mesenchymal stem cells (hMSC), the precursors of bone-forming osteoblasts [46].

In most of these composites, HAp nanoparticles are used as one component, while the many methods used to prepare such composites (e.g., solid-state reactions, wet-chemical syntheses/precipitation, hydrothermal processes, mechano-

Table 11.2 Examples of pure inorganic nanocomposite scaffold materials.

Component 1	Component 2	Reference(s)
HAp	β-TCP	[44, 45]
Calcium silicate	β-TCP	[47]
Silica	β-Rhenanite (β-$NaCaPO_4$) and $Ca_2P_2O_7$	[48]

chemical, pH shock wave, microwave synthesis) may lead to different crystal morphologies [14].

Although the combination of established inorganic phases to develop new materials for biomedical applications represents a common strategy, the application of sintering and melting processes often yield modified (but monophasic) materials that, in this chapter, are not considered to be nanocomposites. Some examples of true inorganic–inorganic nanocomposites that can be used for tissue engineering are listed in Table 11.2.

Silica and calcium phosphates are often combined to form nanocomposites that can be used not only as implant materials but also as scaffolds for tissue engineering. These pure inorganic nanocomposites, which may consisting of HAp, TCP and silica, are known to be both biocompatible and osteoconductive. Silicon has been shown as a major contributor to bone growth during the early stages of mineralization [49], and this has led to the substitution of silicon for calcium in synthetic HAp. It has also led to the development of a wide variety of silica-based biomaterials, including bioglasses [50], silica-containing cements [51], cyclosilicate-based silica–calcium phosphate nanocomposites [52], nanosilica-fused whiskers and silicon-substituted calcium phosphate powders [53], calcium phosphate HAp nanoparticles within organized silica structures [54], and composites containing silica gel and xerogel [55, 56].

Silicon is the second most abundant element in the earth's crust. When associated with oxygen, it forms silica (SiO_2) which can be either crystalline or amorphous (the latter material is mainly of biogenic origin). Silica, which is mostly formed by diatoms, glass sponges, higher plants and insects in distinct biomineralization processes [18], is also known to be involved in the formation of cartilage and bone in higher animals.

The use of silica in medical applications is mainly characterized by the development of bioactive glasses, which have been shown to bond to living bone tissue [50, 57]. Based on silica, these materials can be differentiated by the inclusion of specific amounts of CaO, P_2O_5, and Na_2O that control their solubility and, therefore, their degradation behavior. It has been reported that silica present in calcium phosphate matrices will both influence and enhance the bioactivity [58, 59]. It has also been shown that a specialized heat treatment can lead to a final product that consists of crystalline nanoparticles embedded in an amorphous silica glass phase; such a material may be denoted as a nanocomposite, in contrast to conventional bioglasses which are termed amorphous materials [60]. Because of their high degradability, the use of such scaffolds for tissue engineering applications appears to be problematic. Notably, the high ion concentrations that result from a rapid dissolution of the material can harm cells during *in vitro* cultivation periods. It is for this reason that bioglasses are mostly used as implant materials.

11.5 Organic–Inorganic Composites

As noted above, the ECM of bone is a typical nanocomposite, composed of HAp nanocrystals embedded in a fibrillar collagen matrix. Due to the outstanding material properties of natural bone, and the importance of this tissue for regenerative therapies and tissue engineering, the development of novel nanocomposite scaffolds has focused on a combination of organic (mostly biopolymeric) and mineral phases. Clearly, such materials can be used only for bone tissue engineering, and not for soft tissues as the mineral phase might cause irritation. Nanocomposites composed of resorbable polymers and inorganic phases are desirable due to the possibility of them acquiring remarkable materials properties. For these materials, the preparation techniques include blending, extrusion, compounding and compression molding [61]. In general, organic components such as biopolymers (e.g., collagen, fibrin, etc.) or synthetic polymers (PLA, PGA, etc.) serve as binding agents, and therefore are added to inorganic components not only to improve the bulk handling characteristics [7] but also to facilitate the preparation of porous scaffolds. In contrast, inorganic components may be added to organic components so as to enhance their mechanical stability. Owing to the numerous possible ways to prepare organic–inorganic composites for tissue engineering purposes, a variety of combinations has been investigated, and these have been reviewed comprehensively by Dorozhkin [62], Rezwan *et al.* [9], and by Murugan and Ramakrishna [14]. Some novel, recently developed organic–inorganic nanocomposite that can be used as scaffolds for the tissue engineering of bone are listed in Tables 11.3–11.5. It should be noted that, as this field has developed rapidly (and continues to develop), the information provided here is far from being comprehensive. Initially, calcium phosphate-based materials are described because they contain the mineral phase of bone, and therefore mimic this tissue better than other inorganic components. Those materials that consist of a defined collagen–HAp nanocomposite closely resemble the biological ECM, and are therefore described in greater detail. Attempts have also been made to mimic biological nanocomposites other than bone; an example of such materials is *nacre*, which is composed of aragonite (calcium carbonate) crystal platelets embedded in a complex organic matrix, so as to achieve novel biomaterials with interesting properties [63].

11.5.1 Mineralized Collagen: Nanocomposites that Mimic the ECM of Bone

The combination of HAp and collagen type I fibrils is intended to mimic the structure and the biological and mechanical properties of natural bone. Hence,

Table 11.3 Calcium phosphate-containing organic–inorganic nanocomposite scaffolds for tissue engineering.

Inorganic phase (calcium phosphate)	Organic phase	References(s)
HAp	Collagen	[64, 65, 14] [66–76]
HAp	Collagen, PLA	[77]
HAp	Collagen, chitosan	[78–80]
HAp	Chitosan	[81–86]
HAp	Chitosan, carboxymethyl cellulose	[87]
HAp	Chitosan, PCL	[88]
HAp	Chitosan, PLA	[89]
HAp	Alginate	[90]
HAp	Alginate, gelatin	[91, 92]
HAp	Gelatin	[93]
HAp	Gelatin, PCL	[94]
HAp	PLA	[95]
HAp	PLGA	[96]
HAp	PC	[97]
HAp	Cellulose	[98]
HAp	pHEMA, PCL	[99]
HAp and montmorillonite	Chitosan	[100]
TCP	PCL	[101]

PC, polycarbonate; pHEMA = Poly(2-hydroxyethyl methacrylate).

Table 11.4 Silica-containing organic–inorganic nanocomposites as scaffolds for tissue engineering.

Inorganic phase	Organic phase	Reference(s)
Calcium silicate	PCL	[114, 115]
Nanosilica	Chitin	[116]
Bioactive glass	PLLA	[117]
Bioactive glass	PCL	[118, 119]
Sol–gel silica	pHEMA	[120]
Sol–gel silica	Collagen	[121, 122]
Sol–gel silica, calcium phosphates	Collagen	[123]

such a biomimetic nanocomposite has been seen by many as an optimal starting material to generate scaffolds for bone tissue engineering. To date, many attempts have been made to synthesize collagen type I–HAp nanocomposites capable of mimicking the ECM of bone. Whereas, some are no more than simple mixtures of both phases, containing HAp crystals or agglomerates on the micrometer scale

Table 11.5 Fiber-containing organic–inorganic nanocomposite scaffolds.

Inorganic phase	Organic (fibrous) phase	Reference
HAp	PCL	[135]
HAp	PHB	[136]
HAp	PLA	[137]
Montmorillonite	PLLA	[138]
HAp	Chitosan (electrospun)	[86]
HAp	Collagen (electrospun)	[139]
HAp	PLGA	[96]
HAp	PLGA, collagen	[140]
HAp	PCL, gelatin	[94]
HAp	PLA, PCL, gelatin	[141]
TCP	PCL	[101]
Bioactive glass	Collagen	[142]

or bundles of insoluble collagen fibers, others have involved the use of optimized conditions in which parameters such as concentration, ion strength, temperature and pH can be controlled. This has led to the formation of defined nanocomposite materials *in situ*, that could be interpreted as true artificial ECMs. These manufactured collagen–HAp composites have been reviewed recently by Murugan and Ramakrishna [14] and Wahl and Czernuszka [102].

As an example of such a defined nanocomposite, one method of preparation will be described in more detail; this is the synchronous biomineralization of collagen, as developed by Bradt and coworkers [103] and recently reviewed by Gelinsky [64]. Briefly, the nanocomposite was created via the mineralization of collagen with HAp nanocrystals during fibril reassembly. Initially, a solution of an acid-soluble collagen type I in dilute hydrochloric acid is mixed with an aqueous $CaCl_2$ solution. The ionic strength of the mixture is adjusted by adding NaCl, and the pH raised to physiological level by adding Tris and phosphate buffer. Warming the mixture to body temperature initiates the collagen fibril reassembly while, simultaneously, the presence of calcium and phosphate ions in the mixture led to the formation of calcium phosphate. According to Ostwald's step rule, amorphous calcium phosphate phases are initially formed, and slowly transformed into nanocrystalline HAp as the most stable phase at neutral pH. The reaction conditions were adjusted in such a way that both processes – collagen fibril reconstitution and HAp formation – occur simultaneously. Hence, the growing collagen fibrils serves as a template for calcium phosphate crystallization, with the consequence that a homogeneous nanocomposite is formed consisting of about 30 wt% collagen and 70 wt% HAp. The mineralized collagen fibrils appear as a cloudy white floating precipitate in the reaction mixture, but no unbound crystals are deposited at the bottom of the flask, and the surrounding solution becomes clear; the latter point verified complete transformation of the initially formed amorphous calcium phosphate phase into crystalline HAp, bound exclusively to the collagen matrix. The product, described as "mineralized collagen

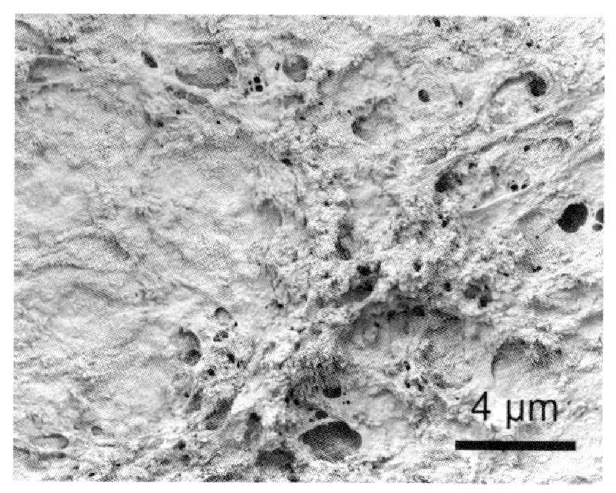

Figure 11.3 Scanning electron microscopy image of reconstituted mineralized collagen I fibrils. An example of an organic–inorganic nanocomposite, mimicking the extracellular matrix of bone tissue on the nanometer scale. The experimental conditions are described in Ref. [72].

fibrils," is isolated by centrifugation and has been used to prepare several types of scaffold material for use in the tissue engineering of bone, or directly as an implant material [64, 72, 104–106].

Full characterization of the nanocomposite, using X-ray diffraction (XRD), Fourier-transform infrared (FT-IR) spectroscopy [103], scanning electron microscopy (SEM), transmission electron microscopy (TEM), and high-resolution TEM (HR-TEM) [72], showed clearly that the mineral phase formed in this reaction was nanoscopic HAp.

A SEM image of the mineralized collagen fibrils, demonstrating homogeneity of the nanocomposite and the close interaction between the mineral phase and the reconstituted collagen fibrils is shown in Figure 11.3.

The dimensions of the HAp nanocrystals are better recognized in TEM images of ultrathin and unstained sections (Figure 11.4a–c). In this case, the mineral phase appeared as dark, needle-like or platelet-like objects with maximum length of approximately 80 nm and a diameter/thickness of 6–8 nm (Figure 11.4c). The collagen could not be identified directly because of its low electron density, and the absence of any additional staining. However, it was assumed that the fibrils, covered with HAp crystals, were located in the lighter, elongated areas of Figure 11.4a and b. In some regions, the HAp crystals appeared to be partly oriented with their long axis (as shown with HR-TEM) as the crystallographic *c*-axis of the hexagonal HAp lattice [72]), parallel to the collagen fibers. In terms of their size, orientation and position of the HAp crystals with regards to the collagen fibrils, the TEM images showed a great similarity to those of fetal human woven bone [107].

As noted above, the preparation of porous scaffolds suitable for tissue engineering applications does not generally require the use of specialized methods, if nanocomposites are employed as the starting materials. Instead, more common

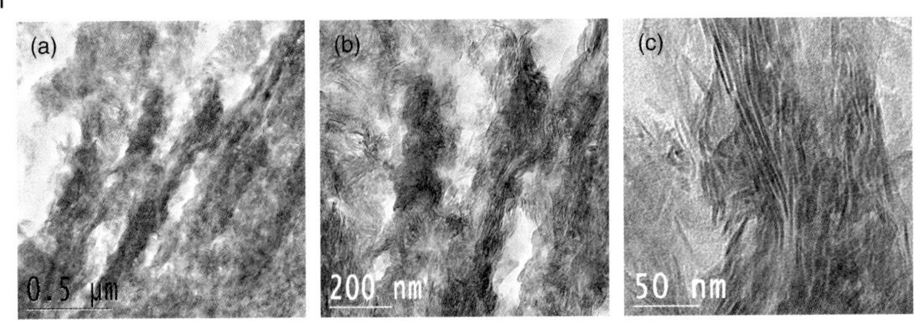

Figure 11.4 (a–c) Transmission electron microscopy images of nonstained ultrathin sections, showing the nanostructure of mineralized collagen fibrils as a biomimetic nanocomposite, at three different magnifications. Images courtesy of Dr P. Simon, Dresden; the experimental conditions are described in Ref. [72].

techniques such as freeze-drying (which is equally effective with monophasic materials) can easily be applied. In the case of mineralized collagen, a porous 3-D scaffold could be produced via lyophilization, followed by chemical crosslinking to introduce covalent bonds between the collagen molecules, making the material stable for use as a matrix for cell cultivation. Notably, these scaffolds could be seeded homogeneously with hMSC [108], could support the osteogenic differentiation of hMSC towards osteoblastic lineage [109], and could also be degraded by osteoclast-like cells (derived from primary human monocytes) *in vitro* [110]. Due to the similarity of this nanocomposite with the natural bone ECM, the scaffold could also be used as an implant material for bone defects [111], and to establish *in vitro* models for the complex cellular processes in bone remodeling [110, 112]. A TEM image of an osteoclast-like cell derived from primary human monocytes, and capable of actively degrading a scaffold of mineralized collagen fibrils, is shown in Figure 11.5. This image was recorded from a co-culture that consisted of osteoclast-like cells and osteogenically differentiated hMSC, following a cultivation period of 21 days [112]. It was also shown that, *in vivo*, this scaffold material is resorbed by osteoclasts when implanted into a bone defect [111].

Despite their great similarity to natural bone ECMs, the properties of scaffolds composed of an artificial collagen–HAp nanocomposite do not equate with the outstanding properties of bone, notably with regards to mechanical stability. Typically, scaffolds applied to tissue engineering must have a much higher porosity compared to bone, in order to allowing homogeneous cell seeding and to permit large constructs. In general, simple methods such as freeze-drying are incapable of creating an hierarchical organization on several length scales, as it is present in compact bone tissue (see Section 11.2). Because of their degradability, however, tissue engineering scaffolds must not be as strong as the natural tissues they are replacing. But, with time the synthetic scaffold material will disappear to be replaced by a newly formed, healthy tissue that inevitably will

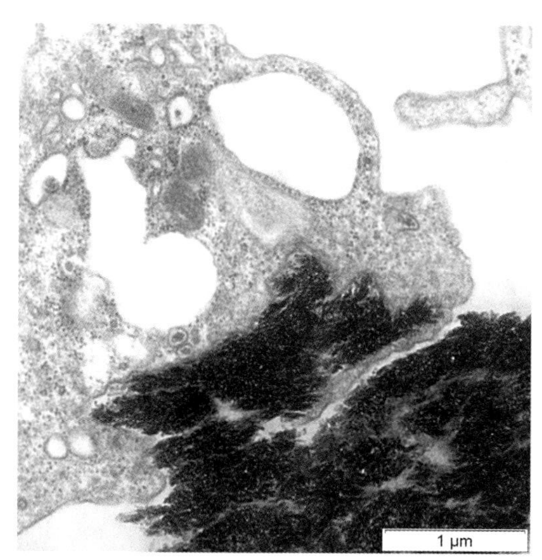

Figure 11.5 Transmission electron microscopy image of an actively resorbing osteoclast-like cell, derived from primary human monocytes and cultivated for 21 days in a co-culture with osteogenically differentiating hMSC. The dark structures at the lower right side of the image are the hydroxyapatite nanocrystals, as parts of the mineralized collagen nanocomposite scaffold (see Figure 11.4). Image courtesy of Dr A. Springer, Dresden; the experimental conditions are described in Ref. [112].

be better adapted to the local mechanical needs than might be any artificial material.

11.5.2
Silica-Based Nanocomposites for Tissue Engineering

Silica and bioactive glasses are biocompatible materials that have emerged as interesting alternatives or supplements to calcium phosphate-based nanocomposites [113]. It has long been recognized that the addition of a particulate bioactive glass to a polymer matrix will modify its degradation kinetics and material morphology. The compressive strength, dimension stability and bioactivity can also be enhanced.

The conventional entrapment of inorganic particles in a polymer network represents a strategy with limited composite characteristics, however. In order to achieve an intimate interaction between the phases, the sol–gel technique has recently been identified as an appropriate method for conjugating inorganic materials with biological systems, owing to the compatibility of the experimental conditions [124]. It is well known that sol–gel routes have more in common with biological processing than with conventional materials processing [125]. Normally, the preparation of silica via this route starts with molecular precursors (alkoxysilanes), the general formula of which is $Si(OR)_4$, where R is an alkyl residue

(C_nH_{2n+1}). The reaction is controlled by the kinetics of the hydrolytic polycondensation that occurs around the silicon atom, through nucleophilic substitution [126]. In this hydrolytic process, H_2O acts as a nucleophile and displaces ROH, resulting in an intermediate soluble form of silica, orthosilicic acid ($Si(OH)_4$). Further polycondensation reactions of the silicic acid lead to siloxane bonds (Si–O–Si) that occur via either hetero- (Si–OH + RO–Si) or homo- (Si–OH + HO–Si) condensation. Under the control of the pH, concentrations, solvent, catalyst, water content, temperature and other parameters, the polymers grow to form either particles or colloids [127, 128], with the viscosity of the solution being steadily increased until the entire solution is gelified and an amorphous 3-D silica network – a hydrogel – is formed [129]. In order to achieve densification of the material, it is first necessary to remove the liquid phase. However, one major problem when transforming silica hydrogels into dry xerogels, is that the capillary forces which occur during drying often exceed the gel's strength [130]; this may cause cracking leading finally to silica fragments (or even a silica powder). An ethanol treatment of the hydrogels, while applying specific temperatures and a high relative humidity during the drying stages, followed by high-temperature sintering processes, has been shown to facilitate the production of crack-free monolithic structures [131].

Novel developments have been based on modifications of the processes described, by introducing organic molecules and taking advantage of their templating activity. Biodegradable PCL–SiO_2 composites prepared via a sol–gel method have shown, after intensive investigation, an enhanced bioactivity with increasing silica content [119]. Such an effect can be explained by the increasing numbers of silanol groups present on the material's surface, and which are known to provide nucleation sites for apatite formation when immersed in a simulated body fluid (SBF). The mechanical properties of the composite are continuously tunable, from ductile–tough (pure PCL) to hard–brittle (pure silica), depending on the organic : inorganic ratio.

Collagen fibrils have also been recognized as an appropriate organic matrix for the fabrication of organic–inorganic nanocomposites, and not only with calcium phosphate phases [132]. The possibility of low-temperature processing when using a sol–gel technique means that the collagen's structure is preserved, retaining its biological activity, while additional components of an organic or inorganic nature can be incorporated. The reactivity of both components is based on the electrostatic interaction of the negatively charged silica species and positively charged amine groups of the collagen at neutral or near-neutral pH values. Thus, silicic acid will be preferentially polymerized at these sites and form nanoparticles at the surface of the collagen molecules or fibrils. Additional phases of lower reactivity (e.g., calcium phosphate particles) will be physically embedded into the forming matrix (Figure 11.6).

Depending on the collagen used, indivisible composite networks exhibiting characteristic structures are formed intermediately as a nanocomposite hydrogel [122]. Further drying at high relative humidity resulted in shrinking and densification caused by the capillary forces. By using this method, and without any heat treatment, a monolithic macroscopic nanocomposite xerogel was produced that

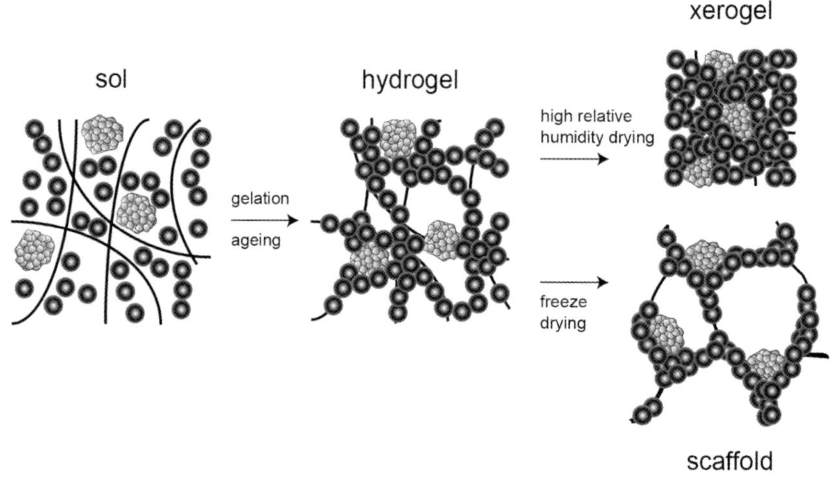

Figure 11.6 Schematic representation of hydrogel formation, starting with a sol of silica precursors (dark gray dots), collagen fibrils as an organic template (black lines), and an additional phase (e.g., calcium phosphate; light gray particles). Drying at high relative humidity yields nanocomposite xerogels, whereas freeze-drying results in porous nanocomposite scaffolds.

exhibited bone-like mechanical properties [123]. Conventional freeze-drying of the nanocomposite hydrogel resulted in highly porous scaffolds.

Subsequent studies using SEM revealed a distinct influence of collagen morphology on the structure of the composite xerogels [122]. The application of tropocollagen resulted in a fine-structured gel surface, similar to that of a pure silica gel. Macroscopically, the tropocollagen-based xerogels were obtained as translucent flake-like fragments, because the capillary forces that occurred during drying exceeded the strength of the gel. Because of this limitation, this composition appeared not to be suitable for the fabrication of porous tissue engineering scaffolds. The incorporation of collagen fibers or fibrils resulted in microporous xerogels that were obtained as homogeneous cylindrical solid monoliths (Figure 11.7a). The proportion of collagen present had a significant effect on the xerogel's structure and its fracture surface, and proved to be a vital parameter with regards to the mechanical properties of the composite xerogels (Figure 11.7b). Whilst pure silica is brittle, a distinct deformability occurred due to the presence of an organic matrix. The compressive strength of the composite (70% silica, 30% collagen) was about fivefold higher than that of pure silica, the corresponding strain at ultimate strength was enhanced about threefold, and the bioactivity was slightly reduced because the high collagen levels caused a decrease in the number of surface silanol groups. The addition of calcium phosphate as a third component caused a lowering of the mechanical strength, but had a positive effect on the bioactivity of the xerogels (Figure 11.7c) [123].

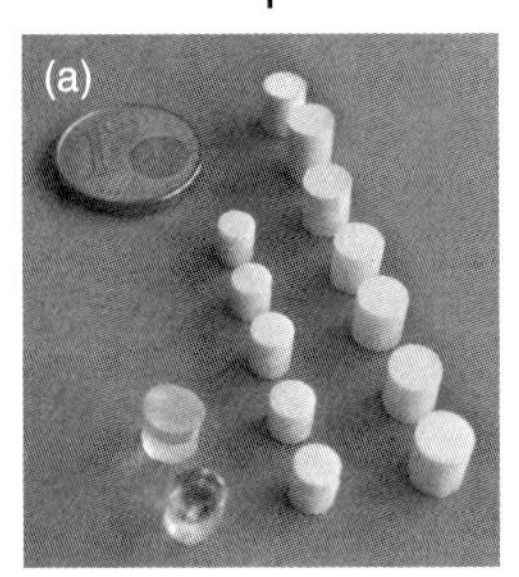

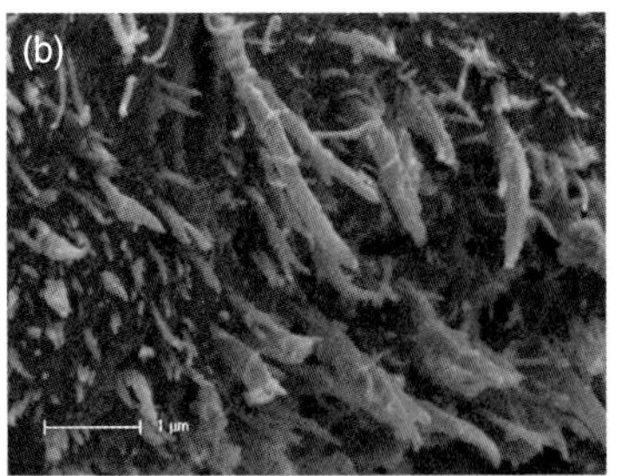

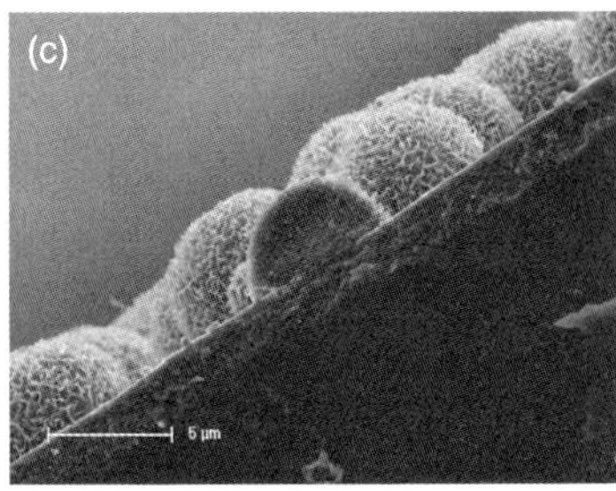

Figure 11.7 (a) Photograph of pure silica xerogels (left row), nanocomposite silica–collagen xerogels (middle row), and nanocomposite silica–collagen xerogel with an additional calcium phosphate phase (right row); (b) Scanning electron microscopy image of a fracture surface, showing the nanocomposite character of the material with collagen fibrils embedded in the silica matrix; (c) Immersion in simulated body fluid results in the rapid growth of a surface apatite layer, demonstrating high bioactivity.

11.5.3
Organic–Inorganic Nanocomposites Containing Fibers

During the past few years, fiber-spinning technologies have attracted increasing attention in regenerative medicine. For example, electrospinning can be used to produce thin polymer fibers down to the nanometer scale, while established adjoining processing technologies such as weaving, stitching, and knitting allow the creation of macroscopic and 3-D medical products [133]. These technologies have also enabled the fabrication of constructs capable of fulfilling most of the requirements of an ideal scaffold system. The advantages over conventional foam-like scaffolds are the higher surface area-to-volume ratio and the interconnectivity of the pore system; both features support the adhesion of cells and the diffusion of nutrients inside the scaffold. Nanofibers which mimic the topography of natural ECMs are known to promote the proliferation and differentiation of attached cells [134]. As noted above, the main disadvantage of electrospun nanofibers is their insufficient mechanical stability, and it is for this reason that porous 3-D scaffolds consisting solely of these fibers cannot be fabricated.

Until now, synthetic and natural polymers have been applied successfully to the tissue engineering of bone, cartilage, heart valves, bladder, and liver. Yet, the unfavorable degradation of synthetic polymers, coupled with the inadequate mechanical strength of natural polymers, has limited their application. The aim of incorporating inorganic phases into nanofibrous polymer matrices is to mimic the structure of natural bone, while providing additional advantages. Notably, the combination is characterized by improved mechanical properties, favorable cell responses, and a controlled degradation of the polymer (the latter effect is due to the inorganic phase buffering the typical acidic degradation products of the polymers [96]). To date, extensive studies have been conducted using HAp-modified polymer fibers comprising poly(hydroxybutyrate) (PHB), PLA, poly(lactic-*co*-

glycolic acid) (PLGA), PCL, collagen, and gelatin. An overview of the fiber-containing organic–inorganic nanocomposites that have served as scaffolds for tissue engineering is provided in Table 11.5. Nanocomposites with biopolymers such as collagen which, by nature, have a fibrous morphology were described earlier in the chapter (see Section 11.2.2).

11.6 Nanocomposites Containing Carbon Nanotubes (CNTs)

During recent years, nanocomposites containing CNTs have intensively been investigated, with many reports having been made describing the application of such nanocomposites for biomedical and tissue engineering purposes (see the reviews of Harrison *et al.* [143], Saito and coworkers [144], and Edwards *et al.* [145]). Bearing in mind that the main function of a tissue engineering scaffold is to provide a temporary tissue replacement which, following degradation, will generate the space for a newly formed, healthy tissue, the incorporation of nondegradable components such as CNTs would appear to be problematic. This problem seems all the more valid because CNTs have been recognized as potentially toxic in humans, in that they are barely removed by macrophages and/or other phagocytic cells, due to their high aspect ratio and a strong tendency to aggregate [146–149]. Although some research groups have not identified any major problems when implanting CNT-containing nanocomposites (e.g., in a bone defect of an animal [150]), the long-term effects of these materials *in vivo* remain unknown. Consequently, it is essential that further studies are conducted with the aim of assessing the possible risk(s) of using CNT-containing materials as scaffolds or biomaterials. In particular, animal studies with a long-term follow-up must be carried out to determine the distribution of CNTs through the body, following the degradation of the resorbable part of the composite. With regards to the temporary requirement of tissue engineering scaffolds, the major factors related to the incorporation of CNTs – namely, to improve the mechanical properties and electrical conductivity – are of no significant relevance. In fact, in the present authors' opinion, CNT-containing nanocomposites will not be approved for medical applications so far as the basic material is biodegradable (which is a main prerequisite for a tissue engineering scaffold). Nonetheless, when safely embedded in nondegradable matrices, CNTs might offer the opportunity to improve the properties of permanent implants, thus opening new possibilities for medical therapies.

11.7 Summary and Outlook

Today, nanocomposite development is a rapidly growing area of research that will surely open up a host of applications. Bearing in mind that mother Nature also uses nanocomposites as ECMs – and therefore as biological scaffolds – it is clear that this represents a promising approach for the synthesis of biomimetic nanocomposites from which scaffolds may be manufactured. But – not any material combination can be used for tissue engineering applications – it is only those that can be adapted to specific needs, as defined by the type of tissue, the defect site, and its size. With nanocomposites, the acquisition of material properties that can barely be realized with monophasic materials will surely lead to the development of novel composites that might revolutionize tissue engineering applications.

References

1 Sen, M.K. and Miclau, T. (2007) Autologous iliac crest bone graft: should it still be the gold standard for treating nonunions? *Injury*, **38** (Suppl. 1), S75–80.

2 Williams, D. (2004) Seasonal fantasies in scaffolds. *Medical Device Technology*, **15**, 8–10.

3 Brittberg, M., Lindahl, A., Nilsson, A., Ohlsson, C., Isaksson, O. and Peterson, L. (1994) Treatment of deep cartilage defects in the knee with autologous chondrocyte transplantation. *New England Journal of Medicine*, **331**, 889–95.

4 Thieme, S., Ryser, M.F., Gentsch, M., Brenner, S., Stiehler, M., Roelfing, J., Navratiel, K., Gelinsky, M. and Roesen-Wolff, A. (2009) Stromal cell derived factor-1 alpha directed chemoattraction of transiently CXCR4 overexpressing bone marrow stromal cells into functionalized three-dimensional biomimetic scaffolds. *Tissue Engineering Part C: Methods*, **15**, 687–96.

5 Hutmacher, D.W. (2001) Scaffold design and fabrication technologies for engineering tissues–state of the art and future perspectives. *Journal of Biomaterials Science – Polymer Edition*, **12**, 107–24.

6 Hutmacher, D.W., Goh, J.C. and Teoh, S.H. (2001) An introduction to biodegradable materials for tissue engineering applications. *Annals of the Academy of Medicine, Singapore*, **30**, 183–91.

7 Chan, C.K., Kumar, T.S., Liao, S., Murugan, R., Ngiam, M. and Ramakrishnan, S. (2006) Biomimetic nanocomposites for bone graft applications. *Nanomedicine*, **1**, 177–88.

8 Karageorgiou, V. and Kaplan, D. (2005) Porosity of 3D biomaterial scaffolds and osteogenesis. *Biomaterials*, **26**, 5474–91.

9 Rezwan, K., Chen, Q.Z., Blaker, J.J. and Boccaccini, A.R. (2006) Biodegradable and bioactive porous polymer/inorganic composite scaffolds for bone tissue engineering. *Biomaterials*, **27**, 3413–31.

10 Polykandriotis, E., Arkudas, A., Euler, S., Beier, J.P., Horch, R.E. and Kneser, U. (2006) Prevascularisation strategies in tissue engineering. *Handchirurgie, Mikrochirurgie, Plastische Chirurgie*, **38**, 217–23.

11 Chaikof, E.L., Matthew, H., Kohn, J., Mikos, A.G., Prestwich, G.D. and Yip, C.M. (2002) Biomaterials and scaffolds in reparative medicine. *Annals of the New York Academy of Sciences*, **961**, 96–105.

12 Chevalier, E., Chulia, D., Pouget, C. and Viana, M. (2008) Fabrication of porous substrates: a review of processes using pore forming agents in the biomaterial field. *Journal of Pharmaceutical Sciences*, **97**, 1135–54.

13 Bonfield, W. (2006) Designing porous scaffolds for tissue engineering. *Philosophical Transactions Series A, Mathematical, Physical, and Engineering Sciences*, **364**, 227–32.

14 Murugan, R. and Ramakrishna, S. (2005) Development o nanocomposites for bone grafting. *Computer Science and Technology*, **65**, 2385–406.

15 Kickelbick, G. (2007) Introduction to hybrid materials, in *Hybrid Materials. Synthesis, Characterization, and Applications* (ed. G. Kickelbick), Wiley-VCH Verlag GmbH, Weinheim, pp. 1–48.

16 Watari, F., Yokoyama, A., Gelinsky, M. and Pompe, W. (2008) Conversion of functions by nanosizing – from osteoconductivity to bone substitutional properties in apatite, in *Interface Oral Health Science 2007* (eds M. Watanabe and O. Okuno), Springer Japan, Tokyo, pp. 139–47.

17 Liu, Y.Y., Liu, D.M., Chen, S.Y., Tung, T.H. and Liu, T.Y. (2008) In situ synthesis of hybrid nanocomposite with highly order arranged amorphous metallic copper nanoparticle in poly(2-hydroxyethyl methacrylate) and its potential for blood-contact uses. *Acta Biomaterialia*, **4**, 2052–8.

18 Ehrlich, H., Heinemann, S., Heinemann, C., Simon, P., Bazhenov, V.V., Shapkin, N.P., Born, R., Tabachnick, K.R., Hanke, T. and Worch, H. (2008) Nanostructural organization of naturally occurring composites – Part I: silica-collagen-based biocomposites. *Journal of Nanomaterials*, Article ID 623838.

19 Ehrlich, H., Janussen, D., Simon, P., Bazhenov, V.V., Shapkin, N.P., Erler, C., Mertig, M., Born, R., Heinemann, S., Hanke, T., Worch, H. and Vournakis, J.N. (2008) Nanostructural organization of naturally occurring composites – Part II: silica-chitin-based biocomposites. *Journal of Nanomaterials*, Article ID 670235.

20 Bäuerlein, E. (2009) *Handbook of Biomineralization. Biological Aspects and Structure Formation*, Wiley-VCH Verlag GmbH, Weinheim.

21 Mann, S. (1996) *Biomimetic Materials Chemistry*, Wiley-VCH Verlag GmbH, Weinheim.

22 Fratzl, P. (2007) Biomimetic materials research: what can we really learn from nature's structural materials? *Journal of the Royal Society, Interface*, **4**, 637–42.

23 Mann, S. (2001) *Biomineralization – Principles and Concepts in Bioorganic Materials Chemistry*, Oxford University Press, Oxford.

24 Fratzl, P. and Weinkamer, R. (2007) Nature's hierarchical materials. *Progress in Materials Sciences*, **52**, 1263–334.

25 Rho, J.Y., Kuhn-Spearing, L. and Zioupos, P. (1998) Mechanical properties and the hierarchical structure of bone. *Medical Engineering and Physics*, **20**, 92–102.

26 Weiner, S. and Wagner, H.D. (1998) The material bone: structure-mechanical function relations. *Annual Review of Materials Science*, **28**, 271–98.

27 Ottani, V., Martini, D., Franchi, M., Ruggeri, A. and Raspanti, M. (2002) Hierarchical structures in fibrillar collagens. *Micronesica*, **33**, 587–96.

28 Rosen, V.B., Hobbs, L.W. and Spector, M. (2002) The ultrastructure of anorganic bovine bone and selected synthetic hydroxyapatites used as bone graft substitute materials. *Biomaterials*, **23**, 921–8.

29 Gao, H., Ji, B., Jager, I.L., Arzt, E. and Fratzl, P. (2003) Materials become insensitive to flaws at nanoscale: lessons from nature. *Proceedings of the National Academy of Sciences of the United States of America*, **100**, 5597–600.

30 Trentz, O.A., Hoerstrup, S.P., Sun, L.K., Bestmann, L., Platz, A. and Trentz, O.L. (2003) Osteoblasts response to allogenic and xenogenic solvent dehydrated cancellous bone *in vitro*. *Biomaterials*, **24**, 3417–26.

31 Kneser, U., Stangenberg, L., Ohnolz, J., Buettner, O., Stern-Straeter, J., Mobest, D., Horch, R.E., Stark, G.B. and Schaefer, D.J. (2006) Evaluation of processed bovine cancellous bone matrix seeded with syngenic osteoblasts in a critical size calvarial defect rat model. *Journal of Cellular and Molecular Medicine*, **10**, 695–707.

32 Meyer, S., Floerkemeier, T. and Windhagen, H. (2008) Histological osseointegration of Tutobone: first results in human. *Archives of Orthopaedic and Traumatic Surgery*, **128**, 539–44.

33 Badylak, S.F. (2007) The extracellular matrix as a biologic scaffold material. *Biomaterials*, **28**, 3587–93.

34 Badylak, S.F., Freytes, D.O. and Gilbert, T.W. (2009) Extracellular matrix as a biological scaffold material: structure and function. *Acta Biomaterialia*, **5**, 1–13.

35 Wang, J., Zhou, W., Hu, W., Zhou, L., Wang, S. and Zhang, S. (2009) Collagen/silk fibroin bi-template-induced biomimetic bone-like substitutes. *Journal of Biomedical Materials Research Part A*, [Epub ahead of print] DOI 10.1002/jbm.a.32602.

36 Xiao, X., Liu, R. and Huang, Q. (2008) Preparation and characterization of nano-hydroxyapatite/polymer composite scaffolds. *Journal of Materials Science Materials in Medicine*, **19**, 3429–35.

37 Lekakou, C., Lamprou, D., Vidyarthi, U., Karopoulou, E. and Zhdan, P. (2008) Structural hierarchy of biomimetic materials for tissue engineered vascular and orthopedic grafts. *Journal of Biomedical Materials Research Part B, Applied Biomaterials*, **85**, 461–8.

38 Millon, L.E. and Wan, W.K. (2006) The polyvinyl alcohol-bacterial cellulose system as a new nanocomposite for biomedical applications. *Journal of Biomedical Materials Research Part B, Applied Biomaterials*, **79**, 245–53.

39 Millon, L.E., Guhados, G. and Wan, W. (2008) Anisotropic polyvinyl alcohol-Bacterial cellulose nanocomposite for biomedical applications. *Journal of Biomedical Materials Research Part B, Applied Biomaterials*, **86B**, 444–52.

40 Teo, W.-E. and Ramakrishna, S. (2009) Electrospun nanofibers as a platform for multifunctional, hierarchically organized nanocomposite. *Computer Science and Technology*, **69**, 1804–17.

41 Sill, T.J. and von Recum, H.A. (2008) Electrospinning: applications in drug delivery and tissue engineering. *Biomaterials*, **29**, 1989–2006.

42 Kumbar, S.G., James, R., Nukavarapu, S.P. and Laurencin, C.T. (2008) Electrospun nanofiber scaffolds: engineering soft tissues. *Biomedical Materials*, **3**, 034002.

43 Ashammakhi, N., Ndreu, A., Nikkola, L., Wimpenny, I. and Yang, Y. (2008) Advancing tissue engineering by using electrospun nanofibers. *Regenerative Medicine*, **3**, 547–74.

44 Ramay, H.R. and Zhang, M. (2004) Biphasic calcium phosphate nanocomposite porous scaffolds for load-bearing bone tissue engineering. *Biomaterials*, **25**, 5171–80.

45 Lin, Y., Wang, T., Wu, L., Jing, W., Chen, X., Li, Z., Liu, L., Tang, W., Zheng, X. and Tian, W. (2007) Ectopic and in situ bone formation of adipose tissue-derived stromal cells in biphasic calcium phosphate nanocomposite. *Journal of Biomedical Materials Research Part A*, **81**, 900–10.

46 Arinzeh, T.L., Tran, T., McAlary, J. and Daculsi, G. (2005) A comparative study of biphasic calcium phosphate ceramics for human mesenchymal stem-cell-induced bone formation. *Biomaterials*, **26**, 3631–8.

47 Zhang, F., Chang, J., Lin, K. and Lu, J. (2008) Preparation, mechanical properties and *in vitro* degradability of wollastonite/tricalcium phosphate macroporous scaffolds from nanocomposite powders. *Journal of Materials Science Materials in Medicine*, **19**, 167–73.

48 El-Ghannam, A., Ning, C.Q. and Mehta, J. (2004) Cyclosilicate nanocomposite: a novel resorbable bioactive tissue engineering scaffold for BMP and bone-marrow cell delivery. *Journal of Biomedical Materials Research Part A*, **71**, 377–90.

49 Carlisle, E.M. (1980) Biochemical and morphological changes associated with long bone abnormalities in silicon deficiency. *Journal of Nutrition*, **110**, 1046–56.

50 Hench, L.L. and Polak, J.M. (2002) Third-generation biomedical materials. *Science*, **295**, 1014–17.

51 Fu, Q., Zhou, N., Huang, W., Wang, D., Zhang, L. and Li, H. (2004) Preparation and characterization of a novel bioactive bone cement: glass based nanoscale hydroxyapatite bone cement. *Journal of Materials Science Materials in Medicine*, **15**, 1333–8.

52 Gupta, G., Zbib, A., El-Ghannam, A., Khraisheh, M. and Zbib, H. (2005) Characterization of a novel bioactive composite using advanced X-ray computed tomography. *Composite Structures*, **71**, 423–8.

53 Xu, H.H., Smith, D.T. and Simon, C.G. (2004) Strong and bioactive composites containing nano-silica-fused whiskers for bone repair. *Biomaterials*, **25**, 4615–26.

54 Diaz, A., Lopez, T., Manjarrez, J., Basaldella, E., Martinez-Blanes, J.M. and

Odriozola, J.A. (2006) Growth of hydroxyapatite in a biocompatible mesoporous ordered silica. *Acta Biomaterialia*, **2**, 173–9.

55 Korventausta, J., Jokinen, M., Rosling, A., Peltola, T. and Yli-Urpo, A. (2003) Calcium phosphate formation and ion dissolution rates in silica gel-PDLLA composites. *Biomaterials*, **24**, 5173–82.

56 Radin, S., El-Bassyouni, G., Vresilovic, E.J., Schepers, E. and Ducheyne, P. (2005) *In vivo* tissue response to resorbable silica xerogels as controlled-release materials. *Biomaterials*, **26**, 1043–52.

57 Hench, L.L. (1991) Bioceramics: from concept to clinic. *Journal of the American Ceramic Society*, **74**, 1487–510.

58 Patel, N., Best, S.M., Bonfield, W., Gibson, I.R., Hing, K.A., Damien, E. and Revell, P.A. (2002) A comparative study on the *in vivo* behavior of hydroxyapatite and silicon substituted hydroxyapatite granules. *Journal of Materials Science Materials in Medicine*, **13**, 1199–206.

59 Borum, L. and Wilson, O.C. (2003) Surface modification of hydroxyapatite. Part II. Silica. *Biomaterials*, **24**, 3681–8.

60 Boccaccini, A.R., Chen, Q., Lefebvre, L., Gremillard, L. and Chevalier, J. (2007) Sintering, crystallisation and biodegradation behaviour of Bioglass-derived glass-ceramics. *Faraday Discussions*, **136**, 27–44, discussion 107–23.

61 Boccaccini, A.R. and Blaker, J.J. (2005) Bioactive composite materials for tissue engineering scaffolds. *Expert Review of Medical Devices*, **2**, 303–17.

62 Dorozhkin, S.V. (2009) Calcium orthophosphate-based biocomposites and hybrid biomaterials. *Journal of Materials Science*, **44**, 2343–87.

63 Waraich, S.M., Hering, B., Behrens, P. and Menzel, H. (2009) Preparation of mother-of-pearl-like materials for biomedical applications by layer-by-layer deposition. *Biomaterialien*, **10**, 152.

64 Gelinsky, M. (2009) Mineralized collagen as biomaterial and matrix for bone tissue engineering, in *Fundamentals of Tissue Engineering and Regenerative Medicine* (eds U. Meyer, T. Meyer, J. Handschel and H.-P. Wiesmann), Springer, Heidelberg Berlin, pp. 485–93.

65 Wahl, D.A., Sachlos, E., Liu, C. and Czernuszka, J.T. (2007) Controlling the processing of collagen-hydroxyapatite scaffolds for bone tissue engineering. *Journal of Materials Science Materials in Medicine*, **18**, 201–9.

66 Ohyabu, Y., Adegawa, T., Yoshioka, T., Ikoma, T., Shinozaki, K., Uemura, T. and Tanaka, J. (2009) A collagen sponge incorporating a hydroxyapatite/ chondroitin sulfate composite as a scaffold for cartilage tissue engineering. *Journal of Biomaterials Science – Polymer Edition*, **20**, 1861–74.

67 Huang, Z., Tian, J., Yu, B., Xu, Y. and Feng, Q. (2009) A bone-like nano-hydroxyapatite/collagen loaded injectable scaffold. *Biomedical Materials*, **4**, 55005.

68 Liu, Y. (2009) Incorporation of hydroxyapatite sol into collagen gel to regulate the contraction mediated by human bone marrow-derived stromal cells. *IEEE Transactions on Nanobioscience*, **9**, 1–11.

69 Sena, L.A., Caraballo, M.M., Rossi, A.M. and Soares, G.A. (2009) Synthesis and characterization of biocomposites with different hydroxyapatite-collagen ratios. *Journal of Materials Science Materials in Medicine*, **20**, 2395–400.

70 Liao, S., Ngiam, M., Chan, C.K. and Ramakrishna, S. (2009) Fabrication of nano-hydroxyapatite/collagen/ osteonectin composites for bone graft applications. *Biomedical Materials*, **4**, 25019.

71 Al-Munajjed, A.A., Plunkett, N.A., Gleeson, J.P., Weber, T., Jungreuthmayer, C., Levingstone, T., Hammer, J. and O'Brien, F.J. (2009) Development of a biomimetic collagen-hydroxyapatite scaffold for bone tissue engineering using a SBF immersion technique. *Journal of Biomedical Materials Research Part B, Applied Biomaterials*, **90**, 584–91.

72 Gelinsky, M., Welzel, P.B., Simon, P., Bernhardt, A. and König, U. (2008) Porous three-dimensional scaffolds made of mineralized collagen: preparation and properties of a biomimetic nanocomposite material for

tissue engineering of bone. *Chemical Engineering Journal*, **137**, 84–96.
73 Liao, S., Ngiam, M., Watari, F., Ramakrishna, S. and Chan, C.K. (2007) Systematic fabrication of nano-carbonated hydroxyapatite/collagen composites for biomimetic bone grafts. *Bioinspiration and Biomimetics*, **2**, 37–41.
74 Yunoki, S., Ikoma, T., Tsuchiya, A., Monkawa, A., Ohta, K., Sotome, S., Shinomiya, K. and Tanaka, J. (2007) Fabrication and mechanical and tissue ingrowth properties of unidirectionally porous hydroxyapatite/collagen composite. *Journal of Biomedical Materials Research Part B, Applied Biomaterials*, **80**, 166–73.
75 Itoh, S., Kikuchi, M., Koyama, Y., Matumoto, H.N., Takakuda, K., Shinomiya, K. and Tanaka, J. (2005) Development of a novel biomaterial, hydroxyapatite/collagen (HAp/Col) composite for medical use. *Biomedical Materials and Engineering*, **15**, 29–41.
76 Tampieri, A., Celotti, G., Landi, E., Sandri, M., Roveri, N. and Falini, G. (2003) Biologically inspired synthesis of bone-like composite: self-assembled collagen fibers/hydroxyapatite nanocrystals. *Journal of Biomedical Materials Research Part A*, **67**, 618–25.
77 Niu, X., Feng, Q., Wang, M., Guo, X. and Zheng, Q. (2009) Porous nano-HA/collagen/PLLA scaffold containing chitosan microspheres for controlled delivery of synthetic peptide derived from BMP-2. *Journal of Controlled Release*, **134**, 111–17.
78 Zhang, L., Tang, P., Zhang, W., Xu, M. and Wang, Y. (2010) Effect of chitosan as a dispersant on collagen-hydroxyapatite composite matrices. *Tissue Engineering Part C, Methods*, **16**, 71–9.
79 Wang, X., Wang, X., Tan, Y., Zhang, B., Gu, Z. and Li, X. (2009) Synthesis and evaluation of collagen-chitosan-hydroxyapatite nanocomposites for bone grafting. *Journal of Biomedical Materials Research Part A*, **89**, 1079–87.
80 Wang, Y., Zhang, L., Hu, M., Liu, H., Wen, W., Xiao, H. and Niu, Y. (2008) Synthesis and characterization of collagen-chitosan-hydroxyapatite artificial bone matrix. *Journal of Biomedical Materials Research Part A*, **86**, 244–52.
81 Chen, J.D., Wang, Y. and Chen, X. (2009) In situ fabrication of nano-hydroxyapatite in a macroporous chitosan scaffold for tissue engineering. *Journal of Biomaterials Science – Polymer Edition*, **20**, 1555–65.
82 Kashiwazaki, H., Kishiya, Y., Matsuda, A., Yamaguchi, K., Iizuka, T., Tanaka, J. and Inoue, N. (2009) Fabrication of porous chitosan/hydroxyapatite nanocomposites: their mechanical and biological properties. *Biomedical Materials and Engineering*, **19**, 133–40.
83 Madhumathi, K., Shalumon, K.T., Rani, V.V., Tamura, H., Furuike, T., Selvamurugan, N., Nair, S.V. and Jayakumar, R. (2009) Wet chemical synthesis of chitosan hydrogel-hydroxyapatite composite membranes for tissue engineering applications. *International Journal of Biological Macromolecules*, **45**, 12–15.
84 Chesnutt, B.M., Yuan, Y., Buddington, K., Haggard, W.O. and Bumgardner, J.D. (2009) Composite chitosan/nano-hydroxyapatite scaffolds induce osteocalcin production by osteoblasts *in vitro* and support bone formation *in vivo*. *Tissue Engineering Part A*, **15**, 2571–9.
85 Thein-Han, W.W. and Misra, R.D. (2009) Biomimetic chitosan-nanohydroxyapatite composite scaffolds for bone tissue engineering. *Acta Biomaterialia*, **5**, 1182–97.
86 Zhang, Y., Venugopal, J.R., El-Turki, A., Ramakrishna, S., Su, B. and Lim, C.T. (2008) Electrospun biomimetic nanocomposite nanofibers of hydroxyapatite/chitosan for bone tissue engineering. *Biomaterials*, **29**, 4314–22.
87 Liuyun, J., Yubao, L., Li, Z. and Jianguo, L. (2008) Preparation and properties of a novel bone repair composite: nano-hydroxyapatite/chitosan/carboxymethyl cellulose. *Journal of Materials Science Materials in Medicine*, **19**, 981–7.
88 Xiao, X., Liu, R., Huang, Q. and Ding, X. (2009) Preparation and characterization of hydroxyapatite/polycaprolactone-chitosan composites. *Journal of Materials Science Materials in Medicine*, **20**, 2375–83.

89 Cai, X., Tong, H., Shen, X., Chen, W., Yan, J. and Hu, J. (2009) Preparation and characterization of homogeneous chitosan-polylactic acid/hydroxyapatite nanocomposite for bone tissue engineering and evaluation of its mechanical properties. *Acta Biomaterialia*, **5**, 2693–703.

90 Despang, F., Börner, A., Dittrich, R., Tomandl, G., Pompe, W. and Gelinsky, M. (2005) Alginate/calcium phosphate scaffolds with oriented, tube-like pores. *Materialwissenschaft und Werkstofftechnik*, **36**, 761–7.

91 Bernhardt, A., Despang, F., Lode, A., Demmler, A., Hanke, T. and Gelinsky, M. (2009) Proliferation and osteogenic differentiation of human bone marrow stromal cells on alginate-gelatin-hydroxyapatite scaffolds with anisotropic pore structure. *Journal of Tissue Engineering and Regenerative Medicine*, **3**, 54–62.

92 Dittrich, R., Despang, F., Bernhardt, A., Hanke, T., Tomandl, G., Pompe, W. and Gelinsky, M. (2007) Scaffolds for hard tissue engineering by ionotropic gelation of alginate – influence of selected preparation parameters. *Journal of the American Ceramic Society*, **90**, 1703–8.

93 Zandi, M., Mirzadeh, H., Mayer, C., Urch, H., Eslaminejad, M.B., Bagheri, F. and Mivehchi, H. (2009) Biocompatibility evaluation of nano-rod hydroxyapatite/gelatin coated with nano-HAp as a novel scaffold using mesenchymal stem cells. *Journal of Biomedical Materials Research Part A*, **92**, 1244–55.

94 Venugopal, J.R., Low, S., Choon, A.T., Kumar, A.B. and Ramakrishna, S. (2008) Nanobioengineered electrospun composite nanofibers and osteoblasts for bone regeneration. *Artificial Organs*, **32**, 388–97.

95 Spadaccio, C., Rainer, A., Trombetta, M., Vadala, G., Chello, M., Covino, E., Denaro, V., Toyoda, Y. and Genovese, J.A. (2009) Poly-L-lactic acid/hydroxyapatite electrospun nanocomposites induce chondrogenic differentiation of human MSC. *Annals of Biomedical Engineering*, **37**, 1376–89.

96 Jose, M.V., Thomas, V., Johnson, K.T., Dean, D.R. and Nyairo, E. (2009) Aligned PLGA/HA nanofibrous nanocomposite scaffolds for bone tissue engineering. *Acta Biomaterialia*, **5**, 305–15.

97 Liao, J., Zhang, L., Zuo, Y., Wang, H., Li, J., Zou, Q. and Li, Y. (2009) Development of nanohydroxyapatite/polycarbonate composite for bone repair. *Journal of Biomaterials Applications*, **24**, 31–45.

98 Fang, B., Wan, Y.Z., Tang, T.T., Gao, C. and Dai, K.R. (2009) Proliferation and osteoblastic differentiation of human bone marrow stromal cells on hydroxyapatite/bacterial cellulose nanocomposite scaffolds. *Tissue Engineering Part A*, **15**, 1091–8.

99 Huang, J., Lin, Y.W., Fu, X.W., Best, S.M., Brooks, R.A., Rushton, N. and Bonfield, W. (2007) Development of nano-sized hydroxyapatite reinforced composites for tissue engineering scaffolds. *Journal of Materials Science Materials in Medicine*, **18**, 2151–7.

100 Katti, K.S., Katti, D.R. and Dash, R. (2008) Synthesis and characterization of a novel chitosan/montmorillonite/hydroxyapatite nanocomposite for bone tissue engineering. *Biomedical Materials*, **3**, 034122.

101 Erisken, C., Kalyon, D.M. and Wang, H. (2008) Functionally graded electrospun polycaprolactone and beta-tricalcium phosphate nanocomposites for tissue engineering applications. *Biomaterials*, **29**, 4065–73.

102 Wahl, D.A. and Czernuszka, J.T. (2006) Collagen-hydroxyapatite composites for hard tissue repair. *European Cells and Materials*, **11**, 43–56.

103 Bradt, J.-H., Mertig, M., Teresiak, A. and Pompe, W. (1999) Biomimetic mineralization of collagen by combined fibril assembly and calcium phosphate formation. *Chemistry of Materials*, **11**, 2694–701.

104 Burth, R., Gelinsky, M. and Pompe, W. (1999) Collagen-hydroxyapatite tapes – a new implant material. *Technical Textiles*, **8**, 20–1.

105 Bernhardt, A., Lode, A., Boxberger, S., Pompe, W. and Gelinsky, M. (2008)

Mineralised collagen – an artificial, extracellular bone matrix – improves osteogenic differentiation of bone marrow stromal cells. *Journal of Materials Science Materials in Medicine*, **19**, 269–75.

106 Gelinsky, M., Eckert, M. and Despang, F. (2007) Biphasic, but monolithic scaffolds for the therapy of osteochondral defects. *International Journal of Materials Research*, **98**, 749–55.

107 Su, X., Sun, K., Cui, F.Z. and Landis, W.J. (2003) Organization of apatite crystals in human woven bone. *Bone*, **32**, 150–62.

108 Lode, A., Bernhardt, A. and Gelinsky, M. (2008) Cultivation of human bone marrow stromal cells on three-dimensional scaffolds of mineralized collagen: influence of seeding density on colonization, proliferation and osteogenic differentiation. *Journal of Tissue Engineering and Regenerative Medicine*, **2**, 400–7.

109 Bernhardt, A., Lode, A., Mietrach, C., Hempel, U., Hanke, T. and Gelinsky, M. (2009) *In vitro* osteogenic potential of human bone marrow stromal cells cultivated in porous scaffolds from mineralized collagen. *Journal of Biomedical Materials Research Part A*, **90**, 852–62.

110 Domaschke, H., Gelinsky, M., Burmeister, B., Fleig, R., Hanke, T., Reinstorf, A., Pompe, W. and Rosen-Wolff, A. (2006) *In vitro* ossification and remodeling of mineralized collagen I scaffolds. *Tissue Engineering*, **12**, 949–58.

111 Yokoyama, A., Gelinsky, M., Kawasaki, T., Kohgo, T., Konig, U., Pompe, W. and Watari, F. (2005) Biomimetic porous scaffolds with high elasticity made from mineralized collagen – an animal study. *Journal of Biomedical Materials Research Part B, Applied Biomaterials*, **75**, 464–72.

112 Bernhardt, A., Thieme, S., Domaschke, H., Springer, A., Rösen-Wolff, A. and Gelinsky, M. (2009)Crosstalk of osteoblast and osteoclast precursors on mineralized collagen – towards an *in vitro* model for bone-remodelling. *Journal of Biomedical Materials Research Part A*, under review.

113 Sakka, S. (2008) Sol-gel technology as representative processing for nanomaterials: case studies on the starting solution. *Journal of Sol-Gel Science and Technology*, **46**, 241–9.

114 Wei, J., Heo, S.J., Liu, C., Kim, D.H., Kim, S.E., Hyun, Y.T., Shin, J.W. and Shin, J.W. (2009) Preparation and characterization of bioactive calcium silicate and poly(epsilon-caprolactone) nanocomposite for bone tissue regeneration. *Journal of Biomedical Materials Research Part A*, **90**, 702–12.

115 Kotela, I., Podporska, J., Soltysiak, E., Konsztowicz, K.J. and Blazewicz, M. (2009) Polymer nanocomposites for bone tissue substitutes. *Ceramics International*, **35**, 2475–80.

116 Madhumathi, K., Kumar, P.T.S., Kavya, K.C., Furuike, T., Tamura, H., Nair, S.V. and Jayakumar, R. (2009) Novel chitin/nanosilica composite scaffolds for bone tissue engineering applications. *International Journal of Biological Macromolecules*, **45**, 289–92.

117 Hong, Z., Reis, R.L. and Mano, J.F. (2008) Preparation and *in vitro* characterization of scaffolds of poly(L-lactic acid) containing bioactive glass ceramic nanoparticles. *Acta Biomaterialia*, **4**, 1297–306.

118 Aho, A.J., Tirri, T., Kukkonen, J., Strandberg, N., Rich, J., Seppala, J. and Yli-Urpo, A. (2004) Injectable bioactive glass/biodegradable polymer composite for bone and cartilage reconstruction: concept and experimental outcome with thermoplastic composites of poly(epsilon-caprolactone-*co*-D,L-lactide) and bioactive glass S53P4. *Journal of Materials Science Materials in Medicine*, **15**, 1165–73.

119 Rhee, S.H. (2004) Bone-like apatite-forming ability and mechanical properties of poly(epsilon-caprolactone)/silica hybrid as a function of poly(epsilon-caprolactone) content. *Biomaterials*, **25**, 1167–75.

120 Costantini, A., Luciani, G., Silvestri, B., Tescione, F. and Branda, F. (2008) Bioactive poly(2-hydroxyethylmethacrylate)/silica gel hybrid nanocomposites prepared by sol-gel process. *Journal of Biomedical*

Materials Research Part B, Applied Biomaterials, **86**, 98–104.

121 Heinemann, S., Ehrlich, H., Knieb, C. and Hanke, T. (2007) Biomimetically inspired hybrid materials based on silicified collagen. *International Journal of Materials Research*, **98**, 603–8.

122 Heinemann, S., Heinemann, C., Ehrlich, H., Meyer, M., Baltzer, H., Worch, H. and Hanke, T. (2007) A novel biomimetic hybrid material made of silicified collagen: perspectives for bone replacement. *Advanced Engineering Materials*, **9**, 1061–8.

123 Heinemann, S., Heinemann, C., Bernhardt, R., Reinstorf, A., Meyer, M., Nies, B., Worch, H. and Hanke, T. (2009) Bioactive silica-collagen composite xerogels modified by calcium phosphate phases with adjustable mechanical properties for bone replacement. *Acta Biomaterialia*, **5**, 1979–90.

124 Carturan, G., Dal Toso, R., Boninsegna, S. and Dal Monte, R. (2004) Encapsulation of functional cells by sol-gel silica: actual progress and perspectives for cell therapy. *Journal of Materials Chemistry*, **14**, 2087–98.

125 Coradin, T. and Livage, J. (2007) Aqueous silicates in biological sol-gel applications: new perspectives for old precursors. *Accounts of Chemical Research*, **40**, 819–26.

126 Corriu, R.J.P. (2001) The control of nanostructured solids: a challenge for molecular chemistry. *European Journal of Inorganic Chemistry*, **2001**, 1109–21.

127 Perry, C.C. (2003) Silicification. *Reviews in Mineralogy and Geochemistry*, **54**, 291–327.

128 Brinker, C.J. and Sherer, G.W. (1990) *The Physics and Chemistry of Sol-Gel Processing*, Academic Press, New York.

129 Iler, R.K. (1980) Isolation and characterization of particle nuclei. *Journal of Colloid and Interface Science*, **75**, 138–48.

130 Scherer, G.W. (1990) Theory of drying. *Journal of the American Ceramic Society*, **73**, 3–14.

131 Zhong, J. and Greenspan, D.C. (2000) Processing and properties of sol-gel bioactive glasses. *Journal of Biomedical Materials Research*, **53**, 694–701.

132 Ono, Y., Kanekiyo, Y., Inoue, K., Hojo, J., Nango, M. and Shinkai, S. (1999) Preparation of novel hollow fiber silica using collagen fibers as a template. *Chemistry Letters*, **28**, 475–6.

133 Jang, J.H., Castano, O. and Kim, H.W. (2009) Electrospun materials as potential platforms for bone tissue engineering. *Advanced Drug Delivery Reviews*, **61**, 1065–83.

134 Laurencin, C.T., Ambrosio, A.M., Borden, M.D. and Cooper, J.A. Jr (1999) Tissue engineering: orthopedic applications. *Annual Review of Biomedical Engineering*, **1**, 19–46.

135 Thomas, V., Jagani, S., Johnson, K., Jose, M.V., Dean, D.R., Vohra, Y.K. and Nyairo, E. (2006) Electrospun bioactive nanocomposite scaffolds of polycaprolactone and nanohydroxyapatite for bone tissue engineering. *Journal of Nanoscience and Nanotechnology*, **6**, 487–93.

136 Ito, Y., Hasuda, H., Kamitakahara, M., Ohtsuki, C., Tanihara, M., Kang, I.K. and Kwon, O.H. (2005) A composite of hydroxyapatite with electrospun biodegradable nanofibers as a tissue engineering material. *Journal of Bioscience and Bioengineering*, **100**, 43–9.

137 Kim, H.W., Lee, H.H. and Knowles, J.C. (2006) Electrospinning biomedical nanocomposite fibers of hydroxyapatite/poly(lactic acid) for bone regeneration. *Journal of Biomedical Materials Research Part A*, **79**, 643–9.

138 Lee, Y.H., Lee, J.H., An, I.G., Kim, C., Lee, D.S., Lee, Y.K. and Nam, J.D. (2005) Electrospun dual-porosity structure and biodegradation morphology of montmorillonite reinforced PLLA nanocomposite scaffolds. *Biomaterials*, **26**, 3165–72.

139 Thomas, V., Dean, D.R., Jose, M.V., Mathew, B., Chowdhury, S. and Vohra, Y.K. (2007) Nanostructured biocomposite scaffolds based on collagen coelectrospun with nanohydroxyapatite. *Biomacromolecules*, **8**, 631–7.

140 Ngiam, M., Liao, S., Patil, A.J., Cheng, Z., Chan, C.K. and Ramakrishna, S. (2009) The fabrication of nano-hydroxyapatite on PLGA and PLGA/collagen nanofibrous composite scaffolds and their effects in osteoblastic behavior for bone tissue engineering. *Bone*, **45**, 4–16.

141 Gupta, D., Venugopal, J., Mitra, S., Giri Dev, V.R. and Ramakrishna, S. (2009) Nanostructured biocomposite substrates by electrospinning and electrospraying for the mineralization of osteoblasts. *Biomaterials*, **30**, 2085–94.

142 Kim, H.W., Song, J.H. and Kim, H.E. (2006) Bioactive glass nanofiber-collagen nanocomposite as a novel bone regeneration matrix. *Journal of Biomedical Materials Research Part A*, **79**, 698–705.

143 Harrison, B.S. and Atala, A. (2007) Carbon nanotube applications for tissue engineering. *Biomaterials*, **28**, 344–53.

144 Saito, N., Usui, Y., Aoki, K., Narita, N., Shimizu, M., Hara, K., Ogiwara, N., Nakamura, K., Ishigaki, N., Kato, H., Taruta, S. and Endo, M. (2009) Carbon nanotubes: biomaterial applications. *Chemical Society Reviews*, **38**, 1897–903.

145 Edwards, S.L., Werkmeister, J.A. and Ramshaw, J.A. (2009) Carbon nanotubes in scaffolds for tissue engineering. *Expert Reviews of Medical Devices*, **6**, 499–505.

146 Aillon, K.L., Xie, Y., El-Gendy, N., Berkland, C.J. and Forrest, M.L. (2009) Effects of nanomaterial physicochemical properties on *in vivo* toxicity. *Advanced Drug Delivery Reviews*, **61**, 457–66.

147 Genaidy, A., Tolaymat, T., Sequeira, R., Rinder, M. and Dionysiou, D. (2009) Health effects of exposure to carbon nanofibers: systematic review, critical appraisal, meta analysis and research to practice perspectives. *The Science of the Total Environment*, **407**, 3686–701.

148 Helland, A., Wick, P., Koehler, A., Schmid, K. and Som, C. (2007) Reviewing the environmental and human health knowledge base of carbon nanotubes. *Environmental Health Perspectives*, **115**, 1125–31.

149 Lam, C.W., James, J.T., McCluskey, R., Arepalli, S. and Hunter, R.L. (2006) A review of carbon nanotube toxicity and assessment of potential occupational and environmental health risks. *Critical Reviews in Toxicology*, **36**, 189–217.

150 Sitharaman, B., Shi, X., Walboomers, X.F., Liao, H., Cuijpers, V., Wilson, L.J., Mikos, A.G. and Jansen, J.A. (2008) *In vivo* biocompatibility of ultra-short single-walled carbon nanotube/ biodegradable polymer nanocomposites for bone tissue engineering. *Bone*, **43**, 362–70.

Index

Nanomaterials for the Life Sciences Vol.8: Nanocomposites. Edited by Challa S. S. R. Kumar

ISBN: 978-3-527-32168-1